1st International Seminar on Earth Sciences and Technology (ISEST 2021)

IOP Conference Series: Earth and Environmental Science
Volume 1047

Online
8 – 9 December 2021

Editors:

Irwan Meilano
Riantini Virtriana
Lamona Irmudyawati Bernawis
Rusmawan Suwarman

ISBN: 978-1-7138-5825-6
ISSN: 1755-1307

Printed from e-media with permission by:

Curran Associates, Inc.
57 Morehouse Lane
Red Hook, NY 12571

Some format issues inherent in the e-media version may also appear in this print version.

Institute of Physics
Dirac House, Temple Back
Bristol BS1 6BE UK

Phone: 44 1 17 929 7481
Fax: 44 1 17 920 0979

techtracking@iop.org

Additional copies of this publication are available from:

Curran Associates, Inc.
57 Morehouse Lane
Red Hook, NY 12571 USA
Phone: 845-758-0400
Fax: 845-758-2633
Email: curran@proceedings.com
Web: www.proceedings.com

TABLE OF CONTENTS

PREFACE

The 2021 1st International Seminar on Earth Science and Technology (ISEST 2021) was hosted by the Faculty of Earth Sciences and Technology, Institut Teknologi Bandung (ITB), Indonesia. The seminar was remotely organized by the Executive Committees from Bandung, Indonesia. ISEST 2021 was successfully held in the form of a virtual conference on December 8th – 9th, 2021 as an attempt to actively support the Prevention of COVID-19 for the health and safety of our participants.

In the modern era, the interest of earth science and technology discipline has shifted following the urge in achieving a more sustainable future. The world is encouraging the future generation to take the present decision into consideration in the hope of maintaining the existence of natural resources, while combating the climate issue. ISEST 2021 served as an international forum, which continued disseminating innovative ideas, discussions, and academic exchanges among researchers. It aimed to integrate sustainability into the field of earth science and technology. The generous support for ISEST 2021 is provided by the Faculty of Earth Sciences and Technology, Institut Teknologi Bandung.

ISEST 2021 hosted contributions on "Earth Science and Technology for Sustainable Development" as its grand theme, owing to the increasing awareness on sustainability. The topic was divided into four main discussions, which consisted of Natural Resources, Earth Data Sciences, Geohazard and Environment and Climate System.

During this session, the opening and welcome speeches were live-streamed from Bandung, Indonesia. It was followed by singing the national anthem "Indonesia Raya". Five virtual conference rooms using the zoom platform were provided for two consecutive days. The discussions presented in each room were held simultaneously in two sessions (morning and evening). The keynote speeches were the first session of the seminar. The second part was the presentation of invited speakers sharing their outstanding research. The session continued with oral presentations from the participants. Each speaker delivered the presentation in 30 minutes including 10 minutes of Q&A session. The pre-recorded presentations were also collected in case any technical issue occurred. A minor internet connection issue arose, but can be managed within seconds, therefore the seminar was overall successfully conducted. ISEST 2021 officially ended with a closing remark from the Vice Dean of the Faculty of Earth Sciences and Technology, ITB, Indonesia.

Notably, over 80 abstracts were delivered in ISEST 2021, surpassing the initial target. The 49 and 56 oral presentations respectively in the first and second day of the seminar captured many fruitful discussions that contributed to the success of the seminar. Furthermore, the seminar was truly global in scope. There were 5 keynote speakers, 19 invited speakers, and 84 speakers coming from 11 countries (Indonesia, United States, UK, France, Australia, Hungary, Japan, Portugal, Malaysia, Singapore, Bangladesh). ISEST 2021 was honored to host world-renowned professionals and scientists as the keynote speakers such as Prof. Phil Cummins from Australian National University, Australia; Prof. Mark Cloos from The University of Texas at Austin, United States; Prof. Josaphat Tetuko Sri Sumantyo from University of Chiba, Japan; Mr. Ron Furness from FIG/IHO/ICA International Board Standards of Competence in Hydrography; and Prof. Hasanuddin Zainal Abidin from Institut Teknologi Bandung, Indonesia to share and exchange their recent research knowledge. Overall, there were 16 talks of Natural Resources, 36 talks of Earth Data Sciences, 23 talks of Geo-hazard & Environment, and 9 talks of Climate System. Of the total number of presented abstracts, almost half of these were included in this proceedings volume, published by the IOP Conference Series: Earth and Environmental Sciences (IOP-EES).

At last, ISEST 2021 was undeniably successful. This international seminar was a breakthrough in the Faculty of Earth Sciences and Technology, ITB. The Committee would like to express sincere gratitude to the distinguished keynote speakers, invited speakers, and all participants involved in the seminar, and special thanks to Prof. Reini Wirahadikusumah (Rector of ITB), Dr. Irwan Meilano (Dean of the Faculty of Earth Sciences and Technology, ITB), Dr. Agus Moch. Ramdhan (Vice Dean of Academic Affairs) and Dr. Mutiara Putri Rachmat (Vice Dean for Resources in the Faculty of Earth Sciences and Technology, ITB respectively for their support in ISEST 2021.

On behalf of the Organizing Committee, we were glad to share this valuable momentum in the latest advanced research in the field of Earth Science and Technology. We truly hope this major success could evolve into a persistent annual conference to connect scholars, scientists and professionals from all corners of the globe.

Best Regards,

Reynaldy Fifariz, Ph.D.
Conference Chair - ISEST 2021
Institut Teknologi Bandung
Indonesia, 8-9th December 2021

IOP Conf. Series: Earth and Environmental Science 1047 (2022) 011001 doi:10.1088/1755-1315/1047/1/011001

1st International Seminar on Earth Science and Technology (ISEST)

Executive Committee

Chair

Reynaldy Fifariz, S.T., M.T., Ph.D.

Secretariat

Gabriella Alodia, S.T., M.Sc., Ph.D.

Treasurer

Ratri Widyastuti, S.T., M.T.

General Affairs

Dr. Idham Andri Kurniawan, S.T., M.T.

Dr. Rendy Dwi Kartika, S.T., M.T.

Logistics

Faizal Ade Rahmahuddin Abdullah, S.Si., M.Si.

Brian Bramanto, S.T., M.T.

Members

Tri Sugiyarti Irianingsih, S.Sos
Endang Ratna K., S.E.
R. Tedi Kusmawan, S.T.
Windiatmoko Sumaryo S.Kom
Hanna Fauziya, S.Pd.
Iwan Pramesti Anwar, S.Si., M.Si.
Riko Adi Wijaya, S.T.
Suci Farissa Nabilah, S.T.
Zainab Sausan, S.T.
Indah M. Ananda, S.T.

Kalingga Titon Nur Ihsan, S.T.
Dimas Candika Alamsyah, S.T.
Muhammad Dzun Nurwinas Saepudin,
S.T. Adib Aulia Syamsu, S.T.
Fatwa Cahya Mustika, S.T.
Dzikri Nahsrul Jabbaaar, S.T.
Muhamad Nur, S.T.

IOP Conf. Series: Earth and Environmental Science 1047 (2022) 011001 doi:10.1088/1755-1315/1047/1/011001

Editorial Board

IOP Conf. Series: Earth and Environmental Science 1047 (2022) 011001 doi:10.1088/1755-1315/1047/1/011001

1st International Seminar on Earth Science and Technology (ISEST)

Steering Committee

Dr. Irwan Meilano, ST, M.Sc.

Institut Teknologi Bandung, Indonesia

Agus Mochamad Ramdhan, S.T., M.T., Ph.D.

Institut Teknologi Bandung, Indonesia

Dr.rer.nat. Mutiara Rachmat Putri, S.Si., M.Si.

Institut Teknologi Bandung, Indonesia

Prof. Dr. Ir. Rubiyanto Kapid

Institut Teknologi Bandung, Indonesia

Prof. Dr. Ir. Eddy A.Subroto

Institut Teknologi Bandung, Indonesia

Prof. Ir. Benyamin Sapiie, Ph.D.

Institut Teknologi Bandung, Indonesia

Prof. Ir. Lambok M. Hutasoit, M.Sc., Ph.D.

Institut Teknologi Bandung, Indonesia

Prof. Dr. Ir. Deny Juanda Puradimaja, DEA

Institut Teknologi Bandung, Indonesia

Prof. Dr.Eng. Nining Sari Ningsih, M.Si.

Institut Teknologi Bandung, Indonesia

Prof. Ir. Hasanuddin Z. Abidin, M.Sc., Ph.D.

Institut Teknologi Bandung, Indonesia

Prof. Ir. Ketut Wikantika, M.Eng., Ph.D.

Institut Teknologi Bandung, Indonesia

Prof. Robert Hall

Royal Holloway University of London, UK

1st International Seminar on Earth Science and Technology (ISEST)

Keynote Speakers and Invited Speakers

Keynote Speakers

Prof. Phil Cummins, *Australian National University, Australia*

Prof. Mark Cloos, *University of Texas, Austin, USA*

Prof. Josaphat Tetuko, *Chiba University, Japan*

Ronald Furness, *FIG/IHO/ICA International Board Standards of Competence in Hydrography, Australia*

Prof. Hasanuddin Z. Abidin, *Institut Teknologi Bandung, Indonesia*

Invited Speakers

Prof. Simon Mathias, *Durham University, UK*

Prof. Masayuki Sakakibara, *Ehime University, Japan*

Prof. Mary Elliot, *Universite de Nantes. France*

Dr. Subrata Sarker Shahjalal, *University of Science and Technology (SUST), Bangladesh*

Dr. Julien Louys, *Australian Research Centre for Human Evolution, Environmental Futures Research Institute, Griffith University*

Dr. Lilian Anne Krug, *Scientific Coordinator of Partnership for Ocean Global Observation (POGO), Portugal*

Prof. Edvin Aldrian, *National Research and Innovation Agency (BRIN), Jakarta, Indonesia*

Dr. Augy Syahailatua, *Oceanography Research Centre – Badan Riset dan Inovasi Nasional (BRIN), Indonesia*

Prof. Benyamin Sapiie, *Head of Geodynamics and Sedimentology Research Group – Institut Teknologi Bandung, Indonesia*

IOP Conf. Series: Earth and Environmental Science 1047 (2022) 011001 doi:10.1088/1755-1315/1047/1/011001

Prof. Lambok M. Hutasoit, *Head of Applied Geology Research Group – Institut Teknologi Bandung, Indonesia*

Prof. Eddy A. Subroto, *Head of Petrology, Volanology, and Geochemistri Research Group – Insitut Teknologi Bandung, Indonesia*

Prof. Nining Sari Ningsih, *Oceanography Research Group – Institut Teknologi Bandung, Indonesia*

Dr. Aswan, *Paleontology and Quaternary Geology Research Group, Institut Teknologi Bandung, Indonesia*

Dr. Totok Suprijo, *Oceanography Research Group – Institut Teknologi Bandung, Indonesia*

Dr. Rusmawan Suwarman, *Head of Atmospheric Sciences Research Group – Institut Teknologi Bandung, Indonesia*

Dr. Wedyanto Kuntjoro, *Head of Geodesy Research Group – Institut Teknologi Bandung, Indonesia*

Dr. Eka Djunarsjah, *Head of Hydrography Research Group – Institut Teknologi Bandung, Indonesia*

Dr. Riantini Virtriana, *Remote Sensing and GIS Research Group, Insitut Teknologi Bandung*

Dr. Rizqi Abdulharis, *Surveying and Cadastre Research Group – Institut Teknologi Bandung, Indonesia*

Peer Review Statement

All papers published in this volume have been reviewed through processes administered by the Editors. Reviews were conducted by expert referees to the professional and scientific standards expected of a proceedings journal published by IOP Publishing.

- **Type of peer review:** Single Anonymous
- **Conference submission management system:** Morressier
- **Number of submissions received:** 48
- **Number of submissions sent for review:** 42
- **Number of submissions accepted:** 37
- **Acceptance Rate (Submissions Accepted / Submissions Received × 100):** 77.1
- **Average number of reviews per paper:** 2.1
- **Total number of reviewers involved:** 30
- **Contact person for queries:**
  Name: Irwan Meilano
  Email: irwanm@gd.itb.ac.id
  Affiliation: Bandung Institute of Technology

IOP Conf. Series: Earth and Environmental Science 1047 (2022) 012001 doi:10.1088/1755-1315/1047/1/012001

Sensing the Terrestrial and Atmospheric Hydrological Dynamic using Satellite Gravimeter and GNSS

Brian Bramanto*, Dina A. Sarsito, Irwan Gumilar, Wedyanto Kuntjoro

Geodesy Research Group, Faculty of Earth Sciences and Technology, Institut Teknologi Bandung, Jalan Ganesha 10, Bandung, Indonesia, 40135

E-mail: brian.bramanto@itb.ac.id

Abstract. Geodetic observation methods, e.g., satellite gravimeter and Global Navigation Satellite System (GNSS), are mainly used for determining the Earth's gravity field and reference system. However, many efforts have been made to adopt the mentioned methods for estimating hydrological dynamics. The satellite gravimeter of the Gravity Recovery and Climate Experiment and its Following-On missions (GRACE/GRACE-FO) have proven to capture the terrestrial hydrological variation. Also, thanks to its signal propagation through the atmosphere medium, GNSS can be used for sensing the hydrology variation in the atmosphere, i.e., troposphere medium. This study aims to explore GRACE/GRACE-FO and GNSS observations to estimate the respective terrestrial and atmospheric hydrological variation in Bandung, Indonesia. Monthly solutions of GRACE/GRACE-FO provided by three agencies were used to estimate the terrestrial hydrological variation. We also used the continuous GNSS site of ITB1 and calculated the atmospheric hydrological series in the form of precipitable water vapor. We found that the terrestrial hydrological series varies approximately ± 20 cm, while the precipitable water vapor ranges between 1 to 5 centimeters. Further, we observed long-wavelength components from terrestrial and atmospheric hydrological variations that correspond to dry and wet seasons. However, we only sensed shorter wavelength components of hydrological dynamic from GNSS observation and not for the GRACE as the estimated hydrological variations were estimated monthly. At the same time, delays were calculated every two hours using GNSS observations. Nevertheless, this study shows the potential uses of geodetic approaches such as satellite gravity and GNSS observations to capture the hydrological dynamic.

1. Introduction

The Global Geodetic Observing System (GGOS) distinguishes geodesy into three aspects, e.g., the geokinematics, gravity field, and Earth rotation [30]. The geokinematics include methods, techniques, and fundamental aspects to determine the figure of the Earth both spatially and as a function of time. The second aspect correlates with the determination of Earth's gravity field describing the mass variations. Finally, the last aspect is related to the planet's rotation and forces affecting the Earth. Together, these three aspects form the reference frames [31].

Presently, several geodetic techniques can be used for addressing those aspects—for example, the Global Navigation Satellite System (GNSS) and satellite gravimeter. GNSS ensures users estimate the figure of the Earth by measuring accurate positions. The accuracy of the estimated position depends on which method that is implemented. When using Real-Time

Kinematic (RTK) positioning method, the achieved accuracy is up to several cm [4, 17, 9]. The accuracy can also be slightly worse at a few decimeters level when utilizing the Real-Time Precise Point Positioning (RTPPP) positioning method [14, 15]. Nevertheless, the highest accuracy (at mm level) is achieved when using static differential positioning, which is mostly implemented for tectonic [29] and land subsidence applications [1]. The Earth's gravity field can be determined using satellite gravimeters. Various studies have been conducted using several satellite gravimeters, e.g., CHAMP (CHAllenging Minisatellite Payload), GRACE/GRACE-FO (Gravity Recovery And Climate Experiment/-Following On) and GOCE (Gravity field and steady-state Ocean Circulation Explorer), to produce geopotential models [3, 12, 23, 22].

Interestingly, we can also obtain other parameters other than the position and gravity field using the mentioned geodetic techniques. An attempt had been made by Long et al. [26] to monitor the terrestrial water storage using the GRACE mission in Texas. They found a large depletion in water storage corresponds with the severe drought. While He et al. [19] used GNSS observations to retrieve the atmospheric water content in the form of precipitable water vapor. Further, they characterized the precipitable water vapor variation due to the super typhoon Mangkut in Hong Kong.

Considering the above, we aim to explore the use of satellite gravimeter (particularly GRACE/GRACE-FO) and GNSS observations to monitor the terrestrial and atmospheric hydrological dynamic. We further explore the behavior of the results. At this preliminary stage, we limit our study area to Bandung, Indonesia.

2. Data and Methods

2.1. GRACE/GRACE-FO Monthly Solutions

GRACE/GRACE-FO mission aims to provide a global-scale of temporal gravity field observation. The GRACE/GRACE-FO twin satellites measure the distance between their satellites by the K-band to estimate the Earth's gravity field. Any mass anomaly beneath the Earth's surface disturbs the satellites and eventually changes the distance between the twin satellites. However, the mass anomaly sensed in the GRACE/GRACE-FO measurements is given in an aggregate form. Therefore, additional measurements are needed to distinguish it into different components. First, the GNSS receiver embedded in the twin satellites and the satellite laser ranging measurements help precisely determine their positions and the orbit location. Next, accelerometers reduce any source of accelerations due to non-gravitational effects. Finally, the two-star cameras determine the relative and absolute orientation of the twin satellites. Detailed explanations can be referred to [21, 28, 34].

In this study, we used the level-2 data of GRACE/GRACE-FO missions, which are given in spherical harmonic coefficients of C_{nm} and S_{nm} for the period 2002 to 2021. These coefficients are provided by several agencies, e.g., the Center for Space Research (CSR), the Geo-ForschungsZentrum (GFZ), and the JPL (Jet Propulsion Laboratory). The solutions from these agencies are given on a monthly basis, with a maximum degree of up to 90. This maximum degree corresponds to approximately 200 km of spatial resolution.

2.2. Equivalent Water Height Estimation

Before computing the equivalent water height from monthly harmonic solutions, we considered applying the following procedures to each monthly solution. First, we replaced the C_{20} coefficients with the coefficients from satellite laser ranging (SLR) observations [27]. Next, we introduced the degree 1 coefficients (C_{10}, C_{11}, S_{11}) as GRACE did not include these coefficients. These coefficients represent the variation of the Earth geocenter, which is mainly due to the seasonal effects [35]. The magnitude of Earth geocenter variation might reach up to ± 10 mm. We also truncated the monthly solution to a degree and order of 60 to reduce the error at smaller wavelengths [36]. The mean value of the coefficients was then removed, and we obtained the residuals of every single month ($\Delta\bar{C}_{nm}$ and $\Delta\bar{S}_{nm}$). Finally, a Gaussian smoothing was applied to the solution because the monthly harmonic solutions are contaminated by the correlated error causing the stripes errors in the spatial domains. The Gaussian smoothing radius of 350 km was used.

The terrestrial water surface, in the form of equivalent water height Δh, can be computed by the following [38]:

$$\Delta h(\theta, \lambda) = \frac{a_E \rho_E}{3\rho_w} \sum_{n=0}^{60} \sum_{m=0}^{n} \frac{2n+1}{1+k_n} \bar{P}_{nm}(\cos\theta)[\Delta\bar{C}_{nm}\cos(m\lambda) + \bar{S}_{nm}\sin(m\lambda)] \tag{1}$$

where θ and λ is the latitude and longitude of the computation point, respectively; a_E is the Earth's mean radius; ρ_E is the density of the Earth; ρ_w is the density of water; k_n is the load love number; $\bar{P}_{nm}$ is the fully normalized associated Legendre functions.

2.3. GNSS Observations and Data Processing

A continuously operating GNSS of ITB1, maintained by the Geodesy Research Group of ITB, was used to retrieve the precipitable water vapor. It used the Leica GRX1200GGPRO GNSS and the LEIAX1202GGPRO GNSS for the receiver and antenna, respectively, classified as geodetic GNSS and antenna. The receiver collected multi-GNSS data, including GPS and GLONASS, with a sampling interval of 1 second. The GNSS observations used in this study cover the period of 2011-2012.

To estimate the delays due to tropospheric disturbances, we used a scientific GNSS data processor of Bernese GNSS software version 5.2 [10]. The differential positioning mode was implemented to ensure an accurate result estimation. The solutions were constrained to twelve International GNSS Service (IGS) GNSS stations near the study area. The tropospheric delays in the zenith direction were estimated every two hours.

Following are several details regarding the processing parameters implemented in Bernese GNSS software: (i) we used the precise orbit and clock products provided by the Center for Orbit Determination in Europe (CODE) [11] to correct the satellite's orbital and clock errors, (ii) the General Mapping Function (GMF) by Boehm et al. [6] was used as a priori information, (iii) the high orders (up to the third-order) of ionospheric delays were estimated by first introducing the global ionospheric map from CODE, (iv) and 15° of elevation mask angle was used to reduce the multipath effects. The Bernese Processing Engine (BPE) was used to estimate the solutions automatically.

2.4. Precipitable Water Vapor Retrieval

Having zenith tropospheric delay (ZTD) estimated, the precipitable water vapor (PWV) can be calculated by subtracting zenith hydrostatic delay (ZHD) from ZTD. The residual is then multiplied by a scale factor of Π. The computation can be mathematically described as follows [40]:

$$PWV = \Pi(ZTD - ZHD) \tag{2}$$

where Π and ZHD are defined as:

$$\Pi = \frac{10^6}{\rho_w R_w (k_2' + \frac{k_3}{Tm})} \tag{3}$$

$$ZHD = (2.2779 \pm 0.0024) P_s / f(\theta, H) \tag{4}$$

with T_m is the weighted mean temperature, while k_2', k_3, and R_w are constants. One can adopt the following value for the constants: (i) $k_2' = 16.48$ K$\cdot$ hPa^{-1}; (ii) $k_3 = (3.776 \pm 0.014)\cdot 10^5$ [K^2

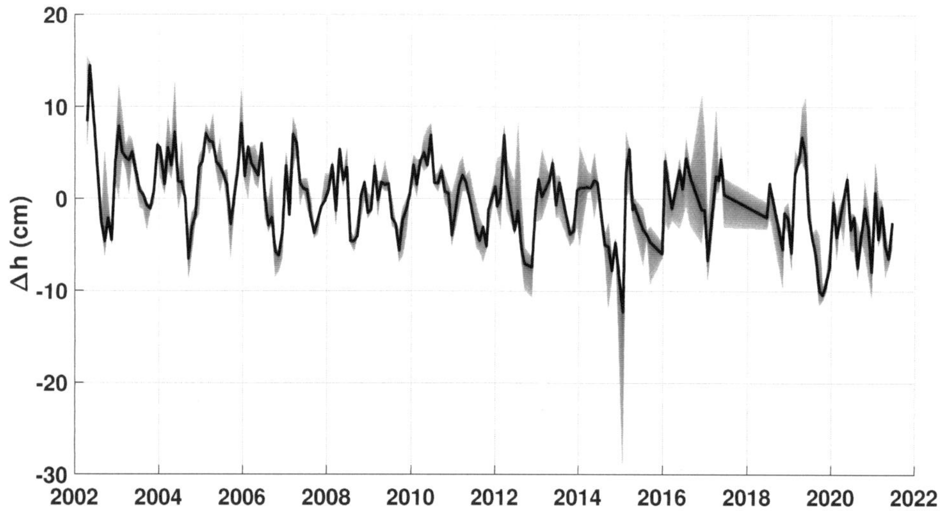

Figure 1: The estimated equivalent water. The black line represents the median solution with its variability marks by the red area.

$\cdot$ hPa^{-1}], and (iii) $R_w = 461$ [J$\cdot$ kg$^{-}1 \cdot$ K$^{-}1$]. P_s is the total pressure and H is the ellipsoidal height.

Since the computation of Π requires complete information about the atmospheric profile data along the zenith direction and we did not measure the required data, we adopted a constant value. Wang et al. [39] reported that Π varies from 0.14 to 0.165. Therefore, 0.15 seems acceptable to be used in this study. While the pressure data for the computation of the hydrostatic delays were obtained based on the Global Pressure Temperature model (GPT) by Boehm et al. [7].

3. Results and Discussion

3.1. Equivalent Water Height from GRACE/GRACE-FO Observations

Figure 1 displays the estimated equivalent water height representing the terrestrial water storage variation at the study area. The median water storage variation variability varies from approximately -12 to 14 cm throughout the observed period. Table 1 lists the basic statistics of equivalent water height from different solutions at the study area. The difference in standard deviation is around 1 cm, indicating that they have a similar variation. However, we notice that GFZ solutions seem to have a larger discrepancy at the beginning of 2015.

We observe an apparent yearly periodicity presents in the time series. The maximum variation likely occurs between December to March. On the other hand, periods between August to October appear to have a relatively low value compared with other periods. Interestingly, this characteristic seems to agree up to early 2014. In late 2014, we obtained a very low equivalent water height, and it lasted until 2015 except for a few months in early 2015. YPTE [41] reported that 2014 and 2015 were two of the hottest years. The oceanic niño index (ONI), used by the National Oceanic and Atmospheric Administration (NOAAA) for classifying warm and cold events, also indicates that those years are considered warm periods. Additionally, Lestari et al. [25] also reported that Indonesia suffered from catastrophic drought at those periods by analyzing three datasets of precipitation, Sea surface temperature, and low-level wind data.

Table 1: Statistics of equivalent water height for each model at study area.

Source	Max (cm)	Min (cm)	STD (cm)
JPL	15.2	-12.3	4.7
CSR	14.4	-12.1	4.2
GFZ	13.4	-28.8	5.2

3.2. Precipitable Water Vapor from GNSS Observations

Here, we show the time series of the estimated tropospheric delays using GNSS observations needed for computing the precipitable water vapor. The tropospheric delays in the zenith direction are shown in Figure 2a. It varies from ~2.2 to ~2.4 meters throughout 2011 to 2012. It should be noted that the delays in GNSS propagation are not only affected by the water content in the tropospheric medium but also caused by other gasses, e.g., N_2, O_2, Ar [24]. Therefore, it is necessary to remove the corresponding 'dry' delays component using Eq. 4 and obtain the wet component of tropospheric delays that varies between 0.1 to 0.4 meters (Figure 2b). Then, we applied the multiplication factor of 0.15 to the data. Figure 2c shows the variability of observed precipitable water content as sensed using GNSS observations. It varies up to 5 cm for the period 2011-2012.

Similar to what we have seen in Figure 1, we also detect an annual periodicity present in precipitable water vapor series. The maximum magnitude lies within the December to March periods, while the minimum one occurs within August to October. However, due to the limited time of observations, we can not further observe any anomalous signature in 2014-2015 as observed using GRACE/GRACE-FO.

3.3. Spectral Content Analysis

To fully explore the data presented in Sections 3.1 and 3.2, we further analyze the spectral content of the data. We used the Lomb-Scargle Periodogram to detect periodic signals. This method allows us to capture the periodicity in random and unevenly sampled data [37]. The spectral content for the equivalent water height series and the precipitable water vapor series are shown in Figure 3. A clear yearly component is detected for the estimated equivalent water height data, but we cannot detect any prominent higher frequency components as the data is given monthly. Meanwhile, we observe a vague yearly periodicity for the precipitable water vapor data. It has a wide range of frequencies for the long-wavelength component indicating that they cannot be resolved or distinguished. This incapability of separating clear annual periodicity frequency with nearby frequencies is due to the limited length of the data. For example, we need at least eleven years of observations to separate the yearly periodicity with a repeating cycle of 365.25 days from the component with a period of 400 days according to the Rayleigh criterion. However, we see clear periodicity components at higher frequencies from precipitable water vapor series, e.g., diurnal, semi-diurnal, and ter-diurnal periodicity components.

Efforts have been made by Sekaranom et al. [33] to monitor the precipitation variability in Indonesia using the Tropical Rainfall Measuring Mission (TRMM) satellites. They successfully classified the climate in Indonesia into three different major climate zones, including monsoonal, equatorial, and anti monsoonal zones (see Figure 4). Our study area is classified as a monsoonal zone with the highest precipitation between December and March. It agrees with what we have seen in both time and frequency domains for the estimated equivalent water height and precipitable water vapor series.

In order to explain the diurnal periodicity presented in the precipitation water vapor series, we further limit our analysis to one-day observation. We select the data on 28th March 2011 as depicted in Figure 5. It has a maximum peak at around 08:00 UTC or 15:00 in local time and

IOP Conf. Series: Earth and Environmental Science 1047 (2022) 012001 doi:10.1088/1755-1315/1047/1/012001

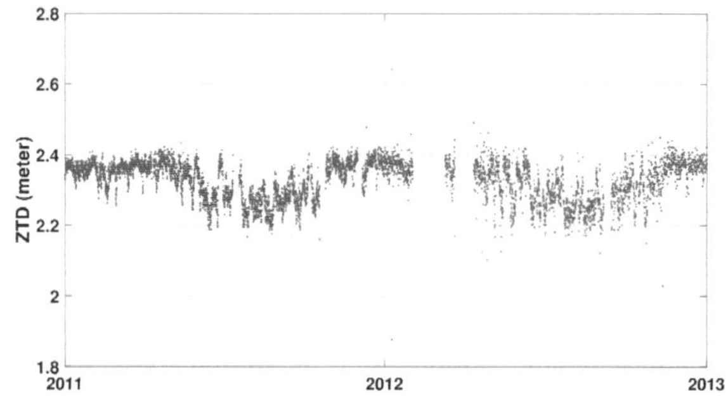

(a)

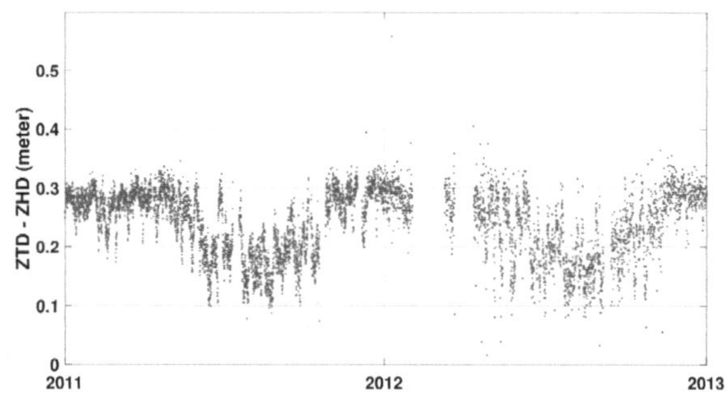

(b)

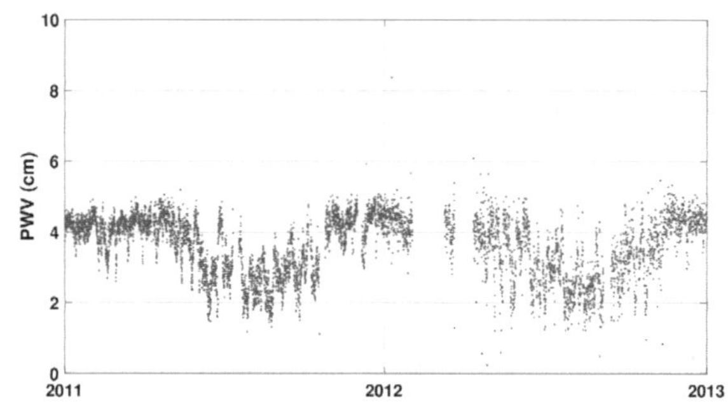

(c)

Figure 2: (a) Zenith tropospheric delays, (b) wet component of tropospheric delays in zenith direction, and (c) precipitable water vapor. Note different units in the y-axis.

has a minimum value at 18:00 UTC or 01:00 in local time. It indicates that the amount of water vapor in the atmosphere relates to the rotation of the Earth, which causes the change between day and night. In daylight, the heat energy from the sun breaks the molecular bonds of water in the liquid state [18, 20]. Eventually, the water transforms into gas, increasing the water content in the atmosphere. However, we leave the causes of semi- and ter-diurnal periodicity unknown and subject to further research.

4. Final Remarks

We present the use of several geodetic techniques, e.g., GRACE/GRACE-FO and GNSS, to monitor the terrestrial and atmospheric hydrological dynamic at Bandung, Indonesia. They are

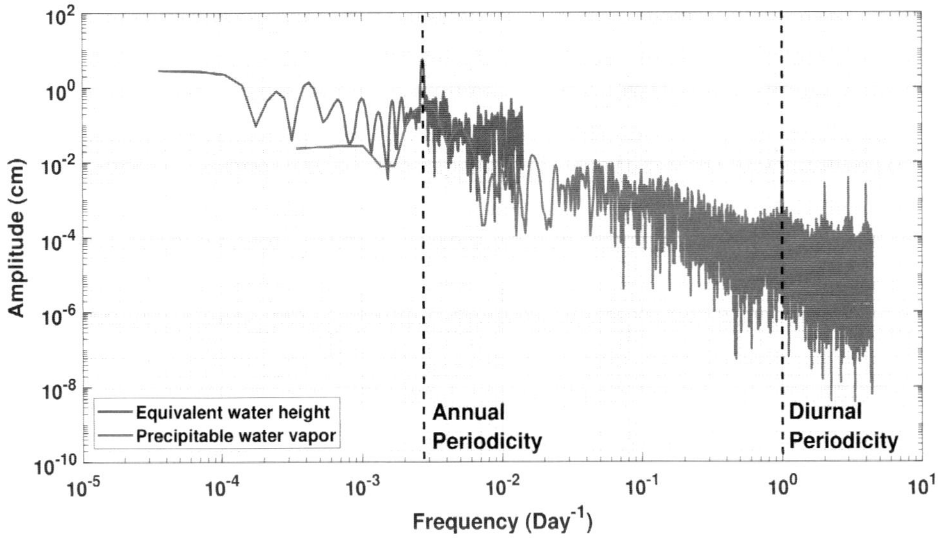

Figure 3: The spectral contents of the equivalent water height and precipitable water vapor.

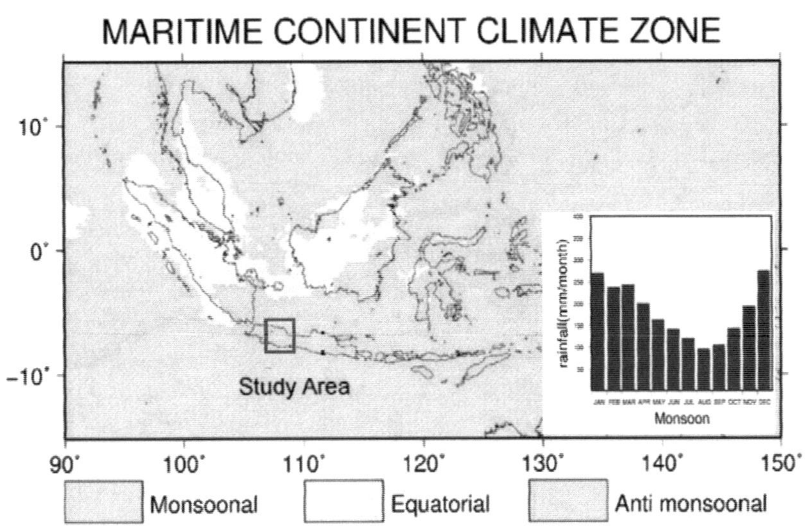

Figure 4: Three major climate zones in Indonesia and the monthly mean precipitation for monsoonal zone (modified from Sekaranom et al. [33]). The red box indicates the study area.

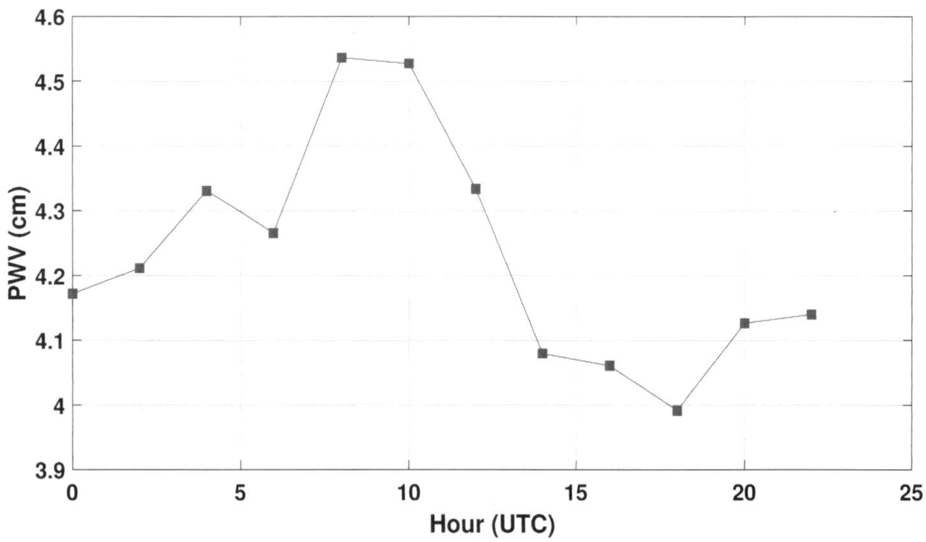

Figure 5: Precipitable water vapor variation on 28th March 2011.

represented in the form of equivalent water height and precipitable water vapor for the terrestrial and atmospheric water storage, respectively. The terrestrial water storage varies between -12 to 14 cm for 2002-2021, and we observe a smaller variation of precipitable water vapor of up to 5 cm throughout the GNSS observations period. Additionally, we observe several periodicity components present in the time series. We capture annual periodicity components that correlate with the monsoonal climate for terrestrial water storage series. While for the atmospheric water storage series, we additionally observe higher frequencies components, e.g., diurnal, semi-diurnal, and ter-diurnal periodicity. The diurnal periodicity relates to the evaporation process. However, the causes of higher frequencies components are still left unknown.

The Geospatial Information Agency of Indonesia (BIG) maintains 137 continuously operating GNSS stations in Indonesia [5]. Considering the significant number of continuously operating GNSS stations in Indonesia, we highlight a huge opportunity of retrieving vast tropospheric studies. For example, monitoring the dynamic of precipitable water content in real-time [19] and robust analysis of tropospheric studies using tomography technique [13], are two of many applications that can be implemented in reality. At a smaller scale or area, one may also consider using a low-cost GNSS receiver and antenna. Bramanto et al. [8] successfully utilized a low-cost GNSS receiver to monitor the atmosphere variability.

Despite the low time resolution of GRACE/GRACE-FO solutions, we stress that long-term monitoring applications using mentioned approach to estimate the secular terrestrial water storage change can be beneficial. Water storage sensed by using GRACE/GRACE-FO solutions is the accumulation of the surface water storage (soil moisture, snow, and plant canopy) and groundwater storage. At present, several global hydrological models have been developed by several agencies, e.g., GLDAS [32], and ERA5 [2]. The model includes the bespoken surface water storage. Eventually, one can observe the groundwater changes after subtracting the surface water storage with the models. This approach can be an alternative to estimate the trend in groundwater changes that are needed for answering several hazard phenomena, e.g., land subsidence [16].

References

[1] Hasanuddin Z. Abidin, I. Gumilar, H. Andreas, D. Murdohardono, and Y. Fukuda. On causes and impacts of land subsidence in Bandung Basin, Indonesia. *Environ. Earth Sci.*, 68(6):1545–1553, 2013.

[2] Clement Albergel, Emanuel Dutra, Simon Munier, Jean-Christophe Calvet, Joaquin Munoz-Sabater, Patricia de Rosnay, and Gianpaolo Balsamo. ERA-5 and ERA-Interim driven ISBA land surface model simulations: Which one performs better? *Hydrol. Earth Syst. Sci. Discuss.*, 22:3515–3532, 2018.

[3] M. J. Amos and W. E. Featherstone. Comparisons of Global Geopotential Models With. *Geomatics Research Australasia*, 79, 2003.

[4] Heri Andreas, Hasanuddin Zainal Abidin, Dina Anggreni Sarsito, and Dhota Pradipta. Study the capabilities of RTK Multi GNSS under forest canopy in regions of Indonesia. *E3S Web of Conferences*, 94, 2019.

[5] BIG. InaCORS BIG: Satu Referensi Pemetaan Indonesia. Technical report, Badan Informasi Geospasial - Indonesia, 2019.

[6] J Boehm, A Niell, P Tregoning, and H Schuh. Global Mapping Function (GMF): A new empirical mapping function based on numerical weather model data. *Geophys. Res. Lett.*, 33(L07304), 2006.

[7] Johannes Boehm, R. Heinkelmann, and H. Schuh. Short note: A global model of pressure and temperature for geodetic applications. *J. Geod.*, 81:679–683, 2007.

[8] Brian Bramanto, Irwan Gumilar, Teguh P Sidiq, Wedyanto Kuntjoro, and A Daniel. Sensing of the atmospheric variation using Low Cost GNSS Receiver. *IOP Conf. Ser.: Earth Environ. Sci.*, 149(012073), 2018.

[9] Brian Bramanto, Irwan Gumilar, Muhammad Taufik, and I Made D A Hermawan. Long-range Single Baseline RTK GNSS Positioning for Land Cadastral Survey Mapping. *E3S Web of Conference*, 94(01022), 2019.

[10] R. Dach, S. Lutz, P. Walser, and P. Fridez. Bernese GNSS Software Version 5.2. User manual. Technical report, Astronomical Institute, University of Bern, 2015.

[11] Rolf Dach, Stefan Schaer, Daniel Arnold, Maciej Sebastian Kalarus, Lars Prange, Pascal Stebler, Arturo Villiger, and Adrian Jäggi. CODE final product series for the IGS. Technical report, Astronomical Institute, University of Bern, 2020.

[12] Serkan Doganalp. An Evaluation of Recent Global Geopotential Models for Strip Area Project in Turkey. *Earth Sci. Res. J.*, 20(3):1, 2016.

[13] Zhounan Dong and Shuanggen Jin. 3-D water vapor tomography in Wuhan from GPS, BDS and GLONASS observations. *Rem*, 10(62), 2018.

[14] I. Gumilar, R. Y. Ananta, B. Bramanto, H. Z. Abidin, Surono, and N. Kishimoto. Initial Performance Assessment of GNSS Augmentation System using Quasi-Zenith Satellite System for Real-Time Precise Positioning Method in Indonesia. *IOP Conf. Ser.: Earth Environ. Sci.*, 767(012021), 2021.

[15] I. Gumilar, B. Bramanto, R. Y. Ananta, D. Haryanto, H. Z. Abidin, Surono, and N. Kishimoto. Performance Assessment of GNSS Augmentation System Using Quasi-Zenith Satellite System for Real-time Precise Positioning Method in Indonesia. *Int. J. Geospat. Environ.*, 8(2), 2021.

[16] I. Gumilar, T. P. Sidiq, I. Meilano, B. Bramanto, and G. Pambudi. Extensive Investigation of the Land Subsidence Impressions on Gedebage District, Bandung, Indonesia. *IOP Conf. Ser.: Earth Environ. Sci.*, 873(012044), 2021.

[17] Irwan Gumilar, Brian Bramanto, Fuad F. Rahman, and I Made D.A. Hermawan. Variability and Performance of Short to Long-Range Single Baseline RTK GNSS Positioning in Indonesia. *E3S Web of Conferences*, 94(01012), 2019.

[18] Jay M. Ham, J. L. Heilman, and R. J. Lascano. Determination of soil water evaporation and transpiration from energy balance and stem flow measurements. *Agric. For Meteorol.*, 52(3-4):287–301, 1990.

[19] Qimin He, Kefei Zhang, Suqin Wu, Qingzhi Zhao, Xiaoming Wang, Zhen Shen, Longjiang Li, Moufeng Wan, and Xiaoyang Liu. Real-Time GNSS-derived PWV for typhoon characterizations: A case study for super Typhoon Mangkhut in Hong Kong. *Remote Sens.*, 12(104), 2020.

[20] J. L. Heitman, R. Horton, T. J. Sauer, and T. M. DeSutter. Sensible heat observations reveal soil-water evaporation dynamics. *J. Hydrometeorol*, 9(1):165–171, 2008.

[21] Pedro Inácio, Pavel Ditmar, Roland Klees, and Hassan Hashemi Farahani. Analysis of star camera errors in GRACE data and their impact on monthly gravity field models. *J. Geod.*, 89:551–571, 2015.

[22] Juraj Janák and Martin Pitoňák. Comparison and testing of GOCE global gravity models in Central Europe. *J. Geod. Sci.*, 1(4):333–347, 2011.

[23] Ramin Kiamehr and L. E. Sjöberg. Comparison of the qualities of recent global and local gravimetric geoid models in Iran. *Studia Geophys. et Geod.*, 49(3):289–304, 2005.

[24] C. Kitpracha, D. Promchot, P. Srestasathiern, and C. Satirapod. Precise Tropospheric Delay Map of Thailand using GNSS Precise Point Positioning Technique. *Int. J. Geoinformatics*, 13(2), 2017.

[25] D. O. Lestari, E. Sutriyono, Sabaruddin, and I. Iskandar. Severe Drought Event in Indonesia Following 2015/16 El Niño/positive Indian Dipole Events. *J. Phys. Conf. Ser.*, 1011(012040), 2018.

[26] Di Long, Bridget R. Scanlon, Laurent Longuevergne, Alexander Y. Sun, D. Nelun Fernando, and Himanshu Save. GRACE satellite monitoring of large depletion in water storage in response to the 2011 drought in Texas. *Geophys. Res. Lett.*, 40:3395–3401, 2013.

[27] Bryant D. Loomis, Kenneth E. Rachlin, David N. Wiese, Felix W. Landerer, and Scott B. Luthcke. Replacing GRACE/GRACE-FO C30 With Satellite Laser Ranging: Impacts on Antarctic Ice Sheet Mass Change. *Geophys. Res. Lett.*, 47(3):1–7, 2020.

[28] Torsten Mayer-Gürr. *Gravitationsfeldbestimmung aus der Analyse kurzer Bahnbögen am Beispiel der Satellitenmissionen CHAMP und GRACE.* Phd thesis, Universität Bonn, 2008.

[29] Irwan Meilano, Rino Salman, Suchi Rahmadani, Qibin Shi, Susilo Susilo, and Eric Lindsey. Source Characteristics of the 2019 Mw 6.5 Ambon, Eastern Indonesia, Earthquake Inferred from Seismic and Geodetic Data. *Seismol. Res. Lett.*, 2021.

[30] Wojciech Pachelski, Małgorzata Paśnicka, Karolina Szafranek, and Agnieszka Zwirowicz-Rutkowska. Geokinematics in GGOS-Different relations and dependences. *Artif. Satell.*, 46(1):9–17, 2011.

[31] Hans-Peter Plag and Michael Pearlman, editors. *Global Geodetic Observing System.* Springer, Berlin, Heidelberg, 2009.

[32] Matthew Rodell, P. R. Houser, U. Jambor, J. Gottschalck, K. Mitchell, C. J. Meng, K. Arsenault, B. Cosgrove, J. Radakovich, M. Bosilovich, J. K. Entin, J. P. Walker, D. Lohmann, and D. Toll. The Global Land Data Assimilation System. *Bull. Am. Meteorol. Soc*, 85(3):381–394, 2004.

[33] Andung Bayu Sekaranom, Emilya Nurjani, M. Pramono Hadi, and Muh Aris Marfai. Comparsion of TRMM Precipitation Satellite Data over Central Java Region – Indonesia. *Quaest. Geogr*, 37(3):97–114, sep 2018.

[34] Samira Sadat Soltani, Behzad Ataie-Ashtiani, and Craig T. Simmons. Review of assimilating GRACE terrestrial water storage data into hydrological models: Advances, challenges and opportunities. *Earth-Sci. Rev.*, 213, 2021.

[35] Sean Swenson, Don Chambers, and John Wahr. Estimating geocenter variations from a combination of GRACE and ocean model output. *J. Geophys. Res.*, 113(B08410):1–12, 2008.

[36] Sean Swenson and John Wahr. Methods for inferring regional surface-mass anomalies from Gravity Recovery and Climate Experiment (GRACE) measurements of time-variable gravity. *J. Geophys. Res.*, 107(B9):2193, 2002.

[37] Jacob T. VanderPlas. Understanding the Lomb–Scargle Periodogram. *Astrophys. J., Suppl. Ser.*, 236(16), 2018.

[38] John Wahr, Mery Molenaar, and Frank Bryan. Time variability of the Earth's gravity field: Hydrological and oceanic effects and their possible detection using GRACE. *J. Geophys. Res.*, 103(B12):30205–30229, dec 1998.

[39] Xiaoming Wang, Kefei Zhang, Suqin Wu, Shijie Fan, and Yingyan Cheng. Water vapor-weighted mean temperature and its impact on the determination of precipitable water vapor and its linear trend. *J. Geophys. Res. Atmos.*, 121:833–852, jan 2016.

[40] Xiaoming Wang, Kefei Zhang, Suqin Wu, Zishen Li, Yingyan Cheng, Li Li, and Hong Yuan. The correlation between GNSS-derived precipitable water vapor and sea surface temperature and its responses to El Niño–Southern Oscillation. *Remote Sens. Environ.*, 216:1–12, 2018.

[41] YPTE. El Nino in 2014 - 2016.

IOP Conf. Series: Earth and Environmental Science 1047 (2022) 012002 doi:10.1088/1755-1315/1047/1/012002

Utilization of Altimetry Satellites Microwave Radiometer for Atmospheric Observation over Indonesia

Fathin Nurzaman, Dudy D. Wijaya, Zamzam A. J. Tanuwijaya, Wedyanto Kuntjoro

Geodesy Research Group, Institut Teknologi Bandung, Jl. Ganesha 10, Lb. Siliwangi, Kecamatan Coblong, Kota Bandung, Jawa Barat 40132 Indonesia

wijaya.dudy@gmail.com

Abstract. Altimetry satellites have an on-board microwave radiometer whose original function is to measure the Tropospheric Wet Delay (TWD) for their main sea level measurement using radar altimeter. The TWD, as it reflects the amount of water vapor in the atmosphere, could be used to derive the Precipitable Water Vapor (PWV) which would be a good parameter to study the atmosphere variability, especially in Indonesia where rainfall is the main variable of its climate. Studies on altimetry-derived PWV are conducted, which include: analyzing their correlation with rainfall observation dataset, and assessing how well they could capture some key atmospheric phenomena in Indonesia. The results show that there are fairly high correlation between the altimetry-derived PWV and the rainfall data in monthly-scale with 0.63 Pearson correlation coefficient. Moreover, they could very well capture some key phenomena in Indonesia such as the monsoon and El Niño Southern Oscillation (ENSO) variation. Thus, a new way of utilizing altimetry satellite data is introduced in this paper: to observe the atmosphere using its water vapor data derived from the microwave radiometer. This open the possibility of utilizing altimetry satellites for both oceanographic and meteorological studies as it could measures both oceanic and atmospheric parameter, with its radar altimeter and microwave radiometer, respectively.

1. Introduction

The atmosphere is holding great influence to everyday lives. Knowledge regarding the atmosphere hence holds great value. Deeper understanding of its characteristics like the variability or long-term trend would improves atmospheric models which would ultimately makes weather forecasts and climate predictions more accurate. With good forecasting we would be able to act accordingly to the state of atmosphere which would let us benefit more from it and even avoid potential disasters. Therefore, to better the understanding of the atmosphere a lot of atmospheric observation has been done. This paper brings up an alternative technology for atmospheric observation, which is the altimetry satellite.

The main focus of altimetry satellite measurement is to observe the world's ocean surface dynamics [1]. Altimetry satellite estimate sea level by measuring the vertical distance to the sea surface with radar altimeter. Some correction parameters are needed for the distance measurement. One of them is the Wet Tropospheric Correction (WTC) to correct the Tropospheric Wet Delay (TWD), which is the path delay caused by the water vapor content in the atmosphere. Water vapor content in the atmosphere have high variability which makes them hard to model. Therefore, to have the needed accuracy an additional measurement that specifically measure the water vapor content is required. For this purpose, altimetry satellite generally has an on-board radiometer measurement to measure the water vapor content in the atmosphere as to provide the precise TWD value.

Altimetry satellites measure water vapor content using an on-board microwave radiometer. Topex/Poseidon and all the follow-on satellites namely Jason-1, Jason-2, and Jason-3 (hereby these four satellites will be referred to as the Jason altimetry mission series by the author), use a microwave radiometer, namely the Topex Microwave Radiometer (TMR), Jason-1 Microwave Radiometer (JMR), Advanced Microwave Radiometer (AMR), and AMR-2, respectively. These radiometers detects emissivity or brightness temperature (TB) at three different frequencies. The main channel detects at 23.8 GHz (21 GHz in Topex/Poseidon) for the water vapor content measurement, the other two channel detects at 34 and 18.7 GHz (37 and 18 GHz in Topex/Poseidon), meant to provide correction on the main water vapor measurement due to non-rainbearing clouds and wind-driven variation respectively [1,2]. The TWD is derived from the three TBs using an algorithm explained in [3]. The uncertainty of the derived wet path delay is less than 1.2 cm [2]. This TWD data, is what going to be analyzed in this paper.

Atmospheric water vapor content is one of the parameter that we could observe to assess the climate characteristics, especially in Indonesia where rainfall is the main variable of its climate. And the fact that Indonesia is mostly covered by sea because it's an archipelagic country means that the spatial coverage of the altimetry data that only available on water-covered surface will still be sufficient. Furthermore, with the TWD retrieval algorithm for near-coastal region has been improved [4] the uncertainty of the data won't be so compromised. These lead us to the possibility of a good atmospheric observation using this technology especially in Indonesia. With the temporal resolution of 10 days and good spatial coverage the altimetry data should be able to capture some global variation which should be useful for atmosphere dynamics study over the Indonesian region.

Study regarding the Precipitable Water Vapor (PWV) retrieval from altimetry satellites' radiometer data has been done by Weng et al. as they try to retrieve PWV from the Jason-2 AMR data using the algorithm to retrieve PWV from Advanced Microwave Sounding Unit A (AMSU-A) data [5]. The PWV is compared with the PWV from AMSU-A and they found high linear correlation between the two (>0.9). Another one is done by Thao et al. as they compare the water vapor from Jason-1 JMR and Envisat MWR (Microwave Radiometer) to the AMSR-E (Advanced Microwave Scanning Radiometer for Earth Observing System) as the reference [6]. They found good agreement on the spatial distribution of water vapor from the radiometers with the correlation between time series is higher than 0.9.

Studies that specifically focus on the utilization of the altimetry satellites' radiometer data for atmospheric observation has not been covered much. Weng et al. did describe the global distribution of water vapor resulted from the altimetry satellite radiometer data but only in general [5]. Though the quality of the radiometer has been showcased in many studies. This is due to the fact that the TWD has always been the primary source of error to the main altimetric measurement. Therefore, improvement on the radiometer instrument along with the TWD retrieval algorithm has been done quite frankly, e.g. [7] and [4]. Varma et al. actually studied the potential of the radiometer data for atmospheric observation in [8]. They found good correlation between the TB from the radiometer with rain-rate estimate from SSM/I satellite, and use the analysis to develop rain-rate estimation algorithm using the TBs from the radiometer. Though at the time of the study, availability of data for the colocation analysis by Varma et al. between altimetry and SSM/I was still so limited.

Currently, altimetry satellite has been operating for around 30 years and GPM (Global Precipitation Measurement), a satellite dedicated for precipitation measurement using radar, has been launched for more than 7 years thus making it possible to do more thorough assessment. As such, this paper will assess the ability of altimetry microwave radiometer for atmospheric observation by comparing the data with a validated precipitation product from GPM. Further assessment are done by analyzing how well the data could capture well-known meteorological phenomena in Indonesia such as monsoon and ENSO (El Niño Southern Oscillation). The methodology of the study along with the data used are given in detail in Section 2. The analysis results are presented in Section 3. Finally the conclusion of the study is given in Section 4.

2. Data and Methods

2.1. Altimetry Wet Tropospheric Correction Data

The author acquire the altimetry WTC data for this study using Radar Altimeter Database System (RADS) version 4 [9]. RADS is a multi-mission altimetry database that offers an easy way to query the data with various criteria specified by user, such as period, missions and area of interest with consistent format.

This paper uses the WTC measured by Jason altimetry mission series with observation period of around 28 years spanning from 1992 to 2021. The spatial distribution of the data can be seen in **Figure 1**. The distance between adjacent pass with the same direction is 315 km in the equator. The resolution along-track or in other word, the space between adjacent footprint is ~5.8 km. Though the spatial resolution of the radiometer instrument itself is different. The TMR and JMR have ~50 km field of view, meaning that the received TB value was actually the averaged value over the field of view area. It got improved to 25 km with Jason-2 AMR and better than 25 km with Jason-3 AMR-2 [7]. The data are smoothed using along-track averaging for every three adjacent footprints to filter out some of the local variation.

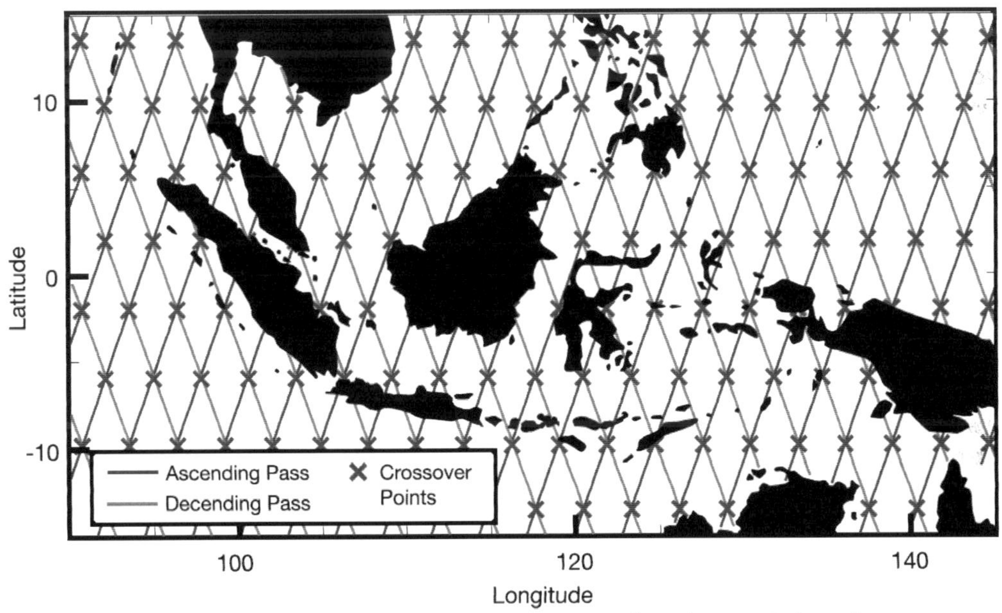

Figure 1 Jason altimetry mission series ground tracks over Indonesia

2.2. Deriving the PWV from Altimetry WTC

PWV is the height of total atmospheric water vapor if it's totally condensed in its vertical column. In this research, PWV is calculated from WTC using the following equations which was derived by Askne & Nordius [10]. Tropospheric Wet Delay (TWD), or the negative of WTC, is the PWV multiplied by a constant K which expressed as follows.

$$TWD = K\ PWV \tag{1}$$

with constant K defined as

$$K = 10^{-6}(\frac{k_3}{T_m} + k'_2)\rho R_v \tag{2}$$

where R_v is the specific gas constant for the water vapor (4.61527 m^3/kg.K), ρ is the density of water (997 kg/m^3), k'_2 = 17 K/mbar, and k_3 = 3.7765 × 10^{-5} K^2/mbar. T_m is the mean temperature of the atmosphere column in Kelvin. The mean temperature wasn't recorded by the altimetry satellite so, alternatively for this research, the T_m value is acquired using an existing model which is the Global Pressure and Temperature 2 wet (GPT2w) developed by Böhm et al. [11].

2.3. *Correlation Analysis with Calibrated Rainfall Data*

To see if the PWV derived from the altimetry could reflect the state of rainfall, the Pearson Correlation Coefficient value between the two is computed. The scatterplot of the two data are also analyzed beforehand to see any indication of correlation between the two. The calibrated rainfall data used in this study is the GSMaP rainfall dataset. Both data are compared in monthly time-scale meaning that the scatterplot and correlation coefficient are computed from the monthly average of both data.

GSMaP is a research project sponsored by Japan Science and Technology Agency (JST) under the Core Research for Evolutional Science and Technology (CREST) from November 2002, which was aimed to develop rainfall retrieval algorithms in order to produce high precision and resolution global precipitation map from satellite data [12]. GSMaP use the GPM satellite, which installed with a state-of-the-art Dual-frequency Precipitation Radar (DPR) and a microwave imager, initiated by NASA and Japan Aerospace Exploration Agency (JAXA). And to further improve the resolution, using the GPM data as core and reference standard, some other satellite data from various missions of partnering countries are assimilated including microwave imager satellites also geostationary microwave imager satellites to improve the temporal resolution, added with ancillary data such as atmospheric condition, sea surface temperature, rain gauge, topographic data, and high-resolution ocean/land flag [13–15].

Validation of the GSMaP_Gauge_NRT V6 Product has been performed in [16] by calculating the Correlation Coefficient (CC) and Root Mean Square Error (RMSE) values of the data with Radar-AMeDAS data as the reference. Radar-AMeDAS is a ground-precipitation radar dataset provided by the JMA which have been calibrated by rain gauges. The comparison are done in the Radar-AMeDAS observation areas, which was 250 km from the radar sites, and within 21 months observation period: April 2017 to December 2018. The results shows that the NRT data yields CC value of 0.61 and RMSE 0.35 mm/h.

2.4. *Identifying Indonesian Atmosphere Phenomena from the Data*

To see how well the data can be used for atmospheric study, the author try to identify from the data two already well-known atmosphere phenomena in Indonesia, i.e. the monsoon and ENSO. The author first analyses the PWV time series data of a single crossover point. Crossover point is used because it has higher temporal resolution as it got passed twice, by both ascending and descending pass, in each repeat period. The temporal variation is analyzed such as the seasonal oscillation which should be the monsoon and the monthly anomaly component which should be affected by ENSO. Spectrum analysis is also conducted to see is there are any wave component captured in the data. Next, in order to analyses spatial distribution or the propagation of PWV, data of a whole region is analyzed. The systematically scattered altimetry data are gridded into a surface and visualized as map. A PWV monthly climatology and monthly anomaly maps are generated for the purpose to identify the spatial variation of monsoon and ENSO.

In the following sub-subsections the mathematical detail of the spectrum analysis and surface gridding which were specifically used in this paper is explained.

2.4.1. Domain transformation using Successive Least Square Spectral Analysis Usual domain transformation such as Fast Fourier Transformation (FFT) can't be done as the sampling of the altimetry data in crossover is not equally spaced (temporally). Therefore, Successive Least Square Spectral Analysis (SLSSA) by Vaníček [17] is used which is the alternative method for domain transformation specifically for unequally spaced data. The method is done by iteratively finding the sinusoidal wave or spectrum that fit best to the data and subtracting that sinusoidal wave from the data before beginning the next iteration to find another spectrum. This subtraction in every iteration eliminate the possibility of spectral leakage in the resulting frequency domain graph.

2.4.2. Surface Interpolation using Tension Spline In analyzing a whole area to study the spatial distribution, the scattered altimetry-derived PWV data is gridded using Generic Mapping Tools (GMT) [18] with its surface module. The GMT perform the gridding using tension spline interpolation method. This method was first introduced by Schweikert [19] that fixes an undesirable characteristic of a normal cubic spline which is the appearing of extraneous inflection point or false local maxima/minima. Tension Spline uses a user-specified Tension Parameter which controls this characteristics. This method compute the gridded values z(x,y) that satisfies the following equation.

$$(1-T) * L\big(L(z)\big) + T * \;(L(z)) = 0 \tag{3}$$

where L is the laplacian factor and T is the user-specified tension parameter between 0 and 1. T value of 1 gives maximum tension which just result in locally linear interpolation and 0 gives "minimum curvature" which would have maximized oscillation and the appearing of local maxima/minima. This paper use 0.25 as the value of T as recommended by GMT.

3. Results

3.1. Correlation between Altimetry-derived PWV and GSMaP Rainfall

The scatterplot between the Altimetry-derived PWV Data with GSMaP rainfall is shown in **Figure 2**. To visualize the relationship pattern more accurately, the density of the scattered points is marked with color. This would reveal the underlying dominant pattern which otherwise wouldn't be apparent with the high amount of point being plotted.

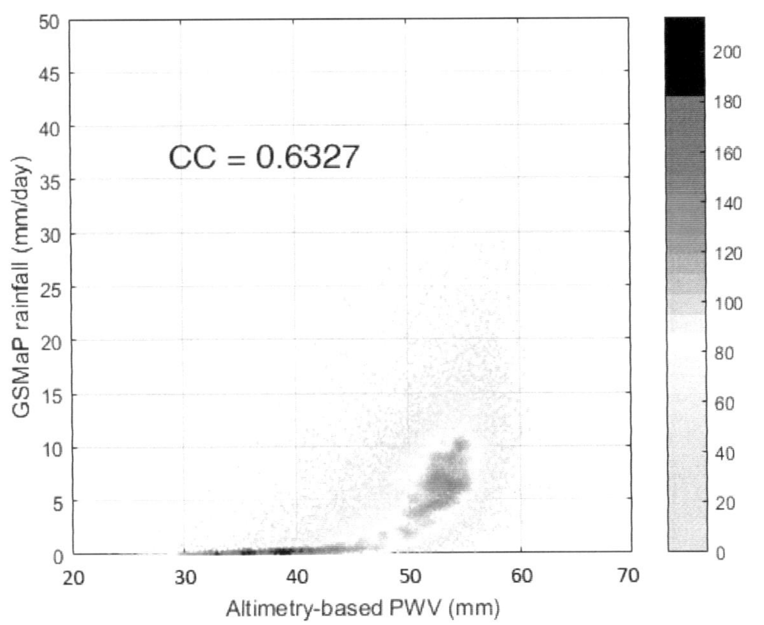

Figure 2. Scatter plot of monthly average Altimetry-based PWV and GSMaP rainfall on all crossover data over Indonesia Region (90°E – 140°E, 9°S – 9°N), the density level which defined as the number of points in the proximity of 0.1 standard deviation is marked with color.

The value of rainfall is lower than the PWV for all points in the scatterplot diagram which explains the fact that not all water vapor fall as rain. The distribution signifies that there are positive correlation between the two data and The Pearson Correlation Coefficient yields 0.63. However there seemed to be a threshold pattern: with less than ~45 mm PWV value, the rainfall value falls flat to zero while above that value the positive correlation begins. For water vapors to condense into water droplets and combine to grow heavy enough to fall as rain, some favorable conditions need to be met which are the saturation of water vapor in the air that depends on the amount of water vapor and the temperature; and the existence of aerosol as the condensation nuclei [20]. The first condition which is helped by the amount of water vapor in the air got materialized in the scatterplot as a pattern of positive correlation over certain threshold. Below ~45 mm of PWV value the chance of air to be saturated by water vapor is low and so the probability of rain is close to zero. And when the PWV value exceed ~45 mm the chance of high saturation starts to rise and so is the probability for rain to happen and the amount of rainfall also increases with the increasing of PWV. Although, the uncertainty to tell how heavy the rain is still high looking at how wide the distribution of the points. This uncertainty is expected because of the other conditions and also the meteorological dynamics: a parcel of air is always moving which means that a water vapor in a certain location might turn into rain only after having transported to another location where the favorable conditions are met.

Another thing that added the uncertainty is the difference in sampling between the PWV and rainfall observations with the rather arbitrary sampling of altimetry data within each month in which was used to compute the monthly mean. The altimetry data sampled only up to 6 data within a month (one ascending and descending per repeat period of ~10 days) while GSMaP data has 1 hour resolution as the result of assimilating data from various satellite mission.

3.2. Atmospheric Water Vapor Characteristics in Indonesia

Evaluating how well the Altimetry-derived PWV data in capturing key atmospheric phenomenon in Indonesia starts with the most common one which is the seasonal variation or the monsoon. **Figure 3** shows the time-series of PWV located in the east of Java Sea (109.1°E, 5.9°S). An oscillation with yearly period is apparent which suggest the change of season between dry and rainy. This oscillation got more highlighted in **Figure 4** in which data of every year got overlayed to each other. The PWV value dropped around DoY 150 to 300 and risen outside that period.

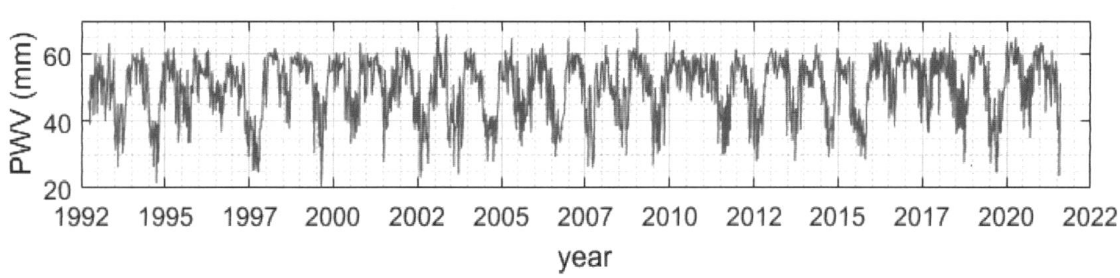

Figure 3 Time series of PWV in a crossover point in Java Sea (109.1°E, 5.9°S).

The period on when the PWV value rise and drop can be seen more clearly in the monthly-mean plot in **Figure 5**. The supposed dry and wet period is marked on the graph and we can see that the PWV value are rising and dropping accordingly to the supposed period which signifies that the annual monsoon variation is well captured by the altimetry radiometer observation. From this graph we could also see how the deviation in each months are across all available year compared to the long-term mean value on that month, or the climatology in meteorological term, which drawn as the red curve in the graph. This deviation from the climatology or, in other word, the anomaly will be identified further to identify the ENSO phenomenon.

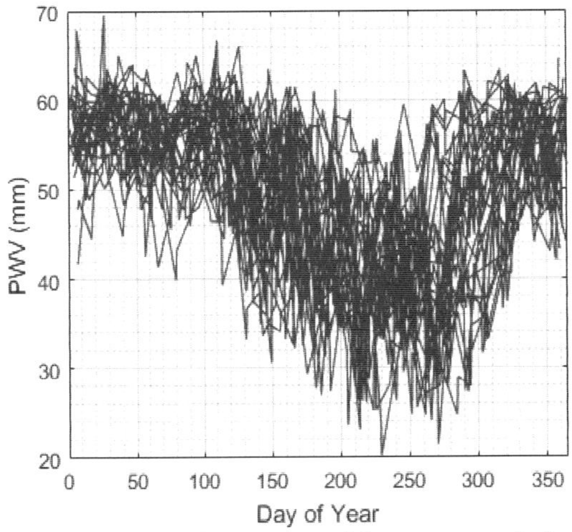

Figure 4 PWV time series with every year being overlayed one another.

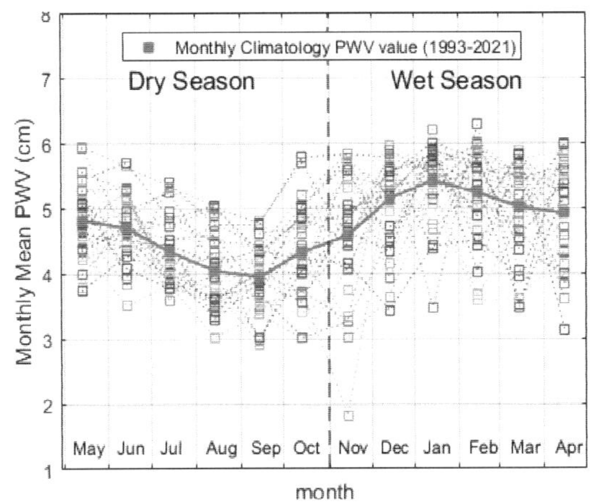

Figure 5 PWV monthly-mean plot for every observed year along with the climatology.

To identify all existing oscillation more thoroughly, the resulted frequency-domain graph by SLSSA is shown in **Figure 6**. From the graph we could see that the annual harmonic oscillation dominates over all other spectrums. There doesn't seem to be any other significant spike beside the annual one which means that the variability of the PWV in the location is dominated by the yearly monsoon variation. However, the frequency-domain graph is only sensitive to oscillation or wave with consistent period like the monsoon. Some anomaly modes which doesn't possess consistent period or only appears occasionally wouldn't be apparent in this graph at least before the dominant variation is filtered from the data.

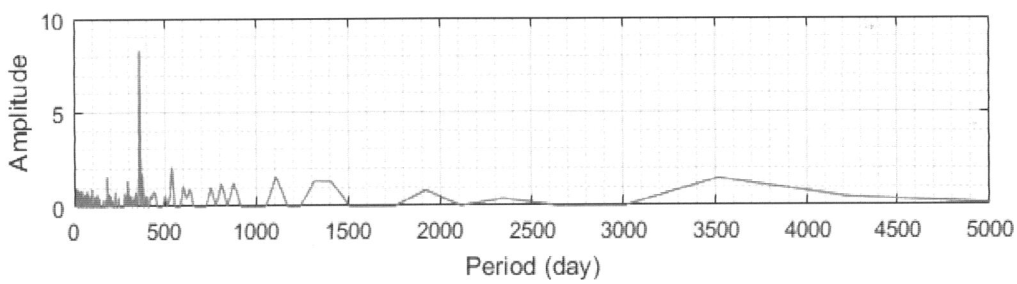

Figure 6 Spectrum analysis of the PWV time-series in a crossover point in Java Sea (109.1°E, 5.9°S).

It is of interest to analyze more locations so we can study the spatial variability of the PWV over Indonesia. The monthly-mean plot of 16 points distributed around the Indonesian Region can be seen in **Figure 7**. All the points were distributed across different latitude and longitude. We could see that the oscillation is weak near the equator and got higher as it goes further from the equator and we could also see the opposing phase of oscillation between the southern and northern hemisphere. Near the equator the semiannual variation is more apparent. This describes the yearly variation due to the shifting of the Earth's inclination towards the Sun which makes the Sun's relative position towards the equator moves. In the equator the sun passes through twice a year which makes the PWV data close to it have semiannual variation. When the sun is more on the north side, during the boreal summer, the northern part of Indonesia got wetter season due to the increasing evaporation of the ocean, while the southern part of Indonesia get their dry season, and vice versa during boreal winter. This will be explained better in the next graph.

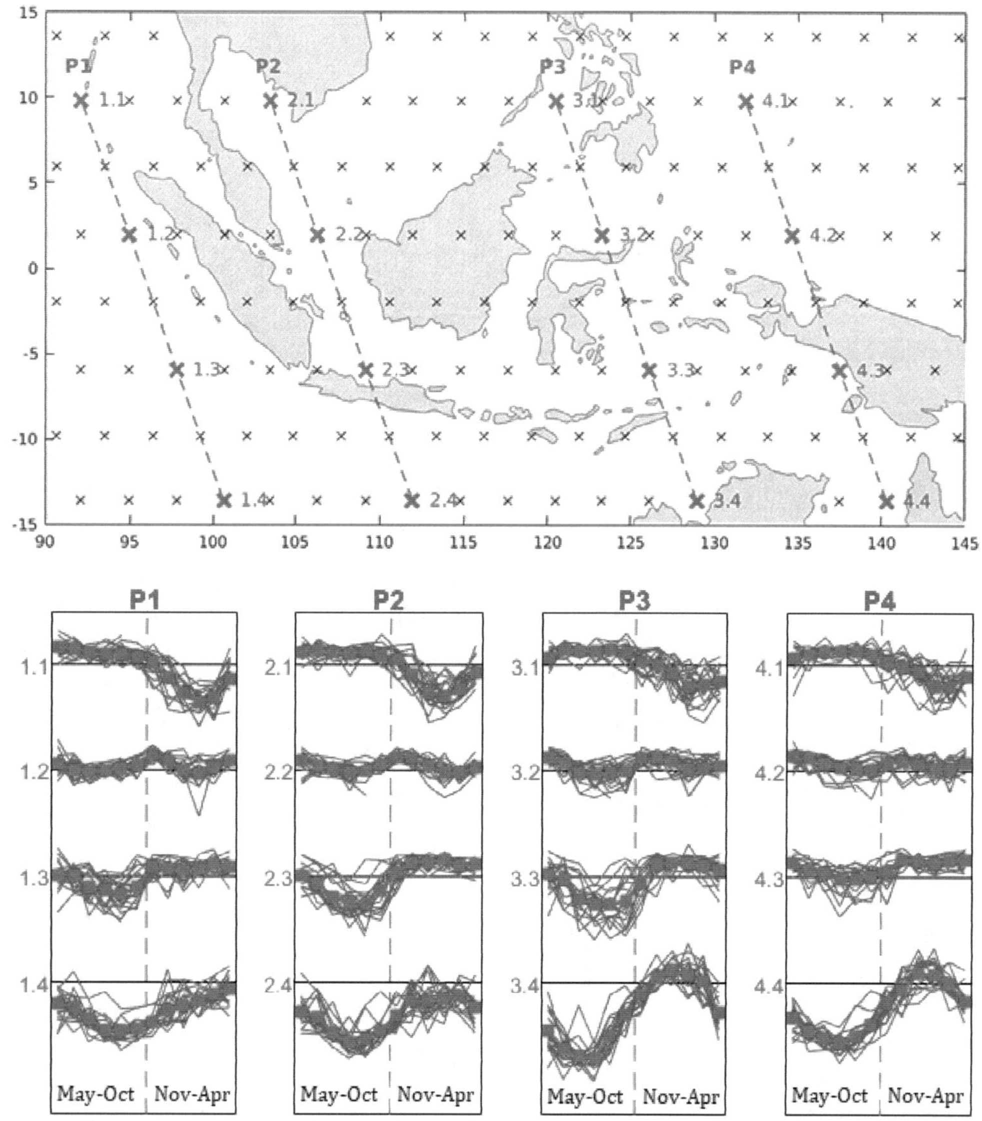

Figure 7 Monthly Mean plot of PWV in 16 crossover points scattered over the Indonesian region (bottom) with the map that shows the distribution of the points (top). The solid red square curve denotes climatology and the grey curves stacked up behind is the monthly-mean of all observed years, the horizontal black line marks PWV value of 40 mm and the separation between adjacent curve in the vertical direction is also 40 mm of PWV as the y-axis unit.

To be able to see the spatiotemporal seasonal variability better, the author use all the computed climatological PWV data over Indonesia and interpolate them into map. This is shown in **Figure 8**. From this PWV climatology map we can see the spatiotemporal variability with more detail. We can see the shifting of a meridional band with high water vapor content which we refer as the Intertropical Convergence Zone (ITCZ). The ITCZ generally follows the relative position of the Sun to the equator as more heating from the sun would cause more evaporation hence forming the ITCZ.

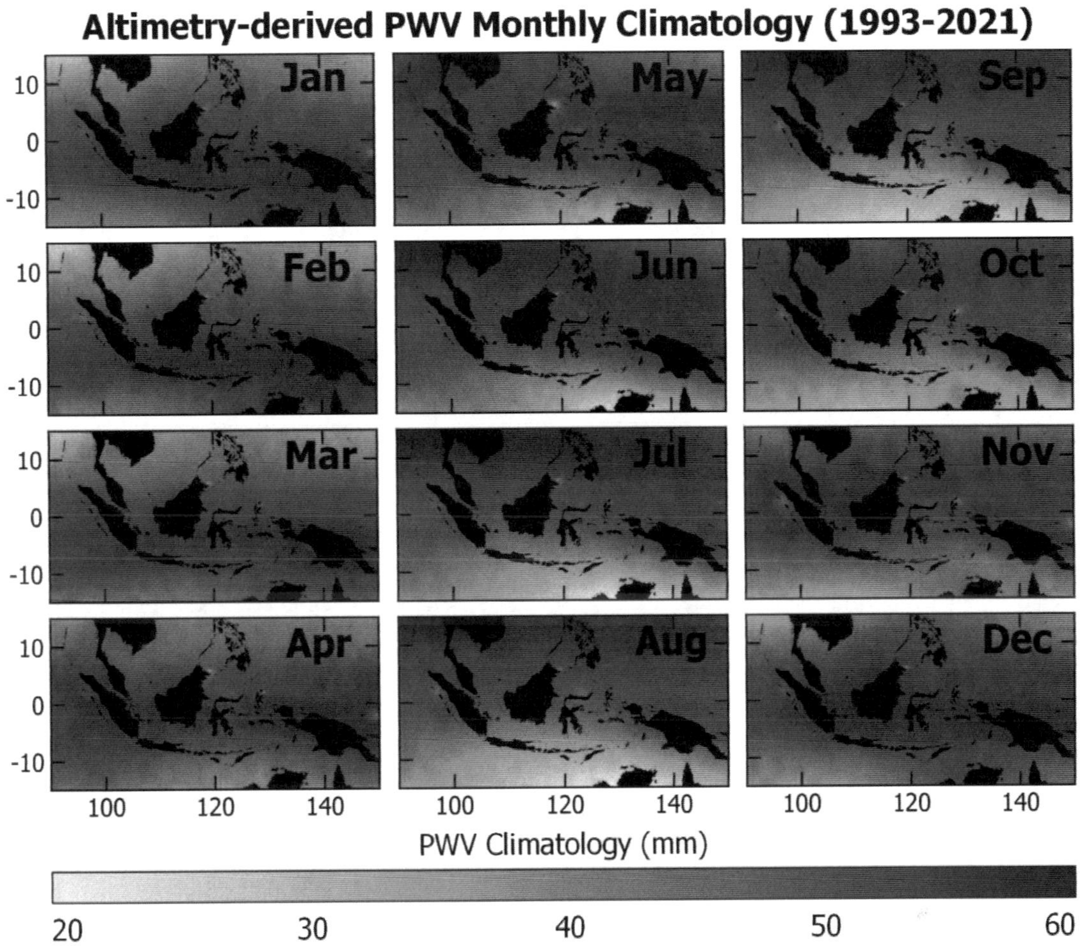

Figure 8 Monthly-Mean Climatology PWV Map over Indonesia

We could also see here the effect of the arid Australian landmass to the climate of the southern part of Indonesia. In austral winter (June-September) as the ITCZ and high atmospheric pressure located slightly northward from the equator, the prevailing wind blows northwest from Australia bringing dry air to some of the southern islands in Indonesia causing dry season. While during boreal winter (Dec-Feb) the prevailing wind blows from the Pacific and South China Sea bringing air that rich in water vapor of the ocean and causes rainy season.

3.3. Impact of ENSO

The impact of ENSO event would be best seen after eliminating the seasonal spectrum from the data thus leaving only the anomaly component. **Figure 9** shows the time-series PWV including the seasonal component which is the computed monthly climatology and in the bottom graph is the anomaly component which computed by subtracting the climatology component from the original data. The author marked three different period with differing ENSO episodes based on the El Niño monitoring index by the NOAA Climate Prediction Center. The three periods are 2010, 2013, and 2015 which respectively are the La Niña, normal and El Niño period. We could noticeably see the different effect on the PWV during these three periods, higher water vapor content during La Niña and lower during El Niño.

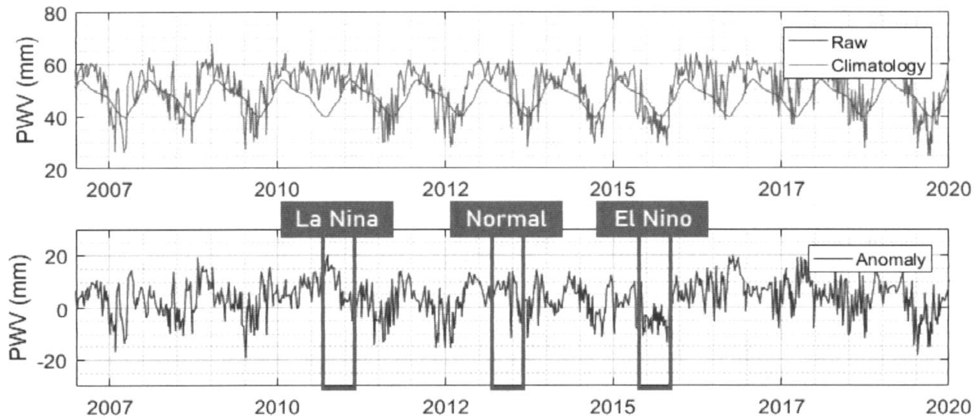

Figure 9 PWV Time-series of a crossover point in the Java Sea (109.1°E, 5.9°S) along with the seasonal component (top) and its anomaly with period of La Nina, normal, and El Nino highlighted (bottom).

To better see the impact of the ENSO on the water vapor distribution over Indonesia the PWV anomaly data over Indonesia are interpolated into maps. The PWV anomaly map of the previous sampled three periods are shown in **Figure 10**. In the figure we could see the impact distribution of the anomaly event. We could see that the southern part of Indonesia got impacted the most during the 2010 La Niña. And during 2015 El Niño the impact are more concentrated on the central islands: around Kalimantan and Sulawesi.

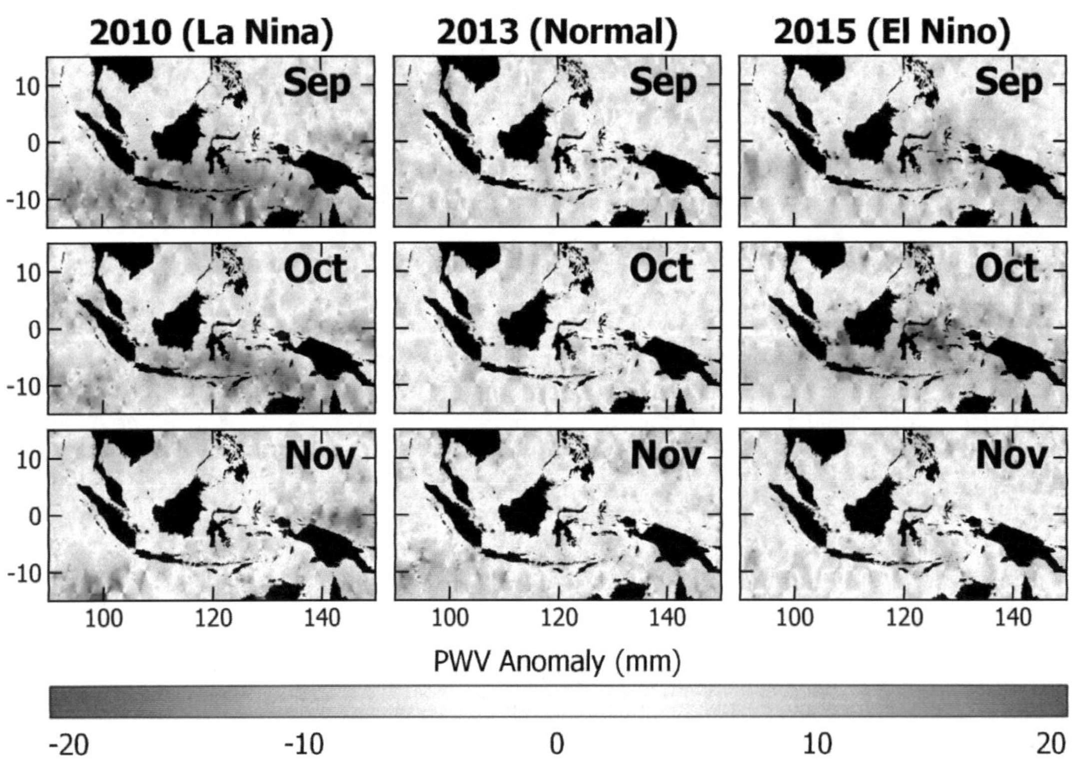

Figure 10 PWV anomaly map during three different condition: extreme La Nina (left), normal (center), and El Nino (right)

4. Conclusion

This paper showcased the ability of altimetry satellites for atmospheric observation using their TWD data over Indonesian region. The altimetry satellites' TWD can be a good estimator for rainfall, the main climate variable in Indonesia, as can be seen from the correlation between the derived PWV data with the GSMaP rainfall data, more specifically the GSMaP_Gauge_NRT V6. The Pearson correlation of both data in monthly-scale is 0.63. The altimetry-derived PWV data performance on portraying two of the well-known atmospheric phenomenon in Indonesia, the Monsoon and ENSO, is also shown here. From the time-series graph of a crossover point in East Java Sea the monsoon variation are apparent as an oscillation of PWV value with one year period. Even better from the generated monthly climatology map, which let us see the spatial distribution of the water vapor, we could see how the dry air came from Australia landmass that causes the dry season in the southern part Indonesia. Furthermore, from the computed monthly anomaly value, the impact of ENSO is portrayed as the overall increasing of PWV value during 2010 La Niña and lowering during 2015 El Niño. The distribution of the impact can be seen in the generated monthly anomaly map.

This results show the potential of using the altimetry satellites' water vapor radiometer data for atmospheric observation. This brings out the possibility of utilizing altimetry data for observing the coupling of ocean and atmosphere as it could measures both oceanic and atmospheric variable: the main sea level measurement, and water vapor from the radiometer which originally used to measure correction.

References

[1] AVISO/Altimetry 1996 AVISO User Handbook for Merged TOPEX/POSEIDON products AVI NT-02-101, Edition 3.0

[2] Jason-1 Products Handbook CNES (AVISO+): SALP-MU-M5-OP-13184-CN, JPL (PODAAC): JPL D-21354

OSTM/Jason-2 products handbook CNES: SALP-MU-M-OP-15815-CN, EUMETSAT: EUM/OPS-JAS/MAN/08/0041, JPL: OSTM-29-1237, NOAA/NESDIS: Polar Ser. J

Jason-3 Products Handbook CNES: SALP-MU-M-OP-16118-CN, EUMETSAT: TBD, JPL: TBD, NOAA/NESDIS: TBD

[3] Keihm S J, Jansses M A, and Ruf C S 1995 TOPEX/Poseidon Microwave Radiometer (TMR): III. Wet Troposphere Range Correction Algorithm and Pre-Launch Error Budget *IEEE Trans. Geosci. Remote Sens.* **33**

[4] Brown S 2010 A novel near-land radiometer wet path-delay retrieval algorithm: Application to the Jason-2/OSTM Advanced Microwave Radiometer *IEEE Trans. Geosci. Remote Sens.* **48**

[5] Weng F, Yu W and Sun N 2011 Retrieval of the total precipitable water vapor and cloud liquid water path from Jason-2 AMR observations *91th AMS Annual Meeting, Seattle, WA, 23-27 January 2011*

[6] Thao S, Eymard L, Obligis E and Picard B 2014 Trend and variability of the atmospheric water vapor: A mean sea level issue *J. Atmos. Ocean. Technol.*

[7] Maiwald F, Montes O, Padmanabhan S, Michaels D, Kitiyakara A K, Jarnot R, Brown S T, Dawson D, Wu A, Hatch W, Stek P and Gaier T 2016 Reliable and Stable Radiometers for Jason-3 *IEEE J. Sel. Top. Appl. Earth Obs. Remote Sens.* **9**

[8] Varma A K, Gairola R M, Kishtawal C M, Pandey P C and Singh K P 1999 Rain rate estimation from nadir-looking topex/poseidon microwave radiometer (tmr) for correction of radar altimetric measurements *IEEE Trans. Geosci. Remote Sens.* **37**

[9] Scharroo R, Leuliette E, Naeije M, Martin-Puig C and Pires N 2016 RADS version 4: An efficient way to analyse the multi-mission altimeter database *European Space Agency, (Special Publication) ESA SP* vol SP-740

[10] Askne J and Nordius H 1987 Estimation of tropospheric delay for microwaves from surface weather data *Radio Sci.* **22**

[11] Böhm J, Möller G, Schindelegger M, Pain G and Weber R 2015 Development of an improved

empirical model for slant delays in the troposphere (GPT2w) *GPS Solut.* **19**

[12] Okamoto K, Ushio T, Iguchi T, Takahashi N and Iwanami K 2005 The Global Satellite Mapping of Precipitation (GSMaP) project *International Geoscience and Remote Sensing Symposium (IGARSS)* vol 5

[13] Aonashi K, Awaka J, Hirose M, Kozu T, Kubota T, Liu G, Shige S, Kida S, Seto S, Takahashi N and Takayabu Y N 2009 Gsmap passive microwave precipitation retrieval algorithm: Algorithm description and validation *J. Meteorol. Soc. Japan* **87 A**

[14] Mega T, Ushio T, Matsuda T, Kubota T, Kachi M and Oki R 2019 Gauge-Adjusted Global Satellite Mapping of Precipitation *IEEE Trans. Geosci. Remote Sens.* **57**

[15] Yamamoto M K and Shige S 2015 Implementation of an orographic/nonorographic rainfall classification scheme in the GSMaP algorithm for microwave radiometers *Atmos. Res.* **163**

[16] Kubota T, Aonashi K, Ushio T, Shige S, Takayabu Y N, Kachi M, Arai Y, Tashima T, Masaki T, Kawamoto N, Mega T, Yamamoto M K, Hamada A, Yamaji M, Liu G and Oki R 2020 Global Satellite Mapping of Precipitation (GSMaP) Products in the GPM Era *Advances in Global Change Research* vol 67

[17] Vaníček P 1969 Approximate spectral analysis by least-squares fit - Successive spectral analysis *Astrophys. Space Sci.* **4**

[18] Wessel P, Luis J F, Uieda L, Scharroo R, Wobbe F, Smith W H F and Tian D 2019 The Generic Mapping Tools Version 6 *Geochemistry, Geophys. Geosystems* **20**

[19] Schweikert D G 1966 An Interpolation Curve Using a Spline in Tension *J. Math. Phys.* **45**

[20] Ahrens C D and Henson R 2018 *Essentials of Meteorology: An Invitation to the Atmosphere 8th Ed.* (Boston, MA: Cengage Learning) chapter 4

3D Modelling of Meteorite from Astomulyo Village, Lampung, Indonesia by Close Range Photogrammetry (CRP) Methods

D G Harbowo[1,5,a], J H Sitorus[2], L K Agustina[2], R Muztaba[3,4,5], T Julian[1], H L Malasan[3,4,6].

[1] Geological Engineering, Institut Teknologi Sumatera, Wayhui, Lampung, Indonesia
[2] Geomatics Engineering, Institut Teknologi Sumatera, Wayhui, Lampung, Indonesia
[3] Atmosphere and Planetary Science, Institut Teknologi Sumatera, Wayhui, Lampung, Indonesia
[4] Observatorium Astronomi ITERA Lampung, Institut Teknologi Sumatera, Lampung, Indonesia
[5] Doctoral Student, Institut Teknologi Bandung, Bandung, West Java, Indonesia
[6] Astronomy, Institut Teknologi Bandung, Bandung, West Java, Indonesia
[a] correspondent author: danni.gathot@gl.itera.ac.id

Abstract. This research was conducted to 3D model our meteorite collection based on photo-capturing. This meteorite are from Astomulyo, Central Lampung, Indonesia. It entry the earth's atmosphere and was found on January 28, 2021. The chance of object degradation nearly appears over time, so It is very urgent to obtain the 3D model of the object for further research necessary. 3D modelling was carried out using a non-metric camera to conduct the close-range photogrammetry method. It applies to modelling the objects based on photo data using the concept of collinearity by using a central projection system on the camera to record. The collinearity will produce an object point between the camera center and the object cloud point. Hence, this process forms the object model from the appropriate image projection, the tie point with the triangular irregular network model. As a result, it produces the DTM (Digital Terrain Model) object formation using stand-alone photogrammetric software, raster graphics, and 3D modelling software.

Keyword: 3D modelling. Astomulyo, Meteorite, Close Range Photogrammetry

1. Introduction

Meteorite encounters are extremely infrequent events. The scientific study of meteorites is a critical and affordable way to comprehend the existence of extra-terrestrial celestial bodies' properties such as physical and chemical properties [1–3]. This study can provide comprehensive perspectives on near-future space exploration. Furthermore, the shape and dimensions of the meteorite are also essential in scientific investigations [4,5]. It could uncover its parent body and its process while entering the earth's atmosphere. However, firstly meteorites are encountered, ideal conditions are often not achieved [6,7]. Rapid natural weathering and unintentional violence can ensue, degrading the object's authenticity, including its shapes. That is why we tried to do 3D modelling to record meteorites' detailed shape and size. This study attempts to apply the close-range photogrammetry (CRP) method.

This study applies this method to meteorites found in Astomulyo, Punggur, Central Lampung, Indonesia. This meteorite fell in the local's rice field (Coordinate: 4°59′ 58.9″ S, 105° 15′ 11.4″ E) on January 28th, 2021, around 21:30 to 22:00 WIB (UTC + 07.00) [3], see **Figure 1**. The possibility of degradation of this object could be disturbing further investigation. It also causes weaken the scientific and historical value of this object. So important to take a step to record and archive the documentation of this object as well as possible. This study attempts to approach object documentation in a 3D model.

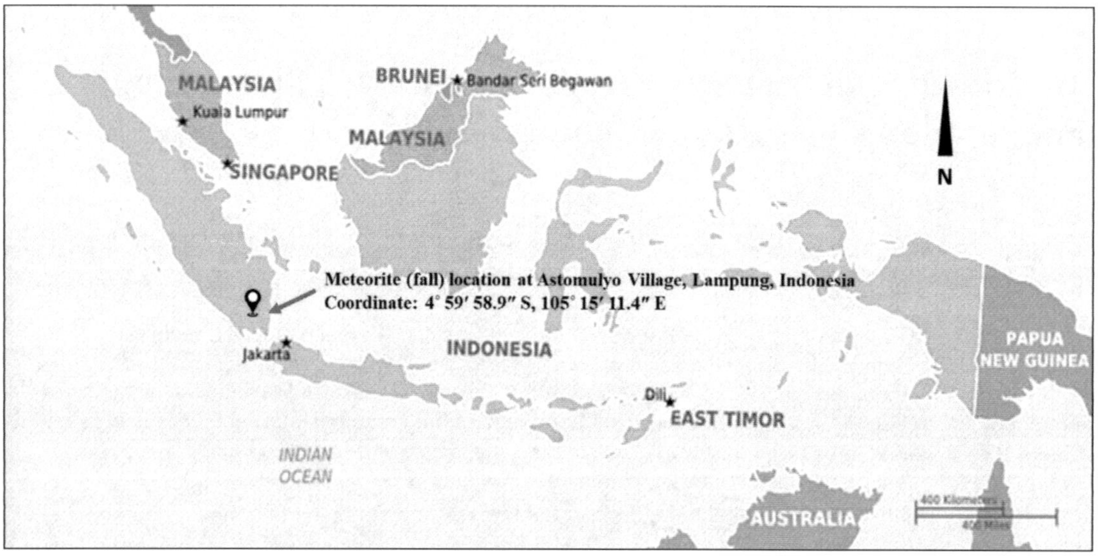

Figure 1. Location of the meteorite fall

Efforts to do 3D modelling on meteorites can be conducted with several methods such as 3D laser imaging and close-range photogrammetry methods [8–10]. The close-range photogrammetry (CRP) method obtains reliable information of the object by recording, measuring, and interpreting the photographic image in the camera's distance to the object of interest relative near-close: shape, dimension, and surface texture covered by this method [11,12]. Collinearity are applied to take photos up to 100 mm distances to calibrate perfectly (**Figure 2**). The CRP method combined photos taken from various angles, and the object must be stationary or move its initial position while taking a picture of the objects [13]. Advantageously, this method can be conducted by non-metric cameras and applied to document several museums artifacts and archeological objects [14,15].

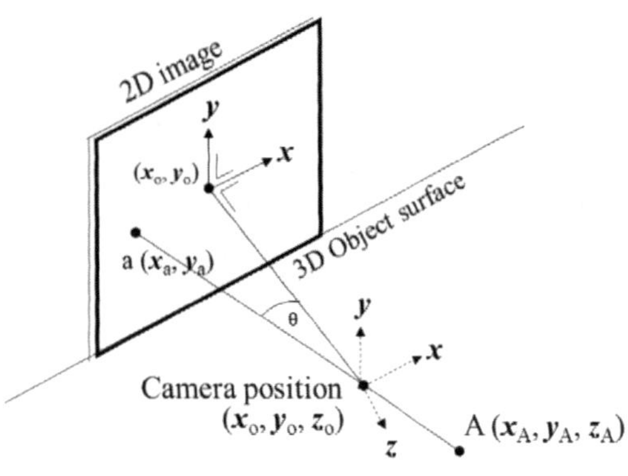

Figure 2. Illustration of the image capturing in this study based on collinearity in close-range photogrammetry (CRP) method.

2. Method

3D object modelling is conducted by observing directly on the object and doing a physical description of each observable feature. Sketches and notes are necessary to record some features that may be missed from the image processing process. Several steps must be carried out consecutively, i.e., image capturing based on parallelism and collinearity (**Figure 2**); photo marking and referencing; image matching; point-cloud building, mesh building, texture building, and size adjustment (**Figure 3**). These method is refer to some researcher practical [10,12–14].

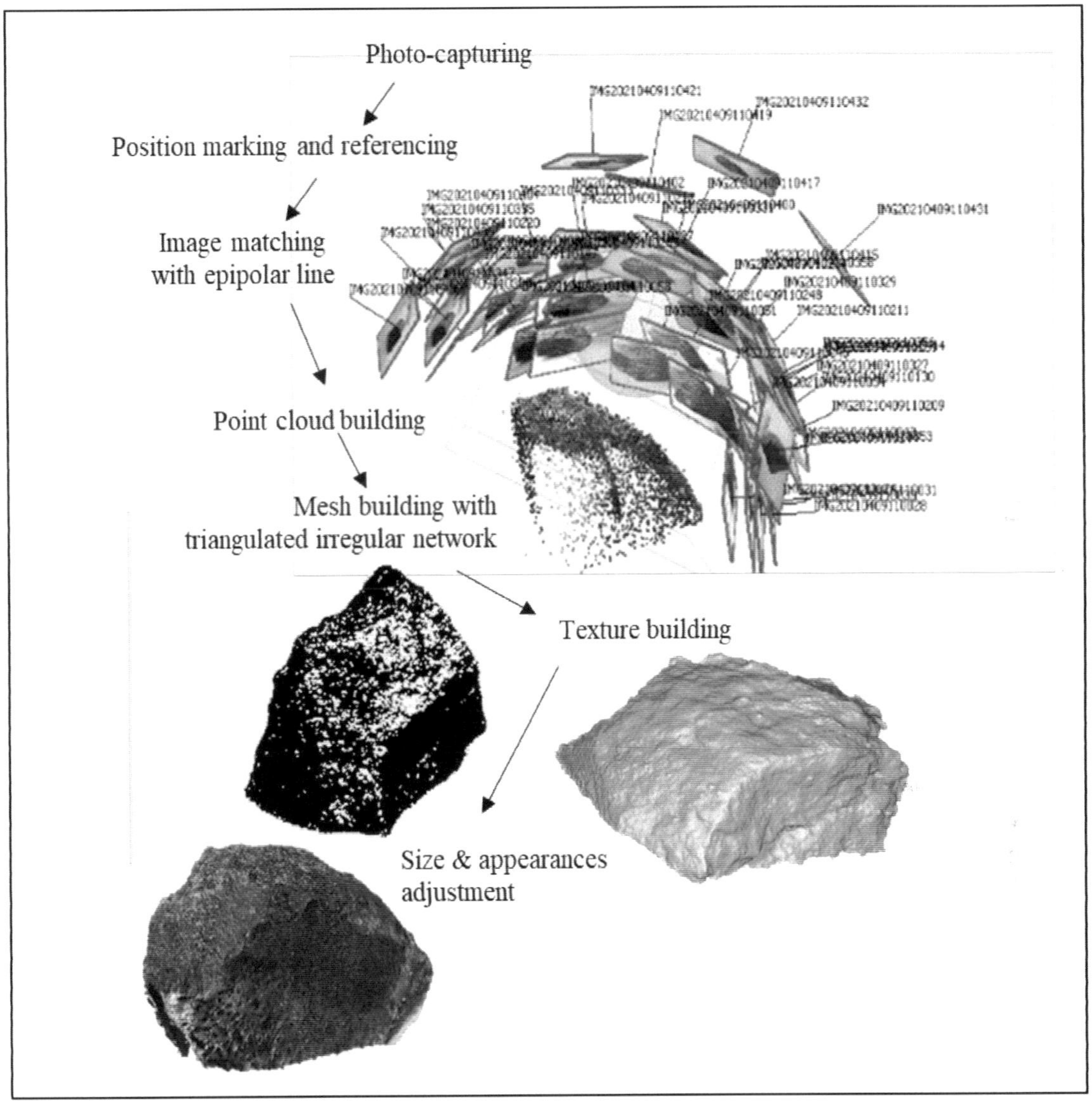

Figure 3. The procedure of close-range photogrammetry (CRP) method in this study.

Firstly, the objects' images were taken on using a camera with specifications of size 142.24 cm x 106.68 cm; 4032 x 3024 pixels, 72 dpi; focal length 4 mm; exposure bias +0.5 step; and exposure time 0.25 s. Distance of camera and the object less than 30 cm, the photo is taken simultaneously with the comparator, from various angles of capture without changing the object and comparator position. Then image process uses stand-alone photogrammetric software.

Secondly, marking and referencing the image are accomplished by marking the points of objects to identify the same point on different photos. Then image matching is conducted by merging overlapping photos to construct the tie point of the epipolar line; this process refers to the adjusted alignment of epipolar geometry. The point-cloud build is carried out by marking the camera position and calculating the group of tie points; it is combined into a pack of dense cloud points to construct a TIN (triangulated Irregular network) model of the modified DTM (digital terrain model). These tie points are processed by connecting them into triangular polygons. Further, mesh building is reconstructed using these triangular polygons based on its dense-cloud data to produce the whole polygonal model of the object. The texture is formed by adjusting the 3D model object's actual appearance and covering the whole object based on its position. It uses raster graphics and 3D modelling and design software. Last, the size setting adjusts based on the actual size (Astomulyo-meteorite) that has been recorded previously.

3. Result and Discussion

The number of photos used in the study was 121, with 48 primary photos forming a tie point, with proper collinearity. It is optimal to employ manual coordinate options adjusted with camera angle and marker references—the image marking and referencing process is optimum with overlay value of 20%. Image matching conditions are obtained 16762 tie points in matching grayscale degrees. These points are grouped in a dense cloud with sufficient quality by calculating the point of color and its confidence value (0.95). A digital terrain model was successfully formed based on the point cloud building with an error value of 0.56 pix; this number is relatively minimal than the size of photo pixels. Optimal mesh building is attained on arbitrary (3D) surfaces type options with face count interpolation up to 149596 points by considering its vertex color. Then, the texture is successfully formed by mosaic blending mode with a texture size/count of 4096. Finally, this model size is adjusted to the actual measurement, which is 16 x 23 x 25 cm (**Table 1**). Final result of this method represent in **Figure 4**.

Table 1. Result of 3D model profiling

Profile	value
Object dimension	16 x 23 x 25 cm
Number of photos	121 photos
Number of primary photos	48 photos
Photo overlay (%)	20 %
Effective tie points	16762 point
Confidence value	0.95
Point cloud error value	0.56 pix
Face count interpolation	149596 points
Texture size/count	4096

Generally, the benefit of the close-range photogrammetry method is rather practical in documenting objects in 3D; the process of taking pictures/photo capturing could be applied using a calibrated non-metric camera [16]. It is relatively reliable for quicker/short time 3D documentation needs. The devices used are relatively uncomplicated, such as most recent cellphone and pocket cameras, but surely, higher camera specifications will enhance image processing standards and quality. The exact distribution of image focus is crucial; many images with badly focus/blur on the photo frame will cause difficulties in the marking and referencing process and image matching, primarily if the automatic mode is carried out. Camera distance, angle of capture, and the code number of the image should be noted. The notes are handy for validating the images while image matching is carried out. In this context, the collinearity condition states that the vector from the perspective center to a distinct point on the image is scaled from the perspective center to the corresponding object point [12,17]. The density of the point cloud is also crucial for processing the TIN (triangulated irregular network) model and constructing the DTM (digital terrain model) model. A loose point cloud will cause the texture to be less detailed and not representative of the actual object. The collinearity produce an object point between the camera center and the object cloud point. Hence, this process forms the object model from the appropriate image projection, the tie point with the triangular irregular network model.

Figure 4. The result of the 3D modelling of meteorite from Astomulyo, Lampung, Indonesia

The close-range photogrammetry method does not require complicated equipment compared to laser imaging [10,16]. In critical conditions in the field, shooting with this method can be more instant, including small-sized objects. This method can be operated by adjusting the camera's capability to take pictures based on its specifications. The close-range photogrammetry method needs to properly hold the stationary position of the object and the angle of the camera. Stationary markers helped as benchmarks. The number of qualified photos and the photos' resolutions is primary roles in successfully implementing the close-range photogrammetry procedure.

4. Conclusion

The close-range photogrammetry method is handy for 3D modelling the meteorite from Astomulyo, Lampung, Indonesia. Parallelism and collinearity are applied in multiple image capturing in various angles, conducted by 121, with 48 primary photos forming a tie point. Then, the comparator/marker plays a role as benchmark and optimum with a minimum overlay of 20%. It generates 16762 tie points optimum for 149596 accounts for mesh building as texture size/count of 4096. It made reliable quality for surface texture processing since the triangular irregular network (TIN) of the object is compatible with the representative result of its digital terrain model.

5. References

[1] Grady M, Pratesi G and Moggi Cecchi V 2015 Atlas of meteorites *Choice Rev. Online* **53** 53-1767-53–1767

[2] Flynn G J, Consolmagno G J, Brown P and Macke R J 2018 Physical properties of the stone meteorites: Implications for the properties of their parent bodies *Geochemistry* **78** 269–98

[3] Harbowo D G, Muztaba R, Malasan H L, Sumardi S, Agustina L K, Julian T, Sitorus J H, Denhi A D A, Sihombing D J P, Mahayu M P and Setyawan D 2021 Meteorite from Astomulyo Village, Central Lampung, Indonesia: Investigation of its chemical properties *IOP Conf. Ser. Earth Environ. Sci.* **882** 012011

[4] Friedrich J M, Weisberg M K, Ebel D S, Biltz A E, Corbett B M, Iotzov I V., Khan W S and Wolman M D 2015 Chondrule size and related physical properties: A compilation and evaluation of current data across all meteorite groups *Geochemistry* **75** 419–43

[5] Borovička J, Tóth J, Igaz A, Spurný P, Kalenda P, Haloda J, Svoreň J, Kornoš L, Silber E, Brown P and Husárik M 2013 The Košice meteorite fall: Atmospheric trajectory, fragmentation, and orbit *Meteorit. Planet. Sci.* **48** 1757–79

[6] Ouknine L, Khiri F, Ibhi A, Heikal M T S, Saint-Gerant T and Medjkane M 2019 Insight into African meteorite finds: Typology, mass distribution and weathering process *J. African Earth Sci.* **158** 103551

[7] Lipka J, Sitek J, Dekan J and Sedlačková K 2014 Analyses of Rumanová meteorite *Hyperfine Interact.* **226** 565–9

[8] Isa M A and Lazoglu I 2017 Design and analysis of a 3D laser scanner *Measurement* **111** 122–33

[9] Percoco G 2011 Digital close range photogrammetry for 3D body scanning for custom-made garments *Photogramm. Rec.* **26** 73–90

[10] Luhmann T, Robson S, Kyle S and Boehm J 2019 *Close-Range Photogrammetry and 3D Imaging* (De Gruyter)

[11] Martínez J, Ordóñez C, Arias P and Armesto J 2011 Non-contact 3D Measurement of Buildings through Close Range Photogrammetry and a Laser Distance Meter *Photogramm. Eng. Remote Sens.* **77** 805–11

[12] Stamatopoulos C and Fraser C S 2011 Calibration of long focal length cameras in close range photogrammetry *Photogramm. Rec.* **26** 339–60

[13] Harahap S J and Handayani H H 2016 Visualisasi 3D Objek Menggunakan Teknik Fotogrametri Jarak Dekat *J. Tek. ITS* **5**

[14] Guery J, Hess M and Mathys A 2018 Photogrammetry *Digital Techniques for Documenting and Preserving Cultural Heritage* (Amsterdam University Press) pp 229–36

[15] Moyano J, Nieto-Julián J E, Bienvenido-Huertas D and Marín-García D 2020 Validation of Close-Range Photogrammetry for Architectural and Archaeological Heritage: Analysis of Point Density and 3D Mesh Geometry *Remote Sens.* **12** 3571

[16] Lee C N and Oh J H 2012 A Study on Efficient Self-Calibration of a Non-Metric Camera for Close-range Photogrammetry *Korean J. Geomatics* **30** 511–8

[17] Abdel-Aziz Y I and Karara H M 2015 Direct Linear Transformation from Comparator Coordinates into Object Space Coordinates in Close-Range Photogrammetry *Photogramm. Eng. Remote Sens.* **81** 103–7

Acknowledgments
We gratefully thanks to Institut Teknologi Sumatera (ITERA) for Collaborative Research Funding Programme in 2021.

IOP Conf. Series: Earth and Environmental Science 1047 (2022) 012004 doi:10.1088/1755-1315/1047/1/012004

Development of Landsat-8 ARD Products for User Service in Indonesia

F Muchsin[1], K Ulfa[1], KA Pradono[1], Hendayani[1], Dianovita[1], KW Veronica[1]

[1] Remote Sensing Research Program – National Research and Innovation Agency (BRIN) of Indonesia

E-mail: kurnia.ulfa@lapan.go.id, fadila.muchsin@lapan.go.id

Abstract. Along with the times and technology development, the number of satellite images produced is increasing and diverse, both in terms of spatial and temporal resolution. In addition, in the last few years, LAPAN has done a mosaic of Landsat data for 16 days. The availability of extensive satellite data makes users desire data ready to use without initial processing to save time and costs. Furthermore, the medium resolution image has a comprehensive enough coverage for applications in the forestry, agriculture, plantation, and marine sectors. Therefore, the need for medium resolution Analysis Ready Data (ARD) increases. The available ARD data has been through the initial processing, including ToA and BRDF corrections. The existing Landsat-8 ARD does not cover Indonesian territory. The Landsat-8 ARD for the Indonesian region was first produced and introduced in this paper. Moreover, the development of ARD data quality is carried out by adding a topographic correction process, and image cropping become tiles with a size of 1-degree x 1-degree, which can be run automatically. This processing system produces ARD Landsat-8 to meet user requirements effectively and efficiently. These advantages are obtained because the pre-processing data load for the user has been reduced, and the user can immediately use the data according to the required scope. The software for making ARD Landsat-8 has been recognized as Intellectual Property Rights. Moreover, ARD Landsat-8 for Indonesian territory has been continuously utilized by the Indonesian Ministry of Environment and Forestry and the Indonesian Ministry of Agriculture.

Keywords: *Landsat-8, analysis-ready data, tile-based*

1. Introduction

The satellite constellation continues to increase, resulting in large volumes of data. On the other hand, storage space availability is not directly proportional to the number of images. Thus, analysis-ready data (ARD) is needed, which users can directly use without pre-processing to save time and cost [1].

ARD data in remote sensing is highly recommended by the Committee on Earth Observation Satellites (CEOS). CEOS defines it as satellite data that has been processed into a data set that meets the minimum requirements (geometric, radiometric, atmospheric correction) and organized into a tile that allows immediate analysis with minimal user effort and interoperability both over time and with the data set. others (https://ceos.org/ard/files/CEOS_ARD_Governance_Framework)[2]. ARD processing and tile structure are also comprehensively described by Dwyer [3].

The need for medium-resolution ARD data such as Landsat-8 and Sentinel-2 is increasing. It is indispensable, especially in extracting information quickly and directly from images and for accurate quantitative analysis of multi-temporal images [4][5]. The existence of Landsat ARD will make it easier to analyze time-series data at a certain level. Such as reflectance TOA, TOA BT, SR, BT, and Pixel band QA can be used in many remote sensing applications and drive new scientific discoveries [6]. Medium resolution images have a broad enough scope to be widely used for global and regional applications in the forestry, agriculture, plantation, and marine sectors.

ARD data with a high level of radiometric accuracy has gone through a complete initial process. Top of Atmosphere (TOA), atmospheric correction, Bidirectional Distribution Function (BRDF), and removal/transfer of fog, clouds, and cloud shadows. This pre-processing can also improve the consistency of data (primarily temporal) being explored, namely cloud and cloud shadow detection and BRDF correction [7]. By-Law Number 21 of 2013 concerning Space and Government Regulation Number 11 of 2018, remote sensing data processing to process primary data includes radiometric correction.

ARD Landsat provided by USGS does not cover Indonesia's territory. Therefore, this research was developed to meet users' needs as utilized by the Indonesian Ministry of Environment and Forestry and the Indonesian Ministry of Agriculture.

The medium-resolution ARD data available at LAPAN has gone through all the previous correction processes. LAPAN's ARD data has been used for various purposes (generally land area applications), including annual composite mosaics, monitoring of rice growth phases, information on forest land changes, and so on. However, the study of the consistency of ARD data for geometric and radiometric aspects and derived geophysical parameters has not been systematically carried out for Indonesia regions.

The provision of ARD data aims to facilitate users in making land cover maps, monitoring land changes and their derivative products, such as geo-biophysical parameters, quickly with consistent results.

2. Material and Methods

2.1 Data and Study data

Landsat-8 OLI level 1TP (terrain precision) was used in this study provided by LAPAN Parepare and Rumpin Ground Station. LAPAN has two Landsat-8 OLI and TIRS receiving by the ground stations. Parepare ground stations in South Sulawesi receive data for most parts of eastern and western Indonesia and Rumpin - West Java to receive data for western Indonesia, which the Papepare earth station does not reach. The data that the earth station has obtained is then processed using an algorithm on the Landsat Processing Generation System (LPGS) with accurate Ground Control Points (GCPs) and Digital Elevation Model (DEM) to achieve geodetic accuracy of 12 meters for the 1TP level (https://www.usgs.gov/landsat-missions/landsat-collection-1).

This study used Landsat-8 data of West Java Province with path/row 121/064-65 and 121/064-65 from January to September 2021, as shown in Figure 1. Each path/row consists of 18 scenes (One month has two adjacent scenes). The entire scene used is 72 scenes. The study area in this paper can be covered by Landsat-8 data, as shown in the following picture:

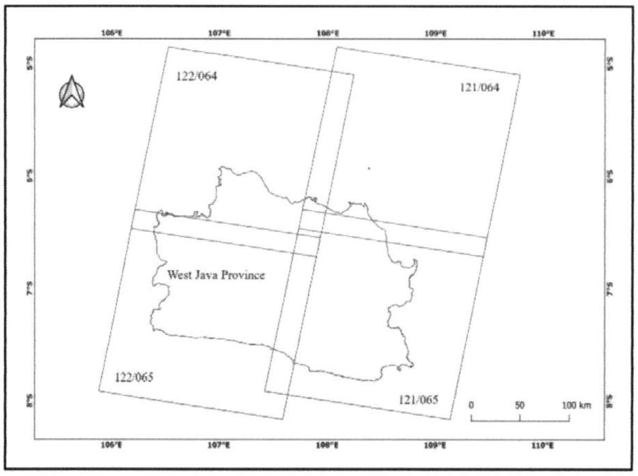

Figure 1. Coordinate location of West Java Province Landsat-8 data with path/row 121/064-65 and 121/064-65

2.2 TOA and BRDF radiometric corrections

Level 1TP data still contains the digital number of objects on the earth's surface, so it needs to be converted into the top of atmosphere (TOA) radiance and reflectance value to derive biophysical parameters, digital classification, and multi-temporal analysis. The conversion process uses equations 1 and 2 [8].

$$L_\lambda = \left(\frac{LX_\lambda - LN_\lambda}{Q_{cx} - Q_{cn}}\right)(Q_c - Q_{cn}) + LN_\lambda$$

$$L_\lambda = G_r \times Q_c + B_r \tag{1}$$

where:

$$G_r = \frac{LX_\lambda - LN_\lambda}{Q_{cx} - Q_{cn}}$$

$$B_r = LN_\lambda \left(\frac{LX_\lambda - LN_\lambda}{Q_{cx} - Q_{cn}}\right) Q_{cn}$$

where:

L_λ = Spectral radiance at the sensor's aperture [W/(m^2 sr μm)]
Q_c = Quantized calibrated pixel value [digital number]
Q_{cn} = Minimum quantized calibrated pixel value corresponding to $LMIN_\lambda$ [digital number]
Q_{cx} = Maximum quantized calibrated pixel value corresponding to $LMAX_\lambda$[digital number]
LN_λ = Spectral at-sensor radiance that is scaled to Q_{calmin} [W/(m^2 sr μm)]
LX_λ = Spectral at-sensor radiance that is scaled to Q_{calmax} [W/(m^2 sr μm)]
G_r = Band-specific rescaling gain factor [W/(m^2 sr μm)]
B_r = Band-specific rescaling bias factor [W/(m^2 sr μm)]

$$\rho_\lambda = \frac{\pi . L_\lambda . d^2}{ESUN_\lambda . \cos\theta_s} \tag{2}$$

where:
ρ_λ = Planetary TOA reflectance [unitless]
π = Mathematical constant equal to ~3.14159 [unitless]
L_λ = Spectral radiance at the sensor's aperture [W/(m^2 sr μm)]
d = Earth-Sun distance [astronomical units]
$ESUN_\lambda$ = Mean exoatmospheric solar irradiance [W/(m^2 sr μm)]
θ_s = Solar zenith angle [degrees]
These parameters above are obtained in metadata from Landsat-8 data.

The next step is to apply a bidirectional reflectance distribution function (BRDF) correction to all bands of the Landsat-8 image to eliminate the effect of illumination due to differences in solar angles [9]. BRDF correction is carried out to eliminate errors caused by the influence of the position of the geometry and the sun, objects, and sensors. For example, the angle of capture and the sun's angle affect the energy received by the sensor. In addition, the results of the BRDF correction make the pixel values more consistent over time and space [10].

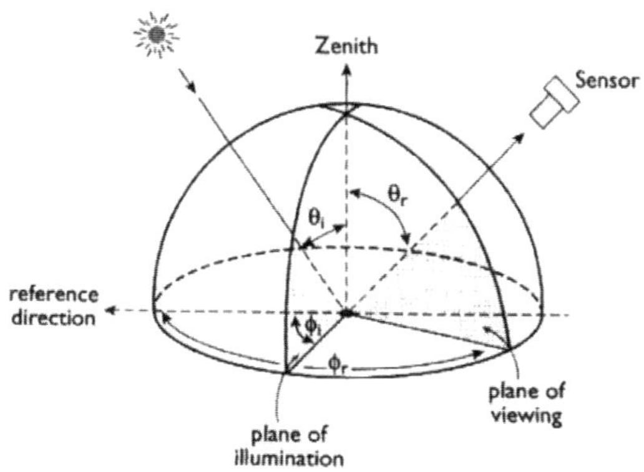

Figure 2. Shooting angle and sun angle effect illustration

$$\rho\lambda = \frac{\rho\lambda'}{\cos(\theta sz)} \tag{3}$$

where:
$\rho\,\lambda$ = Planetary TOA reflectance [unitless]
$\rho\,\lambda'$ = Planetary TOA reflectance [unitless]
θsz = solar zenith angle [degrees]

2.3 Topography/ Terrain Correction

In mountainous areas of the earth's surface, terrain effects can make the pixel value of an object inconsistent, so terrain correction is necessary if the image is to be used for quantitative analysis such as land use and land cover classification. The method commonly used is c correction because it is easy to apply and produces perfect accuracy. The c correction method used in this study is as follows [11][12]:

$$\rho_{H\lambda} = \rho_{T\lambda} \frac{[\cos(\theta_s) + c]}{[\cos(\beta) + c]} \tag{4}$$

$\rho_{H\lambda}$ = Reflectance from horizontal surface (corrected image)
$\rho_{T\lambda}$ = Reflectance over sloped terrain (un-corrected image)
θ_s = Solar zenith angle
β = Solar illumination angle
c = the limiting coefficient, which is the ratio between the intersection point and the gradient (b/m) of the regression equation $\rho_{T\lambda}$ = m Cos (β) + b

The C approach yields better results for the visible bands if the TOA reflectance product is used, but it will lose an advantage if the C normalization is performed on the surface reflectance product [13].

2.4 Scene cropping (Grid ARD)

After getting Landsat-8 data corrected by TOA, BRDF, and topography, a data mosaic is done to get an entire scene of Indonesia with a combination of Red, Green, Blue, SWIR, and NIR channels in VRT format. The virtual format was chosen due to the efficient storage space for data processing. The dataset will then be tilled with the help of the GDAL (Geospatial Data Abstraction Library) using grid Sentinel2 UTM, which is re-projected to geographic with a tile size of 1°x1° (□111 x111 km^2). Next, we need a database of ARD tiles that have been converted into geographic projections as a reference in cropping images. The following is the algorithm applied in cropping Landsat-8 into Grid ARD Indonesia. The cropping process follows the ARD Sentinel-2 grid as shown in Figure 4 by considering that in the future, ARD Landsat-8 will be integrated with Sentinel-2.

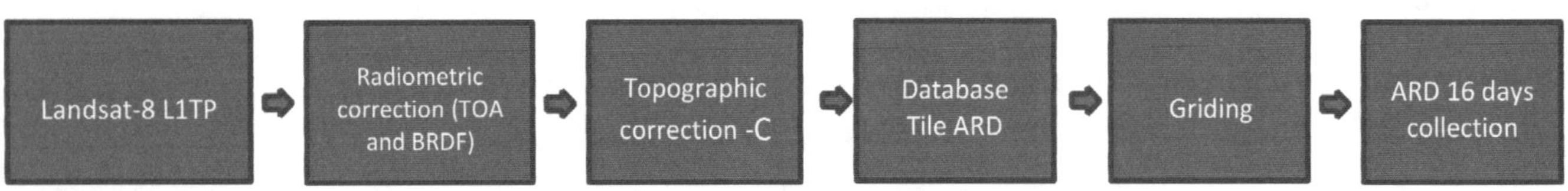

Figure 3. ARD processing flowchart

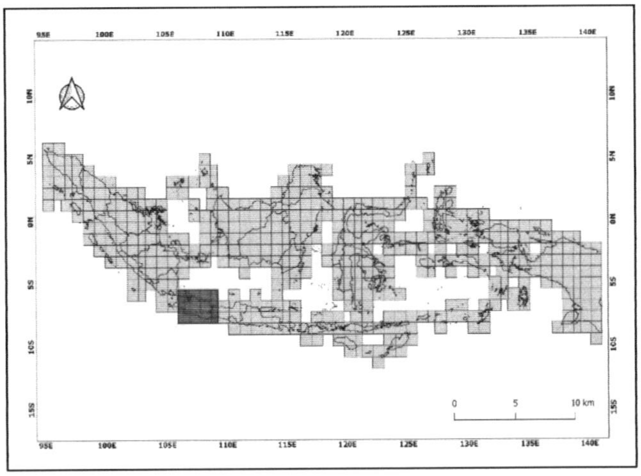

Figure 4. ARD Grid of Landsat-8 (region of Indonesia) refers to the Sentinel-2 grid, with the tile size being 1°x1° (~111 x111 km2). The red box is the coordinate location of West Java Province

The next stage of data collection-16, which has been in ER-Mapper (.ers) format from the results of grid processing, will then be stored in storage to meet user needs. The ER-Mapper format accommodates users because the header files are stored separately. The results of this stage will be visually checked to determine the quality of the data.

3. Result and Discussion

The processing is carried out on the Landsat-8 data acquired for 16 days with 225 scenes of data. These scenes will be processed simultaneously using the python library. The processing takes 12 hours for ToA-BRDF correction, 8 hours for topographic correction, and 4 hours for making virtual mosaics and ARD tiles. This process will generate 485 folders according to the Sentinel-2 UTM grid order. The expert has developed many tile systems. The Sentinel-2 grid is considered to align with the development of the automation system already running at LAPAN. In addition, to make it easier to develop the following method. Each folder contains data in ER-Mapper format that corresponds to its name. The generated .ers data will be added automatically to each appropriate folder in the following processing.

The previous product produced by LAPAN was a Landsat-8 image in the form of a scene corrected by ToA and BRDF. This study adds additional processes, namely topographic correction and image cropping, into tiles referring to the Sentinel-2 UTM grid. The advantages of the image in the form of tiles include having a smaller size than the image in the form of a scene and having gone through a water masking process, namely the removal of sea areas. Furthermore, with this smaller size, ARD data requires less storage capacity and is lighter in processing if information extraction is carried out. This data is very beneficial for users because users can take tiles in certain areas they need. The following is an example of comparing the image in a scene and ARD in the West Java region.

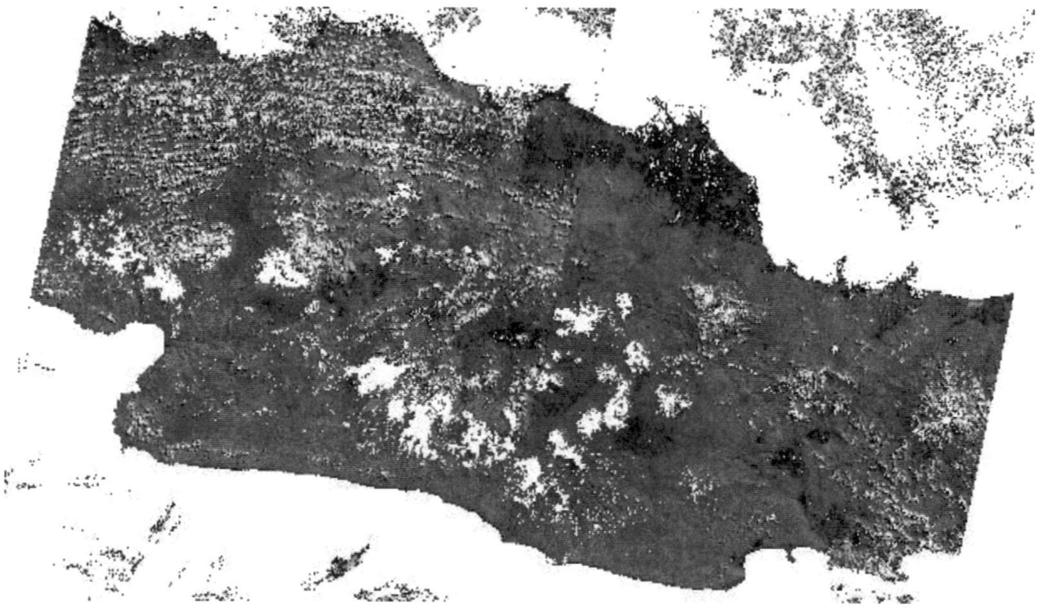

Figure 5. Landsat-8 images of the West Java region displayed per scene (total four scenes): path/row 121/064-65 acquisition date May 27, 2021, and 122/064-65 acquisition date May 27, 2021

Figure 6. Landsat-8 images of the West Java region displayed per tile (total of 14 tiles) from the scenes in Figure 6

The two images above have the same input data. Figure 5 only goes through the ToA and BRDF correction process, while Figure 6 is processed until it becomes an ARD tile. Figure 5 consists of 4 scenes, while Figure 6 consists of 14 tiles. In terms of file size, one ARD tile is 83 KB in size while one Landsat-8 image scene is one GB, so to get an image in a specific AOI, the ARD data is still relatively smaller than the scene data, although the number will be more. The image in Figure 5 is the most explicit scene that has been selected from all the scenes acquired in January – September 2021 in the West Java region. Therefore, the ARD processed image looks more precise than the scene image for a similar region. The ARD data looks clearest because the results of ARD processing have gone through the selection of the most precise image on the same tile from January to September 2021.

4. Conclusion

Landsat 8 ARD data that has gone through topographic and water masking corrections have been generated. With this product, the existing satellite imagery that has been owned so far can be used optimally. These data have several advantages compared to scene data. ARD data has a reasonably small size (100 x 100) km2) and is light in storage. In addition, ARD data also makes it easier for users to extract information according to their needs. Further research can improve the smoothness between ARD tiles by developing a color balancing method.

Acknowledgment

Thanks to Drs. Kustiyo, M.Si. as Group Leader of ARD and LAIPS medium resolution satellite data for his guidance in compiling and improving the programming script for the ARD Landsat-8.

References

[1] D. Frantz, "FORCE-Landsat + Sentinel-2 analysis-ready data and beyond," *Remote Sens.*, vol. 11, no. 9, 2019.
[2] B. Zhong *et al.*, "Analysis ready data of the Chinese gaofen satellite data," *Remote Sens.*, vol. 13, no. 9, pp. 1–17, 2021.
[3] J. L. Dwyer, D. P. Roy, B. Sauer, C. B. Jenkerson, H. K. Zhang, and L. Lymburner, "Analysis ready data: Enabling analysis of the Landsat archive," *Remote Sens.*, vol. 10, no. 9, pp. 1–19, 2018.
[4] A. V. E. Id, D. P. Roy, and A. Kommareddy, "Demonstration of Percent Tree Cover Mapping Using Landsat Analysis Ready Data (ARD) and Sensitivity with Respect to Landsat ARD Processing Level," 2018.
[5] A. V Egorov, D. P. Roy, H. K. Zhang, Z. Li, L. Yan, and H. Huang, "(ARD) Observation Coverage over the Conterminous United States and Implications for Terrestrial Monitoring," vol. 7, 2019.
[6] Z. Zhu, "Science of Landsat analysis-ready data," *Remote Sens.*, vol. 11, no. 18, pp. 3–6, 2019.
[7] S. Qiu, Y. Lin, R. Shang, J. Zhang, L. Ma, and Z. Zhu, "Making Landsat time series consistent: Evaluating and improving Landsat analysis-ready data," *Remote Sens.*, vol. 11, no. 1, 2019.
[8] G. Chander, B. L. Markham, and D. L. Helder, "Summary of current radiometric calibration coefficients for Landsat MSS, TM, ETM+, and EO-1 ALI sensors," *Remote Sens. Environ.*, vol. 113, no. 5, pp. 893–903, 2009.
[9] T. Danaher, X. Wu, and N. Campbell, "Bi-directional reflectance distribution function approaches to radiometric calibration of Landsat ETM+imagery," *Int. Geosci. Remote Sens. Symp.*, vol. 6, no. February, pp. 2654–2657, 2001.

[10] Kustiyo, O. Roswintiarti, A. Tjahjaningsih, R. Dewanti, S. Furby, and J. Wallace, "Annual forest monitoring as part of the Indonesia's National Carbon Accounting System," *Int. Arch. Photogramm. Remote Sens. Spat. Inf. Sci. - ISPRS Arch.*, vol. 40, no. 7W3, pp. 441–448, 2015.

[11] X. Wu, S. Furby, J. Wallace, and F. Park, "an Approach for Terrain Illumination Correction," *Area*, vol. 6121, no. 618, pp. 1–9, 2004.

[12] B. Tan, R. Wolfe, J. Masek, F. Gao, and E. F. Vermote, "An illumination correction algorithm on Landsat-TM data," *Int. Geosci. Remote Sens. Symp.*, pp. 1964–1967, 2010.

[13] R. Richter, T. Kellenberger, and H. Kaufmann, "Comparison of topographic correction methods, "*MDPI: Int. Remote Sens*, pp. 184-196, 2009.

TECTONIC SETTING OF MOUNT AGUNG, BALI: INSIGHT FROM PETROLOGY AND GEOCHEMISTRY ANALYSIS

A Syafitri[1], I G B E Sucipta[2], A N Arifa[1], A Saepuloh[2], and S Widiyantoro[3,4]

[1] Laboratory of SEM-EDS UPP Chevron – Bandung Institute of Technology, Indonesia

[2] Petrology, Volcanology, and Geochemistry Research Group, Geological Engineering, Faculty of Earth Science and Technology – Bandung Institute of Technology, Indonesia

[3] Global Geophysics Research Group, Faculty of Mining and Petroleum Engineering – Bandung Institute of Technology, Indonesia

[4] Faculty of Engineering, Maranatha Christian University, Bandung, Indonesia

Corresponding author: auliasyafitri10@gmail.com

Abstract. One of the longest arcs in the world originated from the complexity of subduction zones is the Sunda Arc that covers 80% of Indonesia's active volcanoes, from the Andaman, Sumatra, Java, and the Lesser Sunda Islands. Previous research in magmatism in Sunda Arc has conceded that the continental crust is in the west and becomes progressively oceanic towards the east. However, recent research has suggested that continental basement is more widespread than previously thought. Therefore, this study aims to re-evaluate the tectonic setting of Mount Agung, Bali, part of the Lesser Sunda Islands. Based on the results of published geochemistry data analysis and our petrological and/or mineralogical data, we found that Mount Agung was influenced by three cogenetic magmas and can be divided into 4 eruption periods, i.e., pre-3200±60 BP, 3200±60 – 1870±40 BP, 1870±40 – 1040±50 BP, and post-1040±50 BP. These calc-alkaline magmas were derived from partial melting caused by the subduction of the Indo-Australian Plate within the Eurasian Plate. It produced basalt to dacite rocks with SiO_2 varying between 51 and 63 wt%. As seen from the spider diagram, Rb, Ba, Th, K, and La – Sm contents are enriched, while Eu – Lu experienced depletion. In addition, the Nb content shows a negative anomaly, which is a characteristic of volcanic products from convergent plate boundaries. Based on the ratio of Zr to Zr/Y, it introduces that Mount Agung is affected by continental arcs. La/Sm to Th/Nb diagram reveals that in the older period (pre-1040±50 BP), the magma differentiation process is subduction-related enrichment, while in the younger period (post-1040±50 BP), there may be a slight influence from the presence of crustal contamination. Thus, these analyses presume that Bali Island has a continental basement (micro-continent basement), which may become the eastern end boundary of Sundaland.

1. Introduction

One of the longest arcs in the world originated from the complexity of subduction zones is the Sunda Arc. The Sunda Arc covers up to 80% of Indonesia's active volcanoes from the Andaman Island, Sumatra, Java, and the Lesser Sunda Islands [1, 2]. It was formed by the Indo-Australian Plate subduction under the Eurasian Plate, which has been continuously activated since the Middle Eocene (around 45 Ma) [2, 3, 4]. This arc is very unique because it contains the transition between continental and oceanic-type arcs [2]. However, due to the lack of basement research, it is still unclear, which part is inherited from the oceanic and the continental crust [2]. Previous research in magmatism has conceded that the continental crust is in the west (Sumatra, Java), through the transition (Bali, Lombok), and becomes progressively oceanic towards the east (Wetar, Banda) [5, 6]. However, recent studies have suggested that continental basement is more widespread than previously thought [2, 7, 8].

This paper presents further work in this direction, particularly in searching the area that is still part of the continental basement. Bali Island, as a transition zone, is an intriguing area to debate. Previous research stated that it is derived from volcanic island arc [5, 9, 10], which is originated from subduction between oceanic crusts [11]. Thus, our study aims to re-evaluate the tectonic setting of Bali. This matter can be identified using the geochemical characteristic of volcanic rocks [12, 13]. Thus, one of the active volcanoes in Bali, Mount Agung provides useful information on interpreting the tectonic environment of this island.

2. Geological Background

Sundaland is the continental core of SE Asia [7, 14, 15] (Figure 1). Its eastern margin is irregular; a boundary has often been drawn through West Java, northeast to Borneo, and then northwest towards South China Sea [7]. Nowadays, further east, including East Java, West Sulawesi, and potentially the several small islands towards Flores and Sumba, is underlain by a continental crust that becomes the edge of Sundaland [7]. The basement of Bali may have varying densities and thicknesses that come from the transition between continental and oceanic types, which may be a product of mélange [4, 5, 16]. The basement of western Bali may originate from Argoland (a fragment of the Australian continental crust), whereas the eastern part may be part of a suture zone that extends to Sulawesi [4, 16] (Figure 2).

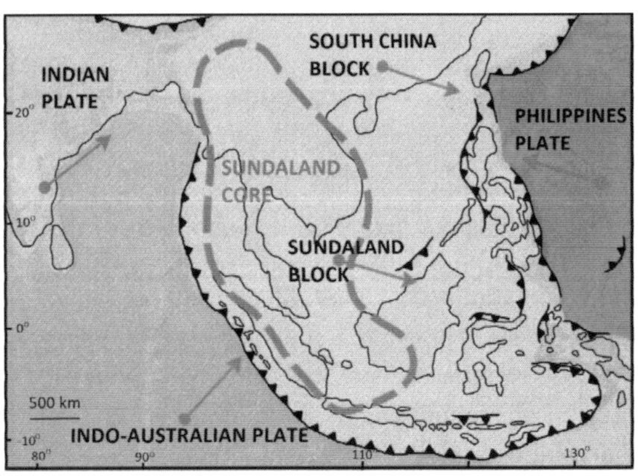

Figure 1. Location of the Sundaland Block as a core in SE Asia [8]. Large arrows represent the direction of plate motions.

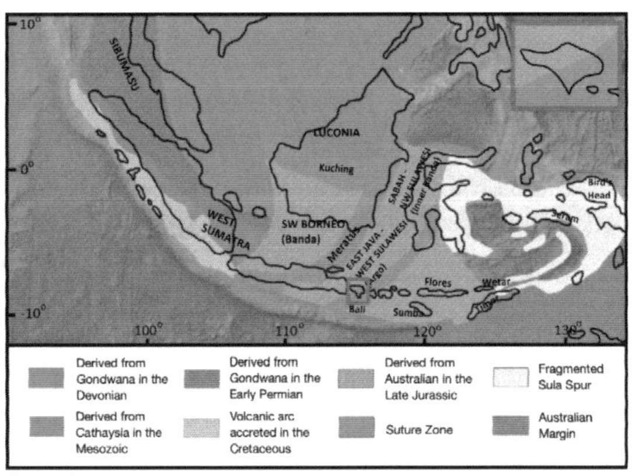

Figure 2. Main arc basement of the Indonesian Crust [7]. Bali Island (shown in box) has a constituent crust originating from the Argoland (a fragment of the Australian Continent Crust in the Late Jurassic Period) in the west and the suture zone in the east [4, 16].

Bali is located in the Lesser Sunda, the eastern part of Sunda Arc, consisting of many small volcanic islands between Java and the Banda Arc (Figure 3). These islands are the product of subduction, collision, and volcanism [16]. The boundary between the Lesser Sunda Islands and the Banda Arc coincides with a change in the present-day plate boundaries between the Indo-Australian and Eurasian plates, from oceanic plate subduction to continental plates collision [16].

The basic stratigraphy of Bali [17, 18] shows that the oldest exposed rock, the Early Miocene Ulakan Formation, is consists of volcanic breccia, lava, and tuff intercalated with carbonate material. Most of the island is covered by Quaternary volcanic deposits originating from the Ancient Volcanoes, Bratan-Batukau-Seraya, and two active volcanoes, Mount Batur and Mount Agung [17, 18]. Mount Agung is situated in eastern Bali. It has one main crater at 3014 meters above sea level. In addition, Mount Agung also has a parasitic cone, Bukit Pawon, on the southeastern slope, with an altitude of 800 meters above sea level. The main crater eruption products consist of 14 lava units, 6 pyroclastic flow units, 3 pyroclastic fall units, and 3 lahars units [19]. Based on the known history, Bukit Pawon only erupted once and produced lava as well as scoria units scattered locally around it [19] (Figure 4).

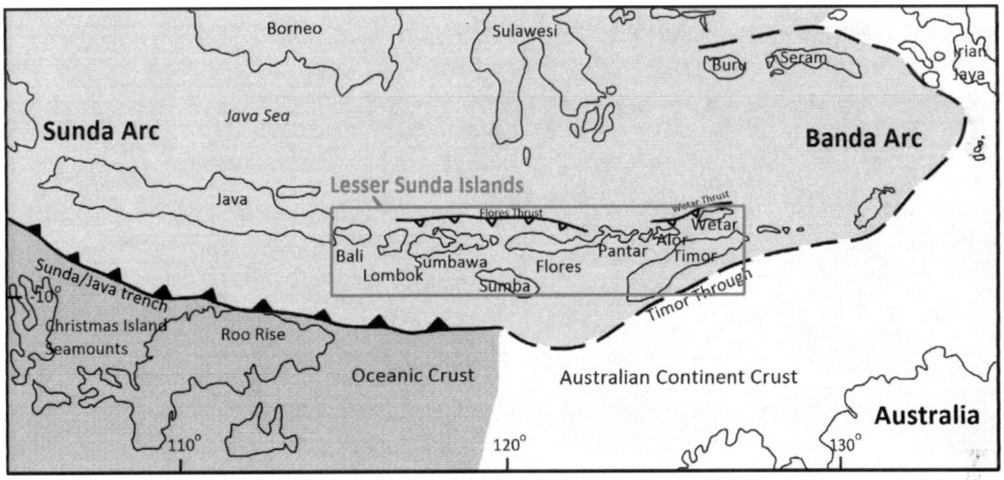

Figure 3. Lesser Sunda Islands between Java and the Banda Arc [2]. The continuity of the trench formed by the subduction of the Indo-Australian Plate under the Eurasian Plate stops at the Sumba Island [16]. That point to the east is influenced by the collision of the Australian Continent with the Banda Arc [16].

3. Sample and Methodology

Well-preserved lava samples were taken from 18 outcrops around Mount Agung (Figure 4). Microscopic observations using a polarizing microscope were conducted on all lava samples to analyze the mineral percentage. Mineral chemistry analysis was carried out on 6 selected samples using Scanning Electron Microscopy-Energy Dispersive X-Ray Spectrometry (SEM-EDS) JEOL JSM-IT-300 of SEM-EDS Laboratory of Chevron-Bandung Insitute of Technology, Indonesia. This test aims to identify the major and minor element content (such as Ti, Al, Fe, Mn, Mg, Ca, Na, K, P, Cr, Ni) of the phenocrysts in the sample. A list of the sample studied in this research is present in Table 1.

As support, geochemical analysis was obtained using published data from Fontjin's [20] and Dempsey's [2] research. Those data are taken from 66 volcanic rock samples (48 samples from [20] and 18 samples from [2]) which consist of major elements (SiO_2, Al_2O_3, Fe_2O_3, FeO, MnO, MgO, CaO, Na_2O, K_2O, TiO_2, P_2O_5) and trace elements (Rb, Sr, Y, Nb, Zr, Cr, Ni, Cu, Zn, Ga, Ba, Pb, Th, U, La, Ce, Nd, Sm). All the supplementary data used in this study are present in Table 2.

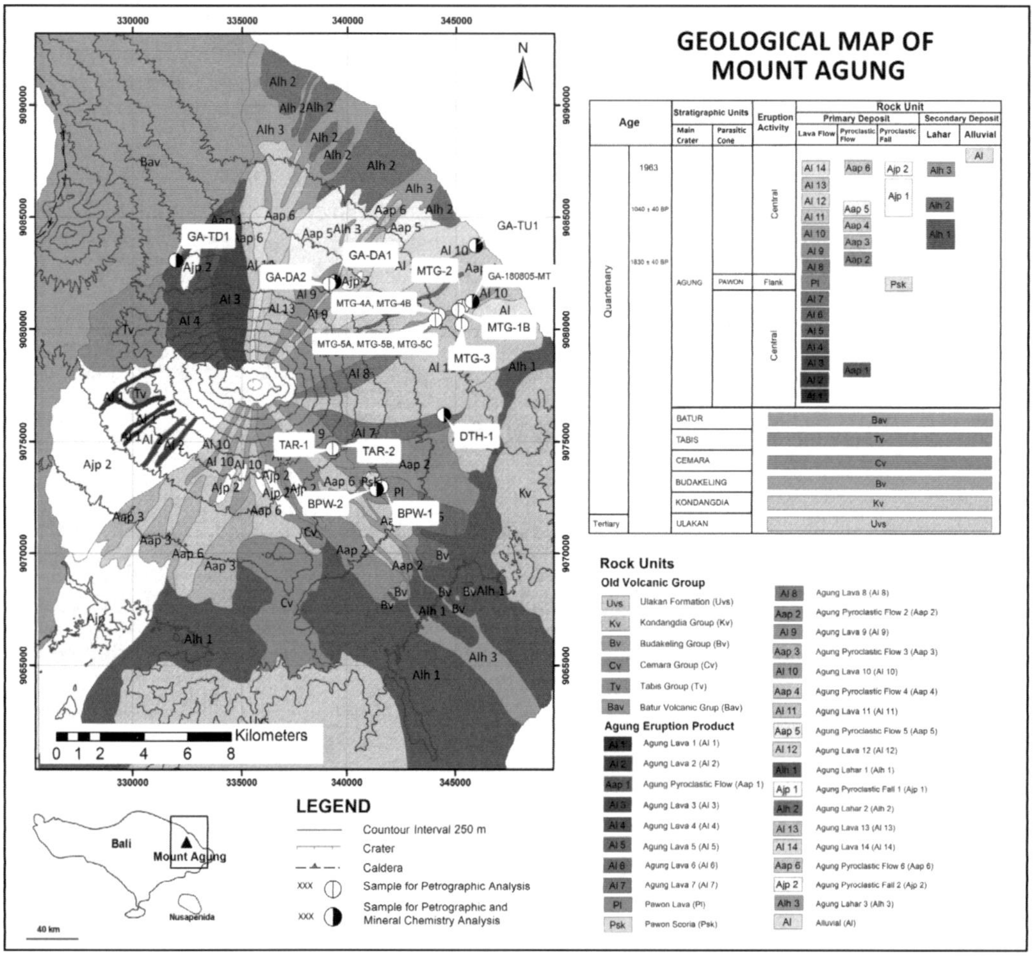

Figure 4. Geological map of the study area [19]. It includes the sample locations used in this study that consist of 18 lava samples for petrographic analysis and 6 of them for mineral chemistry analysis.

Table 1. List of samples used for microscopic observation and mineral chemistry analysis.

Sample Code	Deposit Type	Analysis		Sample Code	Deposit Type	Analysis	
		Petrography	Mineral Chemistry			Petrography	Mineral Chemistry
GA-TU1	Lava	✓	✓	MTG-5C	Lava	✓	
MTG-3	Lava	✓		TAR-1	Lava	✓	
GA-180805-MT	Lava	✓	✓	TAR-2	Lava	✓	
MTG-1B	Lava	✓		GA-DA1	Lava	✓	✓
MTG-2	Lava	✓		GA-DA2	Lava	✓	
MTG-4A	Lava	✓		BPW-1	Lava	✓	
MTG-4B	Lava	✓		BPW-2	Lava	✓	✓
MTG-5A	Lava	✓		DTH-1	Lava	✓	✓
MTG-5B	Lava	✓		GA-TD1	Lava	✓	✓

Table 2. List of samples used as supplementary data in this research for geochemical analysis. It includes rock geochemistry data from published research by Fontjin [20] and Dempsey [2].

Sample Code	Deposit Type	Analysis	Sample Code	Deposit Type	Analysis
		Geochemistry			Geochemistry
Supplementary Data [20]					
AG022A	Lava	✓	AG006bisE	Scoria	✓
AG012I	Scoria	✓	AG006bisF	Ash Fall	✓
AG008B	Pumice	✓	AG006bisH	Ash Fall	✓
AG016C	Scoria	✓	AG021C	Scoria	✓
AG012E	Pyroclastic Flow	✓	AG021D	Ash Fall	✓
AG028B	Pyroclastic Flow	✓	AG021E	Scoria	✓
AG042B	Pyroclastic Flow	✓	AG021H	Ash Fall	✓
AG001bisH	Pumice	✓	AG021I	Scoria	✓
AG045D	Pyroclastic Flow	✓	AG021K	Scoria	✓
AG042D	Pyroclastic Flow	✓	AG006bisO	Scoria	✓
AG001bisB	Pyroclastic Flow	✓	AG006bisS	Scoria	✓
AG042G	Pumice	✓	AG006bisR	Ash Fall	✓
AG001bisC	Scoria	✓	AG006bisT	Scoria	✓
AG001bisE	Pyroclastic Flow	✓	AG006bisK	Scoria	✓
AG042I	Pyroclastic Flow	✓	AG006M	Scoria	✓
AG042bisD	Pyroclastic Flow	✓	AG006N	Scoria	✓
AG042bisB	Pyroclastic Flow	✓	AG006O	Ash Fall	✓
AG024E	Pyroclastic Flow	✓	AG006R	Scoria	✓
AG024D	Pyroclastic Flow	✓	AG006S	Scoria	✓
AG008G	Ash Fall	✓	AG021B	Ash Fall	✓
AG021L	Ash Fall	✓	AG006bisAB	Scoria	✓
AG008I	Scoria	✓	AG006B	Scoria	✓
AG008L	Scoria	✓	AG006F	Scoria	✓
AG006bisC	Scoria	✓	AG024B	Pyroclastic Flow	✓
Supplementary Data [2]					
AG16	Lava	✓	Agu07	Pyroclastic	✓
Agu18	Lava	✓	Agu10	Lava	✓
Agu20	Lava	✓	Agu12	Lava	✓
Agu21	Lava	✓	Agu13	Lava	✓
Agu22	Lava	✓	Agu15	Lava	✓
Agu23	Lava	✓	Agu03	Pyroclastic	✓
Agu24	Lava	✓	Agu30	Lava	✓
Agu25	Pyroclastic	✓	Agu31	Lava	✓
Agu06	Pyroclastic	✓	Agu33	Lava	✓

4. Results

For better understanding, data and stratigraphic correlation were needed to find the connection between all of the existing samples used in this research. The lava samples of Mount Agung eruption were taken from the field based on the division of rock units in Nasution's geological map [19] (Figure. 4). According to the volcano-stratigraphy of Mount Agung, there are 2 dated rock units, Agung Pyroclastic Flow 2 (Aap 2) unit, dated to 1040 ± 40 BP, and Agung Pyroclastic Flow 5 (Aap 5) unit, dated to 1830

± 40 BP. Similar data are also present in Fontjin's research [20]. Sample AG042bisD and AG042bisD have an absolute age of 1040 ± 50 BP, whereas AG006bisE, AG006bisF, and AG006bisH were dated to 1870 ± 40 BP. These two ages become the basis of the sample correlation hypothesis between all the rock units [19] and the absolute age [20] (Figure 5). In addition, these two ages also become the basis for dividing the eruption periods of Mount Agung used in this study.

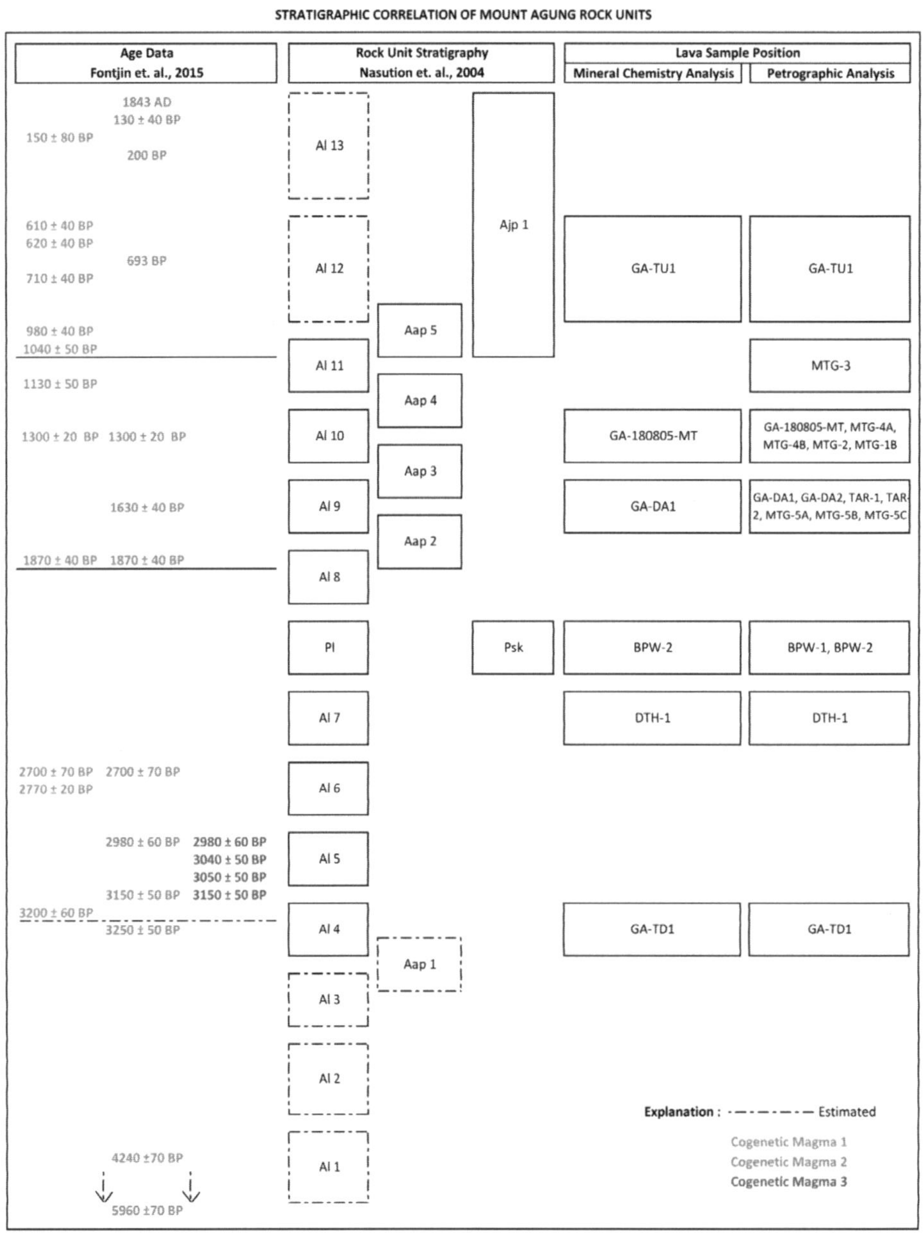

Figure 5. Stratigraphic correlation of age [15], rock units [14], and sample positions representing the rock units used in this study. The cogenetic magma division is obtained based on the sample distribution on the K_2O vs. SiO_2 diagram.

To specify historical activity, the eruption period is detailed using the basis of the geochemistry data plot of Fontjin's [20] and Dempsey's [2] on the K_2O versus SiO_2 diagram, with the assumption K_2O is not affected by the crystal fractionation process. Based on Figure 6, there are three trends of Mount Agung magma differentiation. These trends portray three cogenetic magmas differentiation processes. At the age of 3200 ± 60 BP, there is a sample that has the lowest SiO_2. This sample may be present the closest content of the primitive magma. Rock samples from 3200 ± 60 – 1870 ± 40 BP dominantly had low SiO_2 content. However, rock samples from pre-3200 ± 60 BP had a higher SiO_2. That gives first evidence of the magma evolution process from the pre-3200 ± 60 BP to the 3200 ± 60 – 1870 ± 40 BP period. With those basics, the authors divided Mount Agung's volcanic activity into 4-periods, i.e., the post-1040 ± 50 BP, 1870 ± 40 BP – 1040 ± 50 BP, 3200 ± 60 BP – 1870 ± 40 BP, and the pre-3200 ± 60 BP.

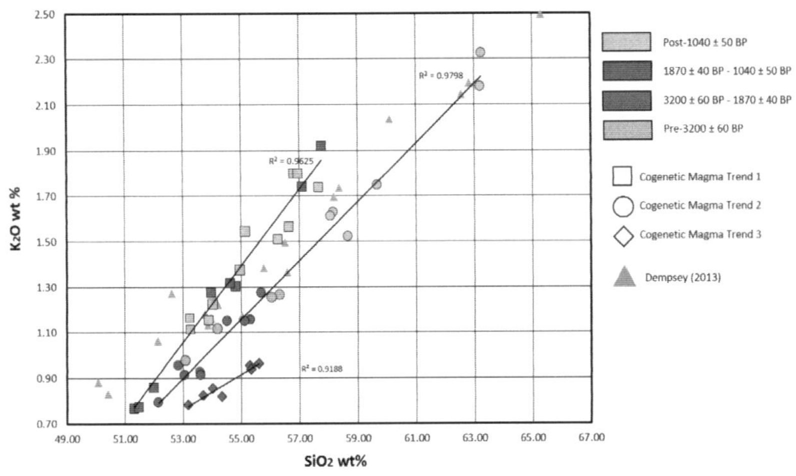

Figure 6. K_2O versus SiO_2 diagram. It shows three cogenetic magma trendlines and the distribution of the samples in every four periods. The data used are from Fontjin's geochemical data [20] and Dempsey's [2] used as a comparison.

4.1. Mineral Composition

Based on microscopic observations of 18 lava samples, all eruption periods produced pyroxene-andesite rocks (name using Williams' classification [21]), which have a porphyritic texture. The phenocrysts contained are dominated by plagioclase, followed by pyroxene, opaque minerals, and olivine.

4.1.1. Plagioclase. Plagioclase is the predominant phenocryst present in all samples. It is found as subhedral crystals that showed varying microtextures. Coarse sieve is abundant followed by glomerocryst, fine sieve, fine-oscillatory zoning, and resorption surface (Figure 7). Based on the An composition [XAn = 100*Ca/(Ca+Na)] taken from mineral chemistry analysis (Figure 8), in pre-3200 ± 60 BP period, plagioclase contains a long-range of An_{38-89}. The core tends to be more calcic than the rim. The core's composition is dominant at An_{70-80} and An $_{85-89}$, while the rim's dominant at An_{50-55}. In 3200 ± 60 – 1870 ± 40 BP, the An content decreased to An_{67-83}. Both core and rim have a wide range of An. In 1870 ± 40 – 1040 ± 50 BP, the An composition generally decreased again to An_{43-88}. In the post-1040 ± 50 BP, plagioclase minerals have An_{38-86}, with the same distribution pattern as the first period. The core is more calcic than the rim. The core's composition is dominant at An_{70-85}, while the rim varies from An_{35-75} but is dominant at An_{65-70}. Based on the scanline test, plagioclase has a decreasing percent mass Ca from the core to the rim (normal zoning) (Figure 8).

4.1.2. Pyroxene. Pyroxene minerals as phenocrysts (Figure 9A and 9B) are present in subhedral–anhedral forms. Locally, pyroxenes are found as cumulocrysts along with plagioclase, olivine, and opaque minerals. There are two types of pyroxene observed in each sample, orthopyroxene and clinopyroxene. Orthopyroxene has enstatite composition, while clinopyroxene has augite composition (classify using [22]). Based on mineral chemistry analysis, In the pre-3200 ± 60 BP period, enstatite has Mg#$_{61-77}$ [XMg# = 100*Mg/(Mg+Fe^{2+})], whereas augite has Mg#$_{73-78}$. In the following period, the Mg# content in enstatite generally increased to Mg#$_{65-76}$, but augite's Mg# decreased to Mg#$_{71-76}$. In 1870 ± 40 – 1040 ± 50 BP period, the Mg# composition in enstatite and augite generally decrease, to Mg#$_{65-75}$ and Mg$_{\#72-77}$ in order. However, viewed from the percentage, the pyroxene minerals are more abundant during this period. In the post-1040 ± 50 BP, the Mg# content in enstatite and augite tend to increase, to Mg#$_{67-72}$ and Mg#$_{67-81}$. Based on the scanline test, both enstatite and augite have a percent mass of Mg that increases from the core to the rim (Figure 10A and 10B).

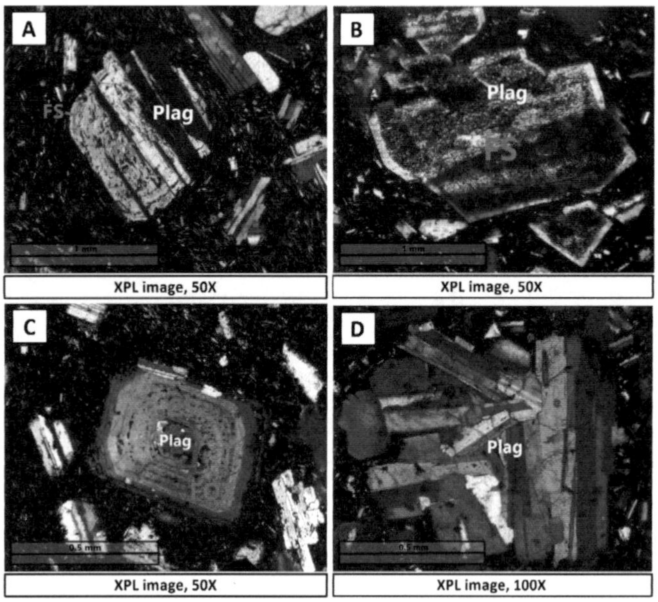

Figure 7. Selected images of optical microscope of plagioclase presented in Mount Agung. (A) Coarse sieve texture in Sample GA-TD1 (pre-3200 ± 60 BP). (B) Fine sieve found in Sample BPW-2 (3200 ± 60 – 1870 ± 40 BP). (C) Fine oscillatory zoning and coarse sieve in Sample TAR-2 (1870 ± 40 – 1040 ± 50 BP). (D) Glomerocrysts found in GA-TU1 (post-1040 ± 50 BP). Detail: Plag = Plagioclase; FS = Fine Sieve; CS = Coarse Sieve; FOZ = Fine Oscillatory Zoning.

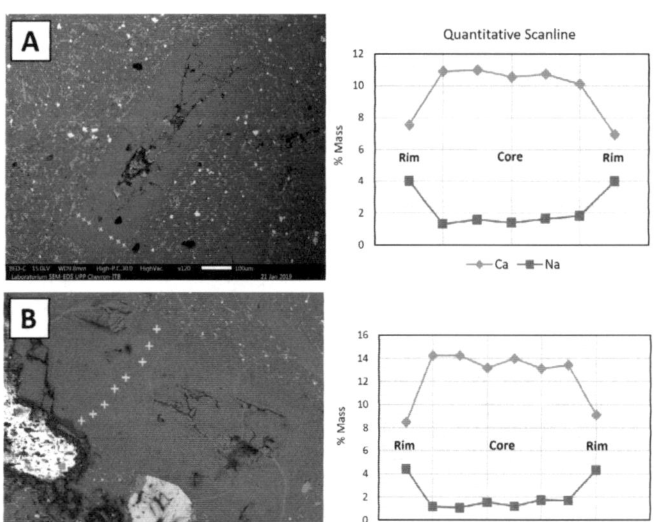

Figure 8. Selected electron images of plagioclase from SEM-EDS mineral chemistry analysis and its quantitative scanline result. Blue +: approximate analyzed points. (A) Plagioclase of Sample GA-TD1 (pre-3200 ± 60 BP period) shows decreasing Ca from the core to the rim. (B) Plagioclase of Sample GA-DA1 (1870 ± 40 – 1040 ± 50 BP) indicates a similar pattern.

IOP Conf. Series: Earth and Environmental Science 1047 (2022) 012005 doi:10.1088/1755-1315/1047/1/012005

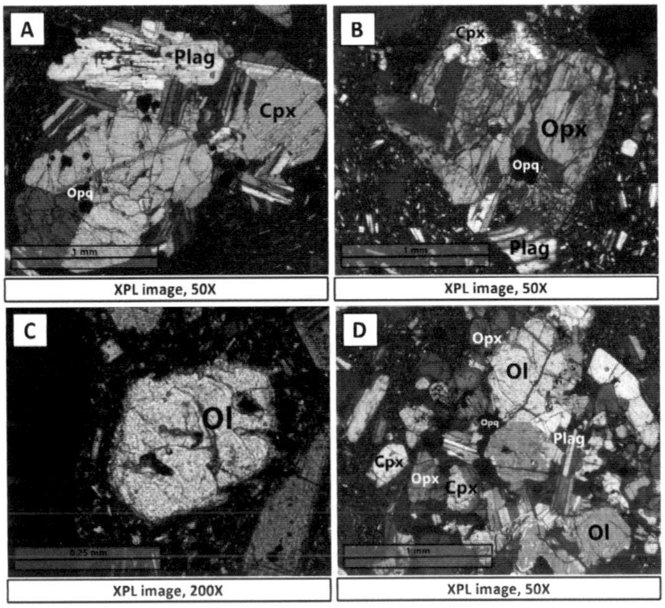

Figure 9. Selected images of optical microscope of minerals presented in the Mount Agung eruption. (A) Clinopyroxene as cumulocrysts of Sample GA-TD1 (pre-3200 ± 60 BP period). (B) Orthopyroxene of Sample BPW-1 (3200 ± 60 – 1870 ± 40 BP). (C) Olivine as microfenocryst in Sample MTG-5A (1870 ± 40 – 1040 ± 50 BP). (D) Olivine, plagioclase, orthopyroxene, clinopyroxene, and opaque minerals cumulocrysts found in Sample MTG-5C (1870 ± 40 – 1040 ± 50 BP). Details: Ol = Olivine; Cpx = Clinopyroxene; Opq = Opaque Mineral; Opx = Orthopyroxene; Plag = Plagioclase.

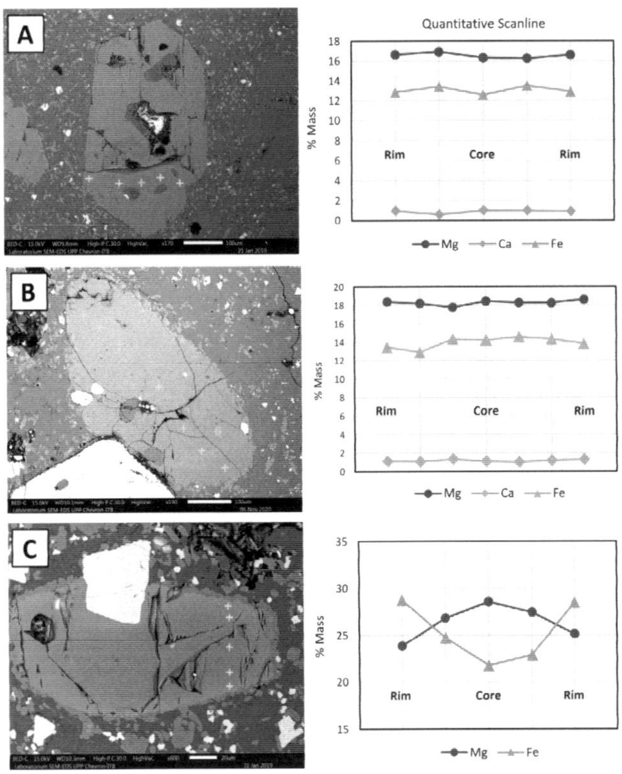

Figure 10. Selected electron images of SEM-EDS mineral chemistry analysis and its quantitative scanline result. Blue +: approximate analyzed points. (A) Enstatite of Sample GA-TD1 (pre-3200 ± 60 BP) generally shows increasing Mg from the core to the rim. (B) Augite of Sample DTH-1 (3200 ± 60 – 1870 ± 40 BP) generally shows increasing Mg from the core to the rim. (C) Olivine with reaction rim of Sample GA-TU1 (post-1040 ± 50 BP) shows decreasing Mg from the core to the rim.

4.1.3. Olivine. Generally, olivine minerals found in all eruption samples are present as microphenocrysts with subhedral–anhedral forms (Figure 9C). Olivine dominantly has a reaction rim composed of pigeonite or enstatite (classify using [22]). Locally, olivine is spotted having a rim that consists of opaque minerals. Based on mineral chemistry analysis, olivine in the pre-3200 ± 60 BP contains Fo_{62-78} [XFo = 100*Mg/(Mg+Fe)] with a more Mg-rich core than the rim. The core's Fo value is dominant at Fo_{75-78}, while the rim is dominant at Fo_{65-70}. In 3200 ± 60 – 1870 ± 40 BP, the Fo content tends to decrease, becoming Fo_{60-71}, but both core and rim have a wide range of Fo. In 1870 ± 40 – 1040 ± 50 BP period, the Fo content slightly decreased to Fo_{59-71}. Locally, olivine is spotted partially oxidized. Then, in the post-1040 ± 50 BP, the Fo composition increase to Fo_{66-76}. Olivine percentage is slightly more abundant compared to the previous period. The core is also more Mg-rich than the rim. The core's Fo content is dominant at Fo_{70-75}, while the rim is dominant at Fo_{65-70}. Based on the scanline test, olivine has a decreasing percent mass of Mg from the core to the rim (normal zoning), which can be caused by the reaction rim (Figure 10C).

4.1.4. Opaque Minerals. Opaque minerals are present as phenocrysts and microphenocrysts with anhedral shapes. Locally, they are found as cumulocrysts along with plagioclase, pyroxene, and olivine (Figure 9D). Based on the mineral chemistry test, the opaque mineral from all eruption periods has a magnetite composition (classify using [23]).

4.2. Geochemical Analysis

Bulk geochemical analysis was carried out using supplementary data from published journals by Fontjin [20]. As a comparison, Dempsey's [2] data was also used in this study, especially to see the data clusters distribution suitability.

4.2.1. Normative Minerals. Normative minerals are the number of minerals present in the norm expressed as their molecular proportion [24]. This paper used the CIPW calculation excel sheet program written by Hollocher [25]. Based on Table 3, all samples from each period are characterized by the presence of quartz and hypersthene that shows saturated to highly saturated silica magma. Hypersthene minerals are found the highest in 3200 ± 60 – 1870 ± 40 BP and 1870 ± 40 – 1040 ± 50 BP. However, the quartz in 1870 ± 40 – 1040 ± 50 BP is slightly higher. It suggests that the magma saturation level reach the lowest in 3200 ± 60 – 1870 ± 40 BP. The quartz minerals are most abundant in the youngest period which implies that the silica saturation in magma is on the highest level compared to the other periods.

Table 3. Percentage of normative minerals present in each period of Mount Agung eruption. The data used are geochemistry data of Fontjin [20].

Eruption Period	Pre-3200 ± 60 BP	3200 ± 60 – 1870 ± 40 BP	1870 ± 40 – 1040 ± 50 BP	Post-1040 ± 50 BP
Deposite Type	Scoria, Pyroclastic Flow	Scoria, Ash Fall	Scoria, Ash Fall, Pyroclastic Flow	Pyroclastic Flow, Pumice, Scoria, Ash
Normative Minerals (wt%)				
Quartz	2.92 - 6.34	1.35 - 6.92	0.39 - 8.26	2.76 - 15.57
Plagioclase	59.99 - 67.66	56.61 - 72.14	55.35 - 68.3	55.1 - 59.82
Orthoclase	5.73 - 7.45	4.61 - 7.68	4.55 - 11.46	6.68 - 10.7
Corundum	0	0 - 0.64	0	0
Diopside	1.22 - 6.66	0 - 8.36	3.13 - 6.43	0.52 - 8.49
Hypersthene	13.85 - 21.34	13.53 - 24.61	13.66 - 25.49	10.48 - 19.98
Ilmenite	1.44 - 2.07	1.42 - 2.22	1.58 - 2.03	1.12 - 1.73
Magnetite	1.09 - 1.59	1.07 - 1.83	1.15 - 1.59	0.87 - 1.48
Apatite	0.53 - 0.65	0.35 - 0.65	0.37 - 0.65	0.53 - 0.63
Zircon	0.01	0.01	0.01 - 0.03	0.01 - 0.03

4.2.2. Rock Classification. Eruption product is a good indicator for determining the origin magma content. Using plotted data on Le Bas diagram [26] of total alkali ($Na_2O + K_2O$) against SiO_2 (Figure 11), the eruption products from each period are as follows: (1) The pre-3200 ± 60 BP consist of basaltic-andesite rocks with SiO_2 53 – 56 wt%; (2) 3200 ± 60 – 1870 ± 40 BP consists of basalt to basaltic-andesite rocks with SiO_2 51 – 56 wt%; (3) 1870 ± 40 – 1040 ± 50 BP consists of basalt to andesite rocks with SiO_2 53 – 58 wt%, but more dominantly in SiO_2 51 – 58%; (4) The post-1040 ± 50 BP consists of basaltic-andesite to dacite rocks with SiO_2 53 – 63% wt%.

4.2.3. Magma Affinity. Magma series or magma affinity can interpret the volcano-tectonic setting [11]. The magma affinity is obtained by plotting the geochemical data on the K_2O vs. SiO_2 [27]. The data plot result (Figure 12) shows that all samples are from the calc-alkaline magma series. According to Wilson [11], it is a feature of convergent plate boundaries (subduction).

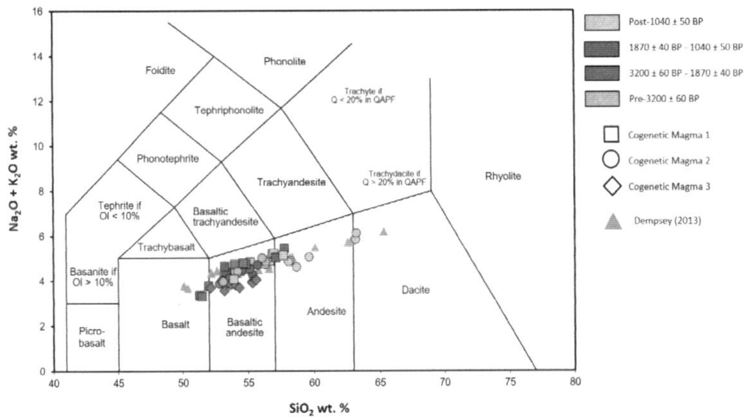

Figure 11. The result of Fontjin's data [20] plotting on the $Na_2O + K_2O$ vs. SiO_2 diagram [26] which determines the geochemical rock name. Dempsey's data [2] is used as a comparison.

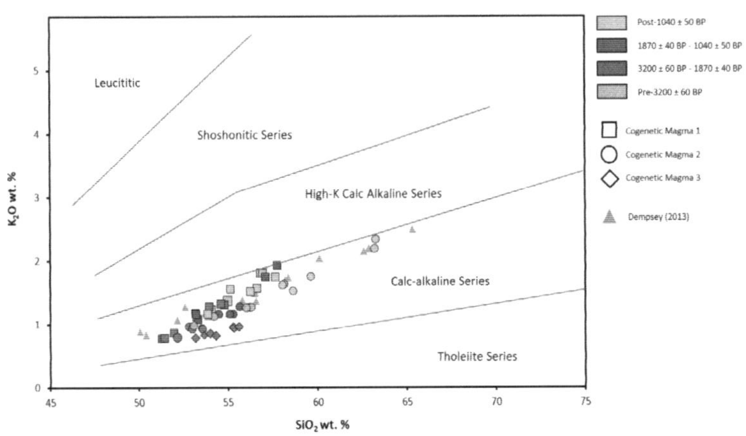

Figure 12. Distribution of Fontjin's data [20] on the K_2O vs. SiO_2 diagram [27] which indicates the magma series of Mount Agung eruption products. Dempsey's data [2] is used as a comparison.

4.2.4. Harker Diagram and Crystal Fractionation. Harker diagrams show the change of major elements to SiO_2 [28, 29]. This diagram can provide magmatic activity interpretation, which includes the evolution and magma differentiation. Crystal fractionation is characterized by a curved pattern due to the depletion of certain compatible elements from the magma solution that depends on the mineral phase undergoing crystallization [30]. Based on Figure 13, crystal fractionation occurred during Mount Agung's magma differentiation. It represents from the TiO_2, MgO, Fe_2O_3, and CaO diagrams. In the TiO_2 diagram, the steep gradient in the older period (pre-1040 ± 50 BP) indicates the fractionation of magnetite. The MgO and Fe_2O_3 diagrams show the fractionation of olivine and/or pyroxene, which is also predominant in the older period (pre-1040 ± 50 BP). Next, the CaO diagram exhibits the plagioclase fractionation, and compared to other minerals, it predominantly happened in the younger period (post-1040 ± 50 BP) instead.

Crystal fractionation can also be seen from the log K_2O against the log Zr/Nb diagram [2] (Figure 14). Based on the previous Harker Diagram, there are differences in the minerals undergoing fractionation. This interpretation is also supported by Dempsey's research [2], which stated that there are two fractionation vectors during the magma differentiation process. The cause can be identified from plagioclase minerals. During the differentiation process, plagioclase fractionation can be temporarily suspended if the magma is under high-pressure conditions or has a high water content [2, 31]. Figure 15 shows data distribution against the two trendlines of plagioclase fractionation [31]. It can be seen that the pre-3200 ± 60 BP, 3200 ± 60 – 1870 ± 40 BP, and 1870 ± 40 – 1040 ± 50 BP periods, in general, followed the curved line indicating that the plagioclase fractionation was restrained. In contrast, the post-1040 ± 50 BP period, in general, followed the other line that shows the occurrence of plagioclase fractionation. These differences correspond to Dempsey [2] and Geiger [32], which said that Mount Agung is possibly controlled by the presence of a high pressure-deep magma reservoir that can withstand plagioclase fractionation and also a shallow magma reservoir with lower pressure.

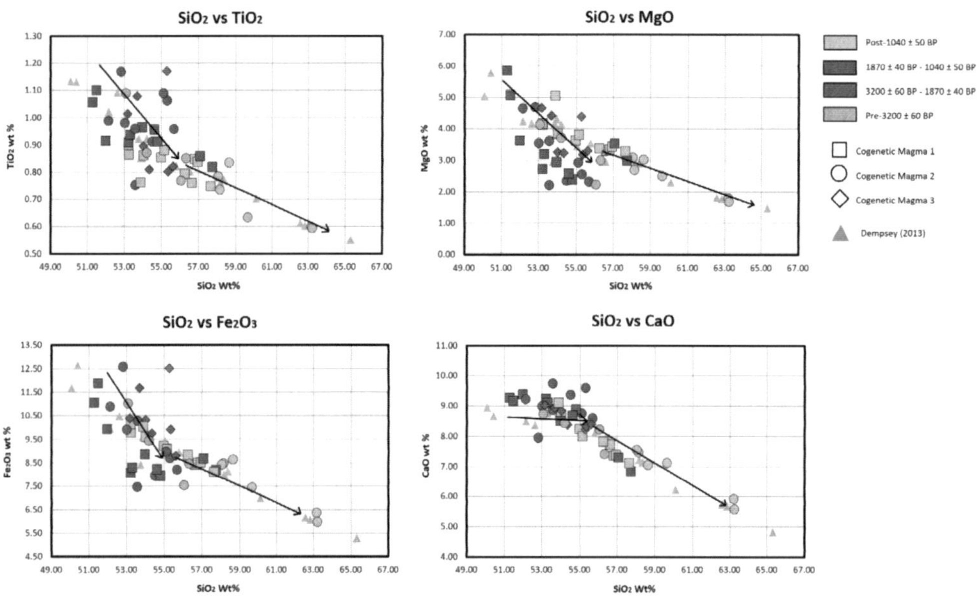

Figure 13. Selected Harker diagrams show the distribution of the major elements using the SiO_2 as a differentiation index. The black arrow indicates the trend of certain minerals depletion. The geochemistry data plotted is from Fontjin [20] and Dempsey's data [2] is used as a comparison.

4.2.5. Spider Diagram and Trace Element. Discovering the tectonic setting can be done by plotting trace elements that have been normalized with the primitive mantle onto a spider diagram [33]. Normalization is used to eliminate the Oddo-Harkins effect on the graph. From Figure 16, it can be seen that the Rb, Ba, Th, and K (LILE) and La – Sm (LREE) contents are enriched, while Eu – Lu (HREE) experienced depletion. According to Dirk [34], these features indicate the properties of a subduction zone. In addition, the Nb content also shows a negative anomaly which is also a characteristic of convergent plate boundaries [11]. Thus, it can be inferred that Mount Agung comes from a volcanic associated with a convergent plate boundary (subduction).

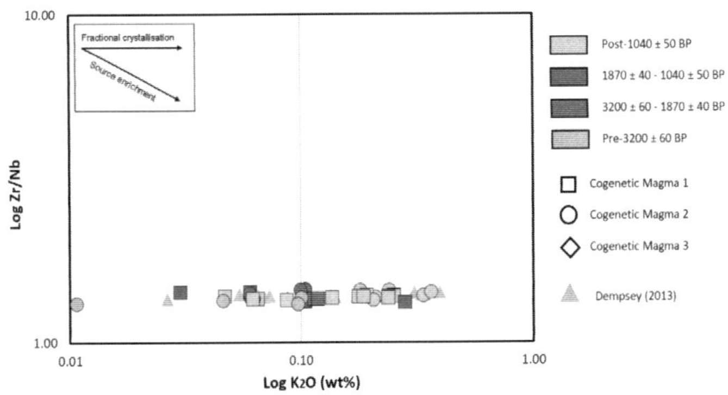

Figure 14. Diagram of log K_2O log vs log Zr/Nb which distinguishes crystal fractionation with primitive mantle enrichment [2]. The rock geochemistry data used is Fontjin's [20] with Dempsey's data [2] as a comparison.

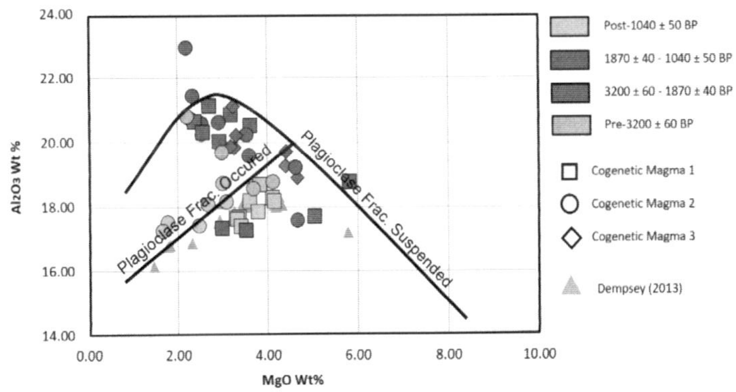

Figure 15. Diagram of MgO against Al_2O_3 with two trends of plagioclase fractionation [2, 31]. The geochemistry data used is from Fontjin [20] with Dempsey's data [2] as a comparison.

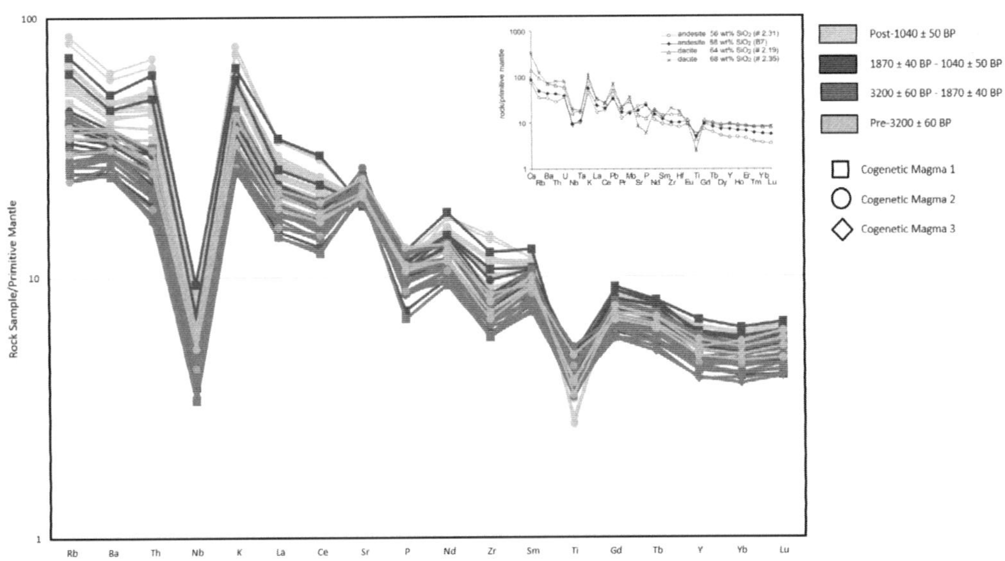

Figure 16. Spider diagram of Mount Agung eruption trace elements. The rock geochemistry data comes from Fontjin [20]. The sample was normalized against the primitive mantle [32]. A similar diagram of the Mount Batur eruption products [35] was used as a comparison.

There are two types of subduction, island arcs and active continental margins [11]. Those two types can be distinguished using the ratio of Zr to Zr/Y from the Pearce [36]. Based on Figure 17, Mount Agung originates from the subduction of oceanic crust beneath continental crust. Next, the La/Sm to Th/Nb diagram can specify the differentiation type influenced by its tectonic setting [37, 38]. Figure 18 shows that in the older period (pre-1040 ± 50 BP), the magma differentiation process is subduction-related enrichment. Whereas in the younger period (post-1040 ± 50 BP), there may be a slight influence from the presence of crustal contamination.

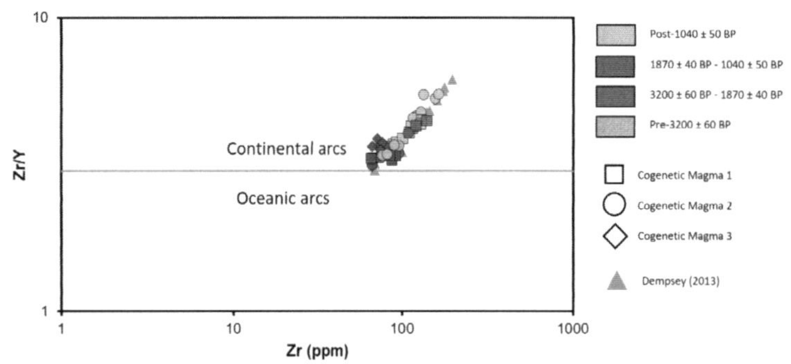

Figure 17. Diagram of Zr vs. Zr/Y which differentiate continental and oceanic arc [36]. The rock geochemistry data is from Fontjin [20]. Dempsey's data [2] is used as a comparison.

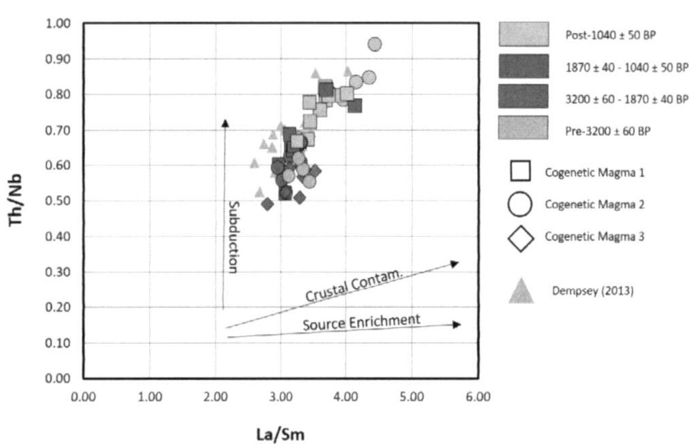

Figure 18. La/Sm vs. Th/Nb diagram with several differentiation vectors [37, 38]. The rock geochemistry data is from Fontjin [20]. Dempsey's data [2] is used as a comparison.

5. Discussion

5.1. Nature of Primitive Magma Source

Magma source interpretation of Mount Agung can be acquired by looking at the Fo content in olivine phenocrysts [39]. Magma originating directly from the mantle source has fosterite content more than Fo_{88} [11, 40]. The Fo in olivine from Mount Agung eruption ranged from Fo_{60-78}. It proves that the primitive magma of Mount Agung originates from the partial melting process instead. The SiO_2 content of the primitive magma of Mount Agung can be interpreted from the plot diagram of K_2O against SiO_2 (Figure 6), assuming that the K_2O content is not affected by the crystal fractionation process and can represent the actual content of the magma. Based on the diagram, three trendlines come from one point at lower SiO_2. Therefore, it indicates that the three cogenetic magmas may originate from the same source. By extending the trendline, the primitive magma has a SiO_2 content of ~50 wt%.

5.2. Magma Chamber Condition

Plagioclase minerals, mentioned in several studies, may be an efficient tool for understanding the dynamic magmatic process due to their high sensitivity to temperature-pressure (T-P) conditions, water content, and melt composition changes [41, 42, 43, 44]. The wide range of cores and diversity in An zoning imply a compositional change of host magma or change in the crystallization conditions that suggest a diverse magma chamber system [44]. Based on mineral chemical analysis, Mount Agung eruption produces a wide range of An zoning, ~An_{38-89}. As stated in the previous Harker diagram and crystal fractionation analysis, Mount Agung is possibly controlled by the presence of a deep and shallow magma chamber [2, 32, 45]. The deep high-temperature magma chamber feeds basaltic magma to the shallower lower-temperature magma chamber [43]. During this process, decompression, heat addition, magma mixing, rapid undercooling, or magma mixing may occur simultaneously [43]. Thus, these processes will produce disequilibrium plagioclase microtextures, such as zoned, sieved (coarse and fine sieve), and dissolved (rounded zone, resorption surface) phenocryst [43] that are all present in every period of Mount Agung eruption.

5.3. Tectonic Implication

The results of this study suggest that Bali Island is interpreted as having a basement that supports the recent studies about the distribution of continental blocks and fragments in SE Asia including Sundaland [7, 8]. Our research found that Bali is underlain by continental crust. This is shown from the Zr vs. Zr/Y diagram indicated that Bali was formed as a result of the continental arc or active continental margin. It is different from previous research [5, 9, 10], which says that Bali was formed due to island arcs or as known as having the oceanic basement. The distinctive feature of island arcs generally is a linear or arcuate chain of small islands [11], which in this case is the Lesser Sunda Islands. The active continental margin, on the other hand, is broadly similar to the island arc, although the passage of magma through the thick continental crust added complexities [11]. This case is not found in Bali, because the crust underlain this island is micro-continent, which may be derived from the Australian continent and suture zone [4, 16].

The formation of Bali Island was deducted in the reconstruction of the Southeast Asian tectonic model [4]. Approximately 15 million years ago, the subduction of the Indian Ocean with Eurasia caused the Australian continent to dock with Sundaland, resulting in the formation of a suture zone [4] (Figure 2). Then, this subduction event generated partial melting, which became the source of volcanic arc magma whose eruptions then slowly formed Bali Island as it is now. This can be proven by the oldest deposits that form Bali Island, the early Miocene Ulakan formation, which is a volcanic rock deposit [17, 18]. The volcanic arc activity was divided into three cycles, i.e., the first cycle occurred in the Miocene, the second cycle occurred in the late Pliocene which was marked by the appearance of more alkaline lava, and the third cycle occurred in the quarternary period [14]. The last cycle formed several volcanoes in Bali and one of them is Mount Agung.

Based on the magma evolution of Mount Agung in each period, it can be seen that the K_2O and SiO_2 content in magma is generally increasing. This development, seen from the major element diagrams, was caused by magma differentiation in the form of crystal fractionation of olivine-pyroxene-magnetite (pre-1040 ± 50 BP) while the younger period was in the form of plagioclase fractionation. However, in the post-1040 ± 50 BP period, crystal fractionation was not the only differentiation process that happened. There was also crustal contamination, which is shown from the La/Sm and Th/Nb trace element diagrams (Figure 18). Based on the spider diagram (Figure 16), post-1040 ± 50 BP also has a higher trace element content of K, Sr, Ba, Zr, and Th than in other periods. It also suggests that crustal involvement [11] indeed took place in the younger period. Lesser Sunda Island, where Bali resides, coincides with a change in the plate convergence type from subduction to collision [16]. This intimates that the Lesser Sunda Islands are slowly will more likely to become affected by the collision of the Australian continent with the Banda arc, which may explain why the continental crust influence is greater in the younger period.

6. Conclusion

From the result and discussion, it can be concluded that:

(1) Mount Agung's primitive magma originated from partial melting and it has approximately 50 wt% of SiO_2 content. This magma underwent a differentiation process in multiple magma chambers.

(2) The tectonic setting of Bali Island is an active continental margin. The trace element analysis confirms that Bali Island has a continental basement, which is more precisely a micro-continent basement and this confirms the recent studies of Sundaland [7, 8]. This evidence suggests that Bali Island is still included in Sundaland and may be part of its eastern boundary.

References

[1] Hall R and Smyth H 2008 Cenozoic arc processes in Indonesia: identification of the key influences on the stratigraphic record in active volcanic arcs *Geol. Soc. of America Special Pub* **436** 27–54

[2] Dempsey S R 2013 Geochemistry of volcanic rocks from the Sunda Arc *Ph.D. Thesis Durham University*

[3] Tregoning P, Brunner F K, Bock Y, Puntodewo S S, McCaffrey R, and Genrich J F, Calais E, Rais J, and Subarya C 1994 First geodetic measurement of convergence across the Java Trench *Geophys. Research Letters* **21** 19 2135–2138

[4] Hall R and Sevastjanova I 2012 Australian crust in Indonesia *Australian J of Earth Sci* **59** 827–844

[5] Hamilton W 1973 Tectonics of the Indonesian region *Geol. Soc. Malaysia* **6** 3–10

[6] Wheller G E 1986 Petrogenesis of Batur caldera, Bali, and the geochemistry of Sunda-Banda arc basalts *Ph.D. thesis University of Tasmania*

[7] Hall R 2014 The origin of Sundaland, *Proc. of Sundaland Resources MGEI Annual Conv.* (Indonesia)

[8] Metcalfe I 2017 Tectonic evolution of Sundaland *Bul. of the Geol. Soc. of Malaysia* **63** 27–60

[9] Katili J A 1975 Volcanism and plate tectonic in Indonesian island arcs *Tectonophysics* **26** 165–188

[10] Purwandono A F, Bonte D, Utami P, Pramumijoyo S, Harijoko A, Beekman F, and van Wees J D 2019 Tectonic and compositional variation in Flores Island, Indonesia: implication for volcanic structure and geothermal occurrences *European Geothermal Cong.* (Netherland)

[11] Wilson M 1989 *Igneous Petrogenesis A Global Tectonic Approach* (Netherland: Springer)

[12] Pearce J A and Cann J R 1973 Tectonic setting of basic volcanic rocks determined using trace element analyses *Earth and Planetary Sci. Letter* **19** 290–300

[13] Wibowo B, Mellawati J, and Susiati H 2011 Kajian evolusi geokimia dan kaitannya dengan tingkat bahaya Aunung Api Muria terhadap Tapak PLTN Muria *Jurnal Pengembangan Energi Nuklir* **13** 1

[14] van Bemmelen R W V 1949 *The Geology of Indonesia (General Geology of Indonesia and Adjacent Archipelagoes)* vol. 1A (The Hague: Government Printing Office)

[15] Hutchison C S 2007 *Geological Evolution of South-East Asia 2nd Edition* (Malaysia: Geological Society of Malaysia)

[16] Minarwan 2012 Tectonic models of the Lesser Sunda Islands *Berita Sedimentologi* **25** 8–15.

[17] Purbo-Hadiwidjojo M 1972 The geologic map of Bali, Indonesia, scale 1:250.000 *Geol. Survey of Indonesia*

[18] Sutawidjaja I S 2009 Ignimbrite analyses of Batur Caldera, Bali, based on ^{14}C dating *Jurnal Geologi Indonesia* **4** 3 189–202

[19] Nasution A, Haerani N, Mulyadi D, and Hendrasto M 2004 Peta Geologi Gunungapi Agung, Bali, Skala 1 : 50.000 *Direktorat Vulkanologi dan Mitigasi Bencana Geologi*

[20] Fontijn K, Costa F, Sutawidjaja I, Newhall C G, and Harrin J S 2015 A 5000-year record of multiple highly explosive mafic eruptions from Gunung Agung (Bali, Indonesia): implications

for eruption frequency and volcanic hazards *Bull Volcanol* **77** 59.
[21] Williams H, Turner F J, dan Gilbert C M 1982 *Petrography: An Introduction to the Study of Rocks in Thin Section 2nd edition* (San Fransisco: W. H. Freeman and Company)
[22] Morimoto N, Fabries J, Fergusson A K, Guizbourg I D, Ross M, Seifert F A, Zussman J, Aoki K, and Gottardi G 1988 Nomenclature of pyroxenes *American Mineralogist* **73** 1123–1133
[23] Buddington A F and Lindsley D H 1964 Iron-titanium oxide minerals and synthetic equivalents. *J. Petrol.* **5** 210–357
[24] Kelsey C H 1965 Calculation of the CIPW norm *Mineralogical Mag.* **34** 276–282
[25] Hollocher K 2004 CIPW norm calculation program *Geology Department Union College*
[26] Le Bas M J, Le Maitre R W, Streckeisen A L, and Zanettin B 1986 A classification of volcanic rock based on the total alkali-silica diagram *J. of Petrology* **27** 745–50
[27] Peccerillo R and Taylor S R 1976 Geochemistry of Eocene calc-alkaline volcanic rocks from the Kastamonu Area, Northen Turkey *Contribution to Mineralogy and Petrololgy* **58** 63–81
[28] Harker A 1909 *The Natural History of Igneous Rocks* (London: Methuen and Co.)
[29] Winter J D 2014 *Principles of Igneous and Metamorpfic Petrology Second Edition* (America: Pearson New International Edition/Pearson Educational Limited)
[30] Novak S W and Bacon C R 1986 Pliocene volcanic rocks of the Coso Range, Inyo County, California *U.S. Geol. Survey Prof. Paper* 1383
[31] Parman S W, Grove T L, Kelley K A, and Plank T 2011 Along-Arc variations in the pre-eruptive H_2O contents of Mariana Arc magmas inferred from fractionation paths *J. of Petrology* **52** 2 257–278
[32] Geiger H, Troll V R, Jolis E M, Deegan F M, Harris C, Hilton D R, and Freda C 2018 Multi-level magma plumbing at Agung and Batur volcanoes increases risk of hazardous eruption *Sci. Reports* **8** 10547
[33] Sun S S, and McDonough W F 1989 Chemical and Isotope Systematics of Oceanic Basalts: implications for mantle composition and processes, magmatism in ocean basins *Geol. Soc. of London* **42** 313–345
[34] Dirk M H J 2008 Petrologi-geokimia batuan Gunung Api Tampomas dan sekitarnya *Pusat Survei Geologi Jurnal Geologi Indonesia* **3** 1 23–35
[35] Reubi O I and Nicholls A 2005 Structure and dynamics of a silicic magmatic system associated with caldera-forming eruptions at Batur Volcanic Field, Bali, Indonesia *J. of Petrology* **46** 7
[36] Pearce J A 1983 *Role of the sub-continental lithosphere in magma genesis at active continental margins* (Chesire: Shiva Publications) p 230–249
[37] Plank T 2005 Constraints from thorium/lanthanum on sediment recycling at subduction zones and the evolution of the continents *J. of Petrology* **46** 921–944
[38] Barresi T, Nelson J L, Dostal J. and Friedman R 2015 Evolution of the Hazelton arc near Terrace, British Columbia: stratigraphic, geochronological, and geochemical constraints on a Late Triassic – Early Jurassic arc and Cu–Au porphyry belt *Can. J. Earth Sci* **52** 466–494
[39] Rogkala A, Petrounias P, Tsikouras B, Giannakopoulou P P, and Hatzipanagiotou K 2019 Mineralogical evidence for partial melting and melt-rock interaction processes in the mantle peridotites of Edessa Ophiolite (North Greece) *Minerals* **9** 120
[40] Arai S 1994 Characterization of spinel peridotites by olivine-spinel compositional relationship: review and interpretation *Chem. Geol.* **113** 191–204
[41] Ginibre C, Wörner G, and Kronz A 2002 Minor- and trace-element zoning in plagioclase: implications for magma chamber processes at Parinacota volcano, northern Chile *Cont. to Mineralogy and Petrology* **143** 300–315
[42] Ginibre C and Wörner G 2007 Variable parent magmas and recharge regimes of the Parinacota magma system (N. Chile) revealed by Fe, Mg and Sr zoning in plagioclase *Lithos* **98** 118–140.
[43] Renjith M L 2014 Micro-textures in plagioclase from 1994-1995 eruption, Barren Island Volcano: Evidence of dynamic magma plumbing system in the Andaman subduction zone *Geosci. Frontiers* **5** 113–126.

[44] Aaltonen M 2018 Crustal contamination recorded by zoned plagioclase in primitive Karoo flood basalts, Luenha River, Mozambique *Master's thesis University of Helsinki*

[45] Syahbana D K, Kasbani K, Suantika G, Prambada O, Andreas A S, Saing U B, Kunrat L, Andreastuti S, Martanto M, Kriswati E, Suparman Y, Humaida H, Ogburn S, Kelly P J, Wellik J, Wright H M N, Pesicek J D, Wessels R, Kern C, Lisowski M, Diefenbach A, Poland M, Beauducel F, Pallister J, Vaughan J D, and Lowenstern J B 2019 The 2017–19 activity at Mount Agung in Bali (Indonesia): intense unrest, monitoring, crisis response, evacuation, and eruption *Scientific Reports* **9** 8848

Acknowledgments

This research is part of Penelitian, Pengabdian Masyarakat, dan Inovasi (PPMI) Institut Teknologi Bandung 2021. The fieldwork was partially supported by the USAID's PEER (Partnership for Enhance Engagement in Research) program with agreement number AID-OAA-A-11-00012.

Spatiotemporal Change of Chl-a Concentration in Jakarta Bay using Sentinel-2 Multispectral Imageries

S Indriyaningtyas[1], F Setiawan[2*], R F Lubis[3], H Bakti[3], W Widyatmanti[4]

[1] Undergraduate School of Faculty of Geography, Gadjah Mada University, Yogyakarta, Indonesia
[2] Research Center for Limnology, National Research and Innovation Agency (BRIN),Indonesia
[3] Research Center for Geotechnology, National Research and Innovation Agency (BRIN), Indonesia
[4] Faculty of Geography, Gadjah Mada University, Yogyakarta, Indonesia

*E-mail: fajar@limnologi.lipi.go.id

Abstract. Jakarta Bay is the estuary of 13 rivers that flows through Jakarta Metropolitan areas. Consequently, nutrients and pollutants accumulate in Jakarta Bay and lead to eutrophication. Therefore, monitoring the water quality of Jakarta Bay is essential to support integrated river coastal management. Chlorophyll a (Chl-a) concentration is one of the essential parameters in describing water environment conditions. For large and dynamic areas, Chl-a concentration distribution monitoring requires multi-temporal remote sensing data. This study examines remote sensing data and methods combined with on-site measurement to get more accurate and effective Chl-a spatiotemporal analysis approaches. We estimate Chl-a concentration of Jakarta Bay during 2016-2021 using selected Sentinel-2A and Sentinel-2B images. We conducted an atmospheric correction using a C2RCC (Case 2 Regional Coast Colour) processor and obtained Chl-a concentration from the embedded Neural Network model. The results showed that the model could retrieve Chl-a concentration with an R^2 of 0.79 and the root mean square error of 31.75 mg/m^3. Time-series results reveal an increase of maximum Chl-a concentration during 2016-2021. Our remotely sensed estimation results well-captured the change of temporal and spatial of Chl- a concentration in the Jakarta Bay. It provides the public and policymakers valuable information to support integrated river coastal sustainable management of Jakarta Metropolitan
Keywords: Chl-a concentration; eutrophication; Sentinel-2; C2RCC processor; Jakarta Bay

1. Introduction

Monitoring of the water environment is required to support integrated river-coastal sustainable management. Policymakers demand comprehensive and continuous data to describe the water environment condition. However, field data collecting required significant time, labor, and financial support. Remotely sensed data could collect spatial and temporal water quality data, especially for large areas. Recently, numerous remote sensing data have become freely available for public use. Multispectral remote sensing data allows us to develop object recognition based on spectral characteristics rather than visual interpretation.

For providing the water quality data from remote sensing data, satellite data must be available, and the models for estimating water quality parameters are needed. In the present study, we used Sentinel-2A and Sentinel-2B data to generate a time series Chl-a concentration data for Jakarta Bay, Indonesia. The Chl-a concentration was selected as the satellite-retrieved water quality parameter in this study because the Chl-a concentration is one of the essential parameters in describing water environment conditions.

Chl-a is a photosynthetic pigment contained in phytoplankton, and its concentration is influenced by water conditions such as temperature and nutrient concentration [2,3]. The presence of Chl-a in water can affect the optical properties [4], enabling it to be monitored using remote sensing sensors. In a clear water environment (Case-1-water), it is assumed that only the Chl-a is present [5]. In inland and coastal waters (Case-2-water) consist of Chl-a and suspended inorganic material or dissolved organic matter. As a consequence, sophisticated atmospheric correction techniques areneeded [6].

The performance of the existing model for Chl-a estimation using remote sensing data need to be asses for the different water environment. Several algorithms worked well in Case-1-water but failed in Case-2-waters, as the influence of Colored Dissolved Organic Matter (CDOM) and Suspended particle Matters [7,8]. The C2RCC (Case 2 Region Coast Color) processor is an atmospheric correction processor with a Chl-a concentration estimation model as one of its processing products. This processor in several studies proved to give good results in estimating Chl-a in various waters in Europe.

However, further studies discussing the use of C2RCC in estimating Chl-a concentration in tropical water environments, such as Indonesia, is still limited. Indonesia has a large coastal area that remains unmonitored as an archipelago country. Remote sensing data (Sentinel-2A ad Sentinel-2B) integrating with C2RCC processor offering practicality to provide a time-series map of Chl-a concentration for Indonesia. Accordingly, our objectives in the present study were to (1) estimate Chl- a concentration in Jakarta Bay using Sentinel-2A images with C2RCC processor, (2) provide spatiotemporal data of Chl-a concentration in Jakarta Bay from 2016 to 2021.

2. Methods

2.1 Research location

This research was conducted in Jakarta Bay, a coastal area in the Northern Jakarta Special Province (Figure 1). Jakarta Bay is the estuary of 13 major rivers that flows through Jakarta Metropolitan areas, namely Citarum, Kali Keramat, Kanal Banjir Timur, Kali Blencong, Cakung, Kali Sunter, Kali Ciliwung, Kali Muara Baru, Kanal Banjir Barat, Cengkareng, Muara Kamal, Dadap, Tahang, and Cisadane. This condition makes Jakarta Bay is vulnerable to eutrophication and water pollution caused by the accumulation of nutrients and pollutants. According to local government data (BPLHD), in 2014, the water quality of Jakarta Bay exceeded the standard seawater threshold as defined by the Ministry of Environment (*Peraturan Menteri Lingkungan Hidup No 51/2004*). Jakarta Bay serves several economic activities, fisheries, commercial harbor, trading, and tourism. Jakarta Bay is a complex ecosystem consisting of mangroves, coral reefs, and estuaries [9].

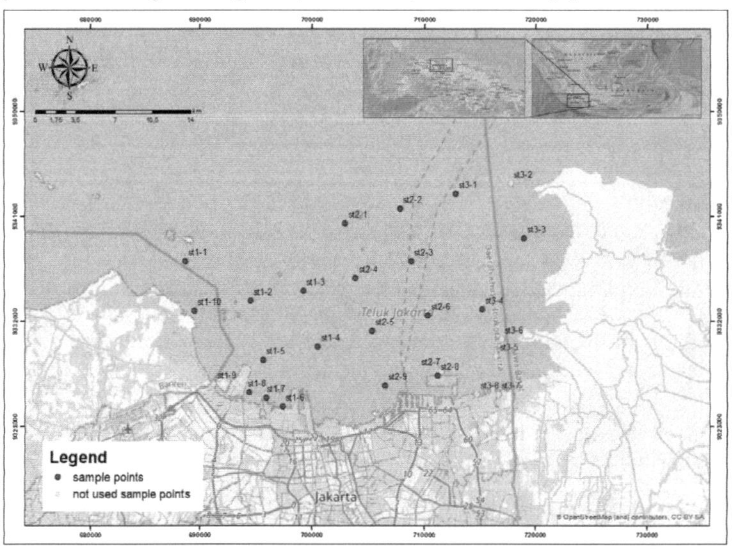

Figure 1. The map of the study area, The Jakarta Bay

2.2 In situ data
We conducted a field survey to collect a water sample to measure Chl-a concentrations from 24 to 26 May 2021. In total, we collected 28 in situ Chl-a concentration data. The sampling points were plotted in Figure 1. The Chl-a concentration was analyzed using the APHA standard method [10]. As the data were used to validate the remotely sensed Chl-a concentration, seven samples (st1-9, st2-7, st3-2, st3-5, st3-6, st3-7, st3-8) as shown using grey colored points in Figure 1 were excluded because it was covered with the cloud.

2.3 Remote sensing data
We collected satellite images from the twin sentinel sensor (i.e., Sentinel-2A and Sentinel-2B). The images were freely downloaded from the Copernicus website (https://scihub.copernicus.eu/). Sentinel-2 Multi-Spectral Imager (MSI) was chosen because it provides higher spatial and temporal resolution than Landsat Imageries. Sentinel-2 has 13 bands with a different spatial resolution classified into 10 meters for visible and Near Infra-Red (NIR) bands, 20 meters for Red Edge 1,2,3, NIR Narrow, and Short Wave Infra-Red (SWIR) bands, also 60 meters for coastal aerosol and water vapor. Sentinel-2 MSI has twin sensors, i.e., Sentinel-2A and Sentinel-2B allow to record the same location every five days. That Sentinel-2 characteristic is suitable as a data source to monitor medium to large scale areas. In addition, specific software for image processing was available for the Sentinel product called "SNAP".

For model validation, we downloaded the Sentinel-2A image acquired on 24 May 2021. To create a time series map of Chl-a concentration, we downloaded Sentinel-2 images acquired in May for the years 2016, 2017, 2018, 2019, 2020, respectively.

2.4 Data processing
Atmospheric correction is crucial in satellite data processing, especially time series analysis. We conducted the atmospheric correction using C2RCC *(Case 2 Region Coast Color)*. C2RCC is a tool for water processing in SNAP software. This processor was initially developed by Doeffer and Schiller [11] from the MERIS Case-2 water algorithm and further developed by Brockman et al. [12] become Case 2 Region Coast Color, which works for several sensors, i.e., OLCI, S2-MSI, Landsat-8, MERIS, MERIS4, MODIS, SeaWiFS, and VIIRS. C2RCC uses an extensive database of radiative transfer simulations inverted by neural networks as basic technology. Besides the atmospherically corrected product, this processor resulted in Chl-a and TSM (Total Suspended Matter) concentration. The unit of Chl-a concentration is in mg/m^3. An investigation by [13] and [14] showed promising results in estimating Chl-a concentration in various types of waters, especially in coastal and inland waters in Europe.

The C2RCC processor requires a processing parameter setting. In this study, we compared three ozone values, i.e., "BMKG" (obtained from Meteorology Agency), "MODIS" (obtained from MODIS data in Giovanni data portal), and "Default." We converted the ozone value from BMKG from 20 ppb into 13 Doxon Units (DU) using Intergovernmental Panel on Climate Change (IPCC) equation. The ozone value from MODIS is 252,875 DU, and the Ozone value by the C2RCC Default is 330 DU. We obtained three Chl-a concentration maps from those three different ozone sources. We calculate the correlation between the measured and estimated Chl-a values (R^2) for the accuracy assessment. The best ozone setting will be used to process the time series data from 2016 to 2021. We provided the workflow of this research in Figure 2.

2.5 Accuracy Assessment
We used the root means square error (RMSE) to assess the models' accuracy. The index is defined as follow:

$$\text{RMSE} = \sqrt{\frac{\sum_{i=1}^{N}\left(X_{esti,i} - X_{meas,i}\right)^2}{n}} \quad (1)$$

where $X_{esti,i}$ and $X_{meas,i}$ are the estimated and measured Chl-a values, respectively; n is the number of samples. We also calculated the correlation between the measured and estimated values (R^2).

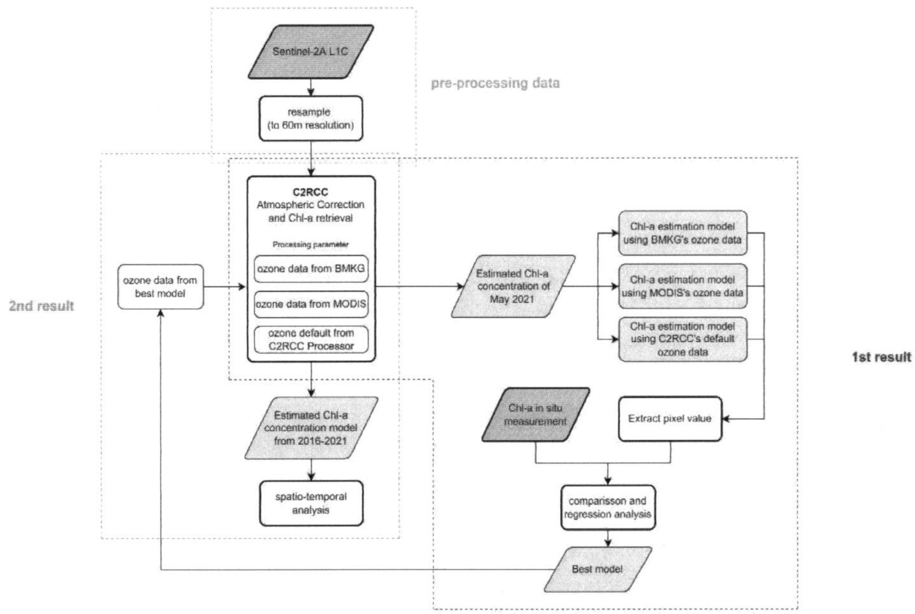

Figure 2. Data Processing Workflow

3. Result and Discussion

This research aims to firstly estimate Chl-a concentration in Jakarta Bay using Sentinel-2A images with C2RCC processor and secondly provide spatiotemporal data of Chl-a concentration in Jakarta Bay from 2016 to 2021. We validate the Chl-a concentration estimation from Sentinel-2A on 24 May 2021 with in-situ Chl-a concentration data acquired from 24 to 26 May 2021. We tested three sources of Ozone data to perform atmospheric correction using C2RCC in SNAP software. We considered the Ozone to describe the atmosphere condition in a specific area. The best Ozone source was selected by comparing Chl-a concentration's estimation with the In situ Chl-a concentration.

3.1 Model Performance

Figure 3 showed that the estimation model could retrieve Chl-a concentration with an R^2 of 0.7933, 0.7453, and 0.6676 with Ozone data from BMKG, MODIS, and C2RCC Default. All models showed an underestimation. We observed higher concentration of Chl-a is spread over the waters close to the mainland and gradually decreases towards the offshore. Higher Chl-a concentration near the shoreline indicates the nutrient transport from the mainland, especially near the estuary.

The model's RMSE values are 31.75 mg/m^3, 31.76 mg/m^3, and 32.32 mg/m^3 using the ozone value from the BMKG data, MODIS data, and with the default ozone value, respectively. Although the model shows an underestimate with a large RMSE value, a quite high R^2 value indicates that the model has succeeded in explaining the data variation with a good proportion of the field data.

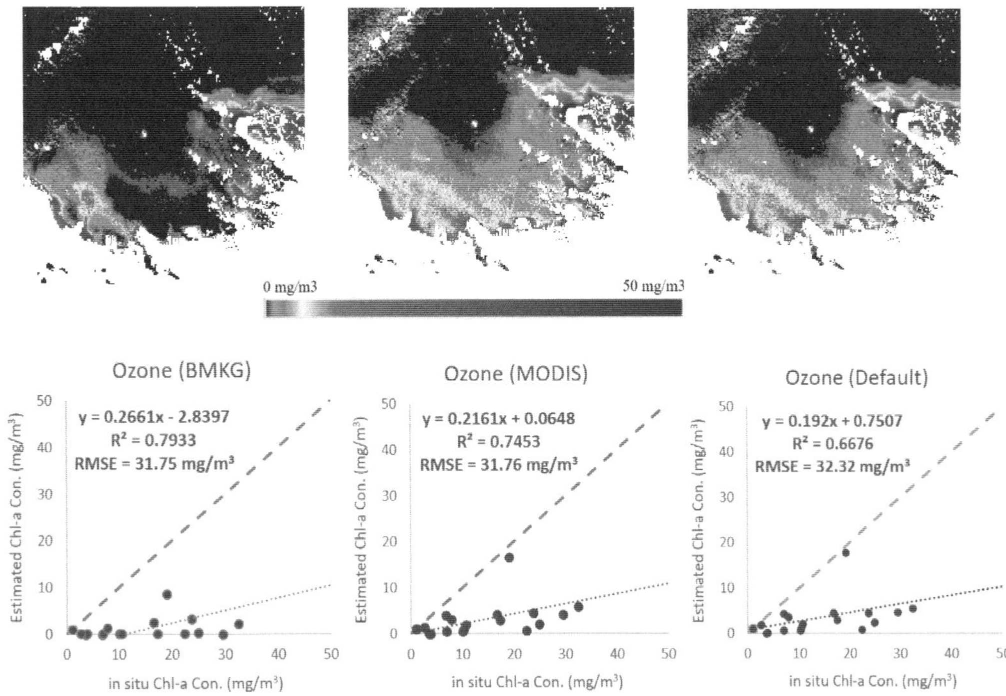

Figure 3. Comparison of the spatial distribution of estimated Chl-a concentration using C2RCC with Ozone from BMKG, MODIS, and Default (upper). Comparison of the in situ-measured Chl-a and the corresponding estimated Chl-a concentration (lower).

3.2 Spatiotemporal of Chl-a concentration

We built a time-series Chl-a concentration map using Sentinel-2 images acquired each May from 2016 to 2021. Those images are dated 20 May 2016, 25 May 2017, 10 May 2018, 25 May 2019, 24 May 2020, dan 24 May 2021, respectively. All images come from Sentinel-2A, except images of 2020 come from Sentinel-2B. As the Ozone data from BMKG resulted in the best Chl-a estimation (using the 2021 image), we further used Ozone data from BMKG to process all time-series images.

The time-series Chl-a concentration map reveals the dynamic spatiotemporal pattern, as shown in Figure 4. In 2016, we observed a high concentration of Chl-a near the shoreline, then gradually decreased towards offshore. We found a similar pattern in 2017, and higher Chl-a dominated the coastal area. In 2018, high Chl-a concentrations were reduced and only scattered in the coastal, and moderate Chl-a concentration scattered over the offshore. In 2019 and 2020, we also identified a similar pattern in 2018. In the 2021 model year, the concentration of Chl-a looks much reduced.

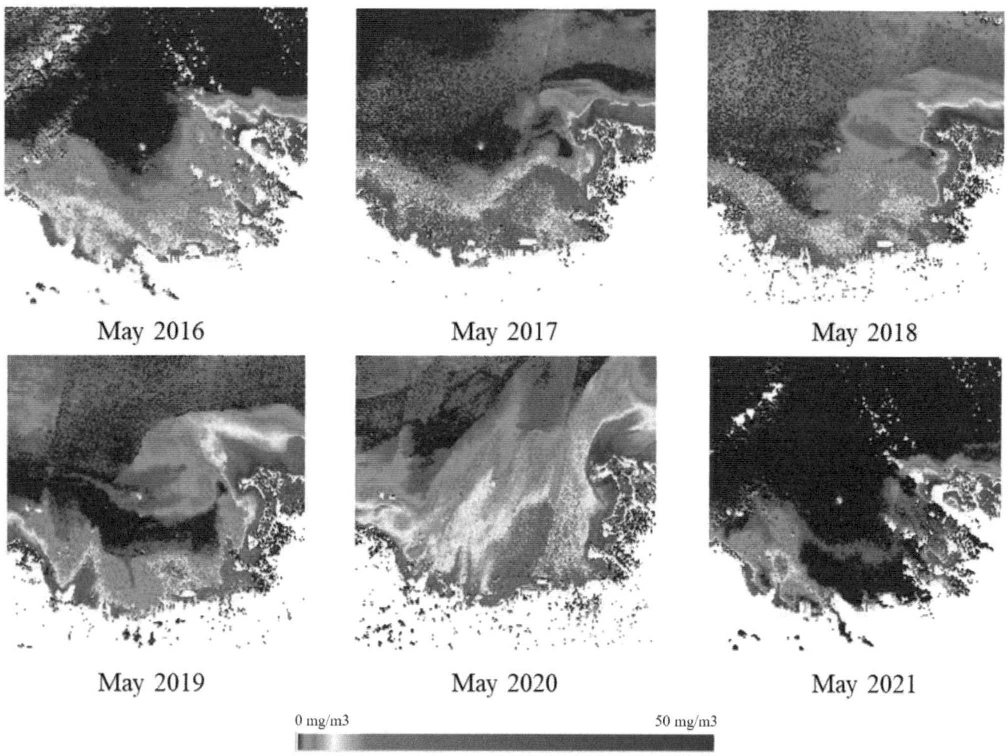

Figure 4. Spatiotemporal of Chl-a concentration on each May during 2016-2021

We calculated the minimum, average, median, and maximum of Chl-a concentration on the time series data, as shown in Figure 5. The minimum value per year is 0, with the maximum value varying from 47.6 to 53.8 mg/m^3. The average value of Chl-a concentration in Jakarta Bay increased from 2016 to 2017, then tends to be stable until 2019 before experiencing an increase in value in 2020 and decreasing again in 2021. In contrast, the maximum value, initially high in 2016 and then decreased in 2017 and 2020. The median value more or less has a stable trend with a slight difference in value each year, with a slight increase in 2020.

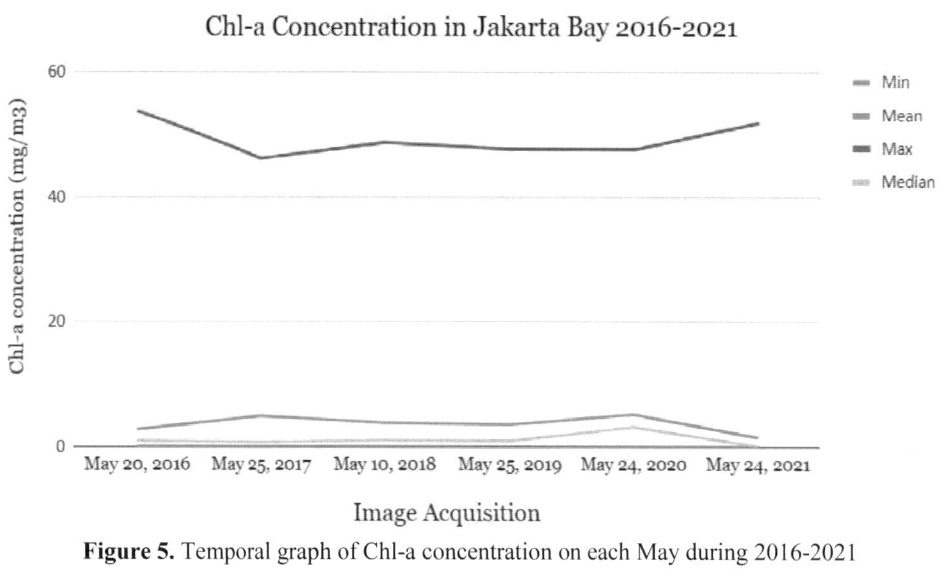

Figure 5. Temporal graph of Chl-a concentration on each May during 2016-2021

Our estimated Chl-a concentration is similar to the data obtained from MODIS images. The Chl-a concentration from MODIS obtained from the NASA Giovanni data portal is provided in Figure 6. Even though we found a similar pattern, there is a slight difference in the average value of the two data sources. Several factors can influence this. The first is the difference in resolution between the Sentinel-2 and the MODIS; Sentinel-2 has 60 m resolution, contrast with MODIS with four km spatial resolution. In addition, the processing parameters to build the spatiotemporal model still use the 2021 data with the best regression in this study. In making the model, processing parameters are also set based on the conditions on each year to obtain more representative results for that year.

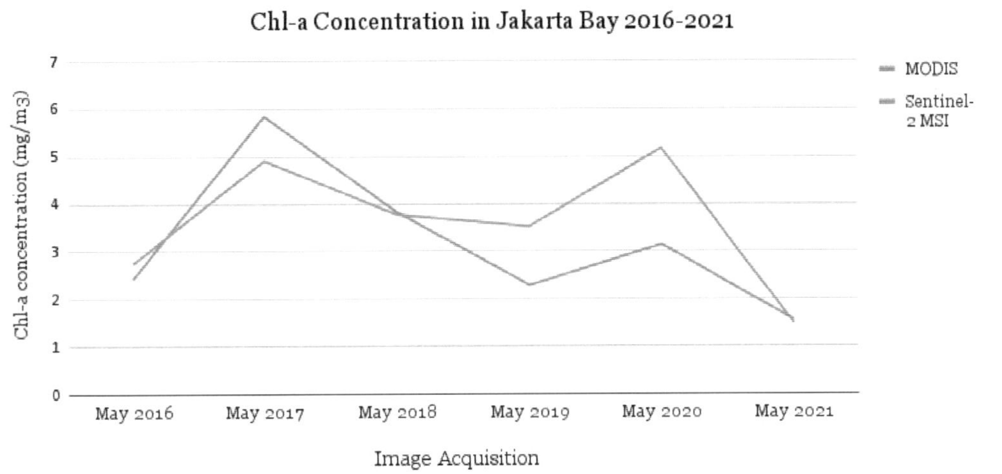

Figure 6. Comparison of average Chl-a concentration derived from MODIS and Sentinel-2 MSI images on each May during 2016-2021

4. Conclusions

This remotely sensed estimation is well-captured the temporal and spatial change of Chl-a concentration in Jakarta bay with the R^2 of 0.79. However, to obtain higher accuracy, the relatively high RMSE value (31.75 mg/m3) should be further investigated, especially related to the tuning parameters. As the increasing trend of Chl-a concentration, consideration is needed to prevent possible related disasters such as alga bloom and eutrophication. This data could provide the public and policymakers valuable information to support integrated river coastal sustainable management of Jakarta Metropolitan.

5. References

[1] Jensen, John R 2014 Remote Sensing of the Environment: An Earth Resource Perspective (2nd edition) Edinburgh: Pearson Education Limited

[2] Wasmund R 1997 Occurrence of cyanobacterial blooms in the Baltic Sea in relation to environmental conditions *Int. Revue Ges. Hydrobiol. Hydrogr.* 82(2), 169–184

[3] Gasiūnaitė Z R 2005 Seasonality of coastal phytoplankton in the Baltic Sea: influence of salinity and eutrophication *Estuarine Coastal Shelf Sci.* 65(1–2), 239–252

[4] Wozniak M, Bradtke K M, and Krezel A 2014 Comparison of satellite chlorophyll-a algorithms for the Baltic Sea *Journal of Applied Remote Sensing* **8(1)**, 083605. https://doi.org/10.1117/1.JRS.8.083605

[5] Gordon H R, Clark D K, Brown J W, Brown O B, Evans R H, and Broenkow WW 1983 Phytoplankton Pigment Concentrations in the Middle Atlantic Bight: Comparison of Ship Determinations and CZCS Estimates *Applied Optics*, **22**:20–36.

[6] Ramsey E W, Jensen J R, Mackey H E, and Gladden J 1992 Remote Sensing of Water Quality in Active to Inactive Cooling Water Reservoirs *Intl. Journal of Remote Sensing* **13(18)**:3465– 3488.

[7] Liew S C, Chia A S, and Kwoh L K 2001 Evaluating the validity of SeaWiFS chlorophyll algorithm for coastal waters *in Proc. 22nd Asian Conf. on Remote Sensing*

[8] Chauhan P 2002 Comparison of ocean color chlorophyll algorithms for IRS-P4 OCM sensor using in-situ data *J. Indian Soc. Remote Sens.* **30(1–2),** 87–94
[9] Puspasari R, Sri Turni H, Regi Fiji A, 2017 Analisis Dampak Reklamasi Terhadap Lingkungan dan Perikanan di Teluk Jakarta *Jurnal Kebijakan Perikanan Indonesia (JKPI)* **9(2)** 85-94
[10] Aderhold J, Davydov V Y, Fedler F, Klausing H, Mistele D, Rotter T, Semchinova O, Stemmer J and Grau 2005 Standard Method for The Examination of Water and Waste Water (Washington D.C: American Public Health Association)
[11] Doerffer R and Schiller H 2007 The MERIS Case 2 water algorithm *International Journal of Remote Sensing* **28(3–4),** 517–535. https://doi.org/10.1080/01431160600821127
[12] Brockmann C, Roland D, Peters M, Kerstin S, Sabine E, and Ruescas A 2016 Evolution of The C2RCC Neural Network for Sentinel 2 and 3 for The Retrieval of Ocean Colour Products in Normal and Extreme Optically Complex Waters *Proc. Living Planet Symposium*
[13] Warren M A, Simis S G H, Martinez-Vicente V, Poser K, Bresciani M, Alikas K, Spyrakos E, Giardino C, and Ansper A 2019 Assessment of atmospheric correction algorithms for the Sentinel- 2A MultiSpectral Imager over coastal and inland waters *Remote Sensing of Environment* **225** https://doi.org/10.1016/j.rse.2019.03.018
[14] Ansper A and Alikas K 2018 Retrieval of Chlorophyll-a from Sentinel-2 MSI Data for the European Union Water Framework Directive Reporting Purposes *Remote Sensing* **11(1).** https://doi.org/10.3390/rs11010064

Acknowledgment

This research is financially supported by the LIPI COREMAP-CTI 2021-2022 project, entitled "Research Priority and Capacity Development Program for Reef Management and Rehabilitation-Reef Triangle Initiative."

Author Contributions

S Indriyaningtyas and F Setiawan are the main contributors to this manuscript.

IOP Conf. Series: Earth and Environmental Science 1047 (2022) 012007 doi:10.1088/1755-1315/1047/1/012007

Utilization of Terrestrial Laser Scanner Technology for Analyzing Shape of Dutch Cave (Bandung, West Java)

G A Jessy Kartini[1,2], A Rizky[2], and F A Rafiq[2]

[1] Doctoral Program of Geodesy and Geomatics Engineering, Faculty of Earth Sciences and Technology, Institut Teknologi Bandung, Jl. Ganesha No. 10, Bandung 40132, Indonesia
[2] Department of Geodetic Engineering, Faculty of Civil Engineering and Planning, Institut Teknologi Nasional, Jl. PH.H. Mustofa No. 23, Bandung 40124, Indonesia

ayujessy@itenas.ac.id

Abstract. Technological developments in the field of surveying and mapping remain in progress. One technology that provides precise and accurate results is the terrestrial laser scanner (TLS). This technology can generate millions of points in a short time and represent objects in 3D. TLS technology has become famous for heritage documentation purposes due to its ability to obtain detailed geometry and information from maintaining historic buildings. This research is in the Dutch Cave, Djuanda Forest Park, Bandung, West Java, to map and document 3D. To produce accurate data, frame measurements were carried out using a total station, GNSS, which ended with acquisition using TLS. The registration method used is the target-to-target method. The goal of this research is to create as-built drawings of the Dutch Cave in analyzing the Dutch Cave's form in greater detail. This investigation determined that the Dutch Cave measures 124.435 meters in length and has a height difference of 69.7 cm between the entrance and the exit. The as-built drawing of the Dutch Cave illustrates that the shape of the cave varies from entrance to exit. There is a cave chamber that widens by 18.8 cm toward the cave's center and narrows by 20.7--32 cm toward the cave's exit. The Dutch Cave has a slope of between 1 and 5 degrees. The Dutch Cave 3D data can be used for a documentation archive for the Djuanda Forest Park.

Keywords: Terrestrial Laser Scanner, Target-to-Target Method, Point Cloud, Heritage, As-Built Drawing

1. Introduction

Cultural heritage is a place where important or historical events occur that can be used as a medium for connecting with the past and can also function as a learning tool and open a view of the importance of a historical event. The Dutch cave as a cultural heritage in the city of Bandung needs to be documented. According to RI Law No. 11 of 2010, to preserve cultural heritage, documentation activities need to be carried out before there are activities that can cause changes to their authenticity. Documentation in preserving cultural heritage includes taking photos and videos, measuring and depicting objects of cultural heritage.

Mapping and modeling the geometry of caves has traditionally been a challenging task, and this is due to limited access and very minimal light conditions [1]. The development of survey equipment technology continues to progress, providing convenience in data acquisition and processing with precise and accurate results, one of which is Terrestrial Laser Scanner (TLS) technology [2,3,4]. The utilization of TLS technology has become famous for documentation of heritage and archaeological sites, geomorphology, paleoclimatology and paleontology, passage stability, and visualization and education application [5]. TLS technology offers a range of advantages concerning traditional cave mapping, including the possibility of measuring high roofs, high operational speed, accuracy, abundant measured points, and construction of realistic 3D models of caves, including cave walls with centimeter resolution [6].

A survey method using TLS produces a set of information in millions of point clouds that already have coordinates for 3D modeling visualization. Point clouds provide significantly more correct geometric completeness and detail [7]. Getting the shape of an object in a 3D model takes several times to stand the tool. Point clouds recorded at each device stand need to be combined in the same coordinate system, commonly referred to as the registration process.

The TLS technology can be used for 3D modeling of cultural heritage sites because it can produce accurate building details. The TLS technology will scan Dutch Cave to document cultural heritage locations, mostly underground. The three-dimensional model created from TLS data is utilized to estimate the Dutch Cave's current state. The goal of this research is to create as-built drawings of the Dutch Cave in analyzing the Dutch Cave's form in greater detail.

2. Data and Methods

2.1. Control Point Data Acquisition

GPS measurements were made using a Trimble 5800 receiver and the differential static technique. Surveys were made for two hours using a thirty-second interval using the radial approach. The distance between the Lembang base station (CLBG) and the Dutch Cave rover is 4 kilometers. Observations were taken at two sites outside the Dutch Cave's entrance, as well as in chosen regions with quite open surroundings, as seen in Figure 1. Trimble Business Center 3.5 was used to handle the data produced by GPS measurements. Table 1 contains the results of GPS data processing with precision in each point.

Figure 1. GPS observation in the front area of the Dutch Cave, left side is measurement in BM1 and right side is in BM2.

Table 1. GPS Measurement Results.

ID	Easting (m)	Northing (m)	Elevation (m)	Solution Type	H. Prec. (m)	V. Prec. (m)
BM1	791521.296	9241561.352	924.533	Fixed	0.018	0.032
BM2	791515.150	9241577.514	924.416	Fixed	0.059	0.086

Electronic Total Stations are carried out to create a basic framework for detailed mapping measurements. Measurement of the basic mapping framework begins by using two tie points located in front of the entrance to the Dutch Cave. There are ten measurement points for the basic framework of an open polygon with a measurement length of 163 meters. The sketch of an open polygon can be seen in Figure 2.

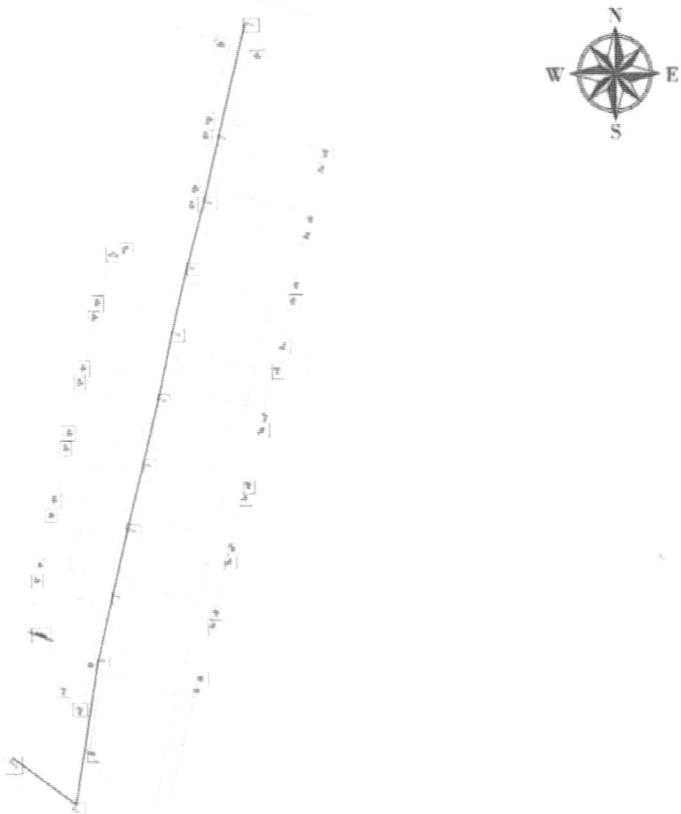

Figure 2. Sketch of polygon measurement, the red line represents a polygon that goes from the entry to the exit gate.

2.2. Point Cloud Data Acquisition

Measurements are made considering several aspects, such as the device's location, the placement of the target, the breadth of the field-of-view or scanning range, the length of storage time, and the resolution used. The total area of the Dutch Cave measurement area is 330 meters. We used six pieces of spherical ball targets and checkerboards that had been previously installed in the outside and inside the Dutch Cave. Trimble TX8 can only display in black and white; thus, photographs for point cloud coloring must be taken using a 360 camera. Trimble TX8 is capable of scanning at 340 meters at a rate of 1 million meters per second. The Trimble TX8 produces an error range of less than 2mm across 2—120 meters.

Figure 3. Spherical target position

Target placement is carried out with the consideration that the target can be scanned or visible from a minimum of two stations so that the point cloud of at least two stations can be registered with a critical point in the form of a target and an overlap between the two stations is 20-40%. The target placement is as much as possible spread out or not in a straight line so that the accuracy of the registered point cloud is evenly distributed, and the target must not change position during the scanning process as in Figure 3.

The density or intensity in the point cloud is adjusted to the object to be scanned. In this study, the thickness of the point cloud using the Trimble TX8 tool at level 2 with a point spacing of 11.3 mm at 30 meters with a storage duration of 11 minutes. The number of scan world is 18 points. The following is an example of a TLS measurement, as shown in Figure 4.

Figure 4. TLS measurement outside Dutch Cave (left) and inside Dutch Cave (right).

2.3. Target-to-target Registration

Target-to-target registration merges several scan world data from the data acquisition process using the target as a reference. The targets used in the data acquisition process can be flat targets and 3D-shaped targets [8]. Flat targets are a reasonably familiar target type found in the data acquisition process, as for the form of flat targets such as checkerboards. 3D shaped targets are another type of target with a spherical shape; there are distinct advantages in using this type of target. The advantage of spherical over flat targets is that they can be easily identified from various storage directions by the TLS instrument compared to flat targets, which are limited by the viewing angle of the TLS instrument. In the data acquisition process in the field, a minimum of 3 targets is needed so that the registration step can be carried out using the target to the target method. The distribution of targets during data acquisition must be optimal to obtain results with high accuracy. The wrong shape of the distribution will impact the poor process of registration results using the target-to-target method.

The registration process is carried out using Trimble Realworks software. The target-to-target registration process is relatively easy, especially if the target points are well planned. The georeferencing process defines the coordinate system on measurement data into the global coordinate system (UTM). Georeference is done using Trimble Realworks software for both registration methods. In this study, three georeferenced control points were used in the front and back of the Dutch Cave. The three georeferenced control points used are spherical ball target objects, namely Target 1, Target 3, and Target 37. The coordinate values of the three points are obtained from the value of the ETS measurement shot, which FBK has processed in Civil 3D 2017 software. The processing uses the Georeferencing tool; then, the three control points were picked manually by entering the coordinate values (x, y, z) and point names. The georeferenced process in the target-to-target model produces an average RMSE of 0.0056 m.

3. Result and Discussion

3.1. Data Acquisition and Processing

The measurement of the Dutch Cave using the Trimble TX8 laser scanner was carried out 18 times scan world at 18 points. Each storage measurement is carried out for 11 minutes with a point spacing of 11.3 mm at 30 meters because the distance of each scan world is not more than 30 meters. At stations 7, 11, 12, and 14, the scanning duration is 7 minutes. In Figure 5, the red dot represents the scan world point, which encompasses the whole research region of the Dutch Cave. The TLS measurement produces raw point cloud data with a maximum of 94,035,667 points and a data size of 12.3 GB.

The target-to-target method has the lowest level of practicality in data acquisition planning because it must determine at least three stable (no shifting) target points that can be seen from two TLS stands (scan world). The other factors that affect this method are the distribution of target points and the type of target. The target placement is spread out as much as possible or does not form a straight line so that the accuracy of the registered point cloud is evenly distributed. This study used two targets: a checkboard and a spherical ball.

When processing registration data, the Trimble Realworks software can automatically detect the target point object with the same name. The RMSE value of this research registration is 6,579 mm. The registration RMSE value refers to the Trimble tolerance reference value in general, with a tolerance value of 8 mm. The target-to-target method uses target point definition, and the software can automatically define the target.

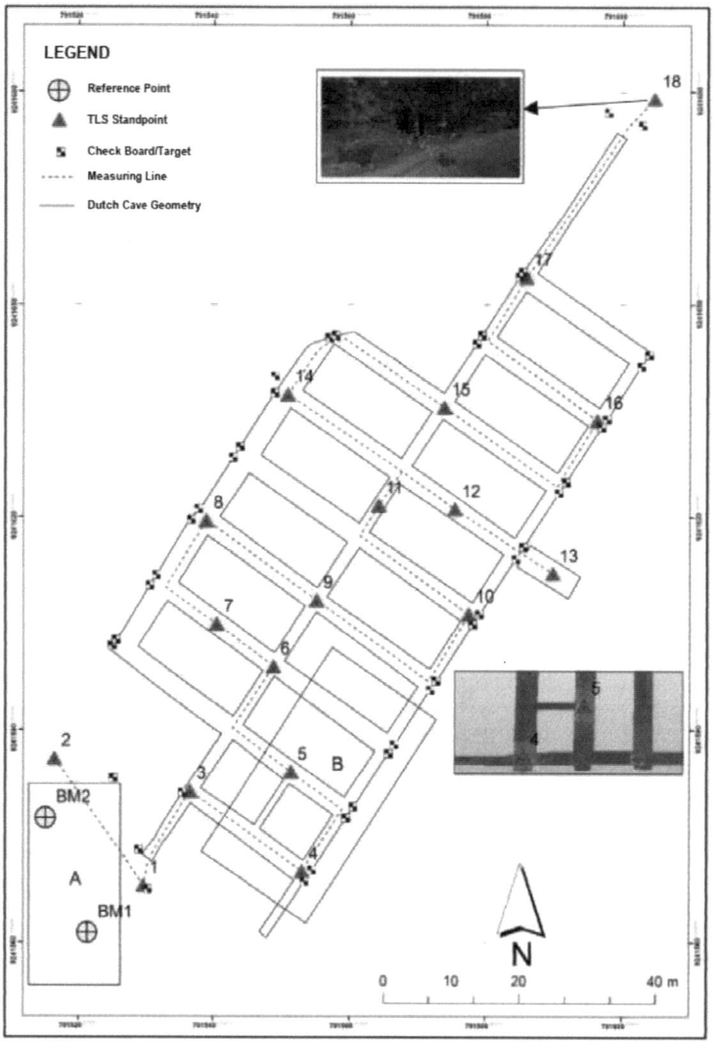

Figure 5. TLS positioning.

In the georeference process carried out using Trimble Realworks software, Target 1, Target 3, and Target 37 are in the front area of the Dutch Cave and selected on the GCP for the ETS measurement. RMSE from the georeferenced results is 0.0057 m. The registration result is still said to be quite good because it reaches the accuracy of the millimeter fraction. The registration result can be seen in Figure 6.

Figure 6. The registration result using target-to-target method before filtering (left) and after filtering process (right).

3.2. Utilization of 3D Point Cloud Data

The main tunnel has a length of 124.435m, measured from the entrance gate to the exit gate of the Dutch Cave. Point cloud TLS measurement results can record points with millimeter accuracy. The surface area of the central aisle increased by 69.7 cm between the entry and exit gates, as we can see in Figure 7. There is a slightly bumpy surface in the main hallway, which is about 40–100m from the entrance gate of the Dutch Cave.

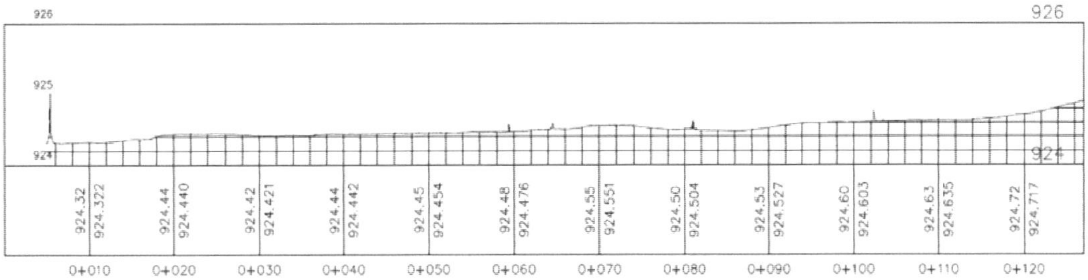

Scale: Horizontal 1:200 Vertical 1:25

Figure 7. Profile of Dutch Cave.

We collect several cross-sectional samples in the central aisle for analysis. The cross-sectional shape sampled is from the entrance to the exit gate on the main corridor. There are 13 cross-sectional forms. The parameters used to analyze the shape of the cross-section of Dutch Cave are width, height from the lowest point, elevation from the east and west sides, the slope of the wall with a reference perpendicular to the tunnel floor. Figure 8 is an example of cross-sectional samples.

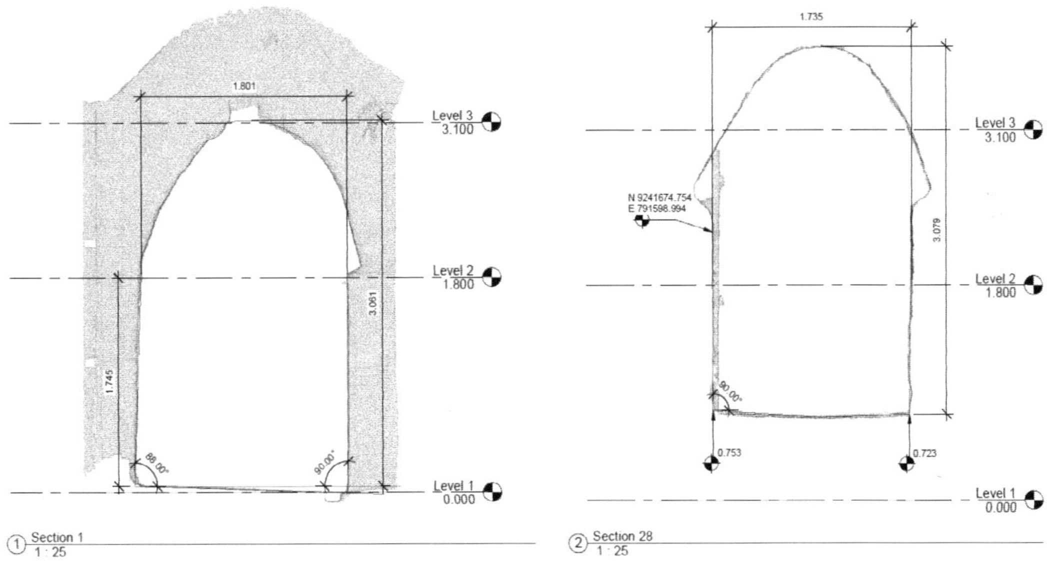

Figure 8. Example of the entrance gate (left) and the exit gate cross-section (right).

There are several results of the cross-sectional analysis that can be seen in Table 1. They widen the side of the cave from the entrance to the cross-sectional point no. 21 by 18.8cm. From the cross-section point no 21 to the exit gate, there is a narrowing ranging from 20.7--32cm. The tunnel height ranges from 2.941—3.179 m, with an average of 3.035 m. The west side of the tunnel surface is 4 cm higher than the east side, and the slope of the tunnel wall is 1°-5°. Table 2 shows the results of the cross-section analysis.

Table 2. Cross-section measurement.

Cross-section Number	Width (m)	Height (m)	Elevation (m)		Slope (°)	
			West	East	West	East
1	1.801	3.061	0.055	0	88	90
3	1.825	3.043	0.075	0.025	91.67	91.67
5	1.956	2.992	0.165	0.078	90	90
7	1.964	3.053	0.133	0.102	90	90
9	1.920	3.018	0.143	0.087	90	91
12	1.940	3.005	0.176	0.116	92	91
13	1.962	3.067	0.179	0.129	90	89
17	1.909	3.047	0.271	0.265	90	90
19	1.917	3.043	0.342	0.269	91.29	91
21	1.989	2.941	0.343	0.32	91	89
23	1.685	2.954	0.359	0.357	93.92	92.87
25	1.669	3.056	0.383	0.383	95	95.58
28	1.735	3.079	0.753	0.723	90	90

4. Conclusion

Technically, the target-to-target method's measurement looks more complicated, but it is easier to process. With TLS capabilities that can reach a millimeter accuracy level, the resulting point cloud data can provide a detailed description of the condition of the Dutch Cave. If there is data about the Dutch Cave from its first construction or discovery, we can observe the changes since it was found. The Dutch Cave 3D data can be used for virtual tourism needs and as a documentation archive for the Djuanda Forest Park.

5. References

[1] Trimmis, K. P. (2018). Paperless mapping and cave archaeology: A review on the application of DistoX survey method in archaeological cave sites. *Journal of Archaeological Science: Reports*, *18*, 399-407.

[2] Jalandoni, A., Winans, W. R., & Willis, M. D. (2021). Intensity Values of Terrestrial Laser Scans Reveal Hidden Black Rock Art Pigment. *Remote Sensing*, *13*(7), 1357.

[3] Büyüksalih, G., Kan, T., Özkan, G. E., Meriç, M., Isın, L., & Kersten, T. P. (2020). Preserving the knowledge of the past through virtual visits: From 3D laser scanning to virtual reality visualisation at the Istanbul Çatalca İnceğiz caves. *PFG–Journal of Photogrammetry, Remote Sensing and Geoinformation Science*, *88*(2), 133–146.

[4] Pennos, C., Lauritzen, S.-E., Lonoy, B., & Tveranger, J. (2018). A 3D model of a cave collapse from Peristerionas cave, northern Greece. *EGU General Assembly Conference Abstracts*, 17322.

[5] Idrees, M. O., & Pradhan, B. (2016). A decade of modern cave surveying with terrestrial laser scanning: A review of sensors, method and application development. *International Journal of Speleology*, *45*(1), 71.

[6] Fabbri, S., Sauro, F., Santagata, T., Rossi, G., & De Waele, J. (2017). High-resolution 3-D mapping using terrestrial laser scanning as a tool for geomorphological and

speleogenetical studies in caves: An example from the Lessini mountains (North Italy). *Geomorphology*, *280*, 16-29.

[7] Sacks, R., Eastman, C. M., & Lee, G. (2004). Parametric 3D modeling in building construction with examples from precast concrete. *Automation in construction*, *13*(3), 291-312.

[8] Reshetyuk, Y. (2009). *Self-calibration and direct georeferencing in terrestrial laser scanning* (Doctoral dissertation, KTH).

Acknowledgments

We want to express our gratitude to PT. GPSLands Indonesia provided the Trimble TX8 to assist with research funding from the Kampus Merdeka Competition Program Research Grant, Geodetic Engineering, Institut Teknologi Nasional Bandung.

Multivariate statistical analysis for sedimentary characteristics in an estuarine environment of the east flood canal river, Semarang

Y Marantika[1], S H Nugroho[2,3], A Kurniasih[1], P S Putra[2,3], Lestari[4], S P A Wibowo[4], M T Kaisupy[4], Amar[2] and Jakah[2]

[1] Departement of Geological Engineering, Universitas Diponegoro, Semarang, Jawa Tengah 50275, Indonesia
[2] Research Center for Geotechnology, National Research and Innovation Agency, Kompleks LIPI, Bandung, Jawa Barat 40135, Indonesia
[3] Departement of Geological Engineering, Institut Teknologi Bandung, Bandung, Jawa Barat 40132, Indonesia
[4] Research Center for Geotechnology, National Research and Innovation Agency (BRIN), Jakarta, Indonesia

Corresponding author: rantika@students.undip.ac.id (YM); sept005@brin.go.id (SHN); anis.kurniasih@live.undip.ac.id (AK)

Abstract. The East Flood Canal River stretches across the eastern part of Semarang, an area that is densely populated with residential and industrial areas. There are many human and industrial activities around this watershed, including the textile industry, food factories, and fish auction sites. Therefore, this river becomes an interesting research object to examine the influence of industrial activities around the river on the characteristics of the sediment. This research was conducted by examining sediment core samples from the estuary of the East Flood Canal. The research aims to reveal the relationship between the characteristics and distribution of sediments and geochemical trace elements. A total of 31 samples of sediment core slices with an interval of 1 cm were analyzed. The granulometric analysis was conducted by Mastersizer 2000, to determine the grain size parameters that were used as the basis for the clustering. Geochemical analysis was conducted by the XRF (X-Ray Fluorescence) to determine the trace elements concentration. The study is interpreted using compositional data analysis associated with a grain size trend analysis (GSTA) and cluster analysis (CA). The GSTA value showed generally dominated by medium to very fine silt. The cluster showed that four lithofacies zones have different sedimentary characteristics. The distribution of the characteristics of the large grain size of the seafloor sediment in the East Flood Canal indicates that as the deposition is farther from the river mouth, the poorer it's sorting. It was recorded that, at a depth of 19 to 27 cm from the surface there was an increase in fine to very fine sand accompanied by an increase in the Fe and Mn which indicated that there was a supply of sediment that occurred from land caused by human activities and flood events.

Keywords: multivariate statistical analysis; sediment characteristics; depositional environmental; East Flood Canal River; Semarang

1. Introduction

The city of Semarang is traversed by several major rivers, one of which is the East Canal Flood River. The East Canal Flood River is a combination of Kali Banger and Kali Tenggang. The East Canal Flood River has a length of 14.25 km with an average discharge of 295.33 liters/second [1]. As for the East Canal Flood river across the eastern city of Semarang which is densely populated and industrial. Many human and industrial activities around the watershed include the textile industry, food factories, and fish auction sites. The amount of human activity around the river will affect the increase in sedimentation processes at the mouth of the river estuary [1].

Geological studies related to sedimentation dynamics in the mouth of East Canal Flood using one sample core of sediment are still very limited. Previous research conducted investigated the distribution of sediments of the riverbed using samples of sediment cores in the mouth of the river and deep sea, there are 10 stations that are sampled sediment, based on analysis of Shepard Triangles and tides. The research found that the type of sediment in the mouth of the river is silt and in the deeper areas contained sand [1]. The granulometric and textural analysis is appropriate to recognize and to give fundamental information on sediment dynamics. The analyses are demonstrated to be of several sedimentary processes. The Granulometric analysis is a good indicator to describe the history of sediment transport and provenance of sediment supply through sediment depositional circum-stances [2]. Some studies have also successfully used the relationship between varying grain size parameters in the interpretation of sedimentary environments [3,4,5].

Statistical analysis is most effective and most useful for the determination of sediments because of facilitates measurement and loss of efficiency generated. The statistical parameters are calculated in phi or mm units and the accuracy of results is depended on the precision of the plot [6]. Multivariate techniques provide automated results based on various sediment variables, avoiding individual sample analysis. In addition, by combining and analyzing the relationship between different variables, sedimentary patterns can be derived and the process is concluded quantitatively rather than qualitatively [7]. Several studies to determine sedimentary facies have used multivariate statistical techniques [8,9,10].

The study describes the relationship between the characteristics and distribution of sediments with the geochemical content of elements. This paper demonstrates a methodological framework for characterizing the distribution of grain sizes in sediments through analysis of compositional data related to grain size trend analysis (GSTA) and multivariate statistical frameworks of cluster analysis (CA).

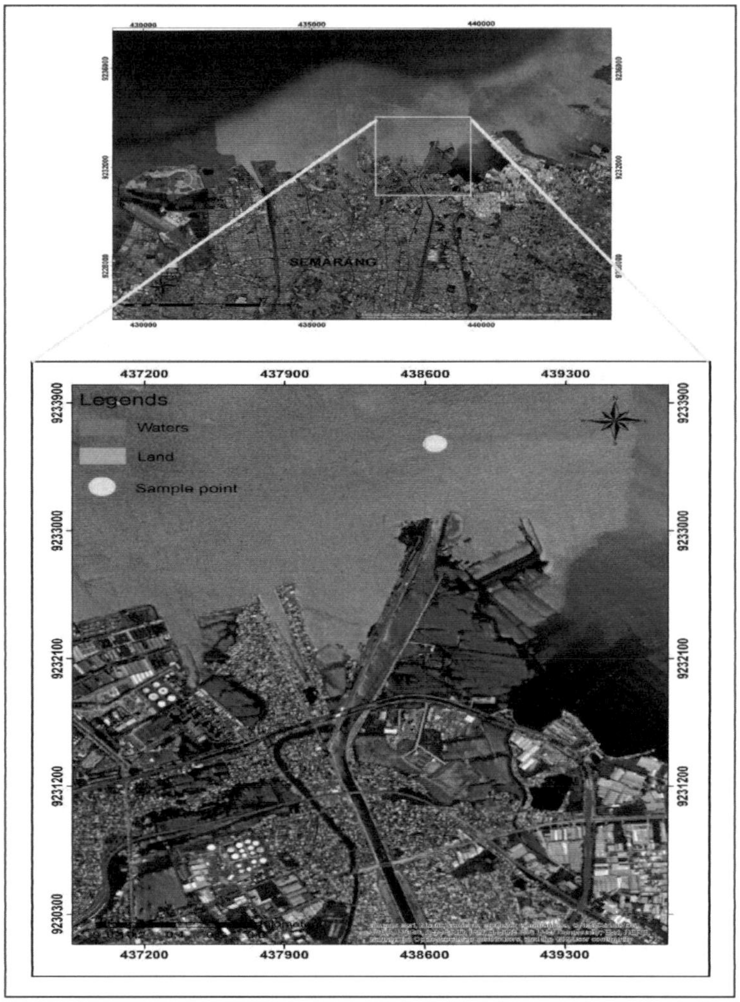

Figure 1. The research site of the mouth of the East Canal Flood River Semarang, and the yellow circle is the point of sampling sediment cores.

2. Geography and Regional Geology

The geographical location of Semarang City is 6° 50' -7 °10' south latitude and 109° 50' -110° 35' east longitude, and is situated between Mount Ungaran (altitude 2000 m) in the south and the Java Sea. The city of Semarang is bordered by Kendal Regency to the west and Demak Regency to the east. The area of Semarang is 37,333.83 ha with 99 sub-districts and 117 sub-districts. The population of the city of Semarang in 1991 was 1,154,536 inhabitants. In addition, industrial locations in Semarang are not located in one particular area but are widely distributed in the east, west, north, and southeast of the city [11].

Based on the regional geological map of Magelang – Semarang [12], the research location is included in the alluvial unit which is composed of alluvial deposits of beaches, rivers, and lakes. The beach deposits consist of clay, silt, and sand with a thickness of 50 m or more. River and lake deposits consist of gravel, sand, and silt with a thickness of 1-3 m. The gravels are composed of andesite, limestone, and small amounts of sandstone.

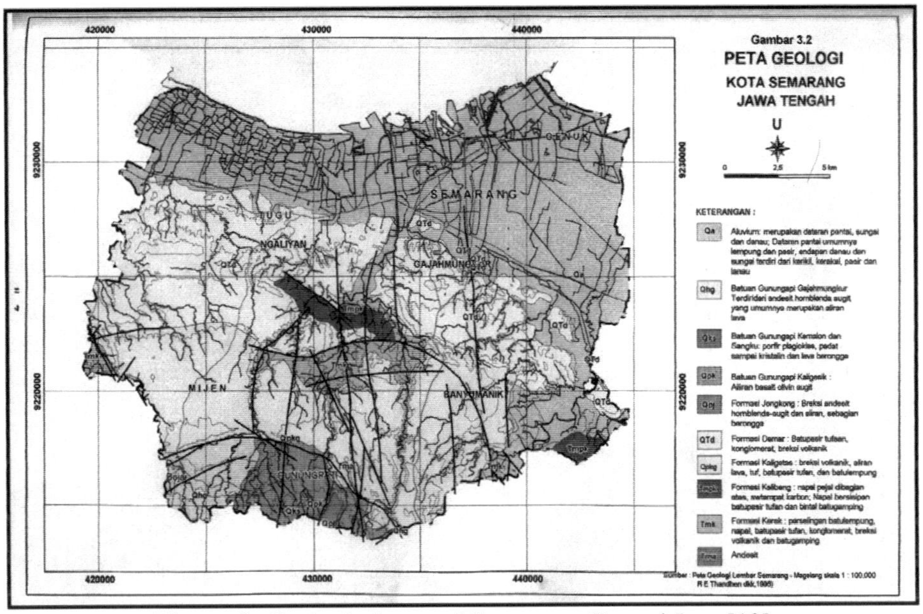

Figure 2. Geological map of Semarang, Central Java [12]

3. Data and Method

One sediment core was taken from the estuary of the East Flood Canal River, Semarang (-6.933050 °S, 110.444631°E) as part of the LIPI's Coremap expedition on March 2021. The sediment sample has a length of 31 cm. The sediment sample was sliced at intervals of 1 cm. A total of 31 sediment samples were used for granulometric analysis and 31 sediment samples were analyzed for XRF.

a. Granulometric analysis

The granulometric analysis is performed using a Mastersizer 2000 tool that aims to determine the value of each grain size parameter such as mean, sorting, kurtosis, and skewness.

b. X-Ray Fluorescence (XRF) Analysis

Geochemical analysis was conducted by the XRF (X-Ray Fluorescence) to determine the concentration of chemical elements. The XRF analysis uses the Thermo NITON XL3t 500.

c. Clustering Analysis (CA)

The clustering analysis used the Past 4.08 software. The value of grain size parameters is used as the basis for grouping sediments through the cluster method to get the lithofacies zones. .

4. Findings and Discussions

The results of a granulometric analysis that has been processed with Gradistate Software are shown in the table below.

Table 1. Grain size parameters of a sediment sample

Code	Mean (μm)	Sorting (μm)	Skewnees (μm)	Kurtosis (μm)
BKT 1	3.583	2.261	0.001	0.947
BKT 2	3.871	2.529	0.02	0.949
BKT 3	4.203	2.81	0.073	1.013
BKT 4	4.019	2.605	0.058	0.971
BKT 5	4.294	2.838	0.05	0.938
BKT 6	5.583	3.24	0.079	0.871
BKT 7	4.831	2.764	0.055	0.95
BKT 8	4.034	2.564	0.059	0.964
BKT 9	3.786	2.463	0.033	0.96
BKT 10	4.621	2.82	0.045	0.949
BKT 11	4.427	2.895	0.042	0.928
BKT 12	4.738	2.923	0.036	0.93
BKT 13	9.538	4.268	0.06	0.852
BKT 14	9.199	4.333	0.103	0.941
BKT 15	7.239	4.095	0.124	0.992
BKT 16	5.378	3.232	0.051	0.935
BKT 17	7.577	4.102	0.116	0.886
BKT 18	7.029	3.935	0.098	0.916
BKT 19	13.3	4.926	0.091	0.731
BKT 20	13.23	4.876	0.078	0.716
BKT 21	11.64	5.143	0.178	0.698
BKT 22	13.64	4.923	0.075	0.705
BKT 23	13.22	5.063	0.096	0.702
BKT 24	11.6	4.606	0.1	0.762
BKT 25	15.42	5.193	-0.042	0.704
BKT 26	12.4	4.821	0.13	0.716
BKT 27	13.39	4.916	0.068	0.721
BKT 28	8.305	3.857	-0.014	0.803
BKT 29	7.318	3.663	0.003	0.797
BKT 30	7.104	3.687	0.027	0.806
BKT 31	4.946	3.279	0.086	0.923

From the results of granulometric analysis obtained grain size parameters such as mean, sorting, kurtosis, and skewness, which are made graphs with Strater software (Figure 3). These parameter values are used as a basis for grouping sediments through the cluster method to obtain lithofacies zones. Grouping analysis uses PAST 4.08 Software and the similarity rate used is 3 to create lithofacies zones.

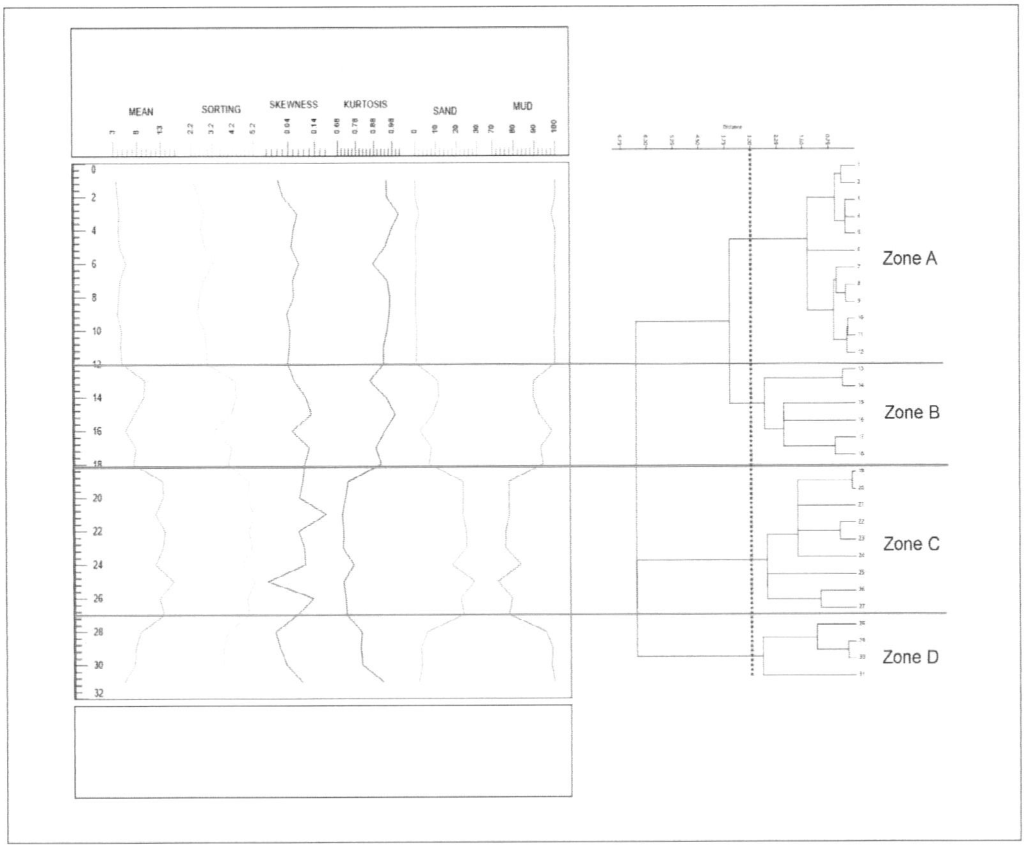

Figure 3. Grain size parameter graph (left) and clustering the lithofacies zones (right)

The cluster showed that four lithofacies zones have different sediment characteristics. Zone A is at a depth of 1 to 12 cm has a grain size of 3.5 - 5.5 µm (fine silt), poor sortation with a value of 2.2 - 3.2 µm (poorly sorted), with skewness (symmetrical), and kurtosis (mesokurtic). Zone B is at a depth of 13 to 18 cm has a grain size of 5.3 - 9.5 µm (fine silt), very poor sortation of 3.2 - 4.3 µm (very poorly sorted), skewness (symmetrical and coarse skewed), and kurtosis (mesokurtic). Zone C is at a depth of 19 to 27 cm has a grain size of 11.6 - 15.4 µm (medium silt), very poor sortation of 3.9 - 5.1 µm (very poorly sorted), skewness (symmetrical), and kurtosis (platykurtic). Zone D is at a depth of 28 to 31 cm has a grain size of 4.9 - 7.3 µm (fine silt), has poor sortation with a value of 3.2 - 3.6 µm (poorly sorted), skewness (symmetrical), and kurtosis (platykurtic).

Differences in grain size occur mainly in zone C, which has a coarser grain size than other zones and could be influenced by several factors. Factors that can affect include the transportation mechanisms, river currents, and the origin of sediment supply [13,14]. Various grain size tends to be coarser showing the sudden change in transportation mechanism, possibly due to fluctuations in water level conditions [15]. The supply of sediment carried by river currents comes mostly from land rather than the sea. It can be indicated that there has been an overflow of river water carrying sediment from the mainland due to flooding in the past. It also indicates that river currents are more dominant than ocean currents so the deposited material is coarser.

In addition to the results of the granulometric analysis in the form of grain size parameters, it is also necessary to be matched with the content of chemical elements in the sample of the research area to be able to ascertain how the sedimentation process occurred.

The XRF analysis used the Thermo NITON XL3t 500 tool can determine 9 chemical elements such as K (potassium), Rb (rubidium), Ca (calcium), Ti (Titanium), Mn (manganese), Fe (iron), Zn (zinc), Zr (zirconium), and Sr (strontium) presented in (Table 2). While the results of the chemical element graph processed with Strater Software are presented in (Figure 4).

Table 2. Geochemical elements of a sediment sample

Code	Zr (ppm	Sr (ppm	Rb (ppm	Zn (ppm	Fe (ppm	Mn (ppm	Ti (ppm	Ca (ppm	K (ppm)
BKT 1	105.7	93.8	23.84	41.44	19888.02	307.53	5966.3	11534.9	5966.3
BKT 2	91.52	114.9	28.07	41.21	19510.67	489.4	2978	17689.8	6810.1
BKT 3	100.12	97.29	25.12	52.74	19012.85	420.47	2742.97	16588.5	6186.9
BKT 4	92.72	99.21	24.29	36.38	18350.55	524.32	2672.75	15465.9	6186.9
BKT 5	82.86	103.55	22.63	41.66	18008.81	420.93	2778.33	17414.9	6093.2
BKT 6	86.33	99.99	23.78	49.86	19691.94	527.88	2770.82	17167.7	6467.6
BKT 7	92.26	102.95	25.95	36.47	18041.43	407.05	2733	17087.9	6927.8
BKT 8	80.03	92.07	26.04	42.71	16293.08	405.11	2442.53	17890.4	6044.4
BKT 9	73.05	95.51	24.81	32.83	16749.35	346.13	2704.92	18468.9	6459.9
BKT 10	92.01	135.93	25.77	37.52	20812.23	441.89	2930.86	19815.3	6320.9
BKT 11	88.57	119.98	22.99	36.18	22878	505.38	3443.69	20243	6722
BKT 12	141.62	184	28.16	60.22	27382.5	545.72	2989.42	20471.2	6938.9
BKT 13	80.6	141.66	25.42	42.24	25621.41	686.27	2835.18	16565.6	5571
BKT 14	85.68	177.91	25.75	48.51	26582.18	735.87	2535.71	19264.7	5452.5
BKT 15	72.08	117.95	19.32	43.19	20895.7	537.35	2395.72	13717.3	5277
BKT 16	86.65	130.24	21.76	47.86	21984.32	519.64	2147.62	14835.1	4798.4
BKT 17	94.41	218.25	22.14	58.33	30781.1	769.38	3542.39	21461.6	5957.1
BKT 18	117.66	241.98	18.96	35.67	38867.03	713.46	3702.7	23298.4	5008.4
BKT 19	103.84	226.06	21.39	64.65	35479.79	677.73	3979.48	22967.8	5933.4
BKT 20	87.67	235.22	17.38	40.32	47954.73	747.76	4444.29	23534.5	4021.3
BKT 21	99.64	243.94	20.12	56.96	46689.38	916.19	4397.84	23363.8	4327.8
BKT 22	111.71	243.22	18.41	48.88	43019.38	796.63	4075.73	23863.9	4667.9
BKT 23	82.12	225.4	24.15	36.5	27864.93	653.73	2766.01	20891.6	5500.6
BKT 24	221.44	233.5	16.9	95.99	63462.82	1070.6	5031.57	19241	3798.7
BKT 25	113.68	236	20.45	43.98	36437.7	798.2	3485.81	25357	4776.4
BKT 26	99.49	255.88	19.71	65.05	51101.76	950.94	4181.89	23342.9	3669.3
BKT 27	88.19	144.37	24.37	56.97	24450.17	666.11	2987.36	15231.2	5547
BKT 28	67.1	125.44	24.65	42.95	21034.68	644.13	2792.3	13522.8	5256.9
BKT 29	90.07	113.84	22.97	58.81	21115.36	596.31	3163.29	15729.4	6392.2
BKT 30	82.04	166.42	25.11	49.43	25405.68	627.23	2928.8	15038.3	5518.8
BKT 31	65.71	94.1	26.03	42.1	22241.78	652.45	2941.25	15722.2	5638.8

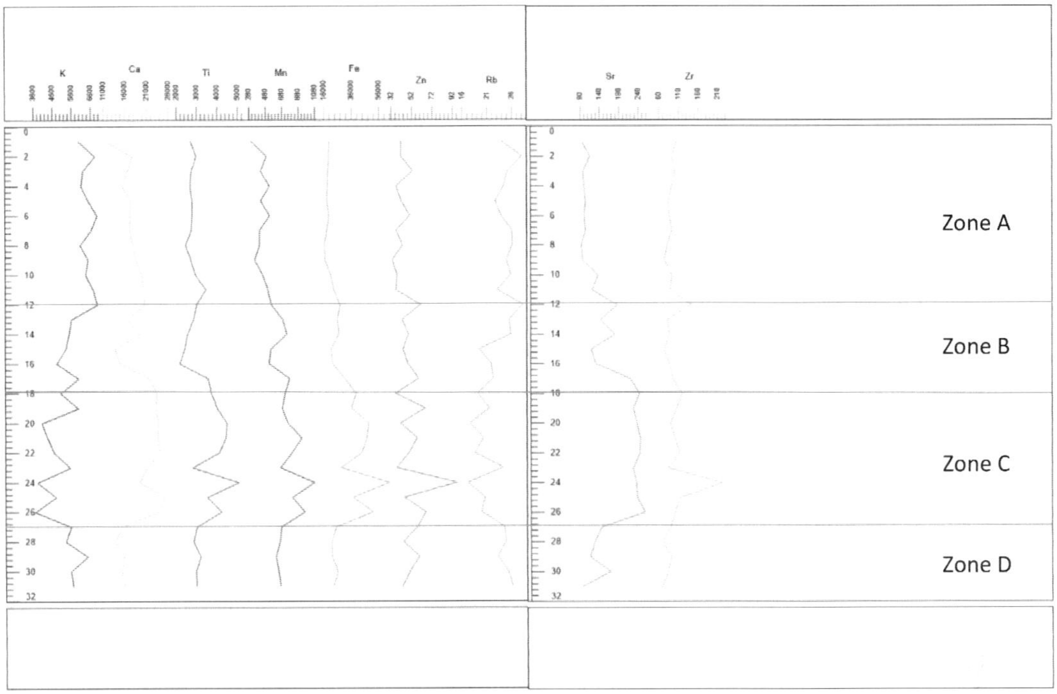

Figure 4. Geochemical elements graph

From the results, the main elements contained in sediments are Fe, Mn, and Ca. Fe ranges from 16293.08 - 63462.82 ppm, Mn ranges from 307.53 - 1070.6 ppm, and Ca ranges from 11534.9 - 23863.9 ppm. In Zone C, there is an increase in elements Fe, Mn, and Ca. The high content of Fe in the estuary is thought to be caused by Fe content derived from several sources, namely apart from the soil also derived from human activities that occur on the mainland, namely the presence of household waste discharge containing iron, water reservoirs from iron, industrial waste deposits and corrosion from water pipes containing iron metal that carried the river flow to the estuary [16].

This is because the research area is close to residential and industrial areas. Pollutants that enter the mouth of the river will be scattered and undergo a deposition process. The absorption of sedimentary metal particles is more dominant because metals have a tendency to bind to iron hydroxide, manganese hydroxide, and organic matter [17]. In addition, the increase in elements Fe and Mn can occur due to the flash flood factor that occurred in the 1990s that brought a lot of material from the headwaters of the river.

While the increase in elements Ca in Zone C, can be indicated from the rock formation of sediment origin containing carbonates [18]. Based on the regional geology of the Semarang area Kerek and Kalibeng Formation composed of limestone can be found at the headwaters of the river, so it can be interpreted that the high Ca element came from these rock formation that was eroded and transported.

5. Conclusion

The characteristics of seafloor surface sediments at the mouth of the eastern canal river flood are dominated by medium to fine silt. In Zone C there is a change in the size of sediment grains accompanied by an increase in elements Fe, Mn, and Ca. Changes in grain size occur because river currents dominated more than tidal currents at the mouth of the river flood so that it deposited a coarser sediment material. The increase in elements Fe and Mn can come from waste pollution from industry and settlements as well as indications of flash floods that had occurred in the past few years. While the increase in elements Ca be indicated by the reworked of older rocks formation of sediment origin containing carbonates.

6. References

[1] Rahmadi, Nanda., Widada, Sugeng., Marwoto, Jarot., Atmojo, Warsito., dan Widiaratih, Rikha. 2021. Studi Sebaran Sedimen Dasar di Perairan Sungai Banjir Kanal Timur Semarang, Jawa Tengah. *Indonesia Journal of Oceanography (IJOCE)* [September] [2021] Vol 03 No **03** : 63 71.

[2] Syvitski, J. P. M. 1991. Factor analysis for size frequency distribution: significance of factor solutions based on simulation experiments. In Principles, methods and application of particle size analysis, ed. J. P. M. Syvitski, 249–63. New York: Cambridge University Press.

[3] Blott, S. J., & Pye, K. (2001). GRADISTAT: A grain size distribution and statistics package for the analysis of unconsolidated sediments. *Earth Surface Processes and Landforms*, **26(11),** 1237 1248.

[4] Folk, R. L. 1966. A review of grain-size parameters. *Sedimentology* **6**:73–93.

[5] Visher, G. S. 1969. Grain size distribution and depositional process. *Journal of Sedimentary Research* **39**:1074–106.

[6] Wachecka-Kotkowska, L., and P. Kotkowski. 2011. Grain-size distribution analysis of Quaternary sediments from the southern part of the Lodz region in Poland: A computational-methods approach. Geologos 17:205–19. doi:10.2478/v10118-011-0012-7

[7] Chambers, R. L., and S. B. Upchurch. 1979. Multivariate analysis of sedimentary environments using grain-size frequency distributions. *Mathematical Geology* **11**:27–43.

[8] Nugroho, Septriono H., Putra, Sulastya P., Yulianto, Eko., dan Noeradi, Dardji. 2017. Multivariate statistical analysis for characterization of sedimentary facies of Tarakan sub-basin, North Kalimantan. *Marine Georesources & Geotechnology*, **36:8**, 907-917.

[9] Oyedotun, Timothy T.D. 2020. Compositional and multivariate statistical analyses for grain-size characterisation of intertidal sedimentary facies in an estuarine environment. *Geology, Ecology,and Landscapes.*

[10] Thoms, M. C., M. E. Parsons, and J. M. Foster. 2007. The use of multivariate statistics to elucidate patterns of floodplain sedimentation at different spatial scales. *Earth Surface Processes and Landforms* **32**:672–86.

[11] Bappedal Propinsi Jawa Tengah. 2002. Laporan program kali bersih tahun 2002. *Laporan Penelitian.* Semarang: Pemerintah Propinsi Jawa Tengah. Hal 13 – 14.

[12] Thanden,dkk. (1996). Regional Geological Map Sheet Magelang – Semarang.

[13] Environmental Impact Management Agency (BAPPEDAL). 1994. *Prokasih (Clean Water Program): Evaluating The Last Four Years and Looking to the Future,* Jakarta, Indonesia.

[14] TICA Study Team, 1994. *Interim Report on Flood Control in BanjirKanal Barat and Babon Rivers,* Semarang, Indonesia p.3-41

[15] Nugroho, S. H., Putra, P. S., & Pamungkas, A. D. G. (2020). Karakteristik Sedimen Holosen-Pleistosen Berdasarkan Ukuran Butir Pada Sedimen Inti Ew17-08, Tepi Baratlaut Paparan Sunda. Jurnal Geosaintek, 6(2), 107-116.

[16] Supriyatini, Endang., dan Endrawati, Hadi. 2015. Kandungan Logam Berat (Fe) pada Air, Sedimen, dan Kerang Hijau (*Perna Veridis*) di Perairan Tanjung Emas, Semarang.

[17] Barreiro, R , Real C. , Carballeria A , 1994. Heavy metal horizontal distribution in surface sediments from a small estuary (Pontedeume, Spain). *Sci. Total Environ.,* **154**: 87 -100.

[18] Takarina, N. (1996). Heavy Metal Contents in Surficial Sediment of Banjir Kanal Barat and Babon Rivers, Semarang, Central Java, Indonesia. Doctoral dissertation (McMaster University).

7. Acknowledgements

The study was partially funded by LIPI's Coremap-CTI 2021-2022 (grant number 17/A/DK/2021) who received by L. and SHN. This study was also partially funded by the Laboratory of Energy Resources, Geological Engineering Department Diponegoro University. We thank to the Deputy of Earth Sciences - BRIN, Director of Research Center for Oceanography - BRIN, Director of Research Center for Geotechnology – BRIN, and Head of Geological Engineering Department Diponegoro University. We are also very grateful to two anonymous reviewers for their useful critical reviews. YM, SHN, AK designed the research concept, prepared and wrote the manuscript. YM contributed to data processing and analysis. PSP and L contributed by providing suggestions and input, as well as review-edited the manuscript. SPAW, MTK prepared the samples and tools in the cruise and A and J in the laboratory. L, SHN, SPAW, and MTK joined the cruise and obtained the samples.

IOP Conf. Series: Earth and Environmental Science 1047 (2022) 012009 doi:10.1088/1755-1315/1047/1/012009

Interaction of Air and Sea above Seamount in the Halmahera Sea

Sobaruddin D P[1], F. Marpaung[2], R.A.B. Putra[3], A. Fahim[4], C.S. Dharma[1*], D.T. Pramono[5], and A. Kristiawan[5]

[1] Indonesian Navy Center for Hydrography and Oceanography Office, North Jakarta, Indonesia
[2] National Research and Innovation Agency
[3] Meteorological, Climatological, and Geophysical Agency
[4] Atmospheric Sciences Research Group, Faculty of Earth Sciences and Technology, Bandung Institute of Technology, Bandung, Indonesia
[5] Center of Research Naval Science and Technology

Email: meteorobo@gmail.com

Abstract. The interaction of sea and air plays a very important role in the early stages of cloud formation. In certain cases, changes in temperature in the sea surface layer and the atmospheric layer closest to the sea will affect the initiation of the formation of water vapor which can become cloud. To monitor the sea-air temperature interaction above Seamount in Halmahera Sea, we took an expedition of Jala Citra-I 2021 Aurora from August 28 to September 9, 2021 using The Indonesian Navy Center for Hydrography and Oceanography research vessel, the Navy's KRI Spica-934. Weather condition was observed using two installed Automatic Weather Stations (AWS) mounted on a vessel, radiosonde in Sorong, West Papua, and Outgoing Longwave Radiation (OLR) data from NOAA (National Oceanic and Atmospheric Administration), while the profile of sea surface temperature (SST) were observed using a 24-hour CTD measurement. Both AWS data were analyzed using two samples Kolmogorov-Smirnov test with h=1 and showed low correlations ($r^2 = 0.2\text{-}0.4$, p-value <0.01) for the parameters of temperature, relative humidity, and wind speed. The hourly SST ranged 29°C – 30°C from 0-50 meter and the Lifting Condensation Level ranged about 939 to 985 mb in the morning on 1 - 2 September. The OLR was a neutral and positive values. These conditions indicated that a strong local effect dominantly created by a strong sea-air interaction in the study area. A warm SST with a strong divergence wind leads a warm air layer developed on the sea surface to rise. A low OLR with a moderate CAPE is enough to support the updraft of water vapor, rising from sea surface and then support convective activity formed in the region. In addition, it shows that the changes activity of convective clouds, developing over warm waters, are dominant mode of diurnal variability. However, investigation of intra-seasonal variability in Halmahera Sea was not clear. Further investigation on spatiotemporal of sea-air interaction on a local scale are needed to capture the phenomena on its season.

1. Introduction

Indonesian waters have an important influence on the global climate. Indonesia is the hub of the thermohaline circulation conveyor belt line or better known as Indonesia Throughflow (ITF). The ITF provides the only inter basin exchange of water at low latitudes from the Pacific to the Indian Ocean. It is important to understand that the dynamics of this region because of the role, it may play in El Nino-

Southern Oscillation (ENSO) development due to its active air-sea interactions [1]. Water from the Pacific Ocean enters the Indonesian seas in the surface layer near the region where the New Guinea Coastal Current (NGCC) meets the Mindanao currents (MC). The NGCC retroflects around Halmahera Eddy into the eastward flowing North Equatorial Counter Current [2]. The NECC plays a key role in shaping the tropical Pacific climate, such as the equatorial asymmetry of the intertropical convergence zone [3], and modulates the warm pool heat budget through the displacement of the western Pacific warm pool in the east–west direction [4]. In addition, the warm pool is caused by annually-long winds that drag masses of tropical warm water into the western tropical Pacific Ocean and then accumulate in the Tropical Western Pacific region that includes the waters of Halmahera and North Papua. These territorial waters also have complex current dynamics that may be caused by complex bathymetric configurations. There are several currents that play a role in transporting water masses in this area such as Mindanao Current, NGCC, Western Equatorial Current, Halmahera Eddy, and Mindanao Eddy [5, 6, 7]. Water mass stratification also plays an important role in many bio-geo-chemicals processes in which the mixed layer regulates interaction with the inner layer and its upper layer.

Although water mass and climate circulation have been successful in explaining many aspects of sea-air interaction in the Tropical Pacific region; a puzzling aspect of the interaction especially above the seamount found to be limited to none. These conditions are due to an insufficient amount of expedition conducted in each region of the Indonesian seas. The expedition of oceanography was recorded in Satal Index 2010 and Halmahera Index 2011, whereas the investigation of the marine meteorology was done during the Indonesia Prima 2017 expedition [8]. The results showed that knowledge of the deep flows and sea-air interaction in each region is scantly available. It has not been (and is still not) possible to show unambiguously that the upper currents compensate to meteorological conditions, and vice versa. On the other hand, Jala Citra-1 2021 "Aurora" expedition was held to explore the air and sea interaction in Halmahera Sea based on the equipment installed on the survey vessel. During the expedition, the investigation about ocean-meteorology was done during Stage II, Jala Citra-1 "Aurora" from August 28 to September 9, 2021. The research was focused on oceanography and meteorology condition above the seamount that was found on previous stage, the Stage I, Jala Citra-I "Aurora" expedition.

2. Data and Method

2.1 Description of the study sites

This study was conducted in the eastern part of Halmahera Sea. It locates 68.9 Nm from the Northeast of Buli, the administration centre of Maba districts, Eastern part of Indonesia, and about 158 nautical miles Northwest of Sorong, the largest city on the Western tip of Papua (Figure 1). The research area has to be accessed with research vessel as the only transportation modes. Field data measurements were carried out on Stage II, Jala Citra-I Expedition 2021 which took place on August 28 - September 9, 2021; The measurement and retrieval of field data were carried out using the Navy's KRI Spica-934.

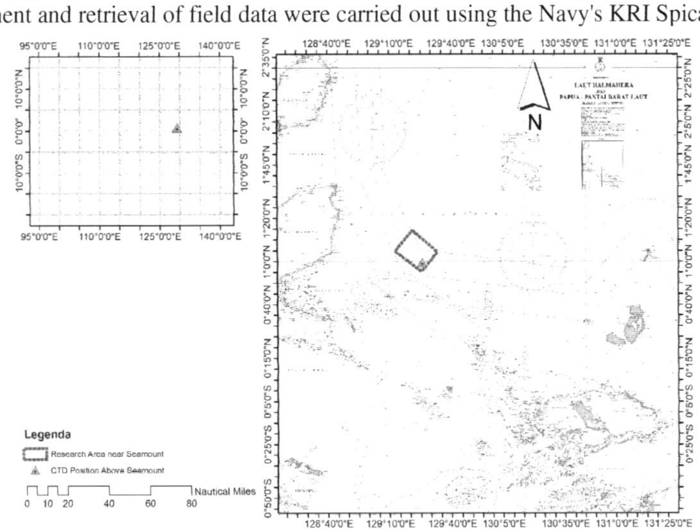

Figure 1. Research Area of Jala Citra-I 2021 "Aurora" in Halmahera Sea, Indonesia.

2.2 Material and Methods

The data used in this study were Automatic Weather Station (AWS) data from the Indonesian Navy Center for Hydrography and Oceanography (Pushidrosal, namely Navy's KRI Spica-934), mounted on board of the research vessel. Two AWS recorded, namely the Vaisala (St.1) and Airmar (St.2) recorded the weather condition. A rain gauge for precipitation measurement only mounted on St.1. St.1 observes with Eastern Indonesia Local Time or Waktu Indonesia Timur (WIT = UTC+9) while St.2 observes in UTC. Due to a time zone difference between both stations, we adjusted the time before further analysis. We also used a rainfall data from the observation, JAXA GSMaP (Global Satellite Mapping of Precipitation), and from NASA Global Precipitation Measurement data (GPM). The GPM measures daily data with 0.1 x 0.1 degree resolution. The GPM was used as a general guide about seasonal rain data in the study area. To monitor the cloud formations around study area, we also used data from radiosonde observation in Sorong, West Papua.

Based on climatological classification, Halmahera region relies on climate zone with non-monsoonal variety, or exhibiting a weak seasonal variety [9]. This generally occurred in areas with climate zone C in Indonesia [10]. Even though the climatological variability was weak, daily pattern of meteorological condition recorded was captured (Figure 2). There are five parameters of meteorological data used in this research, namely temperature, humidity, pressure, rain, wind speed and wind direction that are analysed to investigate sea-air interaction above study area. The Sea Surface Temperature (SST) data is ideally observed with non-isolated bucket, condenser or cooling system intake, and hull conduit or thorough hull sensors. This observation has 0.3 – 1.0°C difference of recorded SST data using a 24-hour CTD measurement. Meanwhile, the air temperature, relative humidity, and surface wind velocity have met the requirements of 0.1°C error measurements, 0.5 ± 0.2 g/kg bias tolerance, and ± 2 m/s with 1-8% variance, respectively [11]. Data observation derived from both stations was then tested its homogeneity using 2-sample Komolgorov-Smirnov and r^2 test.

The SST data collected from the Copernicus Marine Services from 1998-2021 with resolution of 0.083° x 0.083° and the Outgoing Longwave Radiation (OLR) data from NOAA (National Oceanic and Atmospheric Administration) with the resolution of 0.25° x 0.25° were then analysed. Wind data from BMKG Ocean Forecast System (OFS) within the range of the expedition time was also explored thoroughly to determine the interactions.

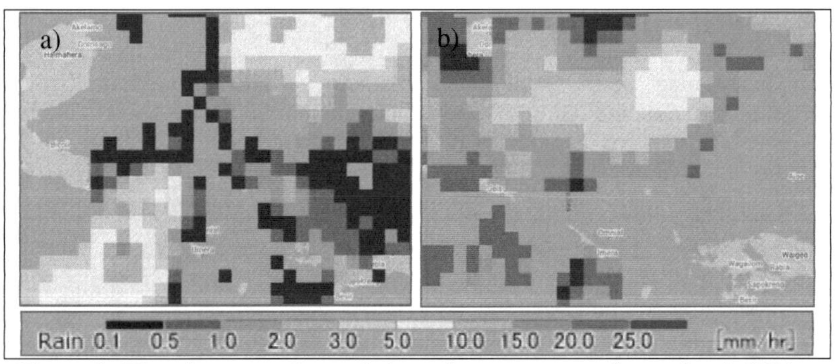

Figure 2. The rain events observed above the seamount on September 1, 2021 at (a) 00 UTC, and (b) 08 UTC in Halmahera Sea. Source: JAXA GSMaP

3. Results and Discussion

Profile of temperature, relative humidity, wind speed and precipitation during the expedition can be seen in Figure 3. It shows that the diurnal pattern of air temperature, relative humidity and wind speed measured from both stations are relatively similar. Although both stations have a similar pattern data recorded, the correlation between data of St.1 and St.2 was low with the correlation with the r^2 values ranges from 0.2 to 0.4. The homogeneity of air temperature with two sample Komolgorov-Smirnov resulting in completely different data characteristics with h=1, and a p-value of < 0.01. Hourly relative humidity and wind speed measured at St.1 have a lower value than hourly data measured at St.2. It also shows that a low relative humidity occurred during midday (10.00 – 14.00 WIT) as increasing air temperature. Hourly wind speed profile ranges between 12 and 15 knot with a high value occurred in the morning (06.00 – 08.00 WIT) and late midday (12.00 – 14.00 WIT).

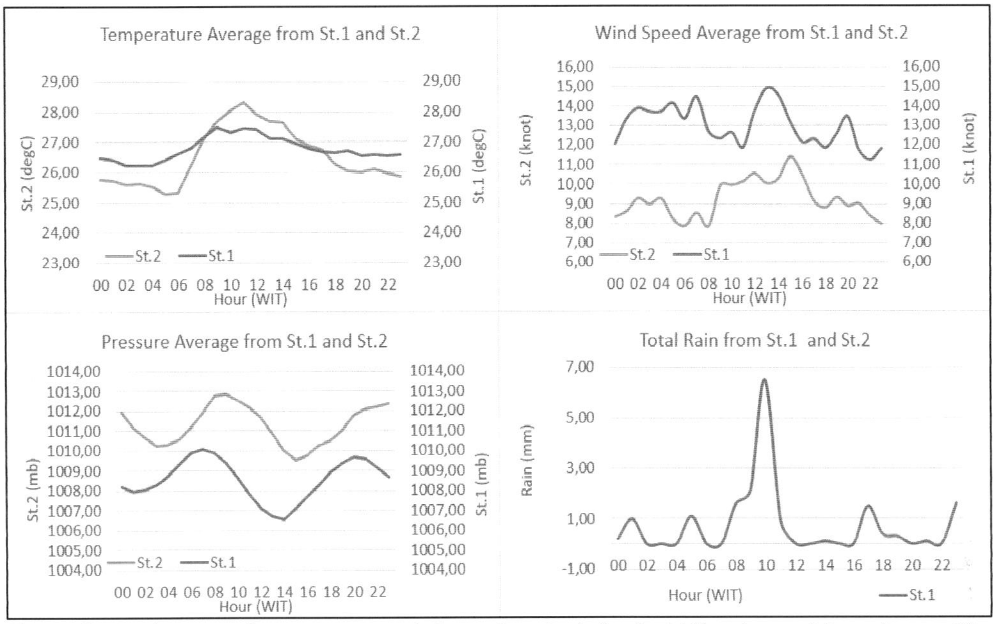

Figure 3. Hourly mean air temperature, wind speed, and relative humidity observed from August 10–September 7, 2021 from both stations, while precipitation only observed in St. 1.

The interaction between sea and air on a regional scale in Indonesia can be observed from the phenomenon of El Nino, La Nina, and IOD and its effects to dry and wet season. On the other hand, in a local scale, the sea-air interaction can be investigated by rainfall event monitoring. Manual observations of the rain events show that on September 1, a slight rain event occurred at 06.06-17.55 WIT with the appearance of Cumulonimbus and Stratocumulus cloud, while on September 2, there were only 30 minutes of rain events from occurring at 12.30-13.00 WIT with the appearance of Cumulonimbus clouds with anvil. Precipitation at 10.00 WIT have a high value, with 6.5 mm of total rain (Figure 3d). The formations of convective cloud in the ocean were more frequent than in the land area. This was mainly caused by a high-water vapor content with a supportive sea surface temperature around midday that can generate rain (Figure 3a, 3c, and 6).

Radiosonde is commonly used as a tool to determine the Lifting Condensation Level (LCL) that the LCL is clearly visible from the base/lowest level of cumulonimbus clouds on daily basis or a cumulus cloud and its formations on an average daily basis. Formation of rain cloud in the study area is shown in Figure 4. It shows that there is a little chance for the formation of rain cloud due to a wide separation between the dew point temperature and air temperature in the morning until afternoon despite of the moderate CAPE value (2486.8 – 2974.4 Joule/Kg) [12]. LCL ranges about 939 to 985 mb at 09.00 WIT. A high LCL of 920 – 898.6 mb generally occurred during a dry season with no rain event and it reaches 953 mb during rain events [13]. It is the contradictions to observed weather condition in the study area. The rain occurred all day on September 1 while the LCL recorded from radiosonde in Sorong was no rain (LCL > 800 mb). This phenomenon might be affected by a strong air-sea interaction. During the early morning, there is a rapid movement and development of cumulus clouds. AWS data show that a high increased air temperature and a high decreased relative humidity with a high SST at 10.00 – 14.00 WIT with a relative high wind speed. It indicated that a formation of cumulus cloud in the study area is strongly affected by the local weather condition due to a strong sea-air interaction.

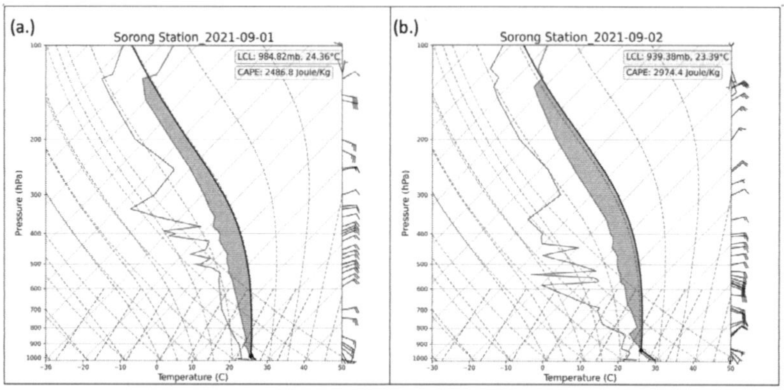

Figure 4. Radiosonde observations from September 1 (a) and September 2 (b), 2021 in Sorong conducted on 00 UTC or 0900 WIT Local Time. Both data shows that in Sorong during that day, was clear with a slight rainy day on the morning until afternoon.

Air movements can be distinguished from the horizontal and vertical movements. The horizontal movements of air are visible by the wind flow. It observed that about 76,3% data from St. 2 were validated with BMKG OFS Wind (Figure 5a) from August 29 to September 7, 2021. On September 1, the wind was blowing from the Southeast direction with the speed of 2-4 knot. In addition, vertical air movements were approached using seasonal wind data from 1998 – 2021, and then calculated using wind divergence on 700 mb heights, as a crucial cloud formation height [14] (Figure 5b). It shows that a strong divergence occurred in the study area while at the Eastern and Southern part of the study area have a weak wind divergence. A dominant wind from Southward resulting in higher air movements and decreasing the possibility of convective cloud formation.

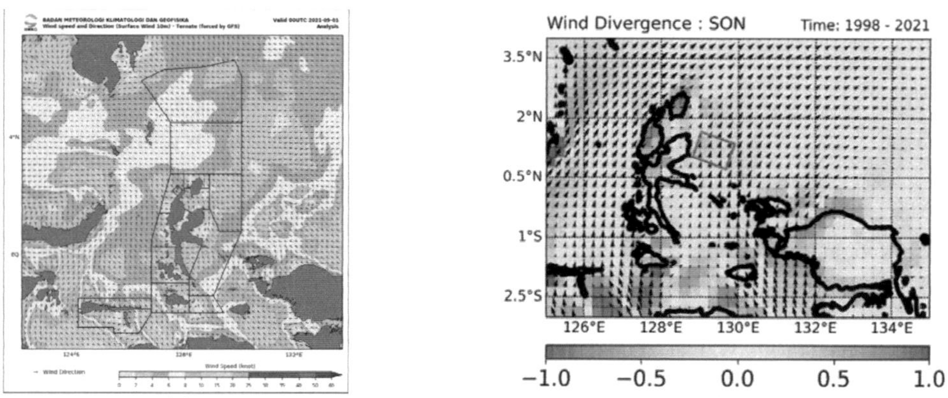

(a) (b)

Figure 5. (a) BMKG OFS Wind Prediction on 1 September 2021, and (b) Seasonal wind divergence pattern on September-October-November periods in Halmahera Sea from 1998-2021.

Radiosonde data and seasonal divergence patterns show a low probability of rain drops occurred in the study area. However, on September 1, rain events occurred for all day. It might be caused by a formed convective cloud with the SST ranged between 29.0°C - 29.8°C (Figure 6). The high SST is a favorable for the formations of convective clouds [15]. Moreover, the anomalies of Outgoing Longwave Radiation (OLR) showed a low value (0-15 w/m^2) (Figure 7). It suggests that there was a positive updraft trend that supported convective activity formed in the region, while on the southern side of the research area there was negative updraft trend that suppress the convective activity.

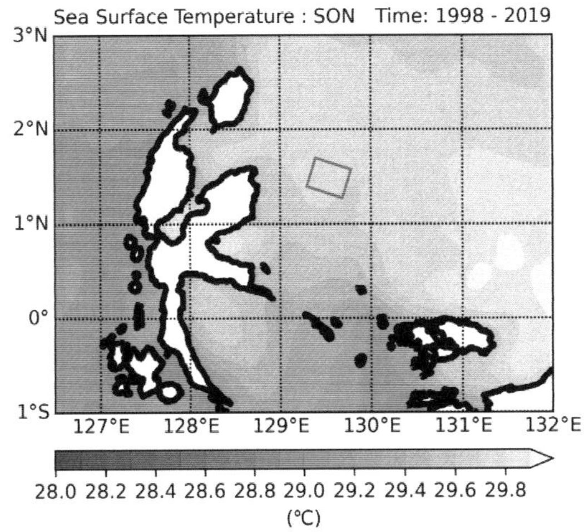

Figure 6. Seasonal Sea Surface Temperature in Halmahera Sea, on September-October-November from 1998 to 2021

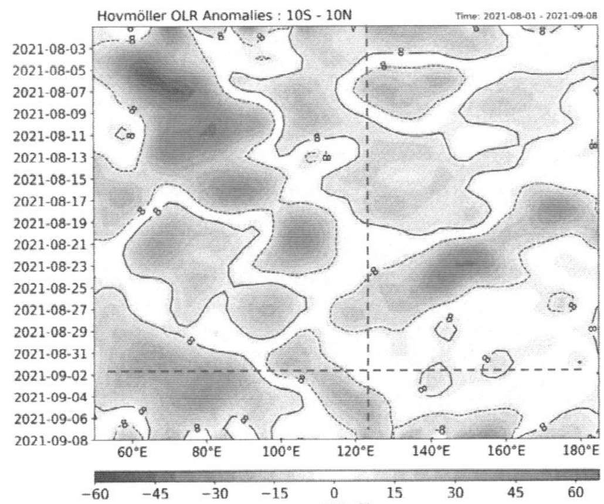

Figure 7. Outgoing Longwave Radiation (OLR) anomalies from August 1 to September 8, 2021.

The analysis of sea-air interaction from the ocean as a base surface also be done using a 24-hour CTD measurement. It shows that the SST has high value ranges between 29°C -30°C (figure not shown) that's warm enough to create vapor as the fuel for convective clouds to be formed. There is a connection between OLR and SST data which is found on the regional scale, the OLR will drive the increase of SST about 1-2 week later [16]. However, in this study area, the effect was direct on one day due to a strong local effect of the interaction of sea and air.

4. Conclusions

The sea-air interaction above the Seamount in Halmahera Sea was investigated using data measurement and observation during the Jala Citra-I 2021 "Aurora" Expedition. During the expedition, it shows that a strong local effect dominantly created by a strong sea-air interaction in the study area. The changes in activity of convective clouds, developing over warm waters in Halmahera Sea, are dominant mode of diurnal variability. During a rainy day of September 1, with a high-hourly mean wind speed of 12-14 knots and a strong divergence wind, a shallow warm air layer developed on the surface of the ocean leads to rising air. This warm layer may be interpreted as a diurnal sea surface temperature of 29.0°C - 29.8°C which favourable contributes for a formed convective cloud. Based on these results, a strong divergence wind also moves the cumulus clouds swiftly (LCL > 800 mb). A low outgoing longwave radiation anomaly with a moderate CAPE are enough to support the updraft of the water vapor in the air, rising from sea surface and then supports convective activity formed in the region.

Acknowledgments
Authors would like to thanks the Chief of the Indonesian Navy, Indonesian Chief of Hydrographer, Head of BMKG, and Head of National Research and Innovation Agency, the ministries, and private companies for the collaboration. This research was part of the Jala Citra-I 2021 "Aurora" Expedition, Funded by Pushidrosal. Authors would also like to give the highest appreciations to all the staff of the Pushidrosal Charting Division, Oceanography and Meteorology Division, the committee of the Aurora Expedition, Commander and all the crew of the Survey Vessel for the support during the expedition.

References

[1] Kessler, W. S., Johnson, G. C., & Moore, D. W. (2003). Sverdrup and Nonlinear Dynamics of the Pacific Equatorial Currents, Journal of Physical Oceanography, 33(5), 994-1008. Retrieved Dec 21, 2021, from https://journals.ametsoc.org/view/journals/phoc/33/5/1520-0485_2003_033_0994_sandot_2.0.co_2.xml

[2] Zhou, H., Liu, H., Tan, S., Yang, W., Li, Y., Liu, X., Ren, Q., & Dewar, W. K. (2021). The Observed North Equatorial Counter current in the Far Western Pacific Ocean during the 2014–16 El Niño, Journal of Physical Oceanography, 51(6), 2003-2020. Retrieved Dec 21, 2021, from https://journals.ametsoc.org/view/journals/phoc/51/6/JPO-D-20-0293.1.xml

[3] Masunaga, H., and T. S. L'Ecuyer, 2011: Equatorial asymmetry of the east Pacific ITCZ: Observational constraints on the underlying processes. J. Climate, 24, 1784–1800, https://doi.org/10.1175/2010JCLI3854.1.

[4] Clement, A. C., R. Seager, and R. Murtugudde, 2005: Why are there tropical warm pools? J. Climate, 18, 5294–5311, https://doi.org/10.1175/JCLI3582.1.

[5] Cresswell G R and Luick J L 2001 Current measurement in the Halmahera Sea J. Geophys. Res. 106(C7) 13953-58

[6] Atmadipoera A, Kuroda Y, Pariwono J I and Purwandani A 2004 Water Mass Variation in the Upper Layer of the Halmahera Eddies Region Observed from a TRITON buoy IEEE

[7] Kashino Y, Atmadipoera A S, Kuroda Y and Lukijanto 2013 Observed feature of the Halmahera and Mindanao Eddies J. Geo. Ress. Ocean. 118 6543-60

[8] Siswanto Siswanto, Kusmanto E, Rahayu P, Florida N, Mcphaden M 2019 Air-Sea Interaction Mechanisms in the Generation of Persistence Low Pressure Area Over the Eastern Indian Ocean Southwest Sumatera Observed During the Indonesia Prima 2017 Field Campaign Environmental Science Jurnal Meteorologi dan Geofisika [in Bahasa Indonesia]

[9] Dang Haowen, Jian Zhimin, Kissel Catherine, Bassinot Franck 2015 Precessional changes in the western equatorial Pacific Hydroclimate: A 240 kyr marine record from the Halmahera Sea East Indonesia Geochemistry Geophysics Geosystems 16 doi:10.1002/2014GC005550

[10] Aldrian E, and R D Susanto 2003 Identification of three dominant rainfall regions within Indonesia and their relationship to sea surface temperature Int. J. Climatol 23 1435-1452

[11] Kagan Boris A 1995 Ocean-Atmosphere Interaction and Climate Modelling British Cambridge

[12] Gettelman A, Seidel Dian J, Wheeler Matthew C, Ross R J 2002 Multidecadal trends in tropical convective available potential energy Journal of Geophysical Research Atmospheres 107(21) doi:10.1029/2001JD001082

[13] Prayoga Ismail, Richard 2020 Profile lifting condensation level (LCL), convective condensation level (CCL) and estimated cloud base height in Makassar region GAW Bulletin Bariri 1 58-66 [in Indonesian]

[14] Bony S, Lau K-M, Sud Y C 1997 Sea surface temperature and large-scale circulation influences on tropical greenhouse effect and cloud radiative forcing Journal of Climate 10 2055-2077 https://doi.org/10.1175/1520-0442(1997)010<2055:SSTALS>2.0.CO;2

[15] Sherwood S C, R Roca, T M Weckwerth, and N G Andronova 2010 Tropospheric water vapor convection and climate Rev Geophys 48 RG2001 DOI 10.1029/2009RG000301

[16] White M, Mohn C. Seamounts: a review of physical processes and their influence on the seamount eco-system. OASIS Report contract; 2004

[17] Napitu Asmi M, Gordon Arnold L, Kandaga P, 2015 Intraseasonal Sea surface temperature variability across the Indonesian seas Journal of Climate 28(22) doi:10.1175/JCLI-D-14-00758.1

IOP Conf. Series: Earth and Environmental Science 1047 (2022) 012010 doi:10.1088/1755-1315/1047/1/012010

Clastic Sediment Characteristics of Gumai Formation: Preliminary Study of Tertiary Rocks in South Sumatra Basin

M Hendrizan[1,2]

[1] Doctoral Program of Earth Sciences, Faculty of Earth Sciences and Technology, Institut Teknologi Bandung, Labtek XI, Jl. Ganesha, 10, Bandung, 40132, Jawa Barat, Indonesia

[2] Paleoclimate and Paleoenvironment Research Group, Research Center for Climate and Atmosphere, Research Organization for Earth Sciences, National Research and Innovation Agency, Jl. Sangkuriang, Bandung, 40135, Jawa Barat, Indonesia

Email: marf001@brin.go.id; m_hendrizan@yahoo.com

Abstract. The depositional environment of Gumai formation is still challenging to be studied. Previous studies of the depositional environment in the Gumai Formation show a debate interpretation of transition, shallow marine, and deep marine environment. The study's objectives are to identify lithology characteristics of Gumai Formation in Baturaja region and determine the potential environment of the Gumai Formation. Miocene clastic sedimentary rocks of Gumai Formation show well-preserved features in Simpang and Muara Dua, South Sumatra. The Gumai Formation discovered in the southern anticline flange, particularly in Muara Dua, composes marl with quartz veins. The opposite flange in the northern part of Simpang region indicates Gumai Formation consists of siltstone, calcareous sandstone, quartz sandstone and sand ball in siltstone layers. Based on the obtained data at Muara Dua and Simpang, the Tertiary rocks of the Gumai Formation are situated in the flange of folded Pre-Tertiary rocks at Garba complex. Gumai Formation in the research area was deposited in transition environment to shallow sea in which the formation in Simpang has shallower marine environment than Muara Dua region.

1. Introduction

Gumai Formation is a hydrocarbon source in several places in the South Sumatra Basin [1]. The formation is considered the maximum transgression phase after deposition of the Baturaja Formation, which contributes to part of the petroleum system as a seal [2-4]. Gumai Formation in Baturaja geological map indicates the formation is rich in benthic foraminifera, including *Nodosaria* sp., *Brizalia*, sp., *Planulina* sp., dan *Uvigerina* sp. [5]. Distribution of that foraminifera disperses in marine paleoenvironment of upper bathyal until lower bathyal in a water depth of 400-2000 m [6-7]. The study of the depositional environment in the Gumai Formation show a debate interpretation of transition [2,8], shallow marine [1,9], and deep marine environment [10-17]. The depositional environment of the Gumai Formation is still challenging to be studied further related to the debate interpretation above. A detailed study about Tertiary rock in Baturaja region is restricted to Baturaja Formation [18]. Gumai Formation in Simpang and Muara Dua area is an ideal location because the location has never been studied in previous studies [1,2,8-17]. The research study was in small rivers and roads in the Simpang and Muara Dua region. This study aims to identify lithological characteristics and determine the Gumai Formation's potential environment of Gumai Formation from its lithological characteristics in Baturaja region.

2. Geological Setting

South Sumatra Basin comprises a group of semi-connected NNW-SSE trending syn-rift basins with a post-rift sequence [12,13]. In the center of the South Sumatra Basin, lithological characteristics are Permian to Carboniferous metamorphic and igneous rocks that crop out in a northwest and southeast trend, including phyllites, slates, argillites, quartzite, gneisses, and granite [13,14]. In the northwest and south of the basin center, the lithology is characterized by Mesozoic metamorphic rocks with local granite intrusion [13,14]. The broad trend of the Permo-Carboniferous and Mesozoic rocks are Mesozoic metasediments and limestone dated as Jurassic or Cretaceous period [13,14]. In the north of the central basin near the city of Palembang is micritic limestone of the Cretaceous Age [13,14].

The South Sumatra Basin was formed initially during Late Eocene rifting which divided into two parts of the Palembang sub-basin to the south and the Jambi sub-basin to the north [17]. The two sub-basins are bounded by rifts that have north-south orientation in the Palembang sub-basin and NE-SW in the Jambi sub-basin [17]. Specifically, the South Sumatra Basin has formed by three major tectonic phases: 1) extension during late Paleocene to early Miocene forming north-trending grabens that were filled with Eocene to early Miocene deposits; 2) relative quiescence with late normal faulting from early Miocene to early Pliocene; 3) basement-involved compression, basin inversion, and reversal of normal faults in the Pliocene to Recent forming the anticlines that are the major traps in the area [19]. A detailed explanation of tectonic processes has been explored further in the previous study [13,17].

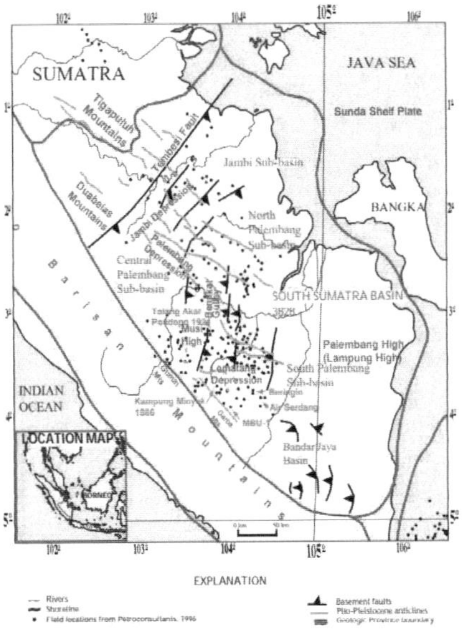

Figure 1. Structural feature of South Sumatra Basin [13]

The grabens and significant faults of the South Sumatra Basin have oriented north-northwest to south-southeast [13]. The fault-bounded Benakat Gulley connects the major basin complexes of the Lematang depression and the Palembang depression (Figure 1) [13,20,21]. The north-south Benakat Gulley is similar to the Bengkalis depression in Central Sumatra, the fault zone that forms the eastern coast of Sumatra, the Sunda and Asri Basins offshore, and the grabens of Northern Sumatra [13,20,21]. A fault zone with a southwest to northeast trend, the Tembesi Fault, forms the northwestern edge of the Jambi Depression (Figure 1).

In South Sumatra Basin, the Lahat and Lemat Formations represent rift deposition [17]. The Lemat Formation is a synonym of the Lahat Formation distinguished by several lithological characteristics [4]. The ages of the Lahat and Lemat formations are given as late Mid-Eocene to Late Oligocene (NP16-

NP24) [1]. The marine sediments of the late transgressive stage are represented in the South Sumatra Basin by Gumai Formation and Baturaja limestone during late Oligocene and mid-Miocene [4,23,24,25]. The Gumai Formation comprises a monotonous series of foraminifer-bearing grey shales and siltstones lenses of tuft [17]. The South Sumatra Basin's maximum transgression in the Mid-Miocene implies to drowned basal Gumai and Baturaja Formation and sealed by marine shales of Gumai Formation (Figure 2).

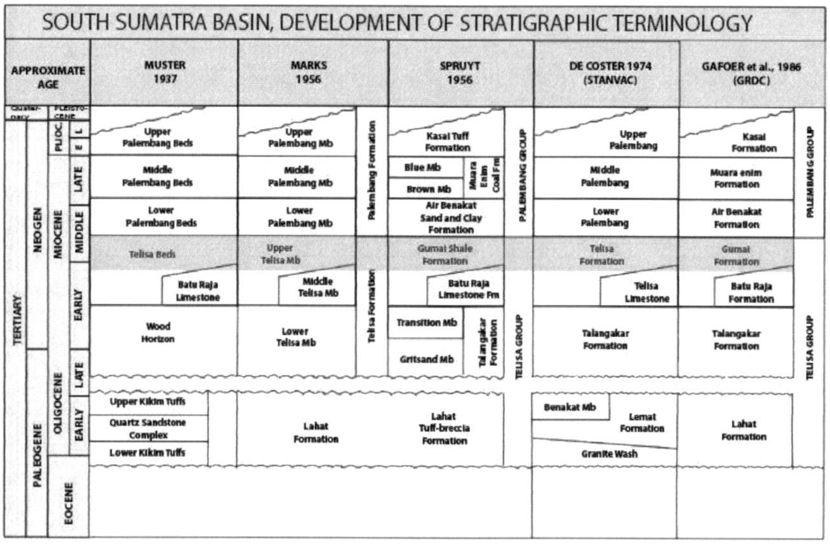

Figure 2. Stratigraphy of South Sumatra Basin [17]

3. Method

The research method in this study includes field data collection in Muara Dua and Simpang region (Figure 3). In Muara Dua and Simpang, field data collection was spotted sampling and systematic sampling in Gumai Formation. Field data collection followed the previous study [18], which is considered a significant component in the clay matrix of the Gumai Formation.

4. Field Data Mapping

Field data collection of Gumai Formation are retrieved in 15 locations at Simpang and Muara Dua region, 2 locations of Gumai Formation are in Muara Dua region, and 13 sites are in Simpang, South Sumatra (Figure 3). THMD-1 crops out greenish claystone, low calcareous, weathered claystone, calcite veins occurrences, strike/dip N N110°E/31°. THMD-2 consists of limestone and platy coral, strike/dip: N120°E/5°. THSP-2 composes calcareous sandstone, quartz sandstone with foraminifera content, strike/dip: N6°E/20°, located in Kembuha River. Outcrop of THSP-4 is characterized by calcareous sandstone in Kembuha River. THSP-5 forms an alternating between quartz sandstone and calcareous sandstone (Figure 4c) in Kembuha River. THSP-6 discovered in the small Kungkilan River consists of claystone with carbon content as fragments and in some places contains sand ball (diameter of 2.5-3 cm) and brownish amber fragments (Figure 4a,d). THSP-7 contains greyish claystone, non-calcareous, in the small Kungkilan River. THSP-8 and THSP-9 consist of greyish claystone, non-calcareous, and the outcrop occurs in the road to small Kungkilan River.

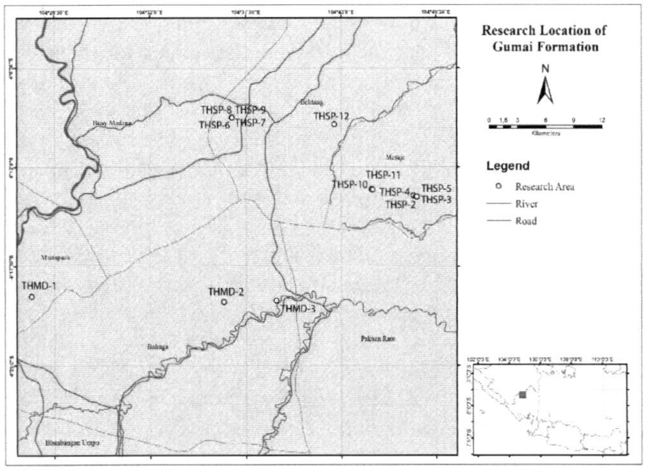

Figure 3. Research Study of Gumai Formation in Simpang and Muara Dua region

5. Discussion

The debate of the depositional environment of the Gumai Formation indicates three depositional settings of shallow marine, deep marine, and transition between marine and terrestrial [2,8,9,10,12,26]. Gumai Formation from West Ogan Komering Block shows the different depositional environment of middle-outer Neritic in JTB-1 well and outer neritic in NRM-1 well based on benthic foraminifera assemblages [9]. The difference is controlled by different structural levels in the South Sumatra Basin. Gumai Formation is most likely deposited on the remnant of local highs formed by the extensional faults [9]. However, further analysis of benthic foraminifer of Gumai Formation should be prepared to constrain better the paleodepositional environment of the Gumai Formation [9].

Other studies mentioned Gumai Formation in South Sumatra Basin was deposited in a deep water environment [10,12,14,26]. Gumai Formation from South Palembang Sub-basin shows the depositional environment of deepwater environment based on well log and seismic analysis [10]. Gumai Formation occurred in the transgressive marine period during the early to middle Miocene and was deposited in the deep marine environment above syn-rift grabens [12,26]. Gumai Formation consists of marine shale, claystone, marls, and fine-grained sandstone, reflecting open marine facies [12]. During great abundance and maximum diversity of planktonic foraminifera, calcareous nannofossil, and benthic foraminifera, Gumai Formation was deposited in a deep water environment [14].

Gumai Formation in Sopa-1 well, South Sumatra Basin is deposited in transition environment [2]. The interpretation is based on geochemical analysis of samples mass chromatogram, which provides a low value of diasterane (terrestrial source) and low values of oleanana (marine origin) [2]. Gumai Formation was deposited in regressive distal delta front as a result of increased sediment load from the Sunda landmass to the northeast; the regressive cycle was periodically interrupted by subtle transgressive events comprising distal delta front shale, distributary mouth bar sands, delta bar sand, channel sands, and interdistributary shale [8]. The process of Gumai Formation deposition in transition between marine and terrestrial initially caused by uplift of Sunda Shield to the east in the Middle Miocene marked the end of the early Tertiary transgression sequence and the beginning of regression sequence that continues to the present day [8].

Previous studies above indicate that the depositional environment of the Gumai Formation could be controlled by the active extensional fault [9]. The fault produced grabens [12,26] that provide Gumai Formation deposited in that tectonic basin. Probably, the spatial distribution of Gumai Formation in low resolution provided debate occurred. Therefore, additional information on Gumai Formation from the new region could strengthen the interpretation of the Gumai Formation environment. Gumai Formation discovered in Simpang composed of siltstone, calcareous sandstone, quartz sandstone, claystone and some places found sandstone with sand ball component in siltstone. Gumai Formation in Muara Dua consists of weathered claystone, and in some areas, quartz veins occurred. Based on data obtained, Gumai Formation in Muara Dua and Simpang indicate Gumai Formation occurred in the flange of Pre-Tertiary rocks. Gumai Formation in Muara Dua and Simpang are likely deposited in transition environment into shallow marine. Gumai Formation in Simpang has shallower marine environment than Gumai Formation in Muara Dua. Illustration model of Gumai Formation in Simpang occurred in lower Gumai and Muara Dua located in the upper Gumai (Figure 5). In lower Gumai, claystone outcrops in THSP-6 – THSP-12 in Simpang were likely deposited in the deltaic system, especially in prodelta due to sand ball, amber, and wood remains fragments in claystone. Transport of a significant part of silt and clay-size material of the river load with abundant plant fragments is typical lithology in prodelta [27]. Non-calcareous characteristics of claystone in Simpang area strengthen little influence marine input in the sediment of THSP6-12. Markers of calcareous sandstone and quartz sandstone in THSP-2 – THSP-5 in Simpang indicate the outcrops probably were deposited as interdistributary channels in the delta plain. In upper Gumai, THMD-1 – THMD-3 in Muara Dua are characterized by low calcareous claystone, and limestone indicates the sediments could be deposited in a shallow sea. However, this interpretation still need to be explored further by micropaleontology analysis quantitatively to strengthen depositional of transition and shallow marine environment of Gumai Formation in Simpang and Muara Dua

Figure 4. Sediments of Gumai Formation in Simpang. A, D. Sand ball fragments in siltstone at small Kungkilan River (THSP-6), B. Alternating quartz sandstone and calcareous sandstone in Kembuha river (THSP-5), C. Claystone in Way Urai River, Way Hering Village.

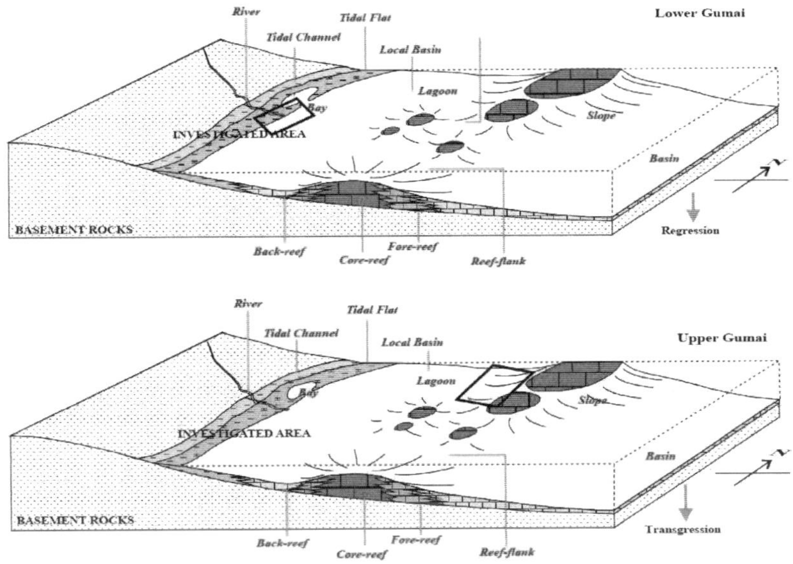

Figure 5. Model illustration of Gumai Formation from modified figure [14] in South Sumatra Basin, A. depositional environment of transition environment in Simpang, B depositional environment of transition environment in Muara Dua.

6. Conclusion

Gumai Formation discovered in Simpang composed of siltstone, calcareous sandstone, quartz sandstone, claystone and some places found sandstone with sand ball component in siltstone. Gumai Formation in Muara Dua consists of weathered claystone and quartz veins occurrence. Gumai Formation in Simpang has shallower marine environment than Gumai Formation in Muara Dua. In lower Gumai, claystone in Simpang was likely deposited in the deltaic system, especially in prodelta due to sand ball, amber, and wood remains as fragments in claystone. Non-calcareous characteristics of claystone in Simpang area strengthen little influence marine input in the sediment at Simpang. Other calcareous sandstone and quartz sandstone attributes in Simpang indicate the outcrops probably were deposited as interdistributary channels in the delta plain. In upper Gumai, low calcareous claystone and limestone in Muara Dua indicate the sediments could be deposited in a shallow sea. However, further analysis, including pollen and benthic foraminifera analysis, should be taken to strengthen the interpretation of transition until the shallow marine environment of the Gumai Formation in Baturaja region.

7. Acknowledgement

The author acknowledges the Alexander von Humbolt Digital Cooperation program to Sri Yudawati Cahyarini ref. no: 3.4-IDN/1158893 for ISEST 2021 conference payment. The author thanks Director of Research Center for Geotechnology for research support. Author thanks Munasri, Tri Hartono, and Nyanjang for their help in the fieldwork.

8. References

[1] Sarjono S, Sarjito 1989 *Proceedings Indonesian Petroleum Association Eighteenth Annual Convention* p 427-428
[2] Patra D H, Noeradi D, and Subroto E 2012 *AAPG International Conference and Exhibition, 20125*
[3] Hartanto, Karsan, Widianto, Eko, and Safrizal 1991 *Proceedings Indonesian Petroleum Association Eighteenth Annual Convention* p 249-267

[4] De Coster G L 1974 *Proceedings of Indonesian Petroleum Association, 3rd Annual Convention* p 77-110
[5] Pardede R, Gafoer S, Budhitrisna N, Ratman N, Suwarna N, Amin, T C, Kusnama G, Burhan G, Baharudin, Mangga A S, and Hartono. *Peta Geologi lembar Baturaja, Sumatera*. Pusat Penelitian dan Pengembangan Geologi. Departemen Energi dan Pertambangan
[6] Van Marle R J 1987 *Marine Micropaleontology* (13) p 97-152
[7] Van Marle R J 1989 *Netherlands Journals of Sea Research 24 (4): 445-457*
[8] Chuanli X, Haofan M, Honggang L, Dongmei L, Xiuli Q, and Benzhong X 2007 *Petroleum Science 4*(2) p19–28
[9] Aswan, Abdurrachman M, Fitriana B S, Mustofa M F, Santoso, W D, Rudyawan A, Rahayu W D, Hamdani A, Rohiman Y, Oo T Z 2017 I*OP Conference Series: Earth and Environmental Science* p 1–9
[10] Fahrudin F, Najib N, Setyawan R, Dennis D, and Fahreza F 2018 *AIP Conference Proceedings* p 020003.1 - 020003.7
[11] Panggabean H, Santy L D 2012 *Jurnal Sumber Daya Geologi 22*(4) p 225–235.
[12] Doust H, Noble R A 2008 *Marine and Petroleum Geology 25* p 103–129.
[13] Bishop G M 2001 *South Sumatra Basin Province, Indonesia: The Lahat/Talang Akar-Cenozoic Total Petroleum System*: USGS Open-File Report 99-50-S p 22
[14] Pannetier W 1994 *Journal of Southeast Asian Earth Sciences 10* (3) p 143–157
[15] Kusnama, Mangga, S A, and Sukarna D 1993 *Geological Society of Malaysia* (33) 2017
[16] Pulunggono, A 1986 *Proc. of 15th Ann. Conv. Indonesian Petroleum Associatio*n. 187- 213
[17] Barber A J, Crow M J, and Milsom, J S 2005, (eds.), Sumatra: Geology, Resources, and Tectonic Evolution: *The Geological Society of London*
[18] Maryanto, S 2014 *Indonesian Journal of Geology 1* p 21–34
[19] Suhendan, A R 1984 *Proceedings Indonesian Petroleum Association Thirteenth Annual Convention* p 63-73
[20] Hutchison, C S 1996 *Southeast Asian Oil, Gas, Coal and Mineral Deposits*: Clarendon Press Oxford
[21] Moulds P J 1989 *Proceedings Indonesian Petroleum Association Eighteenth Annual Convention*, October, 1989 p 217-246
[22] Pulunggono A, Haryo S A, and Kosuma C G 1992 Proceedings *Indonesian Petroleum Association Twenty First Annual Convention* October 1992 p 339-360
[23] Gafoer S, Cobrie T, Purnomo J 1986 *Geologic Map of the Lahat Quadrangle (1012), Sumatra. Scale 1:250 000*. Directorate General of Geology and Mineral Resources, Geological Research and Development Centre, Bandung
[24] Marks P 1956 *Indonesia: Lexique Stratigraphique Internationale.* 3(7a). Congrs Geologique International-Comission de Stratigraphie. CNRS, Paris
[25] Musper K A 1937 *Geologische kaart van Sumatra: toelichting by blad 16 (Lahat), schaal 1:200 000.* Dienst Mijnbouw Nederlandsch Indie
[26] Sitinjak T Y, Tampubolon R A, Pratama A, Nusantara W P, Muis A, and Setiani, H 2020 *Berita Sedimentologi* (45) p 19–38
[27] Einsele G. 1991. *Sedimentary Basin: Evolution, Facies, and Sediment Budget*. Springer-Verlag Berlin, Heidelberg New York

IOP Conf. Series: Earth and Environmental Science 1047 (2022) 012011 doi:10.1088/1755-1315/1047/1/012011

Understanding the Oceanography Condition in Java Sea using Hydrodynamic Simulation for Patimban Port Development

Kholillah Y. Isnaeni[1], Hanif Diastomo[1*], Candidad A. D. S. Nusantara[2], Wiwin Windupranata[2]

[1]Oceanography Research Group, Faculty of Earth Sciences and Technology, Institut Teknologi Bandung
[2]Hydrography Research Group, Faculty of Earth Sciences and Technology, Institut Teknologi Bandung

*Corresponding author: hanifdiastomo@oceanography.itb.ac.id

Abstract. This study's purpose is to gain knowledge and information on the hydrodynamic condition in the Java Sea, North Subang Region. The data was compiled from ECMWF for wind data and TMD for tide constituent data as a hydrodynamic model generator and the field data taken from a field survey in 2020 and 4 tide station BIG (Jakarta (3) and Cirebon (1)). The software developed by Deltares (DELFT3D) was used for hydrodynamic simulation to simulate hydrodynamic conditions (water level and current). A structured rectangular grid was used to construct domain boundary model on the north coast of West Java Province. The bathymetry data from BATNAS (BIG) has been processed to produce a 100 m resolution to accommodate the high-resolution bathymetry data in the coastal areas, especially in Patimban Port. The result shows the water level fluctuation and current movement on the north coast of the West Java Sea during 2020. The model verification shows that the tidal phase already has a similar pattern, especially on the western side of the West Java Sea. RMSE values for stations in Jakarta Sea are 0.2, 0.04, 0.21, and 0.39 with coefficient correlation are 0.94, 0.91, 0.93, and 0.69 for Pondok Dayung, Sunda Kelapa, Kolinamil, and Cirebon respectively. The monsoon seasonal phenomena affected the current condition in the study area. However, the tidal model amplitude needs to be tuned in, so that it can fit with the field data. The study of the hydrodynamic condition is necessary for Patimban Port development which is currently under construction.

1. Introduction

The rapid development in coastal areas caused by the population growth and the need for logistics fulfillment has resulted in the need for infrastructure development such as ports that can accommodate these logistics needs. Currently in Indonesia, the construction of the Patimban Port is being carried out in the Subang area, West Java. This port was built to reduce logistics load by bringing production centers closer and connecting various productive areas in West Java [13]. Patimban Port is the second largest port after Tanjung Priok Port [13] and the most sophisticated port that will be supported by a modern maritime city called Patimban City [14].

The port development and the marine city near the coastal area naturally will have major impact on the oceanographic condition within the area. One of the oceanographic conditions that greatly affect activities at the port is the hydrodynamic conditions in the form of currents and water level. The influence of currents on shipping is very important especially in navigation in shallow areas, including ports and rivers during berthing operations [1]. Water level fluctuation in the port area also important for docking activity in the port. Therefore, hydrodynamics information in the port area is very important for port development and daily port operation.

In assessing hydrodynamic conditions, numerical simulations require wind and tidal inputs to generate water levels and currents. The hydrodynamic simulation research in Java Sea was done by several researcher. The previous research result found that the waters of the Java Sea are strongly influenced by monsoon winds which cause fluctuations in current conditions according to wind conditions (direction and magnitude). The research by [2] focused on Banten Bay, [3] and [4] focused on Jakarta Bay. In addition, [5] and [6] states that the tidal form in the Java Sea is dominated by tidal constants k1 originating from the Pacific Ocean and M2 originating from the Indian Ocean [5][6]0. A further study of tidal conditions in the Patimban area was carried out by [7] which applied a different use of bottom roughness and showed that bottom roughness is not a parameter that affects tidal significance.

In this study, we will conduct a hydrodynamic simulation in the waters of the Java Sea which stretches from Jakarta to Pemalang (Central Java) with the waters of Subang Region (West Java). The modeling will be carried out using DELFT-3D FLOW in 2020. This study aim is to understand the hydrodynamic condition in Java Sea, especially in Patimban port area, Subang Regency, so that this knowledge able to improve the oceanographic information which important for port construction, development and operation that can be effective and environmentally friendly.

2. Materials and Method

2.1. Data and Model Domain

The data used in this study (Table 1) are input data (bathymetry, wind, pressure, and tides) and data for verification (water level). Bathymetry data from BATNAS is processed to produce bathymetric data with a resolution of 100 m. Wind and pressure data are used as input values that change with time and act as generating forces throughout the domain. Tidal constant data is data on an open boundary that allows forcing from the boundary into the domain. Only 5 dominant and represent tidal constituents were used, namely K1, O1, P1, M2, S2. And the last one is tidal station observation data, which is water level observation data at 4 tidal observation stations (3 in Jakarta (Pondok Dayung, Sunda Kelapa, and Kolinamil), and 1 in Cirebon). The water level data is used to verify the model results using the RMSE (Root Mean Square) and CC (Correlation Coefficient) statistical methods. Observational data used for verification have the same range as the model's data (January-December 2020).

Table 1. Details of Data.

No.	Types of Data	Resolution	Source
1.	National Bathymetry (BATNAS)	6 arc-second (185 m)	Geospatial Information Agency (BIG) [8]
2.	Tidal Harmonic Constituents (K1, O1, P1, M2, S2)	1/6° (18.5 km)	Tide Model Driver (TPXO-9 Model) [10]
3.	Wind (u10 and v10) and Pressure	Spatial: 1/4° (27.75 km) Temporal: hourly	ERA 5 ECMWF [9]
4.	Tidal Station Observation Data		Geospatial Information Agency (BIG) [11]

Fig 1. Shows the constructed model domain which stretches from Jakarta to Pemalang (Central Java). It can be seen on the map that there are 4 tidal station observation points (3 in Jakarta and 1 in Cirebon) and 1 observation point in the area of interest, namely the Patimban areas (Subang, West Java). The reasons for the domain selection shown in Fig. 1 are intended to keep the area of interest

away from the boundary so that the area of interest will avoid errors that may occur from the boundary. In general, the error problem in the model is due to open boundary problem. Other reason the model domain is wide, because the model input from wind speed and pressure has big resolution which is less represent for small scale models. Therefore, for the physical process to be fulfilled, in this case using tides as the main driving force (long waves), it requires a large area. The final reason for choosing the domain is the availability of data for verification which in this case uses observational data from tidal stations whose locations are limited (Jakarta and Cirebon) and are not found in the area of interest. The red square in Fig. 1 indicates the area of interest which will be discussed further in section 3.

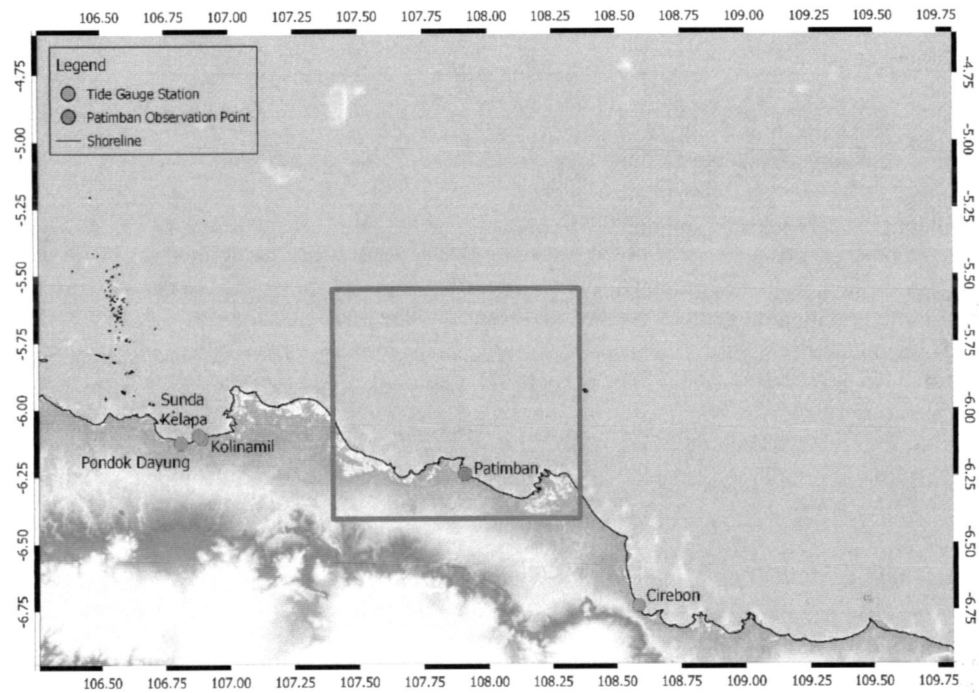

Figure 1. The domain models

2.2. Hydrodynamic Model Setting

2.2.1. Governing Equations. The hydrodynamic simulation in DELFT3D using FLOW module is a multi-dimensional hydrodynamics simulation program that calculates non-steady flow and transport phenomena that result from tidal and meteorological forcing on a rectilinear or a curvilinear boundary fitted grid [12]. For hydrodynamics simulation follows the basic equation of continuum (1) and momentum equation (3 and 4). The continuum equation can be shown in equation 1 [12].

$$Q = \frac{\partial \zeta}{\partial t} + \frac{1}{\sqrt{G_{\xi\xi}}\sqrt{G_{\eta\eta}}}\frac{\partial\left[(d+\zeta)U\sqrt{G_{\eta\eta}}\right]}{\partial \xi} + \frac{1}{\sqrt{G_{\xi\xi}}\sqrt{G_{\eta\eta}}}\frac{\partial\left[(d+\zeta)V\sqrt{G_{\xi\xi}}\right]}{\partial \eta} \quad (1)$$

Q shows the contribution per unit area due to the discharge or withdrawal of water, precipitation, and evaporation.

$$Q = H \int_{-1}^{0}(q_{in} - q_{out})\, d\sigma + P - E \quad (2)$$

Where q_{in} and q_{out} in equation 2 are the source and sink terms of water per unit volume. P and E are precipitation and evaporation, respectively. The momentum equation can be shown in Equations 3 and 4 [12].

$$\frac{\partial u}{\partial t}+\frac{u}{\sqrt{G_{\xi\xi}}}\frac{\partial u}{\partial \xi}+\frac{v}{\sqrt{G_{\eta\eta}}}\frac{\partial u}{\partial \eta}-\frac{v^2}{\sqrt{G_{\xi\xi}}\sqrt{G_{\eta\eta}}}\frac{\partial\sqrt{G_{\eta\eta}}}{\partial \xi}+\frac{uv}{\sqrt{G_{\xi\xi}}\sqrt{G_{\eta\eta}}}\frac{\partial\sqrt{G_{\xi\xi}}}{\partial \eta}-fv = -\frac{1}{\rho_o\sqrt{G_{\xi\xi}}}P_\xi+F_\xi+\frac{1}{(d+\zeta)^2}\frac{\partial}{\partial\sigma}\left(u_v\frac{\partial u}{\partial\sigma}\right)+M_\xi \quad (3)$$

$$\frac{\partial v}{\partial t}+\frac{u}{\sqrt{G_{\xi\xi}}}\frac{\partial v}{\partial \xi}+\frac{v}{\sqrt{G_{\eta\eta}}}\frac{\partial v}{\partial \eta}-\frac{u^2}{\sqrt{G_{\xi\xi}}\sqrt{G_{\eta\eta}}}\frac{\partial\sqrt{G_{\xi\xi}}}{\partial \xi}+\frac{uv}{\sqrt{G_{\xi\xi}}\sqrt{G_{\eta\eta}}}\frac{\partial\sqrt{G_{\eta\eta}}}{\partial \xi}+fu = -\frac{1}{\rho_o\sqrt{G_{\xi\xi}}}P_\eta+F_\eta+\frac{1}{(d+\zeta)^2}\frac{\partial}{\partial\sigma}\left(v_v\frac{\partial v}{\partial\sigma}\right)+M_\eta \quad (4)$$

2.2.2. Model Configuration. As previously mentioned, the model uses a structured grid with the same size except in the area near the shoreline. The area near the shoreline is gridded to have a different grid size because the grid conditions must follow the shoreline. The configuration of the model is shown in Table 2. The simulation duration is 1 year throughout 2020 (January - December) with a spin-up of 2 months (November-December 2019). The aims to set a spin-up time because the model is adjusted to the initial steady-state condition, which means that the water level and current conditions are 0, so that the model requires adaptation time to adjust the given force. Giving spin-up time is also one method to avoid blow-up in the simulation. The model used 1 minute for time step, but the model output is given every 120 minutes interval.

Table 2. Model Configuration.

No.	Types of Data	Resolution
1.	Time Step	1 minute
2.	Grid size	1 km
3.	Duration	1 year (2020)
4.	Forcing	Pressure and wind
5.	Initial condition	Steady-state (uniform value [water level 0 m])
6.	Boundary Condition	Astronomic (44 parts used 5 components: K1, O1, P1, M2, S2)
7.	Output parameters	Water level and current
8.	Output time interval	120 minutes

The model driving force used is only wind and pressure applied to all grids during the simulation with a resolution of 27.75 km so that this value will be interpolated every 1 km to meet the entire model domain which has a resolution of 1 km. At the open boundary, tidal constituent was given as generating force which divided into 44 sections (45 points) on the east, north, and west open boundaries (Fig. 2). Forty four sections were chosen because the data limit from the global model TPXO-9 which has a resolution of 18.5 km. While the TPXO-9 produces 15 components, this study only uses 5 tidal components because these 5 components significantly affect water level conditions in the study area. Such as wind and pressure data will be interpolated along the boundary grid because the input data grid is larger than the boundary grid.

Fig. 2 shows a visualization of the model configuration, which consists of depth data that has been adjusted based on the grid created. The bathymetry in model domain has a depth range of 99.3-4.9 m. In the model, 4 observation points are set as time series outputs for verification data (tide gauge station in Fig. 2). One observation point in the area of interest is useful for time-series analysis of the areas. The output from this hydrodynamic model is water levels and currents, both spatially in entire domain and in a time series for each point of view.

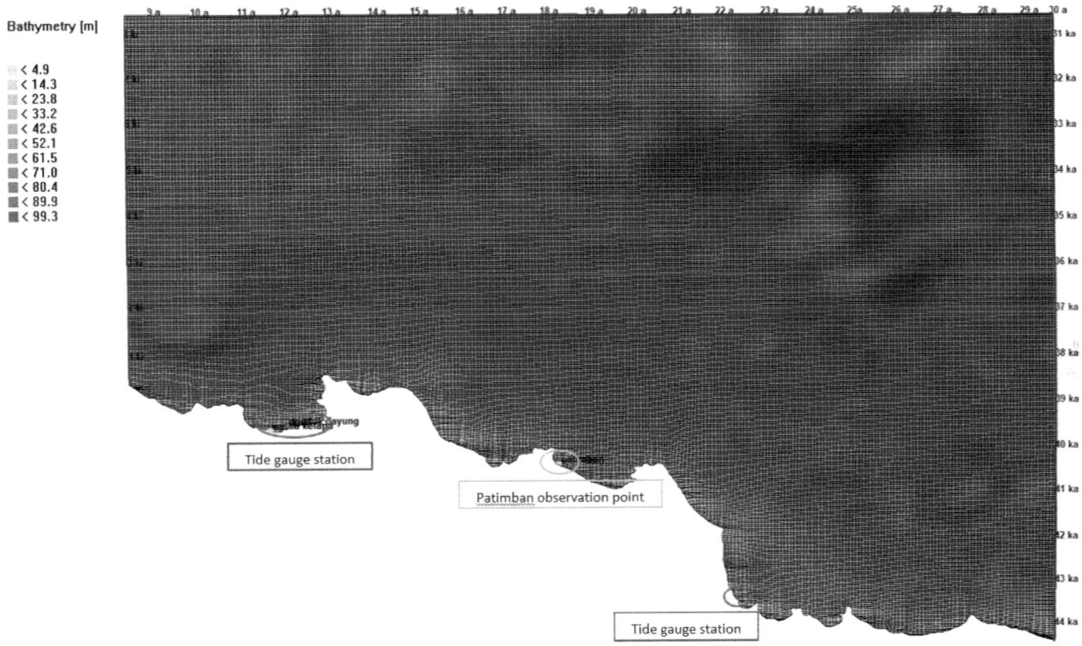

Figure 2. Model domain and model's output and input configuration

2.3. *Verification Method*

In this study, the field data (water level) collected from the BIG tidal observation station was used [13]. The verification method used is statistical methods that are RMSE and CC. RMSE was chosen because it shows how many errors in the model results indicate the difference between the model results and the field data. While CC is used to see the correlation between field data and model data. The formulation of RMSE (5) and CC (6) is as follows.

$$RMSE = \sqrt{\frac{1}{n}\sum_{i=1}^{n}(y_i - x_i)^2} \tag{5}$$

$$CC = \frac{\sum_{i=1}^{n}(x_i - \bar{x})(y_i - \bar{y})}{\sqrt{\sum_{i=1}^{n}(x_i - \bar{x})^2 \sum_{i=1}^{n}(y_i - \bar{y})^2}} \tag{6}$$

Where y_i is field data, x_i is model results, $\bar{y}$ is filed data average, and $\bar{x}$ model results average.

2.4. *Hydrodynamic Model Simulation Flowchart*

Fig. 3 explains the flowchart or algorithm to build this model starting from data collection, pre-processing, running, and post-processing. Pre-processing is a stage from data processing to making a mesh (grid). In data processing, wind and pressure data are processed into a file that can be used as model input. In addition, bathymetric data processing is also carried out and will use in the mesh manufacturing process. After getting bathymetry data for generated mesh, then create a grid and depth file (a mesh) according to the input format in DELFT-3D. The next process is to set up the model and run it. After running the model, verification and initial visualization are carried out to see the suitability of the model with field data or to ensure that there are no errors or blow-ups in the model results.

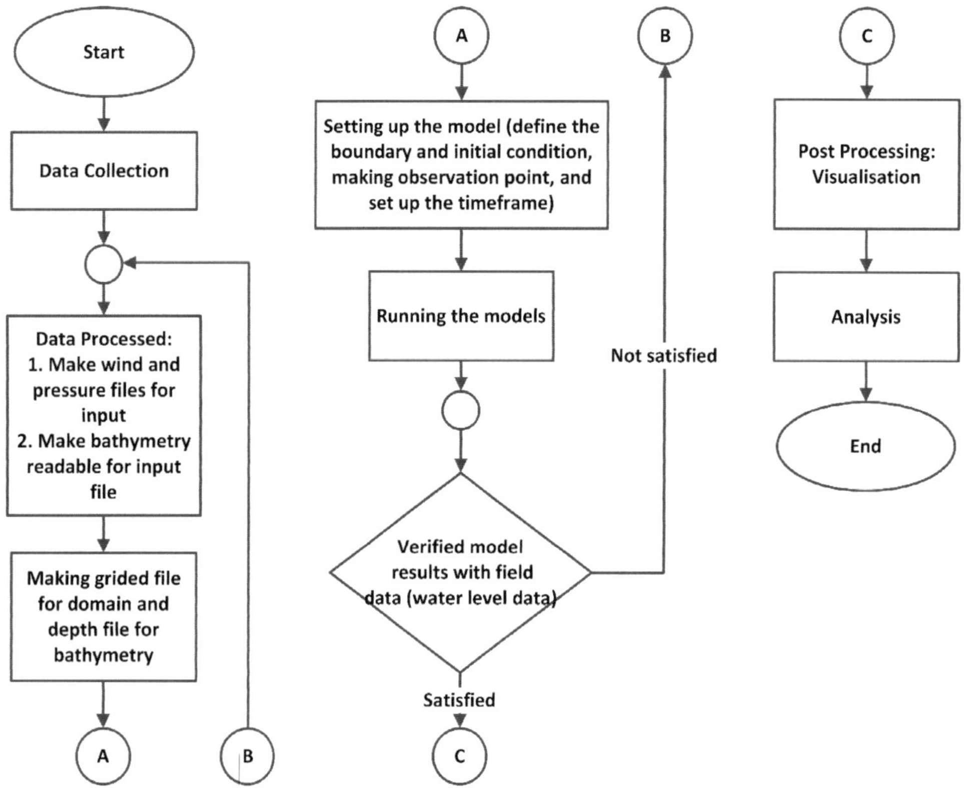

Figure 3. Hydrodynamic Simulation Flow chart

If the verification process is satisfactory, the model can be continued at the post-processing stage in the form of visualization of the model output (water level and current) which can then be continued with the analysis process. However, if the model is not satisfactory, then it is necessary to repeat the data processing this is because sometimes the model error or does not match the results because the input data is not in accordance with the model set up, besides that sometimes it is also necessary to re-create the existing grid, such as reducing or widening the domain area. This process can occur if there is a problem with bathymetric data or sometimes it is enough to reprocess the bathymetric data without having to change the grid to solve the problem. Basically, the main problem of the model usually occurs at the open boundary so that the model setup is the first thing to check if the model results do not match the field data.

3. Result and Discussion

3.1. Model Verification

The verification results show a fit match between the model and the field data, especially for the tide gauge stations located in Jakarta. The best RMSE value is in the Sunda Kelapa area (reaches 0.04) which indicates that the error is very small. From Fig. 4b can be seen that the water level pattern follows the tidal pattern from the field data with the appropriate amplitude. CC for the Sunda Kelapa area is also high, with a value close to 1 (0.91) which means that the relationship between the model and field data have positive correlation (if the value of the field data increases, the model value also increases, if it decreases, the model value will also decrease). These results mean that the model results and field data are all in phase with an RMSE value close to 1.

From the 4 review points, all of them show a tendency to overestimate model values, especially at the Cirebon observation point, which has a large elevation difference between the model and the field data, marked by a fairly large RMSE value, which is 0.39 but still acceptable. In phase, the results in Cirebon still follow the pattern of fluctuations in the field data even though the CC is only 0.69. This fairly high RMSE value is due to bathymetry and coastline problems at the Cirebon Review point. The shoreline is complex and the bathymetry data in Cirebon is steep. In other hand, the review point in the model domain located on the land area which resulted inaccuracy of tidal data.

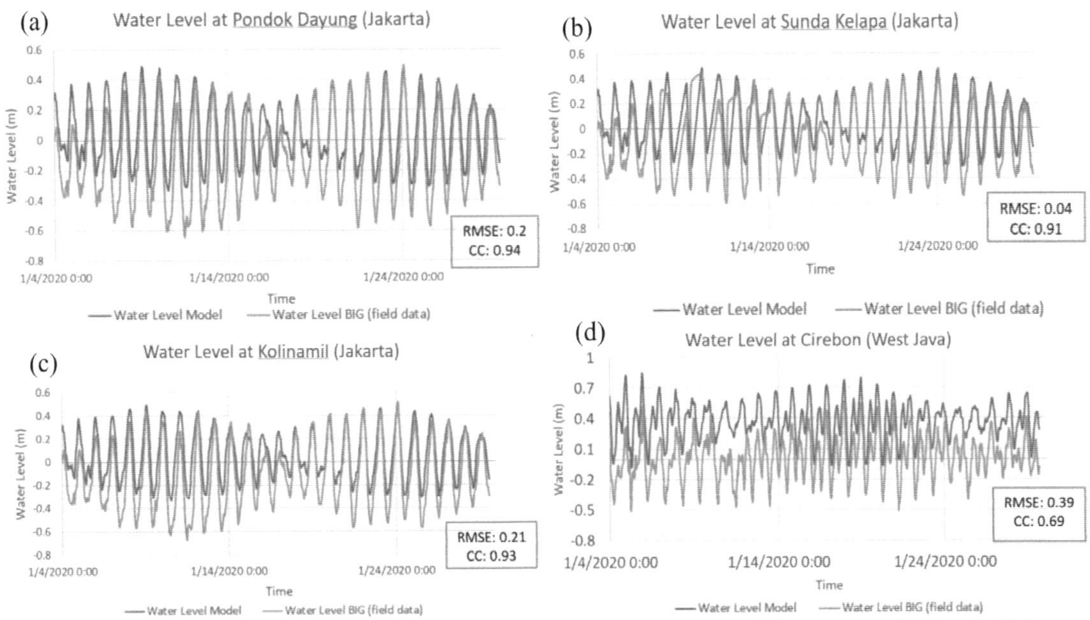

Figure 4. Comparison between model result with field data at (a) Pondok Dayung, (b) Sunda Kelapa, (c) Kolinamil, and (d) Cirebon for water level data

In this study, only 5 tidal components were used (K1, O1, M2, S2, and P1) however, the trend of the model results (water level) was higher than the field data, this indicates that not all tidal components affect the oceanographic conditions in a local area. Tides have many long periods components which values are not representative to be used for models with short simulation durations such as those carried out in this study. Before choosing this model, the author has tried to use 15 tidal components according to the number of TPXO-9 model outputs, but the model results have a very high RMSE and cannot represent the conditions in the field. it can be concluded with the many assumptions used in the model to simplify natural events in numerical calculations. It is necessary to select the most optimal input so that the model can represent real conditions with assumptions that limit it.

3.2. Currents

As mentioned before, ocean currents in the Java Sea area are influenced by the moonsun wind which changes direction seasonally. The model result were differentiate into 4 results to represent moonsun seasons. The output figures namely DJF (December, January, February) or west season represented by conditions on January 15 at 12.00 WIB, MAM (March, April, May) or transition 1 represented by April 15 at 12.00 WIB, JJA (June, July, August) or western season was represented by July 15 at

12.00 WIB, and SON (September, October, November) or transition 2 was represented by September 15 at 12.00 WIB. Due to the influence of the wind, current conditions will also change with seasons alterations. The wind at DJF has a movement to the northeast from the southwest which indicates the west monsoon is blowing. The pattern of movement towards the northeast is clearly visible in the offshore area, while in the coastal area it tends to go north first, especially in the eastern area of Subang (Fig. 5a.). This can be explained because the tendency of the wind to lead to the offshore area so that in the eastern area of Subang in the form of a bay the wind will move north towards the offshore and after leaving the coastal area it will be influenced by the wind that leads to the northeast. Meanwhile, in transitional season 1 in the entire study area, the wind is directed to the north. The movement to the north is due to a change in the direction of the monsoon wind, which will slowly change to the northwest from the southeast. If we compare the wind magnitudes at DJF and MAM, it can be seen that the magnitude is larger during JJA. This can be explained because, during the transition season, it seems as if the wind is weakening due to a direction shifting, which in this case is shifting wind blow origin. (initially from Asia, then reversed from Australia).

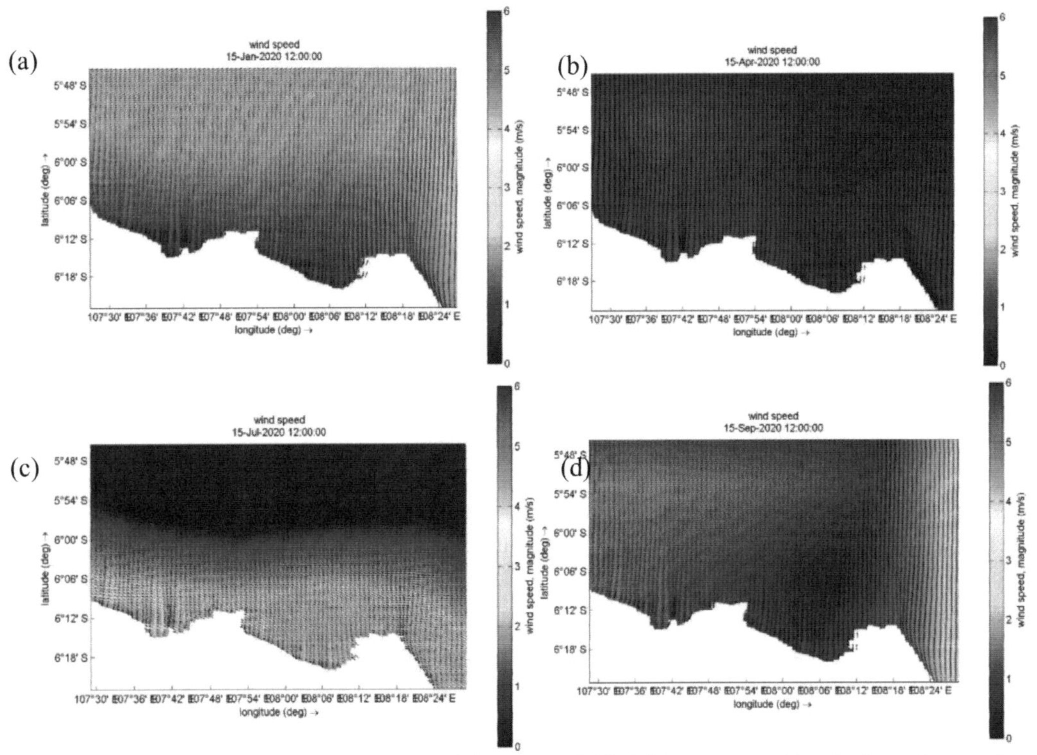

Figure 5. Wind data at (a) DJF, (b) MAM, (c) JJA, and (d) SON seasons. Simulation taken at January-December 2020.

When JJA season, the wind heading to the northwest on the coastal area while heading to west at the offshore area. This pattern is slightly different when the DJF in the bay area the wind is still northward. This is because when southeast monsoon in JJA wind blows from Australia so that in the study area it seems the wind is moving from land to coast with a tendency towards Northwest. In this season, the wind has the highest magnitude compared to other seasons, the magnitude in the Patimban area reaches 3.5 m/s. In transitional season 2, the wind returns to the north, but in the right side of the domain, the wind begins to move to the northeast. This indicates that the west monsoon has begun to dominate again. As the case during transition season 1, the wind magnitude also weakened compared to when JJA.

The occean current pattern occassionally different every month compares with wind pattern. The current during DJF has a high magnitude with a dominant direction to the east. At MAM the current has reversed completely towards the west with a magnitude larger than when DJF. When JJA the current direction is the same as MAM, which is to the west but the magnitude is lower than MAM. In the SON season, the current velocity is the lowest correspondin to the to the lowest wind speed magnitude, and the direction of the current in SON is to the east with the speed about 0.02 m/s. The current pattern which is not in accordance with the wind pattern indicates there are others force that affects the oceanographic conditions in the study area. However, the wind is one of the dominant forces in the area, especially in determining the direction of the current which looks more in line with the wind direction each season when compared to its magnitude.

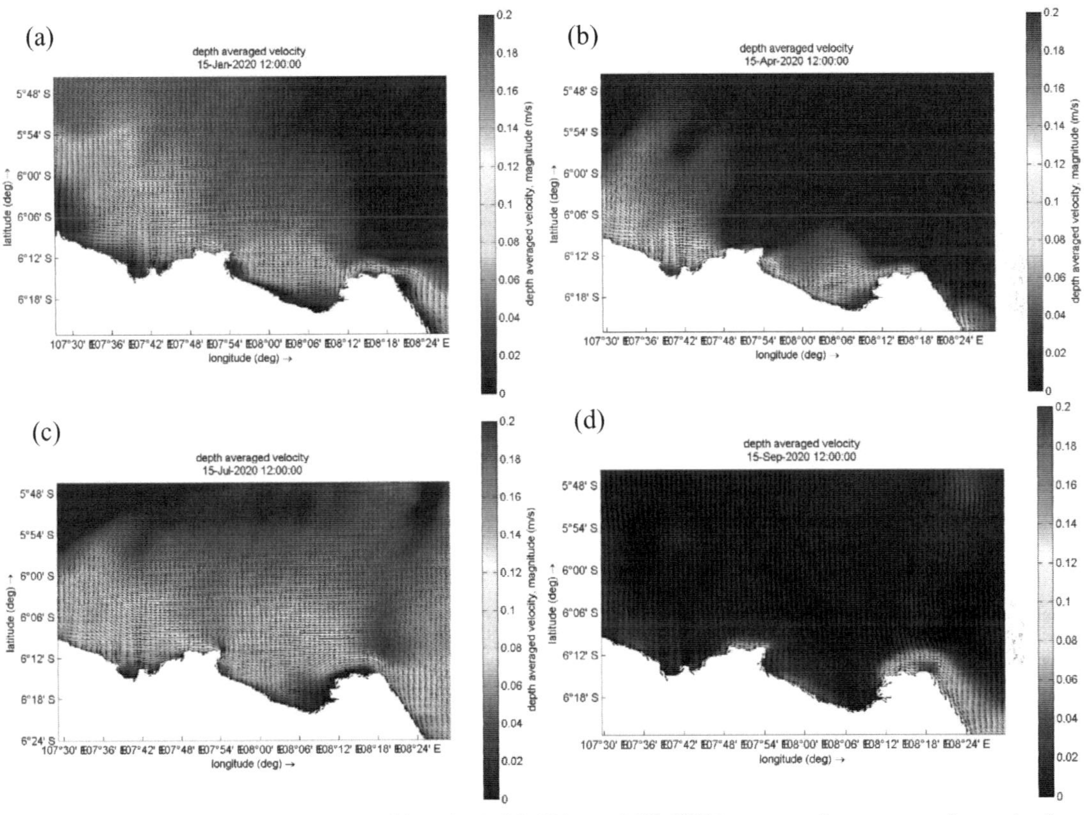

Figure 6. Model results at (a) DJF, (b) MAM, (c) JJA, and (d) SON seasons for currents (magnitude and direction). Simulation taken at January-December 2020.

3.3. *Water Level*

The water level turns out to have seasonal patterns such as wind and current which means that the wind also affects the water level condition. In Fig. 7a when the DJF season water level is low with the range about 0.1 m. This corresponds to the wind magnitude but does not correspond to the current speed because when DJF current has a high speed like other seasons. At MAM the water level rose in the range of 0.3 m or increased 3 times than DJF season. When viewing the wind pattern, when MAM wind speed decreases compared to DJF, the water level should also be lower because the wind speed decreases. However, when the MAM has the highest current magnitude compared to other seasons, it is assumed that in MAM the influence of currents is more dominant than wind events. So, it can be understood that there is a pause between the occurrence of wind, current formation, and water level. In Fig. 7c shows the condition of the water level when JJA which has the highest elevation. The water level is still considered reasonable and not too dangerous for port operational such as docking process because the highest water level is only up to 0.6 m. Furthermore, the water level of SON (Fig. 7d) decreased from that of JJA.

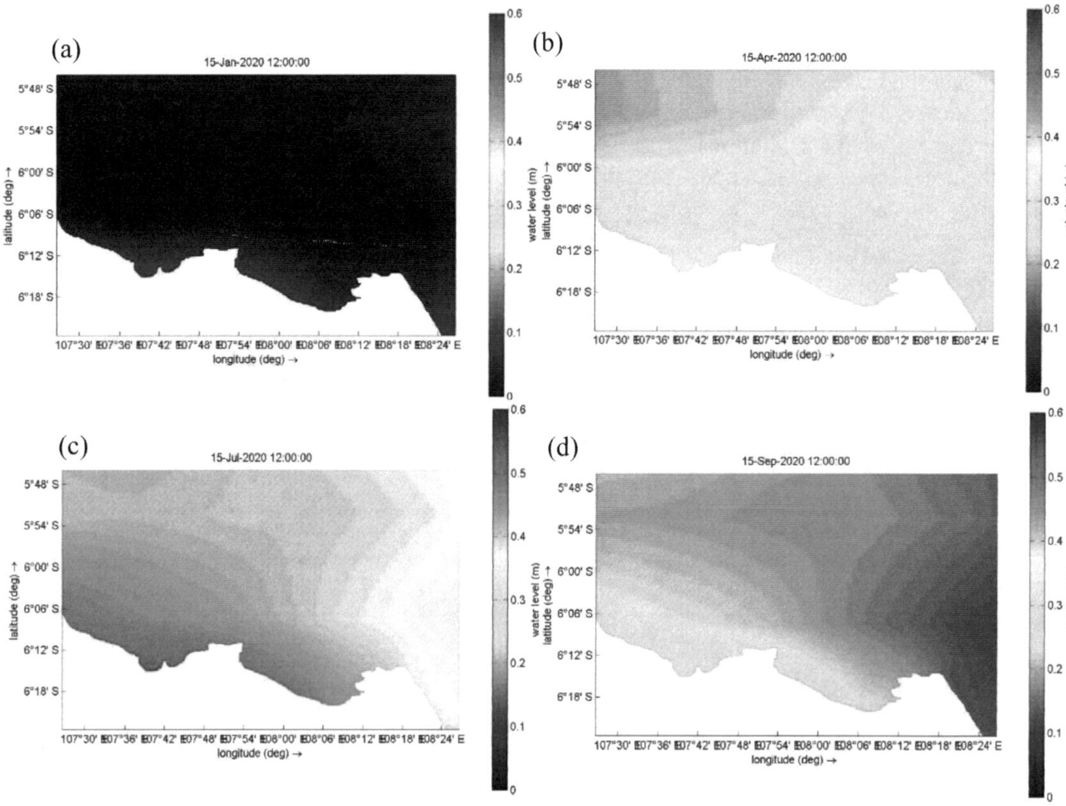

Figure 7. Model results at (a) DJF, (b) MAM, (c) JJA, and (d) SON seasons for water level. Simulation taaken at January-December 2020.

Looking at the pattern that occurs at the water level, it can be concluded that the wind and current affect the water level fluctuation. However, it is not the only things that affect the tidal elevation which be seen with the different patterns between the water level, the current and the wind. In this case, the water level has a more continuous pattern where at the beginning of the year (DJF) with weak wind conditions it causes a slight increase in water level, but when winds became stronger, the water levels get higher. When the wind weakens in the transition season the water level is still high because the wind still does not affect the water level (lag). When the JJA water level is rose due to the strengthening of the wind speed significantly so that the effect of decreasing wind in the previous month is not depicted for this seasonal visualization. Furthermore, after the wind reaches its maximum speed in 1 seasonal cycle (monsoon), the wind will weaken and result in a lower water level during SON. Water elevation and current conditions information is very important in port operations because it is related to safety and stability on ships of various sizes, so the response to oceanographic conditions in the form of water levels and currents will be different.

4. Conclusion and Suggestion

The hydrodynamic simulation carried out in this study can represent the hydrodynamic conditions in the study area because it has a high CC in the range of 0.9 (close to 1) and a good RMSE in the range of 0.2 (close to 0). It is evident that current and water level conditions are influenced by seasonal winds and tides, but seasonal analysis shows that wind dominates seasonal patterns of currents and water level in the study area. The highest current velocity occurs when the MAM season which reaches 0.18 m/s and the lowest occurs when the SON (0.02 m/s). The highest water level occurs at JJA (0.5 m) and the lowest occurs at DJF (0.1 m). Further research is needed with a better resolution, one way is by using a nested grid or using an unstructured grid. Researchers who focus more on the port area and can capture the existing conditions of the port can accommodate a good understanding of the daily patterns of tides that occur in the study area, so the timing of ebb, flood, and slack water can be determined. It determines the best conditions for loading and unloading activities. Other operations, both based on seasonal and daily conditions from the hydrodynamics of the Patimban Port.

Acknowledge

This publication is resulted from a Collaborative Research Scheme funded by the Faculty of Earth Sciences and Technology ITB 2021.

This research was a part of the project titled "Marine Science & Technology Cooperation between Korea and Indonesia (20180319)" and "Ocean and Coastal Basic Survey and Capacity Enhancement in Cirebon, Indonesia (G52440)" funded by the Ministry of Oceans and Fisheries, Korea.

References

[1] Akbar, Z.A., D.G. Pratomo, A.K. Negara & N. Riyadi. (2018). Identifikasi Objek Berdimensi Kecil Menggunakan Sapuan Multibeam Echosounder. J. Chart Datum 4(2): 127-136.
[2] Ahmad, B., U. J. Wisha, & D. Surinati. (2018). Modeling Tidal Current of Banten Bay During Transitional Monsoons 2015-2016. *Jurnal Segara*, 14(2).
[3] Maryan, Y. S., W. S. Pranowo, D. Ardianto, A. Kurniawan, N. B. Sukoco & I Made J. Astika. (2021). Modeling The Hydrodynamic Characteristics of Tidal and Monsoonal Currents in Pondok DayungPort of Tanjung Priok Harbor, Jakarta. *IOP Conf. Series: Earth and Environmental Science*, 925(2021).
[4] Ardiansyah, I. (2020). *Pemodelan Spektrum Energi Gelombang Berdasarkan Variasi Tinggi Pasang Surut di Kawasan Reklamasi Teluk Jakarta (Studi Kasus: Kejadian Badai Hagibis Tahun 2007)*, [M.Sc. thesis]. Institut Teknologi Bandung.
[5] Koropitan, A. F. & M. Ikeda. (2008) Three-dimensional modeling of tidal circulation and mixing over the Java Sea. *J. Oceanogr.* 64, 61-80.
[6] Yusuf M. & T. Yanagi. (2013). Numerical modeling of tidal dynamics in the Java Sea. *Coast. Mar. Sci.* 36, 1-12.
[7] Nuraghnia, A., W. Windupranata, A. R. Hakim, C. A. D. S. Nusantara, & Poerbandono. (2021). Modeling of Tide in the Java Sea Coastal Area Between Jakarta and Cirebon, Indonesia: Bathymetric Data Source and Sensitivity Tests due to Bottom Roughness and Boundary Condition. *IOP Conf. Series: Earth and Environmental Science*, 777(2021).
[8] BIG. (2018). Batimetri Nasional. http://tides.big.go.id/DEMNAS/index.html
[9] ECMWF. (2020). ERA5 Hourly Data on Single Levels from 1979 to Present. https://cds.climate.copernicus.eu/cdsapp#!/dataset/reanalysis-era5-single-levels?tab=form.
[10] OSU TPXO Tide Models TPXO Global Tidal Solutions https://www.tpxo.net/global.
[11] BIG. (2020). *Data Pasang Surut Tide Gauge Sunda Kelapa, Kolinamil, Pondok Dayung, dan Cirebon.*
[12] Deltares. (2011). *Delft3D-FLOW User Manual.*
[13] Kementrian Koordinator Bidang Kemaritiman dan Investasi. *Kemenko Marves Meninjau Kemajuan Pembangunan dan Perkembangan Operasionalisasi Pelabuhan Patimban.* https://maritim.go.id/kemenko-marves-meninjau-kemajuan-pembangunan-perkembangan-operasionalisasi-pelabuhan/ (accessed 20.12.21).
[14] Pemerintah Daerah Provinsi Jawa Barat. *Patimban City Sokong Pengembangan Rebana Metropolitan.* https://jabarprov.go.id/index.php/news/40852/_Patimban_City_Sokong_Pengembangan_Rebana_Metropolitan (accessed 21.12.21).

IOP Conf. Series: Earth and Environmental Science 1047 (2022) 012012 doi:10.1088/1755-1315/1047/1/012012

Registration Strategy of Handheld Scanner (HS) and Terrestrial Laser Scanner Integration for Building Utility Mapping

Ratri Widyastuti*, Asep Yusup Saptari, Arif Rahman

Department of Geodesy and Geomatics, Surveying and Cadastre Research Group, Institut Teknologi Bandung, Bandung 40132, Indonesia

***Correspondence Author:ratri@itb.ac.id**

Abstract. Currently, the need for a 3D model that represents the existing condition of a building is needed, especially for building management. While performed the 3D mapping of building utilities such as pipelines in buildings, the Terrestrial Laser Scanner (TLS) technology couldn't cover the object in the ceiling, so Handheld Scanner (HS) was used to complete the scanned objects. The purpose of this article is conducting the process of mapping between HS and TLS for building pipelines. The integration between two different sensor resolutions and sensor scan coverage requires a separate strategy to generate 3D models for pipeline objects in building. The use of specific target and 3D transformation method could do the registration between two set of point clouds which was obtained by two different technologies. The target object, such as spherical object which is usually used as a tie point between scan results in the registration process, cannot be recorded by HS. The specific targets were pasted on the pipelines in order to be captured by the HS. The specific targets become tie point in the registration process. Meanwhile, the registration process using the Iterative Closest Point (ICP) algorithm cannot be carried out because the two scan results do not meet the overlap percentage standard, so that the 3D transformation would be used in registration process. The result is that the registration accuracy is 0.062 m.

1. Introduction

Building Information Modeling (BIM) is a digital form of building information including the assets that contain three components; technology, processes, and digital information that can be used to improve the operation of assets in buildings [1]. The digital forms can be either two-dimensional (2D) form or three-dimensional form (3D). Generally, the 3D form is used in Architecture, Engineering, and Construction (AEC) design [2]. In construction projects, BIM benefits not only during pre-construction and construction, but also in post-construction by taking the stage in operation control and maintenance scheduling information, replacement parts ordering information [3]. BIM development can be seen as data transformation from the virtual BIM world to the construction site [2]. Consequently low data quality and poor coordination between each user can lead to the inefficient of BIM utilization [4].

Laser technology can be used to provide the real-world data which can record a lot of point at the same time that represent the object of the real world to support the 3D BIM model with high geometry accuracy [5]-[6]. The laser scanning technology principally classifies in twofold based on the data acquisition: moving laser scanner and static laser scanner. The static laser scanner, called Terrestrial Laser Scanner (TLS), is tailored for engineering and monitoring purposes. Although TLS outperforms other technologies in producing 3D model of an object in its class, the level of accuracy depends on several aspects: coverage angle of scanning, the distance between the instrument and the scanned object, and material type of the object itself [7]–[9]. In its applications, TLS can monitor 3D objects by determining point changes in the structure and deformed shape of the structure [10]. However, the performance of TLS coverage in inspection and engineering activities is limited by its application design. In other words, it has difficulty to capture some shadow and narrow areas. Also, this function is compensated by using the Handheld Scanner (HS). Some variants of HS, have better resolution, and require less time for acquisition and processing, but need some improvements in texture compared to the camera acquisition, (e.g. Canon) [11]. However, the HS itself based on the structured laser.

The utility of the building is not very textured and located in the narrowed area, so the TLS couldn't scanned the utility completely. Therefore, the HS technology could cover the limitation of the TLS but another problems was occurred while processing the point cloud from both technologies. The overlapping point clouds was less than 30%, so that the cloud-to-cloud registration couldn't be performed. Another registration method could be used to perform both point clouds, called target-to-target registration, but another problem was occurred, the initial tie point from natural point couldn't be identified because the homogeneous object. Hence, this paper is describing how to overcome those problems.

2. Material and Method

2.1. Material

Data acquisition was done using TLS Topcon G*eodetic Laser Scanner 2000 (GLS-2000*) and HS *Mantis Vision F6 Smart.* The specification is shown in the Table 1.

Table 1. Instrument Specification

Specification	TLS	HS
Scanner Type	Pulsed precise scan technology	Structure Light Scanning
Sensor	Liquid 2-axis tilt sensor	CMOS (Aptina AR134 1.2 M)
Digital Camera	5 MP	1.3 MP
Spot Size	4 mm of 20 m	4 mm of 20 m
Scanning Rate	± 120,000 points/s	± 640,000 points/s
Accuracy	3.5 mm (in range 1 – 150 m)	0.1% - 0.2% of distance
Depth Resolution	3.1 mm; 6.3 mm; 12.5 mm; 25 mm; 50 mm	1 mm
Scanning Range (Target to Device)	> 350 m	0.5 m – 4.5 m

This research used 12.5 mm of depth resolution. The other instrument that used in this research is specific target. This specific target is used as tie point in the registration process of point clouds. The specific target shows in Figure 1. The material of specific target is vinyl paper. The size of the specific target is 7 x 7 cm.

Figure 1. Specific Target

2.2. Method

2.2.1. Scanning Process

Scanning using TLS was performed using two techniques, called fixed scan and free scan. Fixed scan is a scanning process that requires a minimum of two control points, which are as a standing point of an instrument and as a target for back sight measurements. By scanning with this technique, point clouds data will be obtained which can be defined directly in the SRGI-2013 coordinate reference system. The other scanning method, was used to scan objects without having to perform back sight measurements. This technique is more flexible because it does not require a control point as a reference for the placement of the instrument during scanning. That way, the instrument can be placed as needed so that it can reach the entire scanned object. However, the resulting point clouds data from fixed scan method are defined in a local coordinate system so that the geo-referencing process is needed in data processing. The TLS was used to scan by fixed scan and free scan method the object inside the building.

The principle of data acquisition using HS tends to be similar using TLS but the scanning process was done by hand. The other differences is that the data acquisition process was carried out continuously, the data received by the instrument can be directly displayed on the device connected to laptop which already has the *Echo Software*. This software can display scanned data in real time when data acquisition is carried out, so that users know if there are parts of objects that are not recorded. The resulting density values also vary, which is 1-4 mm depending to the scan distance from object to the instrument. The HS was used to scan only the utility object.

In order to get the point cloud in global coordinate, the control point inside the building must be placed and measure first as the reference point. Control points in the building were measured using resection method so that a new control point is obtained. In this case, additional control point inside the building called the *TB-01* was measured from three other control point outside building called *TK-10, B-01* and *B-02*. The illustration of measuring the additional control point shows in Figure 2.

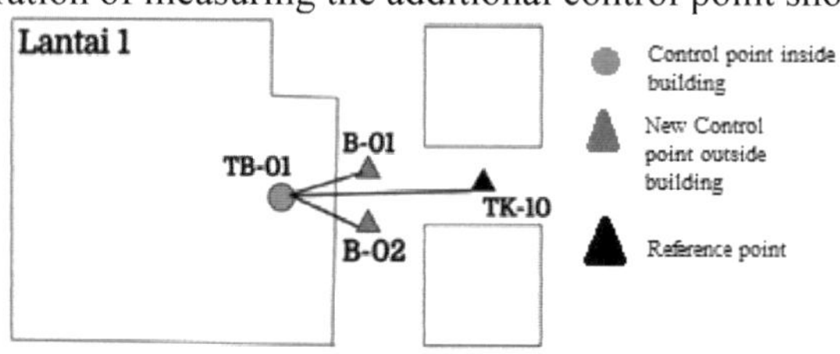

Figure 2. The illustration of measuring the additional control point

2.2.2. Processing Data

The basic processing was done by exporting the data from the device, data filtering and registration process. Exporting data from TLS was carried out using the *Magnet collage software*. This software can do the export data and define the coordinate of the resulted point cloud, in global or in local coordinate. The exported data was in e.57 format. The data filtering was done manually which discarded the unused point cloud. The common registration technique is described below [12]:

- target-to-target registration
  This registration uses targets (natural or artificial) as tie point during the data recording process. To be able to use target-to-target registration, a minimum of three targets from two scans are required to complete the six transformation parameters (three translation parameters and three rotation parameters), more targets will be better because it will be able to minimize errors by using least squares smoothing calculation.
- cloud-to-cloud registration
  This registration technique requires patching of two scans with a patch area of 30-40%. This registration technique uses the Iterative Closed Point (ICP) algorithm. The ICP algorithm builds a transformation matrix by finding a point-to-point correspondence between two point cloud datasets

and aligning the two point cloud datasets accurately through iterations [13]. This algorithm is used so that all scanned point cloud datasets have a minimum distance.

Basically, this research used the basic registration technique to align all point clouds into one, but as mentioned before in Section 1, the problem occurred when the overlap percentage between two dataset of point cloud from TLS and HS less than 30% so that the cloud-to-cloud registration technique couldn't be apply to register the point cloud from HS. The modification of registration process was needed in order to register the point cloud from HS into registered point cloud of TLS. The flow diagram of registration process shows in Figure 3.

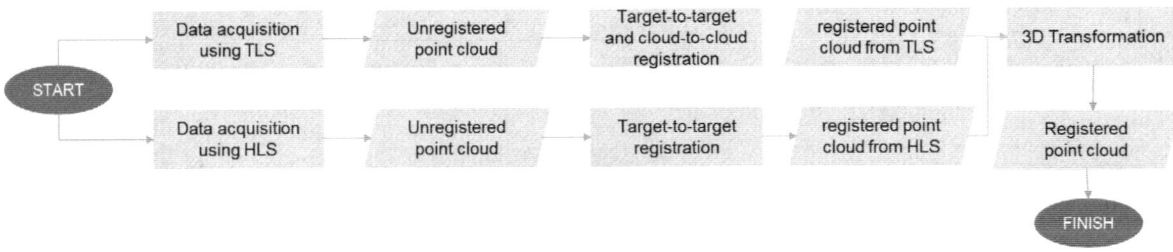

Figure 3. Flow diagram of registration process

Figure 3 shows that the raw data of point cloud from TLS was registered to the referenced point cloud using target-to-target and cloud-to-cloud techniques. The target-to-target technique was done by choosing the initial natural point to align all of the point cloud. The cloud-to-cloud technique was used to closing the gap between point clouds from each scanning. Afterwards, the other unregistered point cloud from HS was registered using target-to-target technique using initial artificial point (specific target). The last step was using the basic 3D transformation between each dataset of point cloud from HS and TLS by choosing the initial artificial point. The output of this research is the registered point cloud of utility of the building.

2.2.3. Data Validation

In order to analyse the quality of the registration, the calculation of Root Mean Square Error (RMSE) between coordinate of point cloud from TLS and HS is needed. Also the calculation of RMSE between coordinate of registered point cloud and referenced coordinate. The referenced coordinate was measured the artificial point (specific target) by Electronic Total Station (ETS). The closer the RMSE value to 0, the better accuracy would be produced.

3. Result and Discussion

The scanning result of point cloud from TLS and HS shows in Figure 4. The Figure 4 (a) shows the unregistered point cloud from TLS which shows the uncompleted scanned pipeline. Figure 4 (b) shows the detailed scanned pipeline by HS scanning.

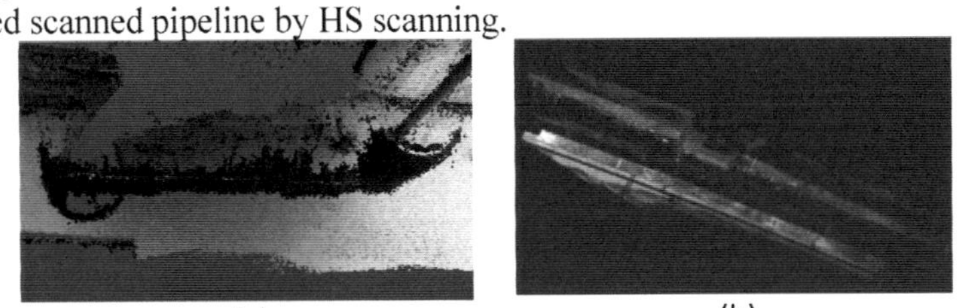

(a) (b)

Figure 4. (a) The unregistered point cloud from TLS; (b) the unregistered point cloud from HS

Figure 4 shows that the data form HS can complete the data form TLS with better resolution. While was doing the registration process of each point cloud dataset from TLS and HS, also produced the RMSE to analyse the quality of the registration process. The RMSE of each registration process of each dataset shows in Table 2.

Tabel 2. The RMSE of each registration process of each dataset

RMSE registration of TLS dataset	RMSE registration of HS dataset
0.002 m	0.002 m

Table 2 shows the satisfactory result for each dataset registration of point cloud. Because of the registration processed of point cloud from HS used the artificial target, this means the artificial target is good enough to help identify the tie points between two point cloud in different scanning position. The point cloud of artificial target shows in Figure 5.

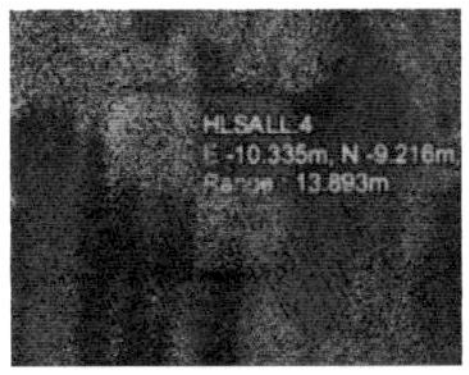

Figure 5. The point cloud of artificial target

Afterwards, the registration process was conducted to align two dataset of point cloud from different instrument. The RMSE was calculated from the residue of the deviation coordinate of artificial target between each dataset. The residue calculation shows in Table 3.

Table 3. The deviation coordinate of artificial target between each dataset

Points	E (m)	N (m)	H (m)	Residue-E (m)	Residue-N (m)	Residue-H (m)
1	788431.007	9237812.810	775.712	0.004	0.005	-0.042
2	788431.289	9237812.846	775.711	-0.001	-0.005	0.044
3	788431.001	9237821.770	775.723	-0.002	0.006	-0.074
4	788431.269	9237821.720	775.721	-0.001	-0.006	0.068
5	788431.008	9237827.796	775.741	-0.001	0.008	-0.066
6	788431.274	9237829.286	775.744	0.003	-0.006	0.070

The quality of registration shows that the RMSEx is 0.002 m, RMSEy is 0.006 m, RMSEz is 0.534 and the RMSE is 0,054 m. The big error occurred in the vertical position. Even though the RMSE of each registration in Table 2 shows satisfactory result, the RMSE of registration between two dataset of point cloud isn't good enough. This means, the handling processed when using the HS was stable for each scanning position but in one scanning position produced an unequal distance so that affects big error in vertical position. This could happen because of the pipeline was located in the ceiling. The result of this research is shown in Figure 6.

Figure 6. The point cloud of the building utility

Meanwhile the analysis of the registration result is discussed above, the position accuracy of the registered point cloud must be also analyse. In order to analyse the position accuracy, the comparison coordinate between artificial point from the registered point cloud and the measurement from ETS must be checked. The RMSE is 0.093 m. The position accuracy is not satisfactory enough because the RMSE is greater than the depth resolution of the instrument. Nevertheless, the HS can complete the incomplete utility scanning by TLS with better resolution.

4. Conclusion

The performance of TLS to coverage building utility mapping is limited. In other words, it has difficulty to capture some shadow and narrow areas, especially when the utility is located in the ceiling. This problem is compensated by Handheld Scanner (HS) to complete incomplete scanned with better resolution. However, the handling process of operating HS is the main factor to determine the result of registration between two different resolutions of scanners. Moreover, the position of each device to capture the specific target as tie point also important.

The registration process could be done by using the target-to-target and cloud-to-cloud registration for each dataset of point cloud but the overlapping area between two dataset is less than 30% causing the cloud-to-cloud registration couldn't be used to register two dataset of point cloud from TLS and HS. In order to overcome the problems, the registration process could be done by registering each set of point clouds from different technology and then do the registration between two sets of point cloud using 3D transformation method.

The RMSE of registered point cloud is 0.054 m which is influenced by big error in vertical position even though the RMSE of registered point cloud of each dataset of point cloud is on millimetre scale. This indicates that the handling process does not only focus on the stability of the use of HS but also must have stability in the vertical line, which mean maintenance the same distance between the sensor and object in every scanning position is also important. To be able to maintain the same distance, the scanning area coverage can be narrowed. Future works could be focuses on the other cases to get the better result of the vertical accuracy.

5. References

[1] S. Mordue, P. Swaddle, and D. Philp, *Building Information Modeling For Dummies*. Chicester: John Willey & Sons, Ltd, 2015.

[2] R. Sacks, C. Eastman, G. Lee, and P. Teicholz, *BIM Handbook A Guide to Building Information Modeling for Owners, Designers, Engineers, Contractors, and Facility Managers*, 3th ed. wiley, 2018.

[3] N. W. Aw and M. A. O. Mydin, "Potential and benefits of building information modeling (BIM) during pre-construction, construction and post construction stage," *Analele Univ. "Eftimie Murgu,"* vol. 24, no. 1, pp. 429–438, 2017.

[4] H. Y. Chong, J. Wang, W. Shou, X. Wang, and J. Guo, "Improving quality and performance of

facility management using building information modelling," *Lect. Notes Comput. Sci. (including Subser. Lect. Notes Artif. Intell. Lect. Notes Bioinformatics)*, vol. 8683 LNCS, no. September 2014, pp. 36–43, 2014, doi: 10.1007/978-3-319-10831-5_6.

[5] R. Staiger, "Terrestrial Laser Scanning Technology, Systems and Applications," in *2nd FIG Regional Conference*, 2003, pp. 1–10.

[6] R. Gupta and H. Chaudhary, "A Literature Review on Low Cost 3D Scanning Using Structure Light and Laser Light Scanning Technology," *Int. J. Innov. Res. Sci. Eng. Technol.*, vol. 3297, no. 7, pp. 15428–15433, 2017, doi: 10.15680/IJIRSET.2017.0607384.

[7] M. Bitenc, D. S. Kieffer, K. Khoshelham, and S. Wavelet, "POTENTIAL AND LIMITATIONS OF TERRESTRIAL LASER SCANNING," vol. XLII, no. June, pp. 10–14, 2019.

[8] K. Tan, W. Zhang, F. Shen, and X. Cheng, "Investigation of TLS Intensity Data and Distance Measurement Errors from Target Specular Reflections," 2018, doi: 10.3390/rs10071077.

[9] R. M. Alkan, "Analysis of the Accuracy of Terrestrial Laser Scanning Measurements," no. November, 2020.

[10] H. S. Park, H. M. Lee, H. Adeli, and I. Lee, "A new approach for health monitoring of structures: Terrestrial laser scanning," *Comput. Civ. Infrastruct. Eng.*, vol. 22, no. 1, pp. 19–30, 2007, doi: 10.1111/j.1467-8667.2006.00466.x.

[11] G. Patrucco, F. Rinaudo, and A. Spreafico, "A NEW HANDHELD SCANNER for 3D SURVEY of SMALL ARTIFACTS: The STONEX F6," *Int. Arch. Photogramm. Remote Sens. Spat. Inf. Sci. - ISPRS Arch.*, vol. 42, no. 2/W15, pp. 895–901, 2019, doi: 10.5194/isprs-archives-XLII-2-W15-895-2019.

[12] M. S. Quintero *et al.*, *Theory and practice on Terrestrial Laser Scanning*, no. June. Flemish Agency of the European Leonardo Da- Vinci programme, 2008.

[13] P. Li, R. Wang, Y. Wang, and W. Tao, "Evaluation of the ICP Algorithm in 3D Point Cloud Registration," *IEEE Access*, vol. 8, pp. 68030–68048, 2020, doi: 10.1109/ACCESS.2020.2986470.

IOP Conf. Series: Earth and Environmental Science 1047 (2022) 012013 doi:10.1088/1755-1315/1047/1/012013

Geochemistry of Gede Volcanic Complex West Java, Indonesia, Compared to Salak Volcano as its Proximity and how The Ciletuh Mélange Complex Lineament Affected It

Erza Winanto*, **Achmad Nazar Abrory, Eric Edwin, Mirzam Abdurrachman**

Department of Geological Engineering, Institut Teknologi Bandung
*Corresponding Author: erza.winanto@icloud.com

Abstract. 'Mélange' has been used to indicate a larger spectrum of chaotic rock bodies formed in various and different tectonic settings (rift–drift cycles, oceanic subduction, continental collision, and intracontinental deformation) by tectonic, sedimentary, and diapiric processes and their mutual interaction. Two well-known mélanges in Java Island are Ciletuh mélange in Sukabumi and another in Karangsambung, Kebumen. Previous studies suggest those mélanges are connected and indicate the Early Cretaceous Sundaland boundary elongated from Ciletuh, Karangsambung, and up until Meratus in Southeast Kalimantan. The Gede Volcanic Complex (GVC) and Salak Volcano of the Sunda island arc (West Java, Indonesia) are Quaternary volcanoes that lie geographically on the Early Cretaceous Sundaland boundary. An indication of crustal assimilation related to the Ciletuh Mélange complex has been observed by previous research. Therefore, this research intended to further investigate whether GVC and Salak are affected by that boundary through an igneous petrogenetic approach.

This research uses geochemical data (major and trace elements) proceeded by various discrimination diagrams to account for the tectonic conditions of the Gede Volcanic Complex and Salak Volcano. Products from both GVC and Salak are all sub-alkalis rocks that came from active continental plate margins. Geochemical analysis shows that one metabasalt sample from Ciletuh Mélange Complex has the same characteristics to anomalies found at GVC by being an island arc tholeiite. Meanwhile, anomalies at Salak could be explained by its location being on top of the Meratus lineament that might contain granitic deposits.

1. Introduction

Mélanges, chaotic blocks of different ages and origin commonly embedded in the various type of matrix (block-in-matrix fabric), represents a significant component of many collisional- and accretionary-type orogenic belts [1]. The type of mélange complex plays a role in the tectonic reconstruction of active margin. The definition of mélange has evolved to cover complex scenarios and is suggested that it should be accompanied by adjectives (tectonic, sedimentary, and diapiric) indicating the relative mélange-forming processes [2].

In the past decades, many research focused on the characteristics of three well-known mélange in Indonesia that correspond to one another; The Ciletuh, Luk Ulo, and Meratus mélange complex. Based on observation of their outcrops and structural configuration, The Ciletuh, Luk Ulo, and Meratus mélange complexes have been regarded as tectonic mélanges [3]. These mélanges have been recognized since the 1970s and are referred to as an indicator of the Early Cretaceous Sundaland zone boundary due to their resemblance and its subduction zone of the southern and southeastern margins of Sundaland [4], [5].

The Sunda arc formed the northward subduction of the Indo-Australian Plate beneath the Eurasian Plate and constructed the western part of the Indonesian subduction zone system [5]. Accordingly, the genesis of volcanic rocks in subduction zone is tortuous due to the multitude of differentiation processes that could occur [6]. Yet, geochemical identification, inventory, and analysis of volcanic products could elucidate the processes occurred beneath the volcano and differentiate one volcano to another.

The Gede Volcanic Complex (GVC) of the Sunda island arc (West Java, Indonesia) is a Quaternary volcanic complex that is geographically located within the Cretaceous subduction zone. It is dominated by the twin stratovolcanoes of Gede (+2,958 m), the only active volcano of the complex, and Pangrango (+3,019 m), an extinct volcano to the northwest. Previous findings show that no significant assimilation of mature continental-type crust with magma beneath the complex due to poor correlation of Sr isotope ratios with indices of differentiation of GVC rocks precludes [6]. This contradicts with the conclusions by Gasparon [7] who suggests the geochemical and isotopic signatures observed in western Sunda arc lavas are affected by upper-crust as the dominant source. Further investigation on a low $\delta^{18}O$ and relatively low Pb and Sr isotopic composition assimilate in West Java (at Gede Volcanic Complex) likely represent more primitive arc rocks or mafic-ultramafic ophiolitic rocks [8] which is found in The Ciletuh Melange Complex [9].

The Salak volcano, +2,211 masl, is a stratovolcano closely located from the GVC. Geographically it is located about 23 kilometers to the west of the GVC area. Previous research concluded that variation in some trace element ratios (K/Rb and Ba/Th) implies crustal assimilation occurring during the fractional crystallization of the magma of the central vent group of Salak Volcano [10]. However, the plot of Pb isotopic ratios versus SiO_2 for volcanic rocks from the Salak shows unclear correlation despite inferred crustal assimilation [8].
Considering the proximity of these two volcanoes, this study aims to investigate further on The Gede Volcanic Complex tectonic setting, petrogenesis, relative to Salak Volcano, and how The Early-Cretaceous Sundaland-zone boundary especially the Ciletuh Mélange Complex lineament affects the geochemical product of both volcanoes. This study combines the newly acquired major and trace element data on volcanic rocks at GVC with previously published geochemical and isotopic data by Handley [6]. Geochemical data of Salak Volcano [10] is also analyzed as a comparison

2. Geological Setting
The Sunda arc elongates from the Andaman Islands north-west of Sumatra, through Java to Flores in the Banda Sea. It is part of the Indonesian subduction zone system that formed by subduction of the Indo-Australian plate beneath the Eurasian Plate at a rate of ~6-7 cm/year [5] in which the tectonic setting described in detail by Hall [11], [12]. A series of Gondwanan lithospheric fragments were thought to be the configuration of the Sunda arc basement and were accreted to the pre-Cretaceous Eurasian Plate (Sundaland) margin in the Cretaceous period [11], [13], [14]. These complexes were exposed in three places in Java, namely The Jiwo Hills and Luk Ulo Area near Karangsambung in Central Java and Ciletuh Area in West Java [9]. The exposed pre-Tertiary rocks in the Ciletuh area consists ultrabasic to basic rocks, metamorphic rocks and sedimentary rocks that are mixed tectonically [9]. Those basic and ultrabasic rocks have a similar age to granite intrusions in the Java Sea, the andesitic to basaltic volcanic rocks which exposed at Cikotok (South Banten area), and the subsurface Jatibarang Formation found in the northern part of West Java [15], [16], [17].

The Ciletuh Melange Complex and Meratus Melange Complex formed a liniament that presumably considered as the southeast limit to Cretaceaous continental crust [5], whereas The GVC and Salak Volcano are located in West Java within the Quaternary volcanic front of the Sunda arc (Figure 1). The location of GVC and Salak Volcano are adjacent to the exposed pre-Tertiary rocks in Ciletuh area and is possible that these rocks impact the geochemical configuration of the products on both volcanoes considering their geological setting.

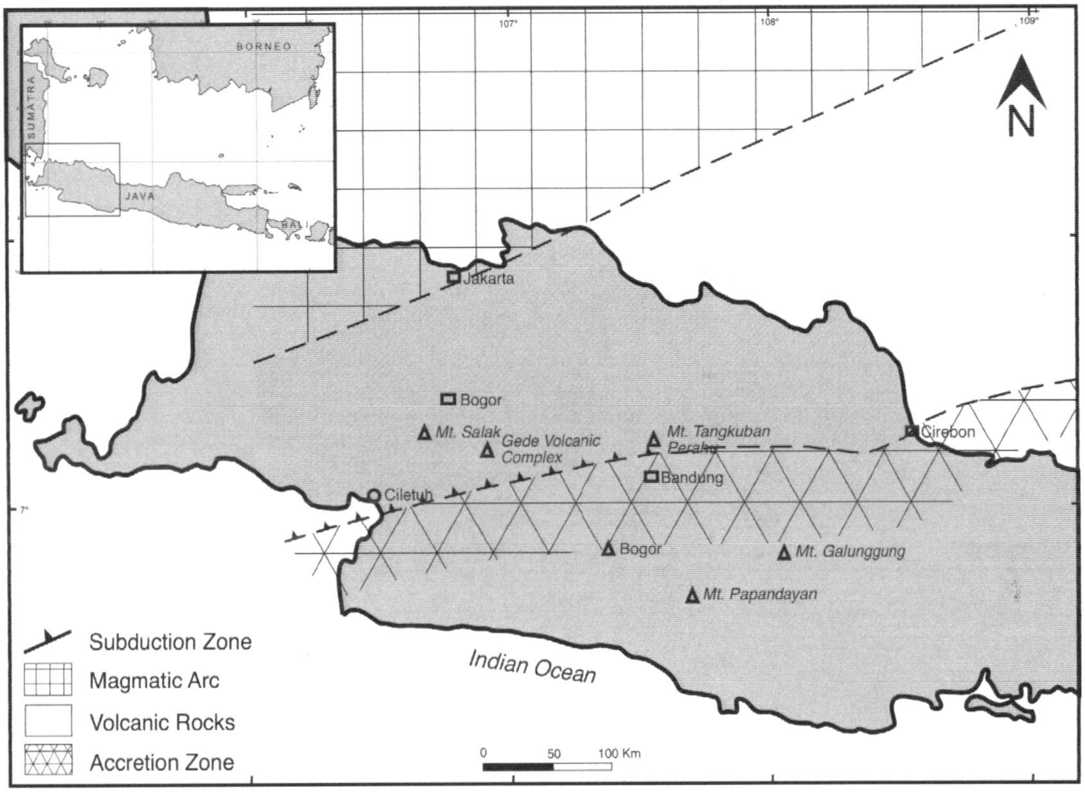

Figure 1. The location of the Gede Volcanic Complex and Salak in Cretaceous volcanism in West Java, modified from Martodjojo [16].

The Ciletuh melange also appears to represent the old subduction zone, whereas eruptive centres in the Java Sea and an area between Cikotok and Jatibarang may represent a magmatic arc (Figure 1). Soejono [18] interpreted that the rocks from Ciletuh showed a specific feature that are found in accretionary prism. Thus, it is assumed that Ciletuh at Early Eocene is part of the outer arc from a bigger island-arc system. Nevertheless, the accretionary prism model previously explained by Karig and Sharman [19], [20], [21] defined the Ciletuh Formation as the "pond" deposits on the inner slope of the subduction trench and the original contact with the underlying melange complex was depositional.

3. Sample Selection and Techniques

Twelve samples used in this study were collected from the lava-flow deposit from GVC and each one represents its source or origins (Figure 2). Then, the samples (Table 1.) were analyzed using XRF (X-Ray Fluorescence Spectrometry) method conducted in the *Laboratorium Pusat Survei Geologi* (Laboratory of Geological Survey Central) in Bandung, Indonesia. A pressed pellet preparation method was used. Laboratory assessment results generally have low LOI indicating the samples were unaltered. The total of major element proportions with ± 2% offset from 100% indicates a good level of accuracy. The lab also validated trace elements, complying with the lab's standard with relatively low standard error. These laboratory assessment values and lack of visible alteration in rock samples and thin sections (except in a few specimens) confirm the samples mostly remain unaltered as they were first deposited.

To facilitate investigation of the tectonic setting of GVC, previously published major and trace elements data from Handley [6] were added as well as samples from Salak [10] for complementary. The new geochemical data for volcanic rocks from GVC are presented in Table 1. All data were later normalized and corrected using CIPW. This procedure employs the distribution of elements into a hypothetical mineral series not necessarily present in the rock, following a crystallization sequence resulting from petrographic and thermodynamic observations. However, the weakness of using this method is that the calculation assumes that the rock is in an anhydrous condition [22].

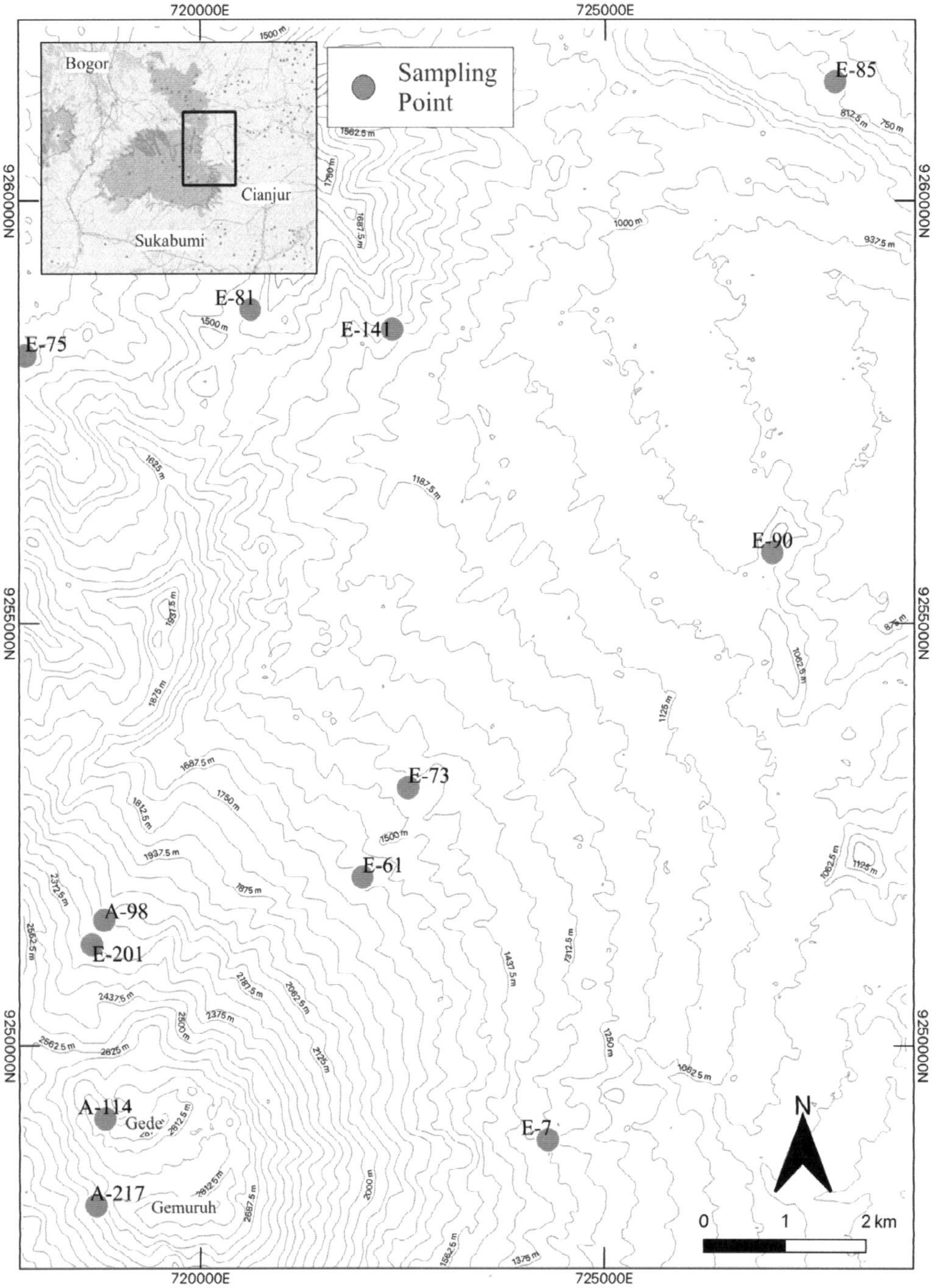

Figure 2. Topographic sketch map of the Gede Volcanic Complex (GVC) showing the distribution of new rock samples.

Table 1. Major and trace element data for Gede Volcanic Complex. These samples were acquired from lava deposits from different eruption centers and then analyzed by XRF and then normalized and corrected using CIPW Norm.

Samples	Olivine Basalt	Porphyritic Basalt		Porphyritic Andesite					Andesite Plagioclase	Pyroxene Andesite		
	E-81	**E-07**	**E-73**	**E-85**	**E-90**	**E-75**	**E-201**	**E-61**	**E-141**	**A-217**	**A-114**	**A-98**
Major Elements												
SiO2	45.37	49.07	48.25	57.43	58.17	53.93	51.65	50.79	51.98	58.80	55.53	56.07
Al2O3	15.94	20.93	19.30	18.73	17.84	22.11	21.28	19.01	24.49	19.09	19.74	18.52
Fe2O3	12.41	8.57	11.87	8.61	7.55	7.78	10.84	10.40	6.13	7.27	9.32	9.32
MgO	7.40	2.51	2.96	1.40	2.84	0.92	1.18	2.67	0.64	0.96	1.67	2.20
CaO	11.01	14.02	8.83	7.14	6.62	8.06	8.26	8.69	5.05	7.68	8.67	7.44
Na2O	1.85	2.41	2.24	3.34	3.45	2.95	2.51	2.21	2.29	3.24	2.76	2.97
K2O	0.83	0.96	0.85	0.94	1.87	0.84	0.85	1.25	1.07	1.55	1.28	1.32
TiO2	0.54	0.55	0.77	0.65	0.73	0.57	0.66	0.66	0.53	0.61	0.72	0.78
MnO	0.18	0.13	0.19	0.17	0.15	0.15	0.18	0.17	0.12	0.12	0.15	0.15
P2O5	0.28	0.18	0.22	0.23	0.22	0.54	0.25	0.25	0.19	0.12	0.14	0.18
LOI	3.94	0.46	3.66	2.40	0.65	2.38	3.09	3.44	7.45	0.57	0.78	1.57
Total	99.74	99.79	99.13	101.04	100.09	100.22	100.75	99.54	99.94	100.01	100.76	100.53
Trace Elements												
V	231	156	253	123	75	87	119	164	63	126	160	170
Cr				23								
Co	53	32	54			25	29	32			25	
Cu	50		66	22	29	48	52	33	45	28	30	33
Zn	63	52	100	53	61	66	74	69	50	64	52	137
Ga	22	28	25		19	28	21	27	25	25	22	25
Rb	0	22	27	34	56	34	38	43	48	63	41	50
Sr	322	384	247	339	356	414	350	297	362	315	268	256
Y	22		20	32				34		25		22
Ba				237	406		277	320	201	293		237
Sx	101	74	149	83	109	565	512	105		154	124	877
Cl			190	224		160	38			194	139	197
S									201			877

Major element contents in wt.%; trace elements concentrations in ppm.

4. Results

4.1. Major and Trace Element Discrimination Diagram

Several geochemical analyses were carried out using discrimination diagrams to classify samples or data into several groups [23]. In the analysis, a bivariate and triangular diagram were made to obtain detailed groupings for all samples. According to Wilson [24], volcanic rocks are divided into two major magma series, alkalic and sub-alkalic. In general, the sub-alkalic magma series can be subdivided into a high-alumina or calc-alkali series and a low-K tholeiitic series.

A discrimination diagram as detailed by Le Bas [25] was used to identify magma series, by comparing total alkali (N_2O and K_2O) against SiO_2 as these elements representing alkali mineral abundance of rocks that correspond to the melt chemical processes. The result shows that magma series for GVC and Salak products are classified as sub-alkaline or tholeiitic rocks (Figure 3.a). Similar to the diagrams made by Ewart [26] where the plotted data point out specific rock names and magma-series by comparing K_2O to SiO_2 (Figure 3.b). From both diagrams, it is shown that both the products from GVC and Salak are defined as calc-alkaline Series, though there are some products from Salak that are considered as high-K calc-alkali.

The trace elements La, Y, and Nb can be usually used to discriminate between different types of mafic to intermediate rocks [27]. Basic-intermediate samples were used in consequence, some felsic samples from Salak have been dismissed in further analysis. Y and Nb have been used before as indicators of the alkalinity index of magmas [28], [29]. In basalt, lanthanum is an element of the Light Rare Earth group which has a strongly hygromagmaphile (incompatible) character, coming just before tantalum and after thorium [30]. Additionally, its behavior during metamorphism is relatively inert [31], [32], [33]. However, special cases like carbonation and epidotization give an exception and furthermore, it is still subject to discussion [34], [35]. Regarding the La/Y ratio (like La/Tb), it directly characterizes the fractionation of light rare earth compared to heavy rare earth and it therefore also represents a good index of alkalinity or "calc-alkalinity" of magmas [31]. Ultimately, the variations of the La/Nb ratio (like La/Ta) were defined to differentiate MORBs [36], [37]. The diagram shows that all samples from both GVC and Salak are stacked within the calc-alkaline to calc-alkaline-tholeiites quadrant (Figure 4). This diagram also shows that all samples lie within one-third of the diagram which is an orogenic domain, thus all products from Salak and GVC are the result of compression (e.g., island arcs, active margin, etc).

Another analysis using ternary diagrams comparing Ti, Zr, and Y by Pearce and Cann [28] is also conducted. Of these elements, Ti, Zr, Y, and Nb have been choosen as they are suggested as being insensitive to processes of alteration [38], insensitive to secondary processes such as weathering and metamorphism, and easy to measure with good reproducibility of results [28]. All samples were screened and two samples from GVC are left out due to alteration. From the diagram, most products from GVC and all from Salak are calc-alkali (figure. 5.a). However, three samples from GVC lie in the C domain. For basic magmas, most crystallizing phases will leave a residual liquid enriched in both Y and Zr [39]. These samples including three from the C domain were later isolated for further analysis using different diagrams to determine their origins.

Another diagram of Ti, Zr, and Sr by Pearce and Cann [28] shows that the samples from both GVC and Salak are clustered in the C section which is defined as calc-alkali although one sample from GVC lies in the B domain, which is an island arc. One sample from Salak lies outside the plotted area (Figure 5. b). However, two observations illustrate the dangers of using Sr for very altered samples: intense albitization could lead to some analyses have very low Sr (and very low CaO), and a possibility indicating the addition of Sr shown by some analyses have high Sr (mostly accompanied by unusually high Rb and K) [28].

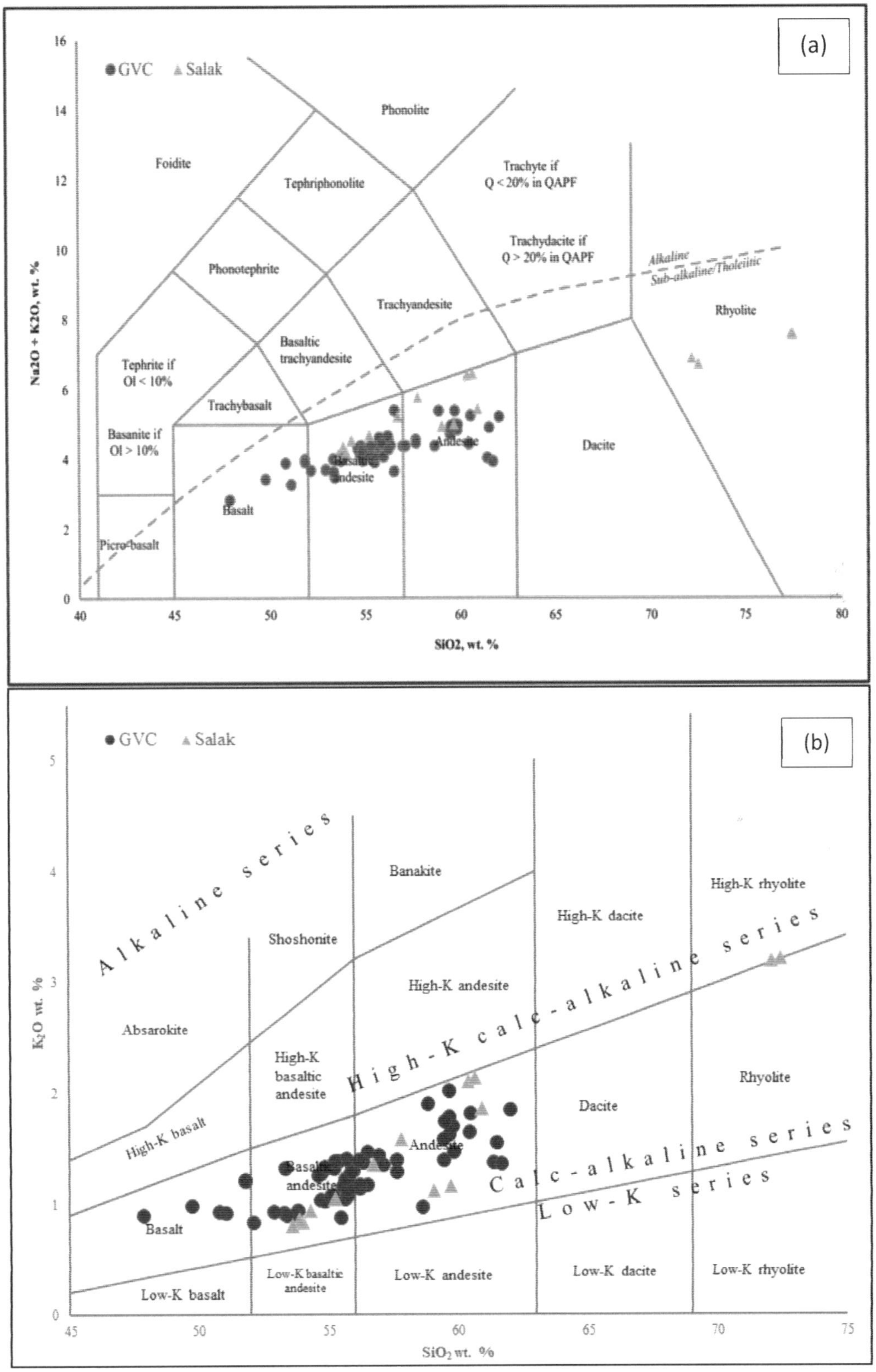

Figure 3. (a) Variation diagram of SiO_2 versus total Alkali (TAS) [25] and (b) Variation diagram of SiO_2 versus K_2O [26] for all samples from GVC and Salak.

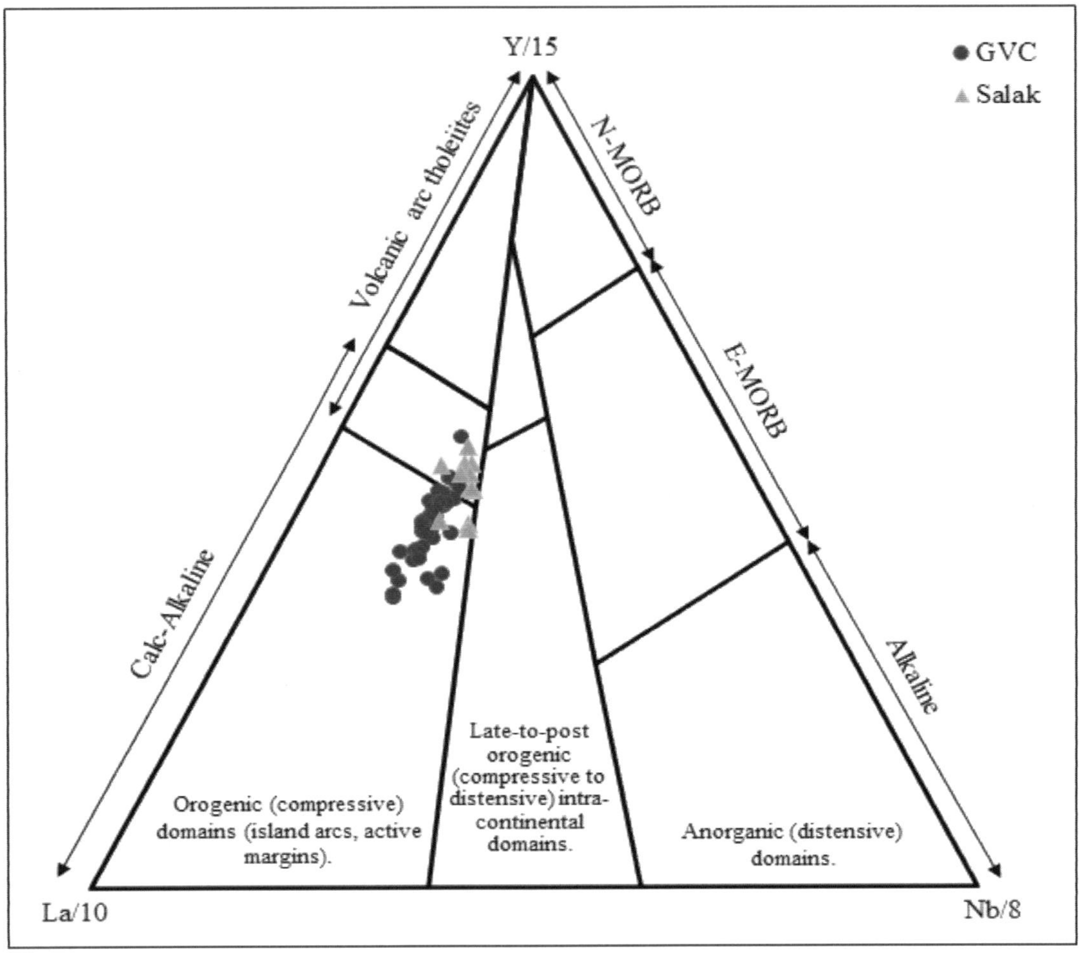

Figure 4. Variation diagram of La, Y, and Nb from Cabanis and Loccelle [27] for all samples from GVC and Salak.

Some analysis was used to further determine the tectonic setting, the one used in this analysis was done by plotting data on several bivariates and triangular discrimination diagrams using major and trace elements. The trace elements such as Ti, Zr, Y, K, Rb, Ba, and Sr, as these elements are relatively immobile, are more effective to be used in tectonic setting discrimination diagrams for volcanic products [23].

The TiO_2–K_2O–P_2O_5 ternary diagram was suggested as a method of discriminating between oceanic and non-oceanic (continental) basalts. Since fractionated rocks cannot be adequately discriminated against, this diagram is effective for non-alkaline "primitive" basalts. Those which have total alkalies $\leq 20\%$ in an AFM diagram are suitable for these analyses [40]. Generally, ocean-floor basalts have consistently low concentrations of TiO_2, K_2O, and P_2O_5 [41], [42]. From Figure 6.a, we can see that all samples from both Salak and GVC are considered to come from non-oceanic environments.

Pearce and Gale [43] created a bivariate diagram by comparing Zr/Y to Ti/Y. The variation in Zr/Y and Ti/Y ratios between basalt suites provides valuable discrimination of within-plate basalts from other magma types is presumably related to long-lived source heterogeneities. Moreover, within-plate basalts can almost all be related to sources with ratios greater than C_3 chondrite, whereas MORB and island arc basalts must almost all be related to sources with ratios less than C_3 [39]. This diagram shows that all samples from GVC and Salak originated from the plate margin environment (figure. 6.b).

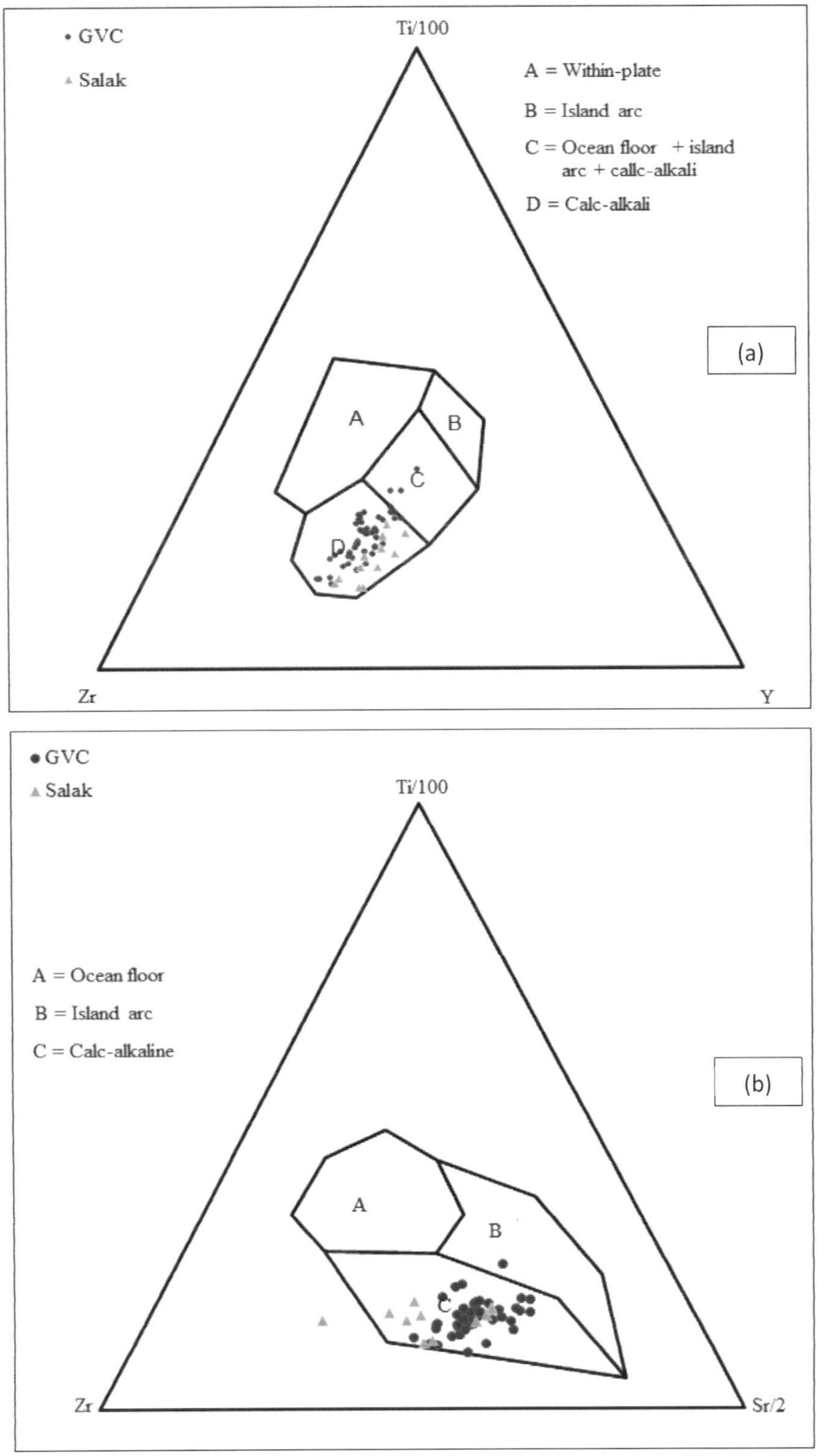

Figure 5. (a) Variation diagram by Pearce and Cann [28] of Ti, Zr, and Y, and (b) Ti, Zr, and Sr for all samples from GVC and Salak.

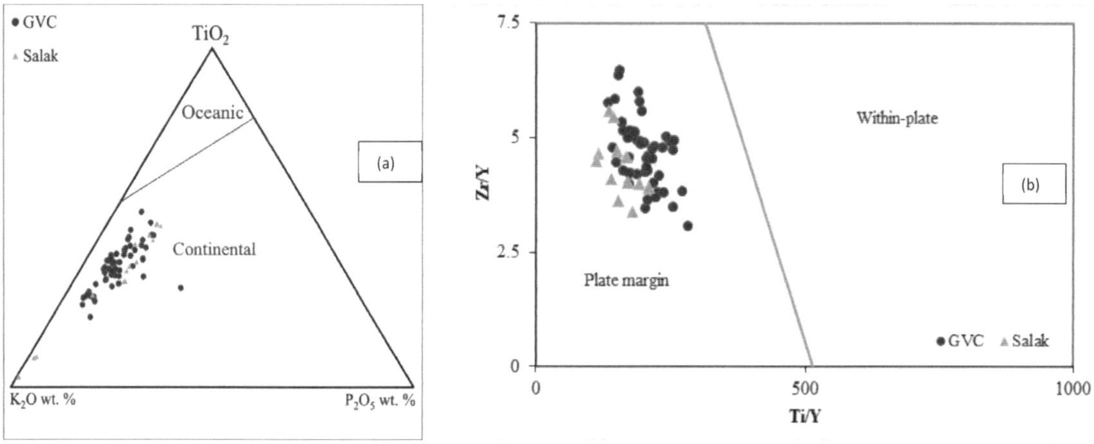

Figure 6. (a) The TiO_2–K_2O–P_2O_5 ternary diagram by Pearce [40] and (b) Variation diagram of Zr/Y to Ti/Y by Pearce and Gale [43] for all samples from GVC and Salak.

Further analysis using another ternary diagram from Pearce [44] were conducted by comparing MgO–FeO(total)–Al_2O_3 (Figure 7.a) to ascertain previous findings. This diagram uses sub-alkaline "basaltic-andesites" (51-56% SiO_2 calculated anhydrous), all samples are not used due to the chemistry of rocks with intermediate silica contents that appears to be most sensitive to different tectonic environments [44]. These three oxides behaviour during alteration is subject to several incompletely studied variables (e.g., submarine weathering, temperature of alteration, degree of crystallinity, and extent of alteration as reflected in the H_2O content of the rocks), and FeO* (total) and Al_2O_3 are likely relatively immobile during the initial stages of submarine alteration whereas the MgO contents have been reported as a variable [45], [46]. From all selected samples, most are clustered within the orogenic field although there are some scattered samples from GVC that lie in the spreading center island field. To account for samples from GVC that are scattered, those samples were later isolated for further analysis.

Accurate determination discrimination diagrams that distinguish among oceanic environments using trace elements requires analysis via techniques such as INAA which are expensive, time-consuming, and not readily available to many researchers [47]. The MgO–$FeO_{(total)}$–Al_2O_3 ternary diagram by Pearce [44] does not discriminate among different oceanic environments, and hence is not directly applicable to some problems such as island arc versus ocean ridge origin of ophiolites. A discriminant diagram from Pearce and Cann [28] (Figure 5.a and 5.b) gave us some uncertainty with a couple of samples being ambiguous. All samples from Figure 5.a that lie within the C domain and outside all domains, and some samples from Figure 7.a that scattered within the spreading center island were later isolated for further analysis.

A ternary diagram using TiO, MnO, and P_2O_5 can discriminate between five petro-tectonic environments of basaltic rocks (45-54% of SiO_2) [47]. Although Pearce [45] noted that major element discriminant diagrams were not very sensitive to crystal fractionation but instead showed differences in primary composition, TiO, MnO, and P_2O_5 relations suggest that fractionation is effective to control of minor element abundances in magmas, although variations such as the great abundance of P_2O_5 in ocean island basalts are primary [47]. The selected GVC samples from the previous analysis were later screened and plotted in this ternary diagram. A sample of metabasalt from Ciletuh by Ikhram [48] that passed selection is also included for comparison.

4.2. Trace Elements Distribution
The distribution of trace elements is often plotted using a spider diagram that is normalized to a series of trace elements from the top crust, lower crust, MORB, or OIB. Samples that are gathered from similar geological settings usually show the same patterns. Thus, we plotted Sr, K, Rb, Ba, Th, Ta, Nb, Ce, P,

Zr, Hf, Sm, Ti, Y, and Yb from GVC's samples following Pearce [49] that is normalized to MORB (Figure 8).

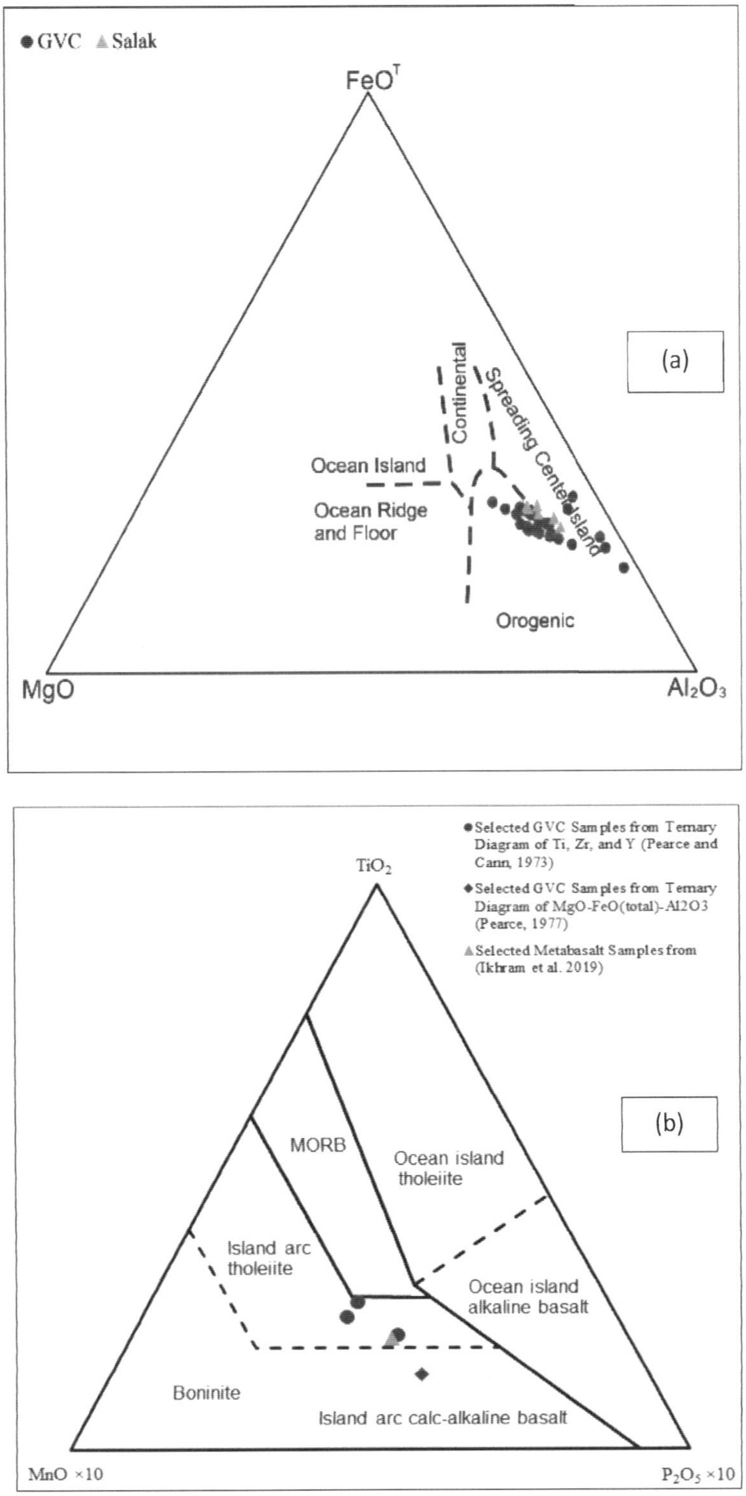

Figure 7. (a) The MgO–FeO (total)–Al_2O_3 ternary diagram by Pearce [44] for samples from GVC and Salak; and (b) Variation diagram of TiO, MnO, and P_2O_5 by Mullen [47] for selected samples from figure. 3.a, 3.b and Figure 7.a.

In this spider diagram from Pearce [49], the elements are divided into two groups based on their relative mobility in aqueous fluids. The mobile elements such as Sr, K, Rb, and Ba are plotted at the left of the pattern, while the remaining elements are immobile. According to Pearce [49], the shape of these patterns is not likely to be greatly changed by fractional crystallization or variable degrees of partial melting. Typically, the arc basalts are characterized by selective enrichment of incompatible elements of low ionic potential (Sr, K, Rb, Ba ± Th) and low abundance of elements of high ionic potential (Ta, Nb, Ce, P, Zr, Hf, Sm, Ti, Y, Yb, Sc, and Cr) relative to N-type MORB [50]. From the diagram, samples from GVC are relatively similar to Salak (Figure 8). Samples from Salak also show a similar trend although there is some significant depletion on P and Ti.

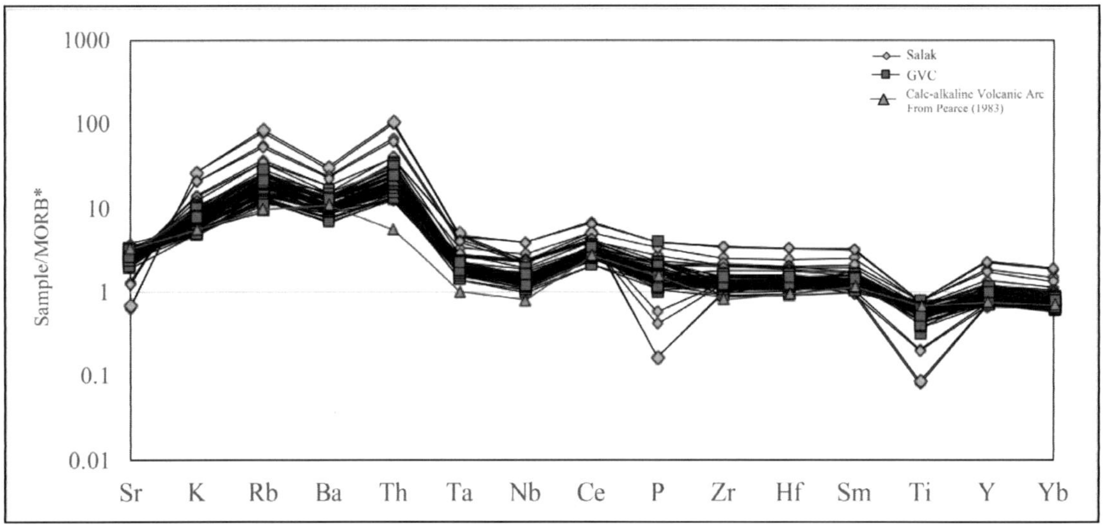

Figure 8. Spider diagrams from whole trace element data from GVC and Salak normalized using constants from Pearce [49].

5. Discussion

5.1. Magma Series

Division in volcanic rocks into two major magma series, alkalic and sub-alkalic, was first proposed by Iddings in 1892 [51] and elaborated by other researchers ever since. A range of basic to acid rock composition is contained in these magma series. The boundary between them in the discrimination diagrams is gradational although it is marked as a solid line. However, these diagrams do not represent the processes responsible for the compositional diversity of magma although it is almost certainly the dominant one [24]. From the binary discrimination diagrams used in this research, samples from GVC and Salak are clustered around in the sub-alkalic domain of the diagrams (Figure 3.a). Generally, the sub-alkalic magma series can be subdivided into low-K tholeiitic and high alumina or calc-alkaline series. Calc-alkali may be further subdivided into low-, medium-, and high-K types as can be seen from (Figure 3.b) based on SiO_2 versus K_2O [26] This diagram shows that four samples from Salak are outlying the rest of the samples with >70 wt% SiO_2. These samples are from (a) pre-Salak, which were suggested as a result of interaction and possible contamination of magma by evolved plutonic rocks, and (b) Western side vent of Salak potentially due to further fractional crystallization [10].

A ternary diagram from Cabanis and Lecolle [27] comparing three trace elements La-Y-Nb. From (Figure 4) we can see that all samples from GVC and Salak are considered the product of an orogenic environment. Moreover, an indication of mixture between different sources or crustal contamination from various sources is shown by the trends of samples being in the middle domain [27]. Negative Nb and Ta have been an indicator of a subduction zone environment due to their immobility relative to other incompatible elements in condition when slab-derived fluids or melts went through separation into two phases or there's a change in its composition [52]. Some models of subduction zone magmatism form imply that fluids leaving the dehydrating slab are initially poor in Nb and Ta due to their low solubility in aqueous solutions in contrast to silicate melts [53], [54]. Accordingly, some samples from GVC and Salak lie between calc-alkaline and volcanic arc tholeiites indicating that some samples are considered as low-K magma series. Conversely, previous diagrams from Ewart [26] (Figure 3.b) shows otherwise due to neither the products of GVC and Salak being considered as low-K series. Another presumption is that there may be contamination from volcanic arc tholeiites.

A Pearce and Cann [28] ternary discrimination diagrams using Ti, Zr, Y, and Sr are used to confirm both GVC and Salak magma series and tectonic setting, considering that all samples are fresh or slightly altered. The first set uses the less mobile elements (Ti, Zr, and Y), and the second set uses an additional element, Sr, which gives better discrimination but is only strictly valid for fresh and slightly altered rock [28]. Most samples from GVC and Salak lie within the calc-alkaline area (Figure 5.a and 5.b), although three samples from GVC lie within the C area (ocean floor basalt/island arc/calc-alkali). Figure 5.b, however, confirms that one sample from GVC is considered as an island arc. However, two samples from Salak lie outside the boundaries with relatively high Zr and low Sr. The consideration needed for using Sr has already been emphasized as it features such as albitization and calcium-depletion in greenschist facies metamorphism [28]. These samples are from the western side vent of Salak [10] and were considered to be slightly altered due to their dull appearance of plagioclase. Another presumption on relatively high Zr concentration from Salak seen in (Figure 5.b) potentially due to continued magmatic evolution from a central vent group-type parent at shallow depth, involving late-stage fractionation of hornblende in the petrogenesis of the western flank SVG rocks [10].

5.2. Tectonic Setting

The magma series in the geochemical analysis of igneous rocks has characteristics that can reflect its tectonic setting [24]. The stages in determining the tectonic setting include identifying the location of the eruption source, whether it is convergent or divergent plate margins, or intra-oceanic or intra-continental within-plate margins. Figure 3.b shows that all samples from GVC and Salak are calc-alkaline. Magma series with calc-alkali characteristics normally has a convergent plate margin tectonic setting [24]. This statment corresponds with Figure 4, which defines all samples from GVC and Salak in an orogenic or compressive domain including island arcs or active margins. However, some samples from Salak and GVC in Figure 4 lie in the interception between calc-alkaline and volcanic arc-tholeiites. Figure 5.a shows that three samples from GVC are defined as sub alkaline (tholeiitic, calc-alkali, or low potassium tholeiites, which later in Figure 5.b that one sample is confirmed as low-K tholeiites. These low-K tholeiites typically occur in young immature arcs, whereas calc-alkaline magmas are typical of more mature arcs and active continental margins [24]. Thus, these occurrences indicate that there might be a contaminant of older arcs, below which align with previous findings from [8].

The TiO_2–K_2O–P_2O_5 ternary diagram by Pearce [40] (Figure 6.a) shows that all samples from both Salak and GVC are considered non-oceanic environment origin. However, this diagram might not be quite effective as not all samples collected from Salak and GVC are "primitive" basalts (and most likely are fractionated rocks). Thus, all analyses are screened through an AFM calculation, and those with total alkalies $\leq 20\%$ came through. In addition, weathering and alteration tend to move oceanic basalts into the non-oceanic field [40] so that altered basalt, which plots in the non-oceanic field, possibly have an oceanic origin.

A binary diagram between Ti/Y and Zr/Y distinguishes within-plate from other magma types by showing a distinctive enrichment in Ti and Zr (but not Y) of within-plate basalts [44]. This diagram (Figure 6.b) clearly shows that all samples from both GVC and Salak are products of plate margin sources. To sum up, combined with previous diagrams and findings, the tectonic setting of both GVC and Salak is presumed as a continental plate margin, which is confirmed by the previous statement that The Sunda arc is part of the Indonesian subduction zone system [5], [11], [12].

From Figure 6.a, 6.b and 4, we can conclude that most samples from GVC and Salak came from continental plate margins. However, these diagrams do not fully yet explain the anomalies in GVC's data seen in both ternary diagrams (Figure 5.a and 5.b). The ternary diagram from Pearce [44] (Figure 7.a) shows that most samples are clustered within the orogenic field, although some scattered samples from GVC lie in the spreading centre island field. Nevertheless, due to the scarcity of data for the spreading centre island field, the defining environment is not well established to account for the indicated domain [44]. The available data suggest that the rocks occurring in this environment are relatively enriched in Fe, and Al compared to otherwise similar rocks from the seafloor and intra-plate islands. This feature could result from high-level fractionation occurring within the volcanoes [44].

Isolated GVC samples from the previous analysis were plotted in a ternary diagram comparing TiO, MnO, and P_2O_5 by Mullen [47] due to samples from GVC that show strange behaviour on previous diagrams that tend to show a more tholeiitic or ocean-related environment. Furthermore, previous research from GVC also stated that samples from GVC are likely represented more primitive arc rocks or mafic-ultramafic ophiolitic rocks [8] and presumably related to the one found in The Ciletuh Melange Complex. Thus, metabasalt samples from Ikhram [48] were also plotted in this diagram for comparison, although only one sample that passed the required criteria. From the diagram (Figure 7.b), we can see that almost all samples selected from GVC are considered as island arc tholeiite, and one sample is island arc calc-alkaline. Moreover, metabasalt found in the Ciletuh ophiolite complex (sample from Ikhram [48]) is also considered as island arc tholeiite, which consolidates the hypothetical idea of this paper.

5.3. Trace Elements Distribution

A complete variety of trace element data is better to obtain the tectonic environment rather than depending on the binary and ternary element diagrams [55]. The idea of understanding the pattern of trace element abundances in any igneous rocks leads us to the concept of spider diagrams [49], [56], [57], [58], [59]. In addition, a Series of trace elements from GVC and Salak were plotted.

The overabundance of LIL cations (Sr, K, Rb, Ba, Th) and depletions of high-field-strength elements are perhaps the most distinctive feature seen in both volcanoes. Those patterns are similar to a sample of calc-alkaline volcanic arc from Pearce [49], with LIL elements tending to be relatively high. This overabundance of LIL elements is explicable in principle by multi-stage enrichment models [57], [60], [61], [62], [64]. From the diagram (Figure 8), it is noticeable that products from Salak are more evolved and diverse than those from GVC. It is presumably due to further fractional crystallization or contamination of magma that occurred at Salak. Further research on Salak has been conducted to elucidate magma evolution and explained by Handley [10].

An indication of island-arc rocks contamination at GVC might be observed in the spider diagram. Chemically, the most distinctive features of the continental-margin volcanic suites compared to those erupted in oceanic island arcs are a relatively higher concentration of K, Sr, Rb, Ba, Zr, Th, and U, higher K/Rb and Fe/Mg ratios, and higher range of Sr, Nd, and Pb isotopic ratios [24]. Contamination from mafic-ultramafic ophiolitic rocks from Ciletuh at GVC was first stated in previous research by Handley [8] by identifying $\delta^{18}O$ and Sr, and Pb isotopic ratios. However, from the spider diagram, this does not affect the trends of both volcanoes (Figure 8) by being similar to that in an active-continental

5.4. Ciletuh Melange Complex Lineament

From Figure 1, it is shown that the Cretaceous magmatic arc, the subduction zone that created the accretionary prism, is located in Ciletuh, and the fore-arc basin has a NE — SW lineament, well-aligned with Cimandiri fault or better known as Meratus Lineament. The deposition of the Ciletuh Formation in Ciletuh was uplifted with a reverse fault that caused the change of the environmental deposition from deep to shallow marine. The Ciletuh formation is part of the Bogor Basin and is well described by Martodjojo [17]. Volcanic rocks, basalts and andesites are also found across West Java with the same age as the accretionary prism. These volcanic rocks are exposed in Cikotok as "old andesite" and found in oil boreholes on land or offshore. This occurrence indicates that these rocks might have an essential role as contaminants under younger Quaternary volcanoes across West Java.

On the other hand, GVC and Salak are the nearest Quaternary volcanoes from the exposed melange complex found in Ciletuh. As previously explained, anomalies found in GVC rocks show similarities to basalt found in Ciletuh. As for anomalies found in Salak previously brought up by Handley [10], there is a possibility of contamination by evolved plutonic rocks or further fractional crystallization.

Granitic rocks with the same age as accretionary prisms (drill hole data from Martodjojo [16]) mainly found beneath the Java Sea indicated that magmatic arc in early Eocene located north-west to West Java. Therefore, the indicated paleocurrent to southeast, south and southwest might be accountable to explain the possibilities of granitic rocks or sediment deposits under Salak. Moreover, this could explain the anomalies found in Salak. However, further geochemical analysis with more accuracy and precision must be conducted to ascertain these findings

6. Conclusions
Products from GVC and Salak are all sub-alkalis, with most of them being calc-alkaline and two samples from Salak are high-K calc-alkaline. Based on several diagrams, these products came from convergent continental plate margins or active continental margins. These findings are in line with previous statements by Hamilton [5] and Hall [11], [12] that the location of GVC and Salak is part of Quaternary volcanism related to the subduction zone. However, certain anomalies are noticeable ternary diagrams using trace elements from Pearce and [28]. Some samples from GVC are shown to be ambiguous between tholeiitic or calc-alkali, and one sample is shown as a product of an island arc. This statement correlates with previous findings that possible crustal assimilation occurs at GVC related to melange or mafic-ultramafic ophiolitic rocks known to outcrop in Ciletuh [8]. A ternary diagram by Mullen [47] proves that meta-basalts found in Ciletuh are similar or nearly identical to one of the products of GVC geochemically. However, the spider diagram pattern of GVC and Salak on rock/MORB did not show significant differences. The pattern indicates that GVC and Salak are products of continental plate margin as it resembles the pattern of a calc-alkaline volcanic arc.

The Cretaceous magmatic arc, subduction and fore-arc basin have a NE — SW lineament, which is well-aligned with the Cimandiri fault or the Meratus Lineament. The subduction zone created an accretionary prism and is composed of various rocks such as basalt and andesite found across West Java. The GVC and Salak are the nearest Quaternary volcanoes from the exposed melange complex found in Ciletuh, and geochemical analysis has shown that anomalies found in GVC are correlated with volcanic rocks found in Ciletuh. Moreover, anomalies found in Salak, previously assumed that contamination from more evolved plutonic rocks could be explained by the occurrence of granitic rocks, came from the Cretaceous magmatic arc found mostly beneath the Java Sea. Therefore, the indicated paleocurrent to southeast might be accountable to explain the possibilities of granitic rocks or sediment under Salak. However, further research and analysis with more accuracy and precision must be conducted to ensure these findings.

Acknowledgments
The author would like to thank Mimi Murdiah, former head of *Taman Nasional Gunung Gede Pangrango,* for her research support. The author would also like to thank all staff at *Taman Nasional Gunung Gede dan Pangrango* and *Pos Pengamatan Gunung Api Gede* for their constructive discussion during the fieldwork and analysis in 2018.

References
[1] Festa A, Pini G A, Dilek Y and Codegone G 2010 Mélanges and mélange-forming processes: a historical overview and new concepts *International Geology Review* **52** 1040–1105
[2] Silver E A and Beutner E C 1980 Melanges *Geology* **8** 32–34
[3] Mukti M M, Aribowo S and Nurhidayati A 2018 Origin of Melange Complexes in the Sunda and Banda Arcs: Tectonic, Sedimentary, or Diapiric Melange *IOP Conf. Ser.: Earth Environ. Sci.* **118** 012003

[4] Katili J A 1973 *Tectonophysics* **26** 165–188
[5] Hamilton W B 1979 Tectonics of the Indonesian Region *U.S. Geological Survey, Professional Paper* **6** 3–10
[6] Handley H K, Macpherson C G and Davidson J P 2010 Geochemical and Sr–O isotopic constraints on magmatic differentiation at Gede Volcanic Complex, West Java, Indonesia *Contributions to Mineralogy and Petrology* **159** 885–908
[7] Gasparon M, Hilton D R and Varne R 1994 Crustal contamination processes traced by helium isotopes: examples from the Sunda arc, Indonesia *Earth Planet Science Letters* **126** 15–22
[8] Handley H K, Blichert-Toft J, Gertisser R, Macpherson C, Turner S, Zaennudin A and Abdurrachman M 2014 Insights from Pb and O isotopes into long-arc variations in subduction inputs and crustal assimilation for volcanic rocks in Java, Sunda arc, Indonesia *Geochimica et Cosmochimica Acta* **139** 205–226
[9] Suhaeli E T 1977 The status of the melange complex in the Ciletuh area, Southwest Java *Proceedings Indonesia Petroleum Association Annual Convention* **6** 241–253
[10] Handley H K, Davidson J P and Macpherson C G 2008 Untangling differentiation in arc lavas: constraints from unusual minor and trace element variations at Salak Volcano, Indonesia *Chemical Geology* **255** 360–376
[11] Hall R 2011 Australia-SE Asia collision: plate tectonics and crustal flow *Geological Society, London, Special Publications* **355** 75–109
[12] Hall R 2002 Cenozoic geological and plate tectonic evolution of SE Asia and the SW Pacific: computer-based reconstructions, model and animations *Journal of Asian Earth Science* **20** 353–431
[13] Metcalfe I and Irving E 1990 Allochthonous terrane processes in Southeast Asia *Philosophical Transactions of the Royal Society of London A* **331** 625–640
[14] Wakita K 2000 Cretaceous accretionary-collision complexes in central Indonesia *Journal of Asian Earth Sciences* **18** 739–749
[15] Arpandi D and Patmosukismo 1975 The Cibulakan Formation as one of the most propective stratigraphic units in the North-West Java Basinal Area *Proceedings of Indonesian Petroleum Association Annual Convention* **4** 181–210
[16] Martodjojo S 2003 *Evolusi Cekungan Bogor, Jawa Barat* (Bandung: ITB Bandung) p 199
[17] Padmosukismo S and I Yahya 1974 The basement configuration of the North-West Java Basinal Area *Proceedings of the Indonesian Petroleum Association Annual Convention* **3** 129–152
[18] Soejono M, S Suparka and S Hadiwisastra 1978 Status Formasi Ciletuh dalam Evolusi Jawa Barat *Geologi Indonesia No. 2* **5** 29–38
[19] Karig D E 1975 Basin genesis in the Philipine sea *Initial Reports of the Deep-Sea Drilling Program* **31** 857–79
[20] Karig D E and Sharman G F 1975 Subduction and Accretion in Trenches *Geological Society of America Bulletin* **86** 377–389
[21] Karig D E, S Suparka, G F Moore and Hehanussa P E 1979 Structure and Cenozoic evolution of the Sunda Arc in the Central Sumatera region *United Nations, ESCAP, CCOP Technical Bulletin* **12** 87–108
[22] Warmada I I, Hendratno A and Sasongko W 2005 Studi komparasi analisis normatif antara metode CIPW dengan metode pemrograman linear LPNORM *Media Teknik Fakultas Teknik Universitas Gajah Mada* **27** 19–24
[23] Rollinson H 1993 Using geochemical data: Evaluation, presentation, interpretation (New York: Longman, Scientific & Technical) p 48–172
[24] Wilson M 1989 *Igneous Petrogenesis* (London: Springer)
[25] Le Bas M J, Le Maitre A, Streckeisen A L dan Zanettin B 1986 A classification of volcanic rock based on the total alkali-silica diagram *Journal of Petrology* **27** 745–750
[26] Ewart A 1982 The Mineralogy and Petrology of Tertiary-Recent Orogenic Volcanic Rocks with Special Reference to Andesitic-Basaltic Composition Range *Andesites: Orogenic Andesites and Related Rocks* ed Thorpe R S (Chichester: Wiley) p 25–95

[27] Cabanis B and Loccelle M 1989 Le Diagramme La/10-Y/15-Nb/8: un outil pour la discrimination des series volcaniques et la mise en evidence des processus de melange et ou de contamination crustale *Compte Rendus de l'Académie des Sciences Series II* **309** 2023–2029
[28] Pearce J A and Cann J R 1973 Tectonic Setting of Basic Volcanic Rocks Determined Using Trace Element Analyses *Earth and Planetary Science Letters* **19** 290–300
[29] Floyd P A and Winchester J A 1975 Magma type and tectonic setting discrimination using immobile elements *Earth and Planetary Science Letters* **27** 211–218
[30] Bougault H, Joronm J -L, M Treuil and Maury R C 1985 Local versus regional mantle heterogeneities: Evidence from hygromagmaphile elements *Initial Reports of the Deep-Sea Drilling Program* **82** 459–482
[31] Cabanis B 1986 Identification des series magmatiques dans les socles metamorphiques sur la base de criteres geologiques, petrographiques et geochimiques *Thèse d'Etat ès Sciences* Université de Paris
[32] Hermann A G, Potts M J and Knake D 1974 Geochemistry of the Rare Earth elements in spilites from the oceanic and continental crust *Contributions to Mineralogy and Petrology* **44** 1–16
[33] Menzies M, Seyfried W and Blanchard D 1979 Experimental evidence of rare earth element immobility in greenstones *Nature* **282** 398–399
[34] Condie K C, Viljoen M J and Kable E J D 1977 Effects of alteration on element distributions in archean tholeiites from the Barberton greenstone belt, South Africa *Contributions to Mineralogy and Petrology* **64** 75–89
[35] Nystrom J A 1984 Rare earth element mobility in vesicular lava during low-grade metamorphism *Contributions to Mineralogy and Petrology* **88** 328–331
[36] Joron J L, Ottonello G, Piccardo G B and Treuil M 1978 Evolution of the upper mantle under the Assab Region (Ethiopia): Suggestions from petrology and geochemistry of tectonitic ultramafic xenoliths and host basaltic lavas *Geol Rundschau* **67** 547–575
[37] Joron J L, Cabanis B and Treuil M 1983 Méthodes d'identification des séries volcaniques anciennes basées sur la géochimie des éléments traces, comparaison avec les séries récentes: exemples d'application *Bulletin Centres Recherches Exploration-Production Elf-Aquitaine* **7** 273–284.
[38] Cann J R 1970 Rb, Sr, Y, Zr and Nb in some ocean-floor basaltic rocks *Earth Planet Science Letters* **10** 7–11
[39] Pearce J and Norry M 1979 Petrogenetic implications of Ti, Zr, Y, and Nb variations in volcanic rocks *Contributions to Mineralogy and Petrology* **69** 33–47
[40] Pearce T H, Gorman B E and Birkett T C 1975 The TiO_2–K_2O–P_2O_5 diagram: A method of discriminating between oceanic and non-oceanic basalts *Earth and Planetary Science Letters* **24** 419–426
[41] Manson V 1967 Geochemistry of basaltic rocks: Major elements *Basalts: The Poldervaart Treatise on Rocks of Basaltic Composition* ed H Hess and A Poldervaart (New York: Interscience) p 215
[42] Cann J R 1971 Major elements variation in ocean-floor basalts *Philosophical Transactions of the Royal Society of London A* **268** 495–505
[43] Pearce J A and Gale G H 1977 Identification of Ore Deposition Environments from Trace Element Geochemistry *Geological Society, London, Special Publications* **7** 14–24
[44] Pearce T H, Gorman B E and Birkett T C 1977 The relationship between major element chemistry and tectonic environment of basic and intermediate volcanic rocks *Earth and Planetary Science Letters* **36** 121–132
[45] Thompson G 1973 A geochemical study of the low-temperature interaction of seawater and oceanic igneous rocks *Transactions of American Geophysics Union* **54** 1015–19
[46] Hart S R, Erlank A J and Kable E J D 1974 Sea-floor basalt alteration: some chemical and Sr isotopic effects *Contributions to Mineralogy and Petrology* **44** 219–230
[47] Mullen E D 1982 $MnO/TiO_2/P_2O_5$: A Minor Element Discriminant for Basaltic Rocks of Oceanic Environments and Its Implications for Petrogenesis *Earth and Planetary Science Letters* **62** 53–62

[48] Ikhram R Hardiyono A, Syafri I Mega A and R M 2018 Petrological Study of Metamorphic Rocks in Melangè Complex of Ciletuh Area, West Java, Indonesia *International Journal on Advanced Science, Engineering and Information Technology* **11** 146–152
[49] Pearce J A 1983 Role of the sub-continental lithosphere in magma genesis at active continental margin *Continental Basalts and Mantle Xenoliths* (Cheshire: Shiva Publication) pp 230–249
[50] Pearce J A 1982 Trace element characteristics of lavas from destructive plate boundaries *Andesites* ed. R.S. Thorpe (New York: John Wiley & Sons) p 525–548
[51] Iddings J P 1892 The origin of igneous rocks *Bulletin of the Philosophical Society of Washington* **12** 89–214
[52] Ionov D A and Hofmann A W 1995 Nb-Ta-rich mantle amphiboles and micas: Implications for subduction-related metasomatic trace element fractionations *Earth and Planetary Science Letters* **131** 341–356
[53] Tatsumi Y, Hamilton D L and Nesbitt D L 1986 Chemical characteristics of fluid phase released from a subducted lithosphere and origin of arc magmas: evidence from high-pressure experiments and natural rocks *Journal of Volcanology and Geothermal Research* **29** 293–309
[54] Saunders A D, Norry M J and Tarney J 1991 Fluid influence on the trace element compositions of subduction zone magmas *Philosophical Transactions of the Royal Society of London A* **335** 377–392
[55] Li C, Arndt N, Tang Q and Ripley E 2015 Trace element indiscrimination diagrams *Lithos* **232** 76–83
[56] Wood D A, Joron J-L, Treuil M, Norry M J and Tarney J 1979 Chemical composition and isotopic ratios of basic lavas from Iceland and the surrounding ocean floor supplement to: Wood D A, Joron J-L, Treuil M, Norry M J and Tarney J 1979 Elemental and Sr isotope variations in basic lavas from Iceland and the surrounding ocean floor *Contributions to Mineralogy and Petrology* **70** 319–339
[57] Sun S -s 1980 Lead isotopic study of young volcanic rocks from mid-ocean ridges, ocean islands and island-arcs *Philosophical Transactions of the Royal Society of London. Series A, Mathematical and Physical Sciences* **297** 409–445
[58] Thompson R N 1982 Magmatism of the British Tertiary Volcanic Province *Scottish Journal of Geology* **18** 49–107
[59] Thompson R M, Morrison M A, Hendry G L and Parry S J 1984 An assessment of the relative roles of crust and mantle in magma genesis *Philosophical Transactions of the Royal Society of London A* **310** 549–590
[60] Best M G 1975 Migration of hydrous fluids in the upper mantle and potassium variation in calc-alkalic rocks *Geology* **3** 429–432
[61] Johnson R W 1977 Distribution and major-element Chemistry of late Cainozoic volcanoes at the southern margin of the Bismarck Sea, Papua New Guinea *Bureau of Mineral Resources, Geology and Geophysics Report* **188** 162–170
[62] Kay R 1977 Geochemical constraints on the origin of Aleutian magmas *Island Arcs, Deep Sea Trenches and Back Arc Basins* ed. Talwani M and Pitman WC (Washington DC: American Geophysical Union) pp 229–242
[63] Dixon T H and Batiza R 1979 Petrology and chemistry of recent lavas in the northern Marianas: implications for the origin of island arc basalts *Contributions to Mineralogy and Petrology* **70** 167–181
[64] Whitford D J and Jezek P A 1979 Origin of late-cenozoic lavas from the Banda arc, Indonesia: Trace element and Sr isotope evidence *Contributions to Mineralogy and Petrology* **68** 141–150

Morphometry Analysis of Sedayu Mountainous Catchment Areas from Detailed Aerial Photos

Tutur[1], Edo Kharisma Army[2,3], Arif Rohman[4*]

[1] Undergraduate Program of Mining Engineering, Faculty of Manufacturing and Earth Engineering, Institut Teknologi Sumatera, Jl. Terusan Ryacudu, Way Huwi, 35365 Lampung Selatan, Indonesia
[2] Research Group of Earth Resource Exploration, Faculty of Manufacturing and Earth Engineering, Institut Teknologi Sumatera, Jl. Terusan Ryacudu, Way Huwi, 35365 Lampung Selatan, Indonesia
[3] Mineral Technology Research Centre, Institut Teknologi Sumatera, Jl. Terusan Ryacudu, Way Huwi, 35365 Lampung Selatan, Indonesia
[4] Geomatic Engineering, Institut Teknologi Sumatera, Jl. Terusan Ryacudu, WayHuwi, 35365 Lampung Selatan, Indonesia

E-mail address: tutur.119370010@student.itera.ac.id

Abstract. Morphometric analysis is a powerful tool to understand the hydrological process quantitatively and the main requirement to determine the hydrological characteristics of a watershed. Morphometric analysis to determine the drainage characteristics of the Sedayu River Sub-watershed, Semaka District, Tanggamus Regency, Lampung Province was carried out using Digital Elevation Model (DEM) data. The primary data used is aerial photography using drones, the location of aerial photos representing watersheds of concern. The catchment area morphometrics were obtained from the extraction of the morphometric characteristics of the combined DEM data from the mountain watershed and the analysis of topographic maps by using ArcGIS software. Based on reports over a period of time, at several points in the area there are frequent river floods that overflow during heavy rains. Morphometric analysis in flood-prone areas with high rainfall is needed. The morphology of the Sedayu area has contrasting variations consisting of several gentle alluvial plains, some of which are Bukit Barisan hills with varying slope variations. The Sedayu watershed in this study is divided into 3 sub-watersheds, namely: sub-watershed A, sub-watershed B, and sub-watershed C. However, only sub-watershed A is the focus of morphometric analysis because its longest river, record flood disasters, landslide records, and has upstream to residential. Morphometric analysis was carried out on parameters such as watershed area (A), watershed length (Lb), main river length (L_n), watershed width (W), drainage density (D_d), bifurcation ratio (R_b), and circularity ratio (R_c). The results of morphometric calculations are the length of the watershed 4.82 km, the length of the main river is Sub-watershed A 8.996 km, the width of the watershed 0.686 km, and watershed area 4.861 km^2 Based on these results can be calculated drainage density (D_d) 2.719 km/km^2 which means Sub-watershed A of a rough type with the inundated surface, permeable sub-soil, and resistant to erosion. Bifurcation ratio (R_b) is 3.333 which means sub-watershed A has a rapid rise and fall in floodwater. Then, the circularity Ratio (R_c) 0.32 which means Sub-watershed A is elongated so that the peak discharge and decrease flow quickly and causes a link with the hydrological response of rainfall-runoff. The results of this morphometric characteristic analysis are the right reference for planning strategies for soil erosion control and soil conservation.

1. Introduction

A watershed is an ideal unit for managing natural resources such as land and water related to natural resource conservation and disaster mitigation in sustainable development [1]. Quantitative analysis of the watershed system is essential so that the watershed's characteristics can be known. The Sedayu watershed is a watershed that often experiences floods and several landslides. It is recorded that an average flood event occurs two times a year in 2020, which is caused by rivers that overflow when rainfall is high. The Sedayu watershed consists of 3 sub-watersheds, namely Sub-watershed A, Sub-watershed B, and Sub-watershed C, which extends through the hills of Bukit Barisan. The Sedayu watershed also traverses areas with high-risk elements, such as the overcrowd Semaka sub-district, Tanggamus, which has a high risk of flash floods. Therefore, the urgency to conduct research increases to solve the problem.

Technological advances in mapping and spatial data processing have led to easier spatial modeling and spatial analysis. Nowadays the development of disaster area mapping technology through remote sensing can be done not only through remote sensing satellites [2] but also through Unmanned Aerial Vehicle (UAV) or so-called drone [3]. Mapping with a drone for mapping disaster areas such as floods and landslides is arising nowadays because the process is fast and can cover most of the, if not all, impact areas. In the phase of a disaster, the drone can help in every phase from prevention to mapping the high-risk area, in the response phase to help provide real-time footage of the evacuation and to help provide a new map (especially terrain map) in the recovery stage [4]. In every phase, the mapping process needs to be executed in a short period. The process of fast mapping the disaster location using drones is commonly called rapid mapping [5]. Combine with Geographic Information System (GIS), we can have powerful tools to help mitigate the flood plain or flood risk area.

Aerial photography technology is a technology that can provide convenience in obtaining fast, accurate, and up-to-date data and information related to objects on the earth's surface. The connection with this research is optimizing aerial photography which then produces an altitude model or DEM (Digital Elevation Model), to analyze the watershed morphometry of the research area quickly, accurately, and up-to-date. Geographic Information System (GIS) is a tool for processing spatial data in the form of maps, thus making it easy to analyze positions relatively close to their distribution [6].

This research activity aims to find out the results of the morphometric analysis of the Sedayu watershed so that it can be a reference for intensive watershed management. This research aims to show data and maps related to the morphometric characteristics of the Sedayu watershed so that it can be determined which sub-watersheds should be prioritized to be managed with natural resource conservation and flood and landslide disaster mitigation. The output of this research can be used for stakeholders in continuing to plan further research activities related to the analysis of the spatial distribution of surface runoff as well as making hydrological study modeling for flood and landslide disaster mitigation using aerial photography technology in this area.

2. Data and Method

2.1. Description of the Research Location

The research location is in the Sedayu watershed, Semaka sub-district, Tanggamus district, Lampung province, Indonesia (Figure 1). This location is considered to have a well-represented morphological type, is located in the mountainous area of Bukit Barisan, has a historical record of floods and landslides worthy of analysis.

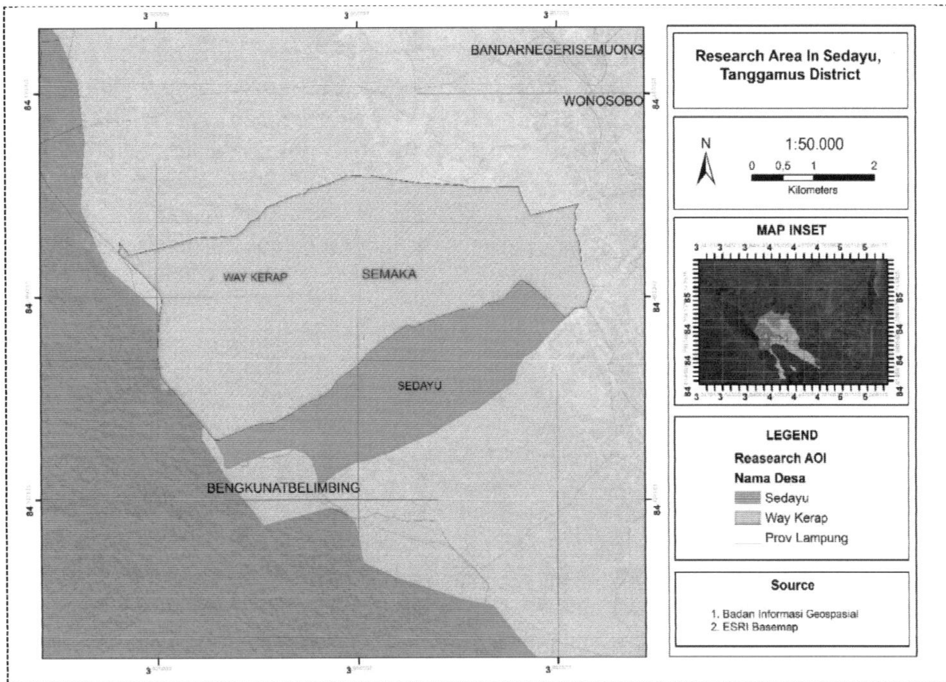

Figure 1. Research Area in Sedayu, Tanggamus District.

A Watershed is an area bounded by natural topography, where all the rainwater that falls in it will flow through a river and out through the outlet on the river [7] (Figure 2). In another word, a watershed is a hydrological unit that describes and uses units of physical-biology and socio-economic activity units for natural resource planning and management. The watershed is limited by the highest points on the earth in the form of a ridge called the stream divide [8].

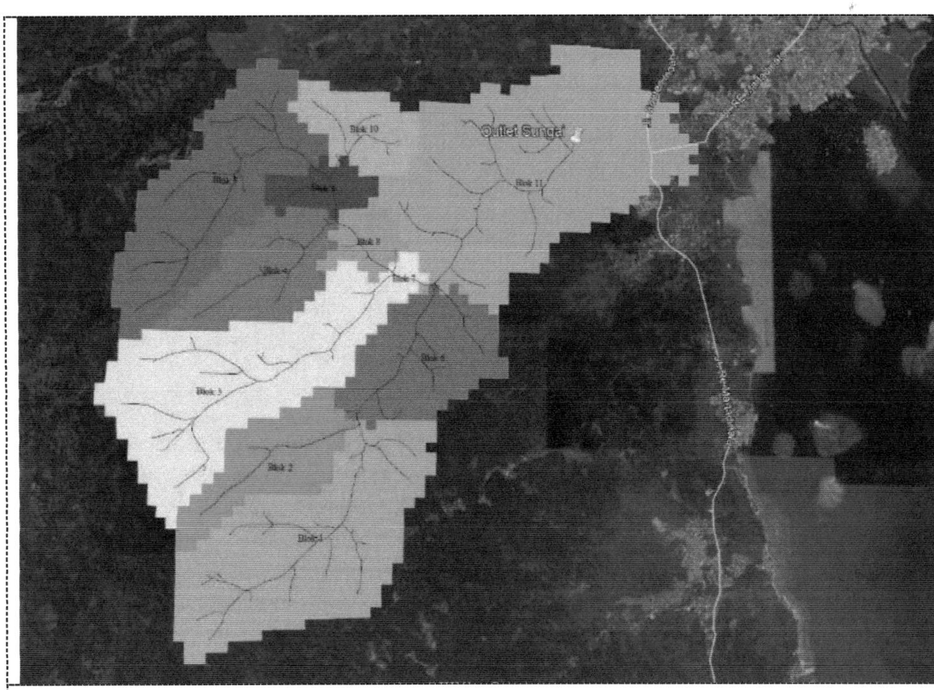

Figure 2. Drainage Basin Example.

2.2. Data Acquisition

Aerial photos are photos obtained from aerial surveys, namely taking aerial photos in certain areas with certain photography rules. A flight path design is made according to the research area, so that the resulting photos are of good quality and according to research needs [9]. The data acquisition using drone requires clear weather. This is due not only to the limited equipment that cannot fly when it is drizzling and windy, but also because the camera requires clear sunlight to produce good photos [10]. Then, there is the ground control point (GCP), which is a field tie point that directs the image to its actual location in the field. Aerial photo processing is carried out using Agisoft Photoscan software to create a Digital Terrain Model (DTM) which will then be processed into a Digital Elevation Model (DEM).

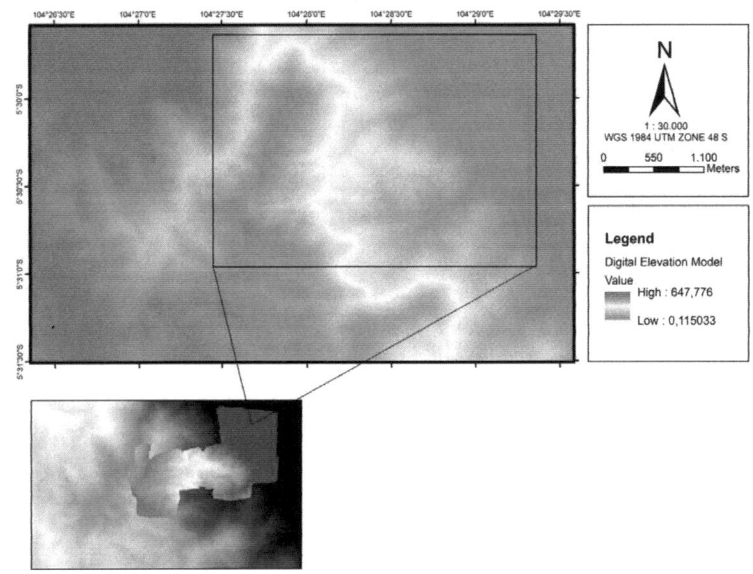

Figure 3. Map Result Combination of DEM Aerial Photography and DEMNAS.

The results of aerial photo processing in the form of an aerial map and Digital Elevation Model (DEM). DEM was obtained from carefully extracted drone UAV data to filter vegetation, buildings. There were problems in the size of the planned area and the resulted area where the resulted area is smaller than the plane area. This is because aerial photos taken by UAV drones cannot be carried out in the area because the drones cannot reach the prepared flight path. It is known that the flyway area is dominated by tall trees and is very steep, so it interfered with the signal. To overcome this, the obtained DEM data is combined with DEMNAS data so that the required DEM area can still be analyzed as seen in (Figure 3).

The data acquisition process using drone includes flight path planning, field orientation and premark installation, premarks measurement, and photos measurement. The flight path planning adapts to the specifications of the UAV used, the UAV used is the DJI Phantom 4 Advanced with specifications:

- Focal Length 35 mm.
- Flight time max 30 minutes.
- Flight Autonomy with Five Vision Sensors.
- Sensor dimension 35.9 x 24 mm.

Planning a flight path using the Drone Deployed application with an accuracy of ground sampling distance (GSD) 3 cm – 5 cm. Footprint covered with the GSD is 205.1 m x 137.1 m with a 3cm/pixel resolution.

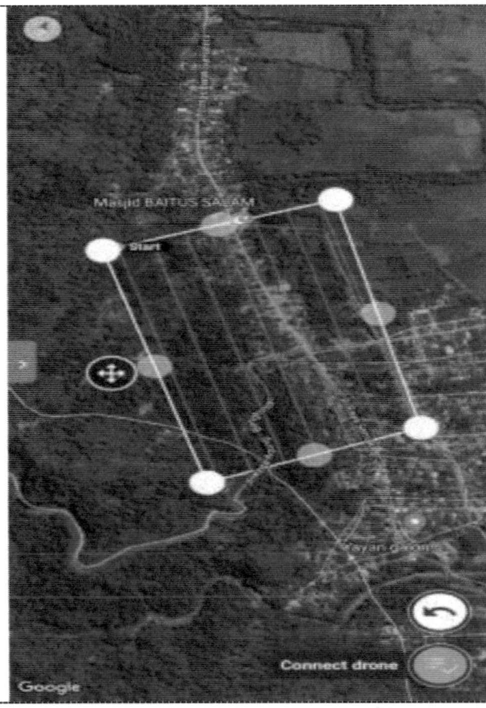

Figure 4. Flight Path.

The control point aims to make the coordinate values in the image the same as the coordinate values in the field. Geodetic GPS and radial method were used to collect the control points. With the Area of Interest (AOI) as seen in figure 5, control points distribution were planned with 9 ground control point (GCP) and 3 independent control point (ICP).

Figure 5. GCP and ICP Planning.

The planned control point is then marked with a size of 30 cm x 10 cm marker (10x GSD premark size) as shown in (Figure 6).

Figure 6. Premark.

Measurement of the GCP and ICP using the static radial method with the closest CORS as a reference point in the Krui area with a GPS observation period of 2 hours. Calibration is done before the drone flies to take aerial photo data. The calibration process on the drone aims to calibrate several sensors on the drone such as the GPS system, Inertial measurement Unit (IMU), altimeter, telemetry and radio system connections.

The photo processing uses Agisoft PhotoScan v1.4.2 Build 6185 software. The general processing process is depicted in (Figure 7).

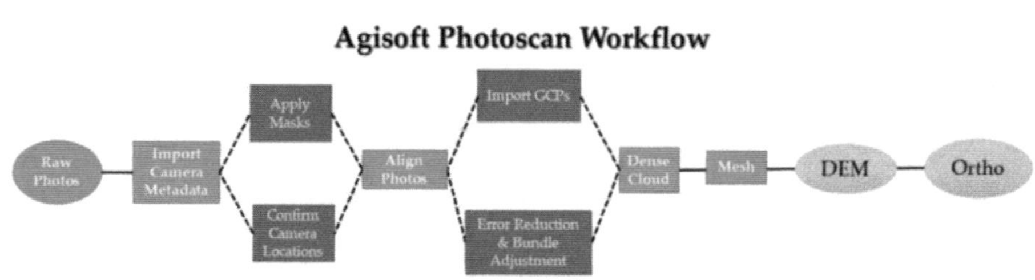

Figure 7. Agisoft PhotoScan Workflow.

The photos that will be processed in the software must first be selected because there are some photos that will not be used. The unused photos are related to the quality of the photos that can be checked from the Agisoft Photoscan.

The output of the processing on the Agisoft PhotoScan software is the Digital Elevation Model (DEM) (Figure 8) and Orthofoto (Figure 9). DEM is divided into 2, namely Digital Terrain Model (DTM) and Digital Surface Model (DSM). The DEM produced in Agisoft PhotoScan software is the Digital Surface Model. The resulting DSM is classified semi-manually and manually to get the ground value to produce a Digital Terrain Model (Figure 8).

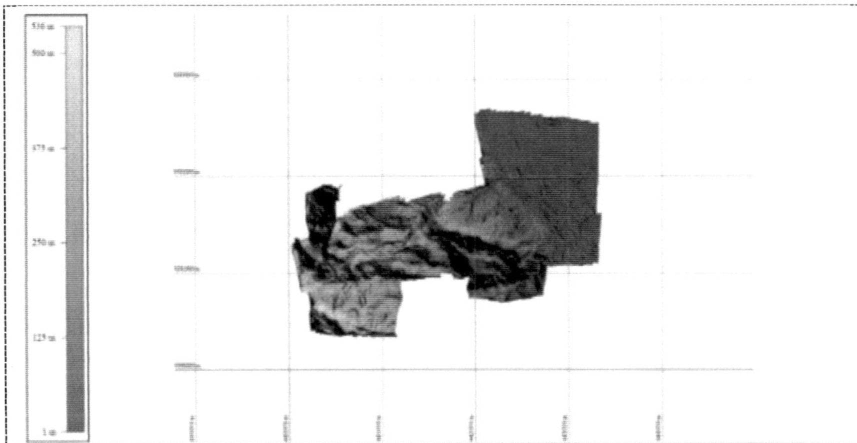

Figure 8. Digital Terrain Model.

Figure 9. Orthophoto.

2.3. Morphometry Analysis

Morphometry is defined as the measurement of shape, which is related to hydrology, first performed by Strahler and Horton between 1940 and 1950 [12]. The measurement parameters in question are watershed area (A), watershed length (L_b), watershed width (W), drainage density (D_d), bifurcation ratio (R_b), circularity ratio (R_c), and watershed shape. The combination of morphometric parameters with human-modifiable factors such as land use, slope, and slope will provide a specific response to the watershed for rainfall, evapotranspiration, infiltration, surface runoff, groundwater content, and river flow behavior [12]. Watershed morphometric characteristics can be used for the emergence of flood events in an area [13].

2.3.1. Area of Watershed. The area of the watershed is the boundary line between watersheds that can separate and divide rainwater into each watershed, if the limit has been determined, area measurements can be carried out. The area referred to in this study is the area of each sub-watershed obtained from morphometric analysis in ArcGIS 10 software. The area of the river is estimated by measuring the area on a topographic map [14].

2.3.2. Watershed Length and Width (L_b and W). The length of the watershed is equal to the flat distance from the mouth of the river upstream along the main river. The watershed width is the ratio between the area of the watershed and the length of the main river. The measurement of the watershed width cannot be carried out directly but by using the following formula [15]:

$$W = A/L_b \tag{1}$$

Where, W is the width of the watershed (km), A is the area of the watershed (km^2), and L_b is the length of the river (km).

2.3.3. Drainage Density (D_d). Drainage density is an index that shows the number of tributaries in a drainage area. Low river density is seen in areas of soil type that are resistant to erosion or highly permeable and when the relief is small [14]. The highest values can occur in soils that are easily roded or relatively impermeable to water, with steep soil slopes and little overgrown vegetation. It is stated as follows [15]:

$$D_d = L_n/A \tag{2}$$

Where, D_d is drainage density (km/km^2), L_n is the total length of all streams in the watershed (km), and A is a watershed area (km^2). Based on the drainage density, watersheds can be graded into the following textures [17]: (a) very coarse (< 2); (b) coarse (2-4); (c) moderate (4-6); (d) fine (6-8); and (e) very fine (> 8).

2.3.4. Bifurcation Ratio (R_b). The flow number law states that the sum of the streams of each successive sequence, A, forms an inverse geometric sequence with the flow order, i, which may be written as:

$$R_b = N_i/N_{i+1} \tag{3}$$

Where, R_b is the branching level index of the river, N_i is the number of river channels for the ith order, and (N_{i+1}) is the number of lion grooves for the ($i + 1$) order. Based on the Strahler method [17], watersheds can be assessed based on the bifurcation ratio with the following conditions (6): (a) $R_b > 3$, the river channel has a rapid increase in flood water level, while the decline is slow; (b) R_b 3-5, the river channel has an increase and decrease in flood water level that is too fast or not too slow; (c) $R_b > 5$, the river channel has a rapid increase in flood water level, as well as the decline, will run quickly.

2.3.5. Circularity Ratio. Circularity ratio (R_c). The ratio of the area of the watershed to the area of the watershed boundary circle, which may be written as:

$$R_c = 4\,\pi\,A/P^2 \tag{4}$$

Where, R_c is the ratio of roundness, A is the area of the watershed (km^2), and P is the perimeter of the watershed (km). Generally, the value of Rc > 0.5 means the watershed is round, Rc < 0.5 means the watershed is elongated [18].

2.3.6. Watershed Shape. River patterns determine the shape of a watershed. The shape of the watershed has an important meaning about river flow, that is, it affects the centralized velocity genre. The watershed form is difficult to express in quantitative terms, the watershed form can be approximated by the circularity ratio shown in (Table 1).

Table 1. The Circularity Ratio ($\boldsymbol{R_c}$) Value Approach to Watershed Shape [18].

R_c	Description
> 0.5	The shape of the watershed is rounded, the peak discharge data is a long time, so does the decline.
< 0.5	The shape of the watershed is elongated, peak discharge comes fast, so does the decline.

3. Findings and Discussion

The results of the delineation, division, and drainage network of the Sedayu watershed based on the combined DEM data from aerial photography and DEMNAS are shown in (Figure 4) below:

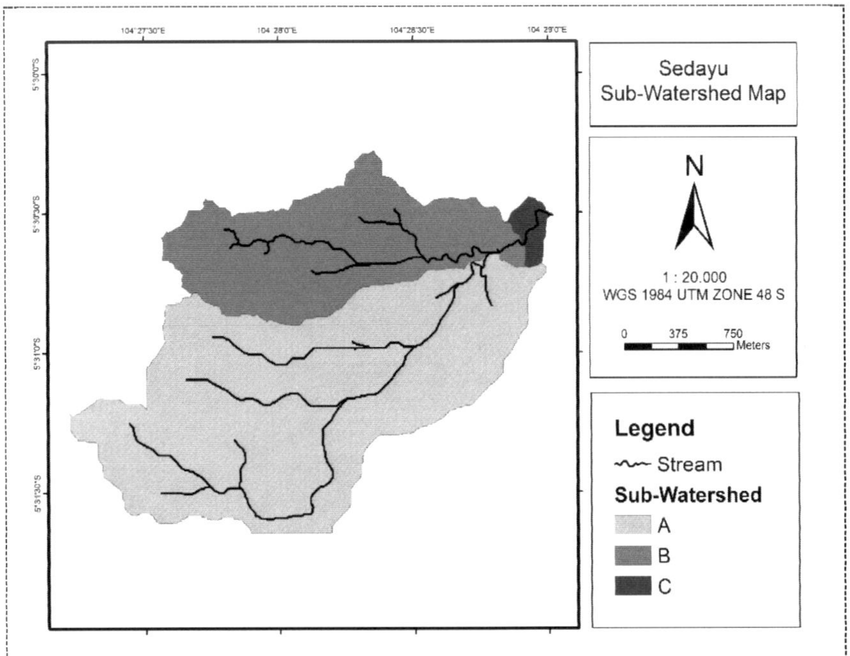

Figure 10. Sedayu River Watershed Drainage Network Map.

Watershed characteristics are physical properties that exist in rivers and are a characteristic of rivers that are described by parameters. The characteristics or physical properties of the watershed such as the length of the main river, the width of the river, and the area of the river are factors that can be analyzed to determine the shape of the watershed. The boundary line between watersheds is a ridge that can separate and divide rainwater into each watershed. The boundary line is determined based on changes in the contours of the image while the area of the watershed can be measured directly using the watershed tool in ArcGIS software.

In this study, the Sedayu watershed is divided into 3 sub-watersheds with the names sub-watershed A, Sub-watershed B, and sub-watershed C as shown in (Figure 10). However, only sub-watershed A has the main flow that will be analyzed for its morphometric characteristics. Sub-watershed A is the river with the longest flow and passes through a path that has a historical record of floods and landslides as can be seen in the (Figure 11). Residents' settlements are known to be upstream of sub-watershed A which is located right in the area of sub-watershed C.

Figure 11. Landslide in the Downstream of the Sedayu Basin From Orthophoto.

Based on the results of the delineation of the boundaries of the Sedayu watershed, it can be known so that the morphometric parameters in sub-watershed A can be calculated and analyzed. The morphometric parameters of the watershed are Area of Watershed (A), Watershed Length (Lb), and Watershed Width (W). In detail, the morphometric parameters of the Sedayu watershed are presented in Table 2 below:

Table 2. Measurement Result of Morphometric Parameter of Sedayu Watershed.

Morphometric Parameters	Measurement Results
Area of Watershed (A)	4.861 km^2
Watershed Length (Lb)	4.820 km
Watershed Width (W)	0.686 km

The results of the above measurements are used as the basis for performing mathematical calculations on other morphometric parameters. So that the calculation value of each morphometric parameter used in sub-watershed A is shown in Table 3:

Table 3. Calculation Results of Morphometric Parameters of Sub-Watershed A Sedayu

Morphometric Parameters	Calculation Results
Main River Length (L_n)	8.996 km
Area of Sub-Watershed A (A)	3.309 km^2
Perimeter (P)	11.392 km
Drainage Density (D_d)	2.719 km/km^2
Bifurcation Ratio (R_b)	3.333
Circularity Ratio (R_c)	0.320

The calculation results show that the value of drainage density is 2,719 km/km^2 which is known from the results of the morphometric analysis to have a coarse type. The low drainage density in sub-watershed A indicates permeable sub-soil under low vegetative cover. Then, it is also known that if the drainage density value is less than 1 mile/mile2 (0.62 km/km^2), then the watershed will experience flooding and if the drainage density value is greater than 5 mile/mile (3.10 km/km^2), then the watershed will often experience drought [19]. Based on these criteria, sub-watershed A is known to be inundated, the soil sub-soil is permeable, meaning it is resistant to erosion.

The bifurcation value from the calculation is 3.333 which is known from the results of the morphometric analysis, that sub-watershed A has a river channel with the increase and decrease of the floodwater level being too fast or not too slow. Then, from the calculation of the circularity ratio value of 0.320, which is known from morphometric analysis, it indicates the shape of the sub-watershed A is elongated so that the peak discharge comes very quickly, as well as its decrease [18]. From the morphometry data obtained it is known that the type of Sub-DAS A is coarse with an elongated shape so that the hydrological response of the rainfall-runoff water relationship is resistant to erosion, flooded surfaces, peak discharges come very quickly, and as well as its decrease.

4. Conclusion

DEM data can be extracted or obtained from detailed aerial photos and DEMNAS data processed using Agisoft PhotoScan software. DEM data is used to extract the morphological and morphometric characteristics of the Sedayu watershed with the help of ArcGIS software so that morphometric analysis can be performed. In this study, Sedayu watersheds were divided into 3 sub-watersheds, with the names sub-watershed A, sub-watershed B, and sub-watershed C. Then the calculation of morphometric parameters is done and the conclusion is that sub-watershed A is the main and longest river with a length of 8.996 km. The result of drainage density (D_d) calculation of 2.719 km/km^2 means Sub-watershed A is of a rough type with the inundated surface, permeable sub-soil, and resistance to erosion. Bifurcation ratio (R_b) 3.333 which means sub-watershed A has a rapid rise and fall in floodwater. Then, the Circularity Ratio (R_c) 0.32 which means Sub-watershed A is elongated so that the peak discharge and decrease flow quickly. The results of these calculations showed that Sub-watershed A studied has a hydrological response of the rainfall-runoff relationship in the water is often flooded, but at peak discharge and the decrease flows quickly.

5. Acknowledgment

This research is supported by Hibah Penelitian ITERA 2021 (ITERA Grants 2021).

6. References

[1] Cahyadi A. Kajian Permasalahan Daerah Aliran Sungai Juwet Kabupaten Gunungkidul Dan Usulan Penanggulanganya. 2017;1–10.
[2] Smith LC. Satellite remote sensing of river inundation area, stage, and discharge: A review. Hydrol Process. 1997;11(10):1427–39.
[3] Feng Q, Liu J, Gong J. Urban flood mapping based on unmanned aerial vehicle remote sensing and random forest classifier-A case of yuyao, China. Water (Switzerland). 2015;7(4):1437–55.
[4] Griffin GF. The use of unmanned aerial vehicles for disaster management. Geomatica. 2014;68(4):265–81.
[5] Tampubolon W, Reinhardt W. UAV data processing for rapid mapping activities. Int Arch Photogramm Remote Sens Spat Inf Sci - ISPRS Arch. 2015;40(3W3):371–7.
[6] Nurfaika. Analisis Karakterist Morfometri Daerah Aliran Sungai Melalui Pemanfaatan Penginderaan Jauh Dan Sistem Informasi Geografi. J Geogr. 2015;1–34.
[7] Eng SM, Dr I. Pelestarian Sumber Daya Tanah dan Air. Penerbit Andi Yogyakarta. 2002.
[8] Hallaf HP. Geomorfologi Sungai dan Pantai. Jur Geogr FMIPA UNM Makassar. 2005.
[9] Lereng PL. Ekstraksi Parameter Topografi sebagai Survei Pragmatik untuk Analisis. 2021;2(1):29–40.
[10] Rohman A, Prasetya DB. Rapid Mapping for Simple Flood Mitigation Using Commercial Drone at Way Galih Village, Lampung, Indonesia. Forum Geogr. 2019;33(1):101–13.
[11] Pidwirny M. Stream morphometry. Fundamentals of physical geography. 2006.

[12] Noges T. Relationships between morphometry, geographic location and water quality parameters of European lakes. Hydrobiologia. 2009;633(1):33–43.
[13] Nugroho S. Respon morfometri dan penggunaan lahan DAS terhadap banjir bandang (Studi kasus bencana banjir bandang di Sungai Bohorok). Taken back from http//sirrma bppt go id. 2009.
[14] Sosrodarsono S. Hidrologi untuk pengairan. 2019.
[15] Seyhan E. Dasar-dasar Hidrologi. Editor, Soenardi Prawirohatmojo. Yogyakarta. UGM Press; 1977.
[16] Krishnamurthy J, Srinivas G, Jayaraman V, Chandrasekhar MG. Influence of rock types and structures in the development of drainage networks in typical hardrock terrain. ITC J. 1996;(3/4):252–9.
[17] Strahler AN. Part II. Quantitative geomorphology of drainage basins and channel networks. Handb Appl Hydrol McGraw-Hill, New York. 1964;4–39.
[18] Soewarno S. Hidrologi Pengukuran dan Pengolahan Data Aliran Sungai (Hidrometri). Nova, Bandung. 1991;825:20.
[19] Linsley RK, Kohler MA, Paulhus JLH. Hidrologi untuk Insinyur, edisi-3. Penerjemah Yandi Hermawan Jakarta: Penerbit Erlangga. 1996.

IOP Conf. Series: Earth and Environmental Science 1047 (2022) 012015 doi:10.1088/1755-1315/1047/1/012015

The Application of Electrical Resistivity Tomography using Wenner-Schlumberger Arrays Configuration to Identify the Geological Structure in Kancah Warm Spring, West Java Indonesia

Patriks Vero Tongkeles[1], Suryantini[1,2], Angga Bakti Pratama[1]

[1]Geothermal Engineering Master's Program, Faculty of Mining and Petroleum Engineering, Institut Teknologi Bandung, Jl. Ganesha 10, Bandung 40132, Indonesia.
[2]Applied Geology Research Group, Faculty of Earth Science and Technology, Bandung, Indonesia.

Email: vero_patriks@yahoo.com

Abstract. Kancah warm spring is located in the southern part of Tangkuban Perahu Volcano, in West Java Indonesia. This warm spring has a temperature 47^0 C and a pH of 3 and it is used as a source of water for nearby tourist attractions. This warm spring has a large discharge, with the source thought to be from subsurface water associated with Tangkuban Perahu Volcanic complex. Knowing the source and geological control of this warm spring is one way to further identify the Tangkuban Perahu geothermal system, especially its permeability. For that reason, this research aims to using Electrical Resistivity Tomography (ERT) method to identify the geological structure that controls the appearance of geothermal manifestation using the resistivity method with Wenner – Schlumberger arrays configuration. The study starts with lineament study using DEM to identify the structure in surface and then determine the geometry of subsurface geological structures are using the electrical resistivity tomography (ERT) method using the Wenner - Schlumberger configuration. with a length 320 meters of each line and the electrode spacing is 10 m. The results of the interpretation of satellite images show that there is a lineament which is interpreted as a fault structure with a NE-SW direction that is right at the location of the discharge of the warm spring. There are 3 - 4 faults identified in each profile, some of the faults identified beneath the Kancah spring and may controlled the warm spring. These faults are a permeable that flows hydrothermal fluids from the subsurface to the surface in the form of warm spring..The lithology beneath of Kancah warm spring are dominated by breccia pyroclastics, clays, and a small amount of andesite igneous rocks.

1. Introduction

The study using ERT (Electrical Resistivity Tomography) was carried out in the manifestation of the Kancah warm spring located in West Bandung Regency, West Java Province. This warm spring has a fluid temperature of 47^0C, the pH value is around 3. This warm spring has a large discharge, with the source thought to be from subsurface water associated with the Tangkuban Perahu Volcanic complex.

Knowing the source and geological control of this hot spring is one way to further identify the Tangkuban Perahu geothermal system, especially its permeability.

ERT also called electrical resistivity imaging, is a geophysical technique used for imaging sub-surface structures such as fault and fracture from electrical measurements made at the surface with an electrode array, in which each electrode is used alternately as a potential and current electrode [1]. This method is a combination of vertical electrical sounding (VES) and constant separation tranvesting (CST) method which is very good for describing subsurface condition[2].

The physical parameter used in this ERT method is resistivity. Resistivity is one of the useful geophysical parameters to predict and identify the presence of geothermal system features. Fluid salinity, alteration minerals, high temperatures, and geological structures are factors controlling the resistivity value of a geothermal system [2]. In the exploration of geothermal systems, the ERT method is very suitable for detecting the position and geometry of small geological structures which usually control the occurrence of geothermal manifestations. Resistivity anomaly caused by geological structures such as faults can be seen with a 2-dimensional resistivity cross-section.

This ERT method has been widely used for the detection of subsurface geological structures such as for the detection of buried fault structures in Shandong China by using a combination of dipole-dipole and Wenner electrode configurations [1], then this ERT method with Wenner-Schlumberger configuration has also been carried out for the identification of subsurface layer [3].

Therefore, based on the above background, the ERT method in this study was carried out to detect subsurface faults that control the emergence of warm spring and subsurface lithology of Kancah warm springs.

2. Electrical Resistivity Tomography

Electrical resistivity tomography is based on the study of the differences in the electrical properties of rocks and minerals in the earth's crust, a kind of geophysical exploration method to solve the geological structure or to find useful minerals by using the spatial distribution of electric field [4]. ERT method in principle works by injecting an electric current into the earth through two current electrodes, causing a potential difference. The results of current and potential differences measurements for each different electrode distance can be used to reduce variations in the value of the resistivity of the layers below the measuring point (sounding point) [2].

The current electrode placement method consists of two, namely the first is depth sounding to determine the vertical variation of the resistivity. This method is known as Vertical Electrical Sounding (VES). The principle of this method is when distance between the current electrodes increases, the electric current will pass through the rock layer with the measurement position in the middle of the electrode arrangement. To reach greater depths, resistivity measurements are made with closer electrode spacing and longer stretches. While the second method is CST or Constant Separation Travesting, also known as electrical profiling, this method produces a horizontal/lateral profile of resistivity variations. The combination of these two methods is called ERT or electrical resistivity tomography, this method combines vertical and horizontal variations and has a high-resolution mapping of the shallow subsurface of the earth [2].

In this study, the ERT method was used to identify and map geological structures and permeability. The arrays configuration used is Wenner-Schlumberger which is a combination of Wenner and Schlumberger. The Wenner-Schlumberger configuration is a configuration with a system of constant spacing rules with a note that the "n" factor for this configuration is the ratio of the distance between the electrodes C1-P1 (or C2-P2) with the space between P1-P2 as shown in Figure 1. If the distance between the potential electrodes (P1 and P2) is the distance between the current electrodes (C1 and C2) is 2na + a. The process of determining the resistivity using 4 electrodes placed in a straight line [5]. Based on the distance between the electrodes, the geometry factor of the Wenner-Schlumberger configuration becomes:

$$K = \pi n(n+1)a \tag{1}$$

The apparent resistivity for the Wenner-Schlumberger configuration is:

$$\rho = \pi n(n+1)a\,\frac{\Delta V}{I} \tag{2}$$

where ρ is apparent resistivity (Ωm), a is electrode spacing (m), I current (mA) and V is potential difference (mV) [5].

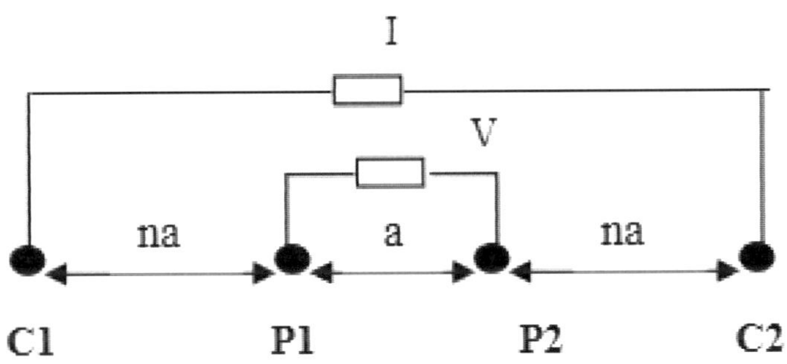

Figure 1. The setting of Wenner – Schlumberger arrays configuration electrode.[5]

3. Overview of the Study Area and Method

3.1. The Study Area

Kancah's location is in the 48S UTM Zone, precisely at 791200 mE and 9264300 mN or ± 16 km from Bandung City and can be reached by car for 30 minutes (Figure 2)

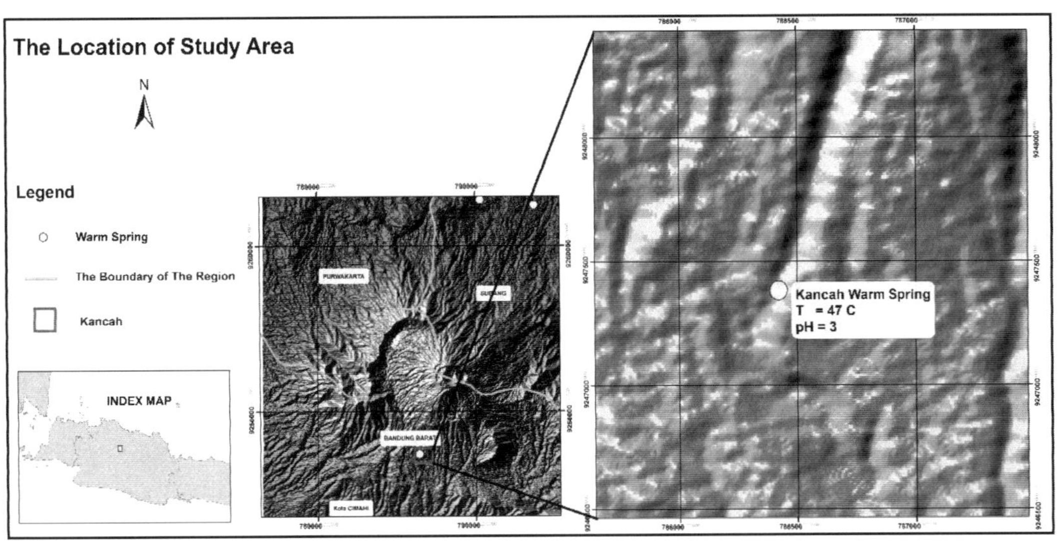

Figure 2. The location of study area

Kancah warm spring is located in southern of Tangkuban Perahu volcanic complex and part of Tangkuban Parahu geothermal system. The Tangkuban Perahu geothermal system is a hydrothermal volcanic system where the heat source is magma under the Tangkuban Perahu volcanic complex and fluids from the surface enter through permeable structures. Based on geological studies, Tangkuban Perahu is a hydrothermal volcanic system with the heat source coming from magma intrusion beneath the surface [6]. Based on the geological map of Bandung in Figure 3, Kancah warm spring is in the south of Tangkuban Parahu volcano, and has a QSL Lithology which is a Sunda andesite unit consisting of volcanic breccia and andesite lava

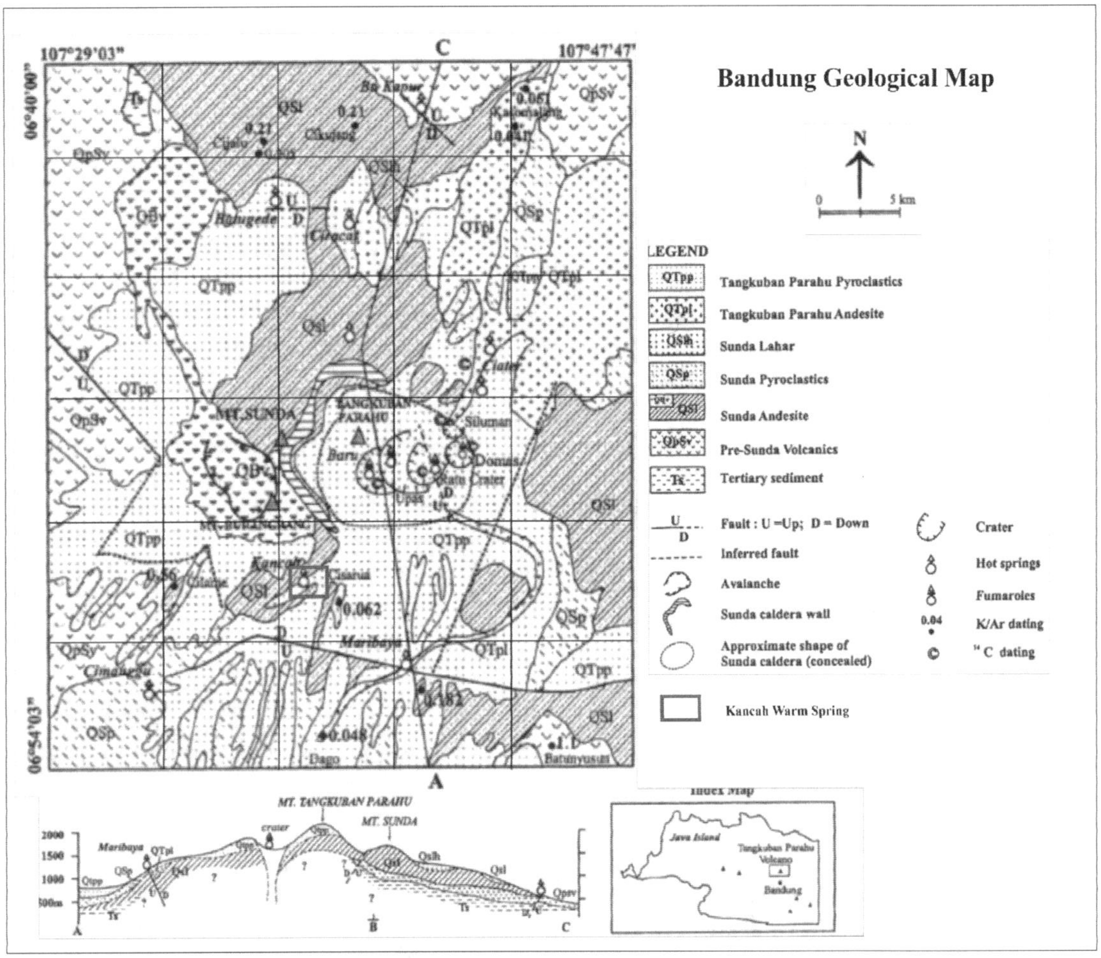

Figure 3. The geology of Sunda and Tangkuban Perahu Volcanic Complex, redrawn after [8]

3.2. Data and Method

To identify the geological control of warm spring in Kancah a literature review has been studied to build initial assumptions and scope of the research. The basic framework is to delineation the geological structures in the surface using lineament study from DEM and identification of geological structures in subsurface using Electrical Resistivity Tomography. This research starting with lineament study to identify the fault based on the surface morphology of satellite. The results of the interpretation of the fault on the surface are the basis for the ERT survey design. The survey design can be shown on Figure 4.

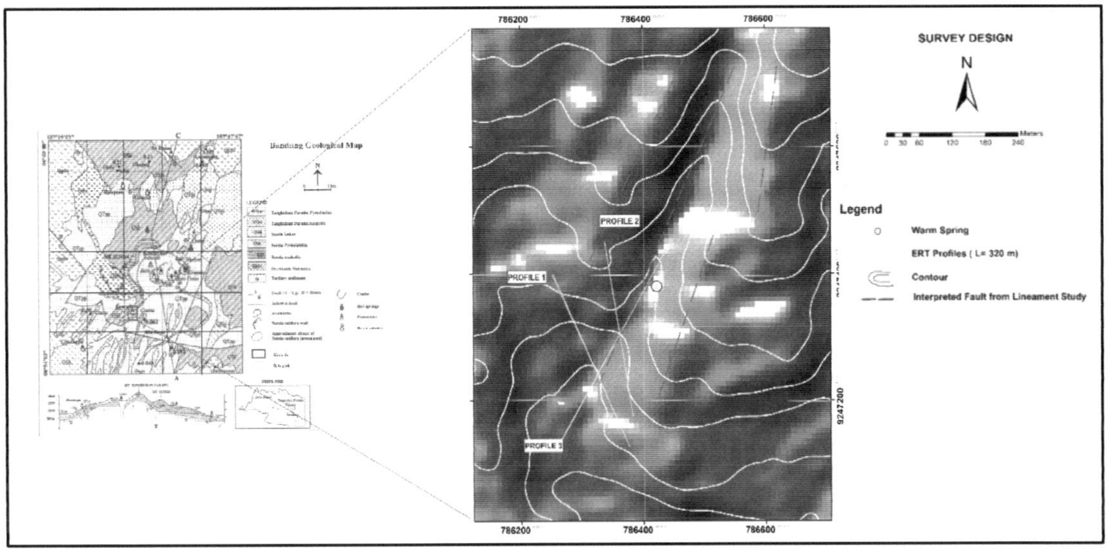

Figure 4. Survey Design of ERT in Kancah warm spring.

Resistivity data acquisition was carried out with 3 measurement lines, line 1 and 2 have the same direction SE-NW, and line 3 perpendicular the two lines in the NE-SW direction. The electrodes used are current and potential totaling 32 with a length of 320 m and the distance between each electrode is 10 m.

Inverse modeling electrode configuration was done on each line. The result of the inversion is 2D resistivity profiles to used for the interpretation of the geological structure and the lithology of Kancah warm spring.

4. Result

The results of the lineament method using DEM (Figure 4) show that there are morphology of valleys and hills which are identified as NE-SW trending fault structures. This structure is probably a permeable structure that control hydrothermal fluids in subsurface. The interpretation of this fault structure becomes the benchmark for the resistivity acquisition of the ERT method.

The ERT acquisition carried out resulted 2D profiles shown in Figure 5. Two-dimensional resistivity profiles were processed by using least-square inversion algorithm to create a model of subsurface resistivity by inverting the apparent resistivity measurements. In this inversion method, the subsurface along the measured profile is divided into a number of rectangular blocks equal in number to the measured apparent resistivity points. During the inversion procedures, the apparent resistivities of the sub surface blocks are estimated and the difference between the observed and calculated data is minimized [3]. 2D resistivity profiles with 320 m length of each line, resulting 209 datum points. The maximum readable depth is 50 to 60 m. The measured resistivity value is in the range of 0.02 Ωm to 2500 Ωm.

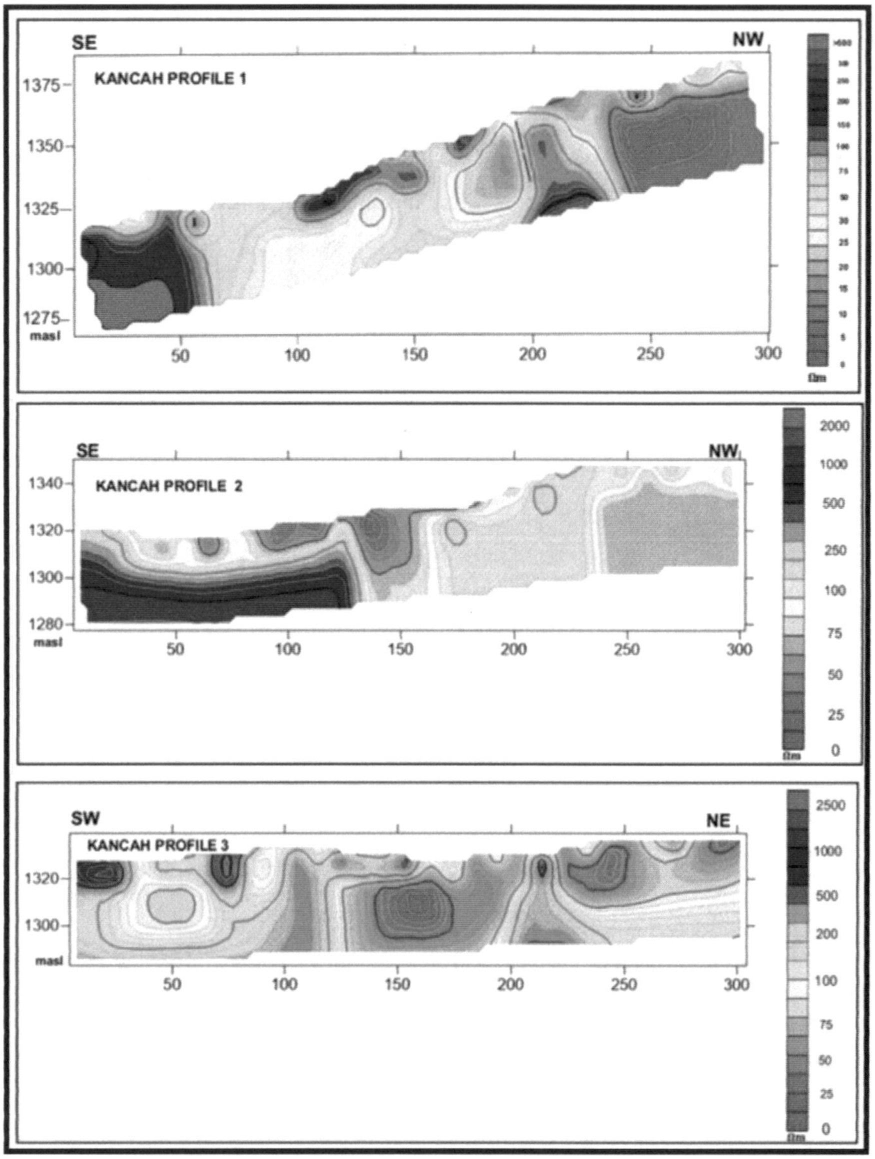

Figure 5. The Uninterpreted Two-dimensional ERT profiles showing the distribution of resistivity with depth

Interpretation is carried out using [7] and [2], and also integrated with Geological Map scale 1.100.000 from [8]. The results of the ERT Profiles interpretation can be seen in Figure 6.

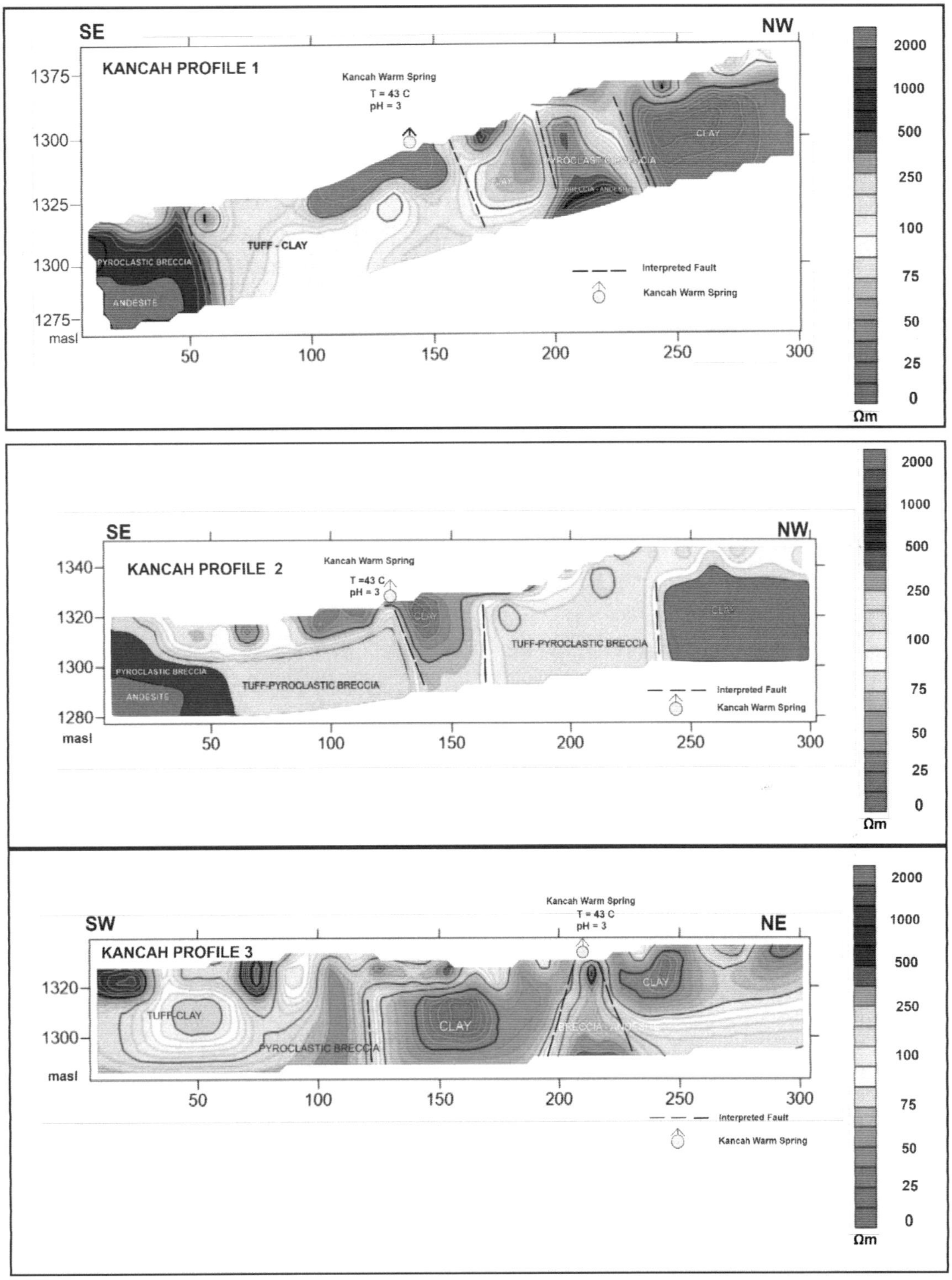

Figure 6. The Interpreted Two-dimensional ERT profiles showing the distribution of Fault and Lithology of Kancah warm spring subsurface

Several faults were identified in the three profiles, these faults were identified based on the anomaly in the cross-section of each profile. In the profiles plot the location of the warm spring with the temperature and pH value.

Based on interpretation of resistivity value with type of the rocks and integrated with geological map, Profile 1 has a SE-NW direction with a depth of 50 m and a line length of 300 meters (Figure 6). In the SE section, the lithology is dominated by pyroclastic breccia with resistivity value around 500 – 1000 Ωm and andesite igneous rocks with resistivity value >2000 Ωm, at a distance of 50 – 200 m, the rocks are dominated by interludes of tuff-clay and clay with resistivity value 75- 150 Ωm while at a distance of 300 m or the NW part of the rock is dominated by clay with resistivity value <50 Ωm and also there are 4 faults identified. Profile 2 (Figure 6) has a SE-NW direction with a depth of 60 m and a line length of 300 m. in the SE section, the lithology is dominated by pyroclastic breccia with resistivity value 500 – 1000 Ωm and andesite igneous rocks with resistivity value >2000 Ωm at depths below 50 m, further to the NW the lithology is dominated by interludes of clay, tuff, and pyroclastic breccias with resistivity value 5 – 500 Ωm. There are 3 faults identified in profile 2. Profile 3 has a SW-NE direction and is perpendicular to profile 1 and profile 2 (Figure 6), the lithology in profile 3 is dominated by interludes of Tuff (75 -100 Ωm), Clay (<50 Ωm), and Pyroclastic Breccias(500 – 1000 Ωm). There are 3 faults identified in this profile.

Several faults were identified right under the warm spring and were the control manifestations of the warm spring. These faults are a permeable that flows hydrothermal fluids from the subsurface to the surface in the form of warm spring.

5. Conclusion

Based on this study, it can be concluded that:

1. Kancah Warm Spring which is located in southern Tangkuban Perahu volcanic complex is controlled by fault, this is evidenced by the results of 2D ERT profiles which show a resistivity anomaly that forms a trend which is interpreted as a fault.
2. There are 3-4 faults identified in each profile, some of the faults identified beneath the Kancah spring and may be control manifestations of the warm spring. This fault is a permeable fault that flows hydrothermal fluids from the subsurface to the surface in the form of geothermal manifestations.
3. The lithology beneath Kancah warm spring are dominated by breccia pyroclastics (500 – 1000 Ωm), clay (<25 Ωm), and a small amount of andesite igneous rocks (>2000 Ωm).

References

[1] Zhu T, Feng R, 2009. The Application of Electrical Resistivity Tomography to Detecting a Buried Fault: A Case Study. Journal of Enviromental and Engineering Geophysics September 2009, Volume 14, Issue 3, pp. 145–151.

[2] Reynolds J M, 2011. An Introduction to Applied and Enviromental Geophysics. United Kingdom.

[3] Jamaluddin, Umar E P, 2017. Identification of Subsurface layer with Wenner-Schlumberger Arrays Configuration Geolectrical Method. Global Colloquium on GeoSciences and Engineering 2017.

[4] Zhang P. 2017. Fault Zone Preliminary Identification Study in Xinchang Block Based on Electrical Resistivity Tomography. 2nd International Conference on Advances in Materials, Mechatronics and Civil Engineering 2017.

[5] Kanata B. 2008. Aplikasi Metode Geolistrik Tahanan Jenis Konfigurasi Wenner- Schlumberger untuk Survey Pipa Bawah Permukaan. Majalah Ilmiah Teknologi Elektro 84(7 2), July 2008.

[6] Saputra M P, Suryantini. 2015. Correlation Study between Volcanic Activities and Thermal Water Changes in Tangkuban Perahu Hydrothermal Prospect, Jawa Barat, Indonesia.

Proceedings World Geothermal Congress 2015 Melbourne, Australia.
[7] Telford, WM., 1990. *Applied Geophysics Second Edition*, Cambridge University.
[8] Nasution A, Kartabinata M N, Sutamingsih E, 2004. Geology, Age Dating and Geochemistry of the Tangkuban Parahu Geothermal Area, West Java, Indonesia.

The Significances of Cretaceous Petrified Wood Fossils from Padangcermin, Lampung in Paleoenvironmental Perspectives

D G Harbowo[1,2,a], R N F A Nahar[1], D Sari[1], T Julian[1], T A Kuswara[1], R A Abimayu[1], R A Lajona[1], S Huzaifah[1].

[1] Department of Geology, Institut Teknologi Sumatera, Wayhui, Lampung, Indonesia
[2] Doctoral Program of Geological Engineering, Institut Teknologi Bandung, Bandung, West Java, Indonesia
[a] correspondent author: danni.gathot@gl.itera.ac.id

Abstract. Reports on the presences of petrified wood fossils in Padangcermin, Lampung, are rare, no strong examination that expose the scientific concern on its paleoenvironment perspective. By the published geological map, these fossils was found in the area that labelled as Cretaceous sedimentary rock (145-66 mya). Recently, the known oldest petrified wood fossil in Indonesia, is from Miocene epoch (23.03-5.33 mya). It is an interesting circumstance to examine. By this study, we are trying to confirm this circumstances by conducting the geological mapping, collecting, and identifying the petrified wood fossils that could be found as *in situ* fossils in the area. This study reveals a new update in paleoenvironmental perspective. The petrified woods in this area are commonly preserved in cherty layer. There are also several fossilized aquatic freshwater plant remains that presences in this area. We suggest the fossilization of the petrified wood fossil in this area is related to high alkaline-freshwater lacustrine. Supposedly, it associated with the Cretaceous terrestrial landmass and freshwater lacustrine environment in southern paleo-Sumatra by paleontological interpretation of data in this study.

Keyword: petrified wood, fossilization, Menanga Formation, paleoenvironment, Lampung.

1. Introduction

Petrified wood is one example of plant fossil. It occurs due to the fossilization of woody plants caused by a series of permineralization and mineral replacement [1,2]. Several wood features would preserved and petrified in million years naturally. Silicification plays a significant role in common petrified wood [3]. Regularly, surface leaching and hydrothermal activity are the main roles in fossil silicification [4,5]. Several cations also significantly affect color of the fossils, such as iron oxide and manganese, which brought reddish to brownish color to the fossils [5]. The presence of petrified wood fossil are significance for geological history interpretation, especially in paleoenviromental perspective. The presences of these fossils are indicate and confirm the existence of terrestrial environment at the past and help to reconstruct the model of paleogeography in a geological time span [6,7]. The Cretaceous paleogeographical model of western Indonesia are puzzling, mainly in the distribution of the landmass in southern Sumatra (Lampung). We expect to contribute in the paleontological data for this circumstances, especially in the petrified wood.

Indonesia is known as an area that has fossilized petrified woods. Most of the recent finding petified wood fossil in Indonesia, found in the Pliocene volcanic deposits (5.33-2.58 mya), including in Sumatra

[8,9]. The oldest petrified wood in Indonesia was recovered in the Miocene sedimentary rock (23.03-5.33 mya), in Kalimantan [10]. Meanwhile, there are only a few well-known studies about petrified wood in Sumatra, particularly in Lampung. Based on the geological map of Tanjungkarang, there are indications of the presence of the older plant fossils [11], which are in the Cretaceous (145-66 mya) sedimentary rock unit, see **Figure 1**. It is much older than the Kalimantan wood fossil. This is an important record worthy of scientific investigation, especially to support a new perspective on paleogeography in western Indonesia during the Cretaceous Period.

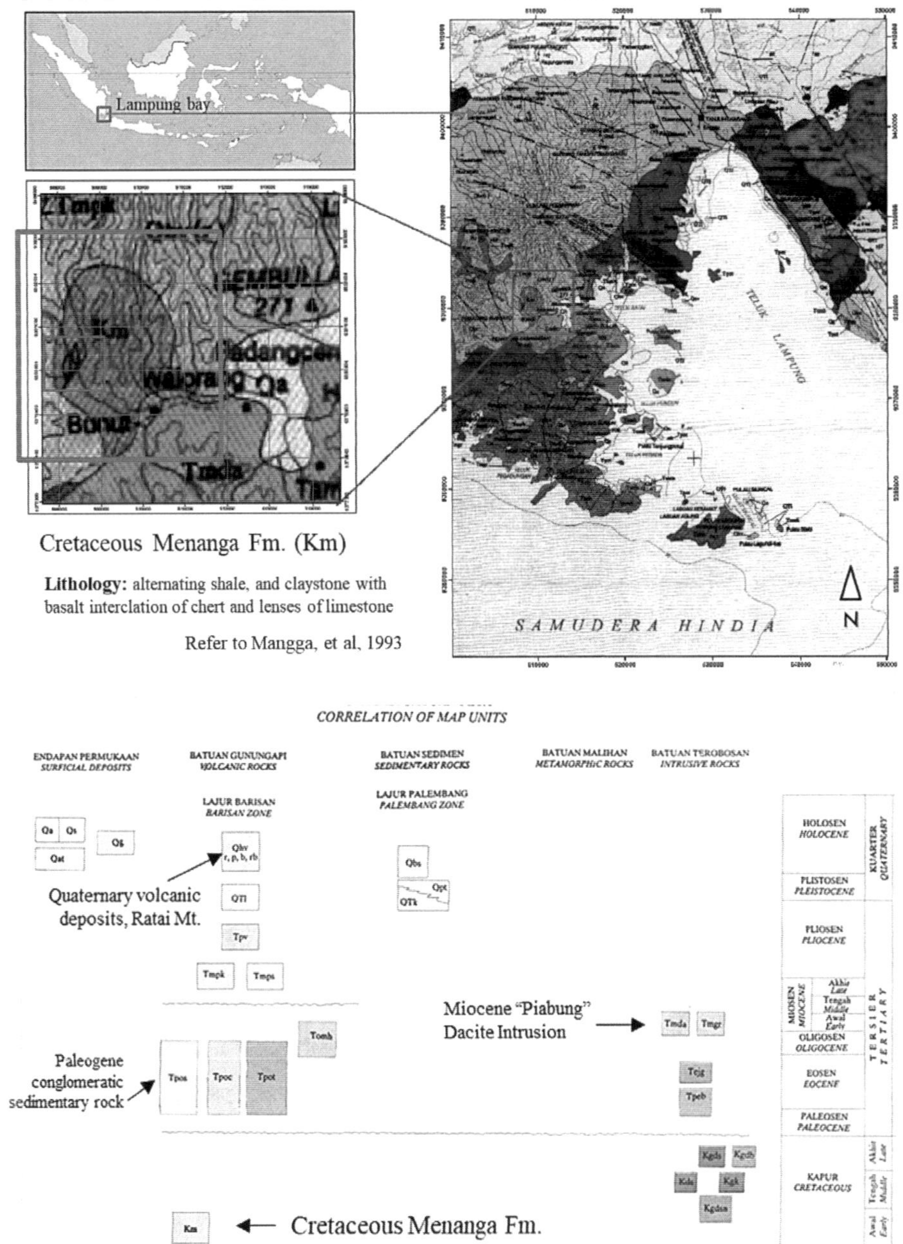

Figure 1. Geological map shows the presence of plant fossil in Cretaceous sedimentary rock in Pesawaran, Lampung, Indonesia referring to [11] - red box shows the location of the study area.

IOP Conf. Series: Earth and Environmental Science 1047 (2022) 012016 doi:10.1088/1755-1315/1047/1/012016

Based on a published geological map of Tanjungkarang, marked the presence of plant fossils with leaf symbols in Menanga Formation, a Cretaceous sedimentary rock. It pointed in Padangcermin, Pesawaran Lampung, at southern of Sumatra. Menanga Formation consists of alternating shale and claystone with basalt intercalation of chert and limestone lenses [11,12]. Menanga Formation is the oldest sedimentary rock unit, followed by Tertiary Hulusimpang Formation, conglomeratic Sabu Formation; Miocene dacite intrusion, and covered by Quarternary volcanic deposit from Ratai Volcanic Complex [11,13,14]. There are not enough descriptions that mention the presences of the plant fossils and its significance of a paleoenvironmental perspective in Padangcermin, Lampung. There are only remains of unknown any fossil that can be found in this area. By this research, firstly, we confirm the presence of the petrified wood in Padangcermin Pesawaran, Lampung by mapping and identifying these fossils finding. Secondly, we determine the features of these fossil samples from to interpret paleoenvironmental condition.

2. Method

We conducted a standard geological mapping in Padangcermin, Pesawaran in Lampung. Located the outcrop using GPS and make a stratigraphy profile. We pointed out and described the outcrop that contain plant fossil, including the petrified wood fossil. The petrified wood fossil that collected are prepared for thin section observation in polarized microscope. The wood fossil anatomy are determined for further examination according to the vessels, ray cells, tracheids, and parenchyma cells.

3. Result

There are about 10 sites that have outcrops containing plant fossils, including fossilized petrified wood (see **Figure 2** and **Figure 3**). Most of the area is strike oriented toward NW-SE, and dip toward NE-SW, with an inclination angel of about 13-18°. The outcrop is located on the north side of the Wayratai River, which flows along the Padangmirin valley. It was oriented W-E, leading to Ratai Bay to the east of the area. The stratigraphic profile shows the facies changes from the lower section to the upper section. This indicates that the lower section is dominated by the alternation of gray claystone and reddish sandstone. Then the middle to the upper section is dominated by a layer of massive reddish-brown chert. We also compared the relative abundances of plant fossil finds in these stratigraphic profiles (**Figure 2**).

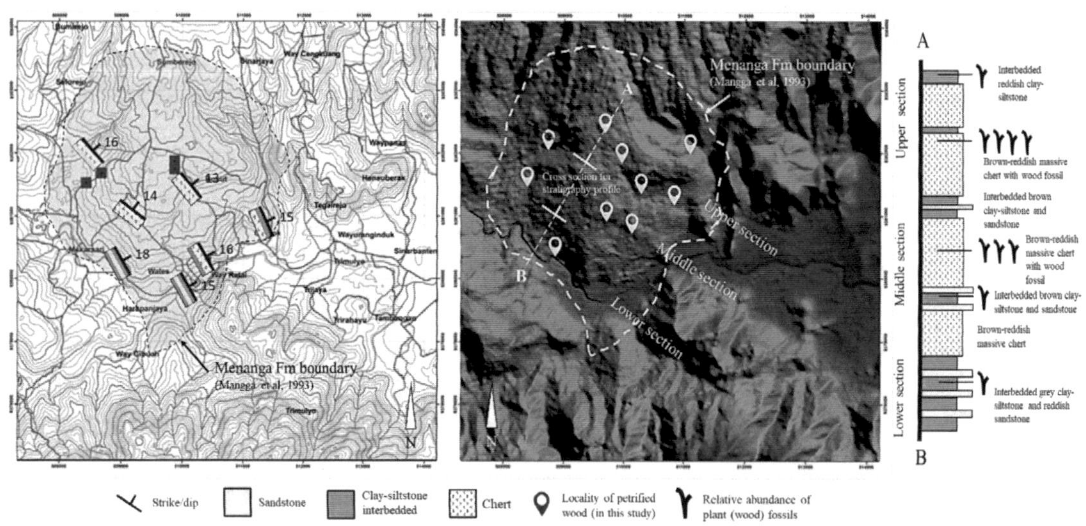

Figure 2. The orientation/strike-dip of sedimentary rock layer on contour map (left), the location of the plant fossils on digital elevation model maps (centre) and its stratigraphic profile (right).

In the lower section, tabular petrified wood fossils (around 33cm x 7cm x 8cm) present in interbedded grey claystone-siltstone (30-20 cm) and reddish sandstone (15-10 cm). The fossils had sharp shape, crusted, and covered by claystone-siltstone. This fossil was largely replaced by silica, having spongy features with a crusted epidermis. These features suggest that these fossils are fossilized freshwater swamp woody plant. In the middle section, there are fossilized branch of woody plant (4-5cm length) that were found inside the chert layer, which are crusted wood fibres. There are also several silicified wood fossils of the trunk and branch of solid woody plants. These fossils were deposited in brown-reddish massive chert layer and between interbedded brown clay-siltstone and sandstone. In the upper section, fossilized plant culms (1-2cm) are abundant in chert layers. These culms had fibrous textures and spongy features. They were considered as graminoid aquatic plant taxa. There are also several fossilized woody trunk and branch fossils. They were found in inside brown-reddish massive chert between interbedded reddish claystone (**Figure 3** and **Figure 4**).

Figure 3. Cretaceous sedimentary rock outcrop in Padangcermin, Lampung, documented in 2021.

By their microscopic observation, the woody fossil are identify as angiosperm woody plants. It was confirmed by the presence of vessels, ray cells, tracheids, and parenchyma cells, and there is no resin duct/canal in these wood fossils. Based on cross-sectional observations, most of the fossils have solitary to multiple vessels with a diameter of 0.1-0.2 mm filled with quartz. It also has aggregate rays, vasicentric tracheid, and longitudinally-banded fusiform parenchyma (**Figure 5**). By comparison of woody anatomy, it is considered a taxa of Juglandaceae.

Figure 4. Plant fossils are collected in each stratigraphy sections, located in Padangcermin, Lampung.

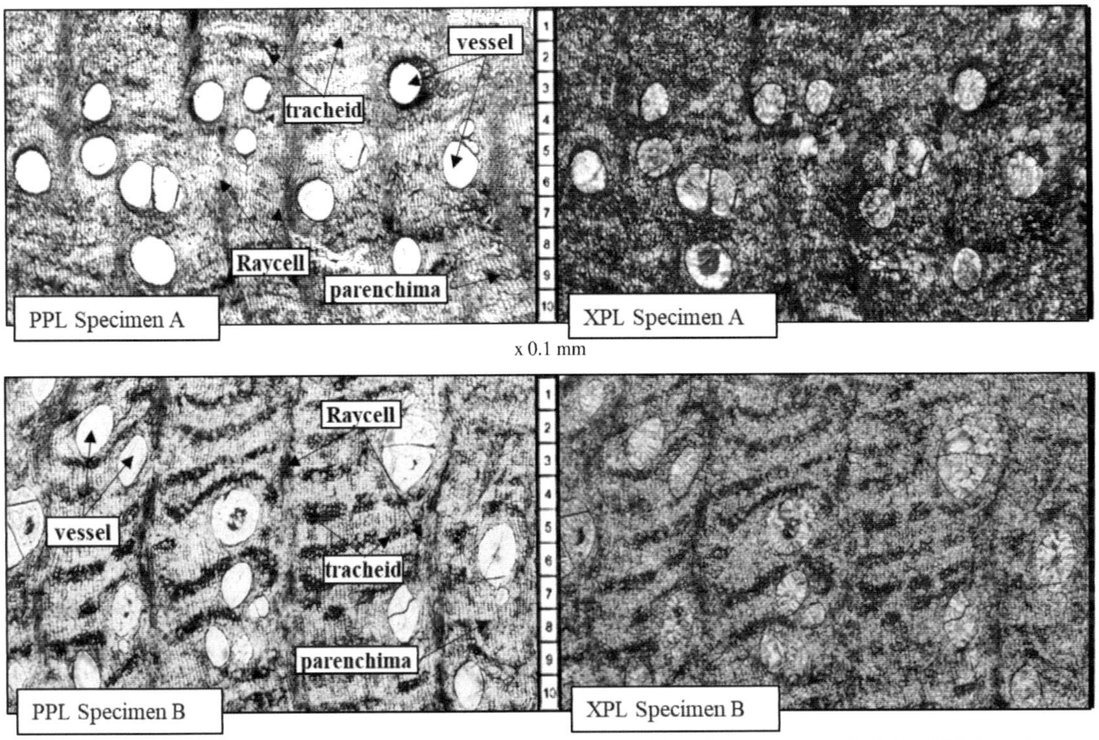

Figure 5. Thin section of selected petrified wood observed in plane polarized light (PPL) and cross polarized light (XPL) from upper section (a) and middle section (b) by the stratigraphy profile in this study.

4. Discussion

The scope of this study is about the presence of petrified wood fossils in Cretaceous sedimentary rock, specifically in Menanga Formation. This formation is mostly exposed in Lampung, mainly in Pesawaran and Tanggamus.. In regional stratigraphy Menanga Formation is grouped with the pre-tertiary formation of Sumatra [11,12,15]. Previously, this formation was interpreted as a deep sea sequence, but it is still arguable due to the fossilization of petrified wood in this formation. It is difficult to explain if such processes poins to a deepwater marine environment. This study suggests other alternative interpretations for Menanga Formation, mainly of those exposed in Padangcermin, Lampung, based on petrified wood fossil.

The presence of petrified wood fossils in these outcrops is *in situ* fossils. *In situ* fossil means the habitat and the fossilization process are located in the same environment. The plant remains were not transported or moved to other environments naturally. This argument is supported by the evidence of the shape of the fossil. These fossils had angular and elongated shapes. Nor is it considered a transported or reworked fossil, it should have rounded and sub-spherical shape from its erosional friction with other fragments during sedimentation transport. These fossils are also clearly found inside the rock layer, meaning rock formation and fossilization were in the same geological time span. Therefore, we suggest that these fossils are *in situ* fossils associated with the formation of these rocks.

The wood fossils were considered as Juglandaceae family based on its microscopic wood anatomies. Juglandaceae is Angiosperms. This family also is known as the walnut tree group. It is generally encountered as trees and shrubs. This family disperse in South East Asia, America, and Eurasia. In Indonesia, the existence of endemic trees of this group is relatively rare [16,17]. Generally, these taxa are associated with freshwater ecosystems, such as rivers and lacustrine [18]. This tree family is known to have existed since the Cretaceous [19]. Based on the aquatic plant fossils identified as graminoid aquatic plants, indicating a freshwater depositional environment surrounded by woody plants such as Junglandaceaea and grassy wetland plants, such as graminoid-aquatic plants occupied the area. It also had stagnant to gently current water columns. Gradually, the remains of these plants accumulated in the lacustrine system and buried underwater over time. Based on these interpretations, we highly suggest that the paleoenvironment of these sections is related to the fluvial-lacustrine environment. Based on the presence of these petrified wood in the chert layer, it seems that silicification had a significant role. These remain are spared from natural decomposition such as organic detritus activities and surficial erosion. Other researchers have reported the geological process of fluvial-lacustrine aquatic plant fossils in the chert layer. Generally, this condition is related to paleoenvironmental conditions with high evaporation and nearby active volcanoes [20–22]. This environment has alkaline fertile soil conditions at fertile to be overgrown by aquatic plants. The plant remains will be deposited in the water column along with the (paleo) soil, then silicified due to leaching from the volcanic material covering them [20,23].

The water column was filled with highly dissolved silicic dioxide. This was likely supplied from a near-volcano activities that deposit the volcanic material in these systems, such as dissolved volcanic ash. This is also occurred by the alkaline conditions in this lacustrine system. These processes condition the silicification process beneath the water column. The evaporation also made this process effective; hotter climate had significant conditions, as exact conditions of Cretaceous warmer paleoclimates [24,25]. These processes shows to the mechanism of chert forming in freshwater environment. So the remains of the plant around the lacustrine could be preserved as silicified wood fossils. This goes direct to the argument that there was already an established terrestrial environment (in the Cretaceous) south of paleo-Sumatra. Other researchers have proposed these paleoenvironment models with the the regional tectonic evolution maps of Southeast Asia model [15,26,27]. This confirms that the paleogeography of the study area is related to the terrestrial environment. Their model also illustrated that the area is close to a southern paleo-Sumatran volcano that was active during the Cretaceous.

5. Conclusion

Based on our discoveries during the geological mapping, we confirmed that indeed petrified woods are found in the Cretaceous Menanga Formation in Padangcermin, Lampung. For the most part, silicification takes main role in their fossilization. Massive chert layers predominates in the area with interbedded claystone-siltstone and sandstone, with several plant fossils presented in these layers. Woody Junglandaceaea and aquatic graminoid plants are commonly discovered in the chert layers. To be more precise, the fossilization processes are related to the chert forming in freshwater environment. The paleoenvironment of this area was related to the alkaline-lacustrine system, provided by enriched-dissolution of silica from a nearby Cretaceous volcanic activities and warmer paleoclimate condition.

6. References

[1] Mustoe G 2017 Wood Petrifaction: A New View of Permineralization and Replacement *Geosciences* **7** 119
[2] Schweingruber F H and Börner A 2018 Fossilization, permineralization, coalification, carbonization and wet wood conservation *The Plant Stem* (Cham: Springer International Publishing) pp 183–92
[3] Ballhaus C, Gee C T, Bockrath C, Greef K, Mansfeldt T and Rhede D 2012 The silicification of trees in volcanic ash - An experimental study *Geochim. Cosmochim. Acta* **84** 62–74
[4] Hofmann A 2005 The geochemistry of sedimentary rocks from the Fig Tree Group, Barberton greenstone belt: Implications for tectonic, hydrothermal and surface processes during mid-Archaean times *Precambrian Res.* **143** 23–49

[5] Mustoe G and Acosta M 2016 Origin of Petrified Wood Color *Geosciences* **6** 25
[6] Bruno D E, Crowley B E, Gutak J M, Moroni A, Nazarenko O V., Oheim K B, Ruban D A, Tiess G and Zorina S O 2014 Paleogeography as geological heritage: Developing geosite classification *Earth-Science Rev.* **138** 300–12
[7] O'shea N, Arthur Bettis E, Zaim Y, Rizal Y, Aswan A, Gunnell G F, Zonneveld J P and Ciochon R L 2014 Paleoenvironmental conditions in the late Paleogene, Sumatra, Indonesia *J. Asian Earth Sci.* **111** 384–94
[8] Oktariani H, Winantris W and Fauzielly L 2018 Fosil Kayu Dryobalanoxylon sp. Pada Formasi Genteng Di Kabupaten Lebak Provinsi Banten Dan Paleofitogeografinya Di Indonesia *Bull. Geol.* **2** 197–205
[9] Oktariani H, Winantris W and Hamzah A 2019 Dryobalanoxylon sp. : Silicified Fossil Wood from Lebak Regency, Banten Province, Indonesia *J. Geol. dan Sumberd. Miner.* **20** 93
[10] Husien N, Wahyuni S, Erwin and Budi A S 2021 Identification of Fossil Wood from Samarinda, East Borneo *Joint Symposium on Tropical Studies* pp 253–7
[11] Mangga, S.A., Amirudin, T., Suwarti, S. and Gafoer S 1993 *Peta Geologi Lembar Tanjungkarang, Sumatera 1: 250,000* (Pusat Penelitian dan Pengembangan Geologi, Departemen Pertambangan dan Energi.)
[12] Barber, A.J., Crow, M.J. and Milsom J 2005 *Sumatra: Geology, resources and tectonic evolution* (Geological Society of London)
[13] Natalia H C, Harbowo D G and Ikhram R 2021 Potensi Geodiversity di Sekitar Kawasan Anak Krakatau-Way Kambas, Provinsi Lampung, Sebagai Kandidat Geopark Indonesia *J. Sci. Appl. Technol.* **5** 47–57
[14] Harbowo D G, Priadi B, Julian T, Amelia R N, Sihombing D J P and Kencana F S 2021 A preliminary study on the element abundance in the Hulusimpang Formation, Way Kalianda, Pesawaran, Lampung, Indonesia *IOP Conference Series: Earth and Environmental Science* vol 882 (IOP Publishing) p 12078
[15] Morley C K 2012 Late Cretaceous–Early Palaeogene tectonic development of SE Asia *Earth-Science Rev.* **115** 37–75
[16] Mandang Y I 1996 Anatomi Delapan Jenis Kayu Kurang Dikenal Dari Suku Flacourtiaceae Sampai Juglandaceae (Wood Anatomy of Eight Lesser Known Wood Species of Flacouriiaceae to Juglandaceae) *J. Penelit. Has. Hutan* **14** 31–45
[17] Song Y, Fragnière Y, Meng H, Li Y, Bétrisey S, Corrales A, Manchester S, Deng M, Jasińska A K and Văn Sâm H 2020 Global biogeographic synthesis and priority conservation regions of the relict tree family Juglandaceae *J. Biogeogr.* **47** 643–57
[18] Song Y-G, Li Y, Meng H-H, Fragnière Y, Ge B-J, Sakio H, Yousefzadeh H, Bétrisey S and Kozlowski G 2020 Phylogeny, taxonomy, and biogeography of Pterocarya (Juglandaceae) *Plants* **9** 1524
[19] Zhou H, Hu Y, Ebrahimi A, Liu P, Woeste K, Zhao P and Zhang S 2021 Whole genome based insights into the phylogeny and evolution of the Juglandaceae *BMC Ecol. Evol.* **21** 1–16
[20] Gihm Y S 2021 Physical and chemical depositional processes when volcanoes meet lacustrine environments: the Cretaceous Imjado Volcanics, Jeungdo, southwestern Korea *Geol. Soc. London, Spec. Publ.* SP520-2021–65
[21] Khosla A and Verma O 2015 Paleobiota from the Deccan volcano-sedimentary sequences of India: paleoenvironments, age and paleobiogeographic implications *Hist. Biol.* **27** 898–914
[22] Harbowo D G and Zahra S 2021 Microscopy Observation of Samosir Formation Paleosoil, Tuktuk Sidaong, North Sumatera, Indonesia. *J. Geosci. Eng. Environ. Technol.* **6** 9–15
[23] Kuma R, Hasegawa H, Yamamoto K, Yoshida H, Whiteside J H, Katsuta N and Ikeda M 2019 Biogenically induced bedded chert formation in the alkaline palaeo-lake of the Green River Formation *Sci. Rep.* **9** 16448
[24] Hay W W 2017 Toward understanding Cretaceous climate-An updated review *Sci. China Earth Sci.* **60** 5–19

[25] Kintisch E 2006 Hot Times for the Cretaceous Oceans *Science (80-.).* **311** 1095–1095
[26] Hall R 2013 The palaeogeography of Sundaland and Wallacea since the Late Jurassic *J. Limnol.* **72**
[27] Clements B and Hall R 2011 A record of continental collision and regional sediment flux for the Cretaceous and Palaeogene core of SE Asia: implications for early Cenozoic palaeogeography *J. Geol. Soc. London.* **168** 1187–200

Acknowledgments
We greatfully thank Institut Teknologi Sumatera (ITERA) 2021 Collaborative Research Programme for financial support. We also thank to Maria Sekar Proborukmi and Mika Rizki Puspaningrum (Institut Teknologi Bandung/ITB) for comments on the manuscript.

Mapping the distribution of natural ecosystems on peatlands through vegetation using the object-based image analysis (obia) method in bangko district, rokan hilir regency, riau.

M.S. Roganda[1] Sigit Heru Murti[2] Wirastuti Widyatmanti[3]
[1]Sungai Panji-Panji, Kubu Babussalam District, Rokan Hilir Regency, Riau, Indonesia
[1]Muhammadslametroganda@gmail.com

ABSTRACT

Expansion of the use of natural ecosystems on peatlands is increasing rapidly in Bangko District, Rokan Hilir Regency, Riau. This utilization tends to cause damage to the natural ecosystem of peatlands which has a major impact on the local environment (in situ) and the surrounding environment (exsitu), so a mapping is needed to detect the presence and ensure the distribution of the natural ecosystem of peatlands. This study aims to (1) map the distribution of natural ecosystems on peatlands through vegetation objects using Sentinel 2A and SPOT 7 using the OBIA method, (2) Compare the accuracy of Sentinel 2A and SPOT 7 in recognizing endemic vegetation types in natural ecosystems on peatlands with using the OBIA method. OBIA is used to detect the presence of peatlands, the detection is carried out through stretched endemic vegetation objects that grow in the natural ecosystem of peatlands. OBIA is carried out in three steps, namely Segmentation, Classification and Accuracy Test. In the classification process, building a ruleset is the key to interpreting the natural ecosystem of peatlands. The results obtained are (1) Sentinel 2A produces a more dominant distribution on the East, North and West edges while in the middle it tends to be less, the total area of natural peatland ecosystems in Sentinel 2A is 94,994 KM^2. In the SPOT 7, the distribution of natural ecosystems of peatlands is more dominant in the eastern part while the western part tends to be small, the total distribution of natural ecosystems of peatlands in the SPOT 7 image is 281,443 KM^2. (2) The accuracy rate of Sentinel 2A is 81.395% with an accuracy of 35 samples from a total of 43 samples of accuracy tests and SPOT 7 of 88.372% with an accuracy of 38 samples from a total of 43 samples of accuracy tests carried out.

Keywords: Natural Ecosystem, Peat, Vegetation, Object-Based Image Analysis (OBIA)

1. Preface

Indonesia is a country that has the 4th largest natural ecosystem of peatlands in the world after Russia, Canada and America (Indonesian Peatland Restoration Agency, 2019). Peat is formed from piles of dead plant debris. The expansion of the use of natural ecosystems on peatlands has increased rapidly in several provinces with large peat areas, such as Riau, West Kalimantan and Central Kalimantan. Between 1982 and 2007 an area of 1.83 million ha was converted or 57% of the total area of 3.2 million ha of peat forest in Riau Province. The rate of conversion of natural ecosystems on peatlands tends to increase rapidly, while for non-peatlands the increase is relatively slower (WWF, 2008). Bangko District, Rokan Hilir District, is one of the sub-districts that contributes to the massive conversion of natural peatland ecosystems in Riau Province. This massive change occurred due to the rapid population growth so that the need for residential, agricultural and plantation land continues to increase.

Increasing the expansion of the use of natural ecosystems on peatlands will certainly threaten and damage the existing ecosystems on peatlands. Damage to the peat ecosystem has a major impact on the local environment (in situ) and the surrounding environment (ex situ). Flooding in the downstream watershed is one of the impacts of the destruction of the peat ecosystem. The magnitude of the impact caused by the destruction of the natural ecosystem of peatlands requires a mapping to detect the presence and ensure the distribution of the natural ecosystem of peatlands in Bangko District, Rohan Hilir Regency, Riau Province, to do this remote sensing technology can be used. Remote sensing output products include images.

The image is the result of sensor recordings carried by the mode. The image becomes the material for analysis related to existing features. In analyzing vegetation, usually using images with medium-high spatial resolution, such as Sentinel 2A and SPOT 7. In addition to spatial resolution, Sentinel 2A and SPOT 7 images also have band compositions that are incredibly supportive for conducting studies on vegetation objects. There are many methods and approaches used in extracting images to obtain the desired information in this case detecting the presence of natural peatland ecosystems. Of the many methods or approaches that have been used, OBIA has advantages over others, this is because the classification process does not only consider spectral aspects but also spatial or object aspects.

1.1 Objective of the study

a. Mapping the distribution of natural ecosystems on peatlands through endemic vegetation objects using Sentinel 2A and SPOT 7 images using the OBIA method.
b. Comparing the accuracy of Sentinel 2A and SPOT 7 images in recognizing endemic vegetation types in natural peatland ecosystems using the OBIA method.

2. Literature Review

Remote Sensing is simply a technique of taking objects on the earth's surface from the air by using sensors. Currently Remote Sensing can be categorized to be growing rapidly. Starting from the addition of new and free images for anyone to access to the increasingly varied uses to be applied or utilized in various themes or fields. Remote Sensing is especially useful in reducing terrestrial survey activities when conducting an inventory and monitoring of natural resources and the environment, one example being mapping the natural ecosystem of peatlands. The natural ecosystem of peatlands is the remains of heaps of dead plants and then decomposed by anaerobic and aerobic bacteria into components that are more stable and have not been seriously damaged by human activities. In addition to the organic substances that make up peat, there are also lesser amounts of inorganic substances.

Terentang is an endemic vegetation (plant) that grows in the natural ecosystem area of peatlands. According to Richter and Dallwitz (2004), *terentang* has the Latin name Campnosperma spp., belongs to the Anacardiaceae family. *Terentang* has various local names, namely *terentang, kelinting, melumut, serentang* (Malaysia), *pauh lebi, tumbus, terentang* (Indonesia), *campnosperm* (Papua New Guinea), *nangpron, huasum sangtrang* (Thailand). In finding the natural ecosystem of peatlands through vegetation, a vegetation index is needed. The vegetation index is the result of the transformation of the spectral values from several remote sensing channels to highlight the spectral values of the vegetation. Practically, this vegetation index is a mathematical transformation that involves several channels at once and produces a new image that is more representative in presenting the vegetation phenomenon (Danoedoro, 2012). The images used in the identification process are Sentinel 2A and SPOT 7.

SPOT is an abbreviation of Systeme Pour I'Observation de la Terre, which is a collaborative project between France, Sweden and Belgium, under the coordination of CNES (Center Nasional D'etudes Spatiales), the French space agency.

SPOT 7 Image Specifications

Resolution	1,5 m Pan/Pansharpened and 6 m Multispectral
Band	Panchromatic dan 4 multispectral bands
wavelength	Panchromatic 597,5 nm, band 1 485 nm, band 2 560 nm, band 3 660 nm and band 4 825 nm
Projection	UTM or Lat/Long
Datum	WGS 84
Format	GeoTIFF
Recording	2013 – now

Source: LAPAN 2019

Sentinel-2A is a European optical imaging satellite launched in 2015. The Sentinel-2A satellite was launched as part of the European Space Agency's (ESA) Copernicus program. Sentinel-2 consists of two constellation satellites, Sentinel-2A and Sentinel-2B, which orbit the poles in a sun-synchronous orbit at an altitude of 786 km. The two identical satellites are 180 degrees apart from each other. The satellite is a medium resolution satellite with a temporal resolution of 10 days for one satellite or 5 days with two satellites. This satellite can be used for operational observations such as land cover maps, land change detection maps and geophysical variables.

Sentinel Image Specification 2A

Resolution	10 m band (B2, B3, B4, B8), 20 m (B5, B6, B7, B8A, B11, B12) and 60 m (B1, B9, B10)
Number of Band	12 band
wavelength	Band 1 0,443 nm, band 2 0,490 nm, band 3 0,560 nm, band 4 0,665 nm, band 5 0,705 nm, band 6 0,740 nm, band 7 0,783 nm, band 8 0,842 nm, band 9 0,945 nm, band 10 1,375 nm, band 11 1,610 nm and band 12 2190 nm
Projection	UTM or Lat/Long
Datum	WGS 84
Format	GeoTIFF
Recording	2015 – now

Source: LAPAN 2019

OBIA is an image classification approach that not only considers the spectral aspect but also the spatial aspect of the object. OBIA is a classification technique that pays attention to the unity of objects based on the hue and texture of pixels. The processing steps in OBIA consist of 3 steps, namely: (Veljanovski.et al., 2011):

IOP Conf. Series: Earth and Environmental Science 1047 (2022) 012017 doi:10.1088/1755-1315/1047/1/012017

a. Segmentation and computation of spectral, geometric, texture, conceptual and temporal attributes.
b. Object (semantics) classification.
c. Post-classification, in the form of verification, elimination of errors and validation of results.

3. Research Method

Broadly speaking, this research consists of 3 steps, namely the first step is preparation including literature study and data collection (first observation), the second step is data processing including field work, image correction, segmentation and classification. The last step includes analysis of the results of data processing and accuracy test. This study took place in a part of Bangko District, Rokan Hilir Regency, Riau Province.

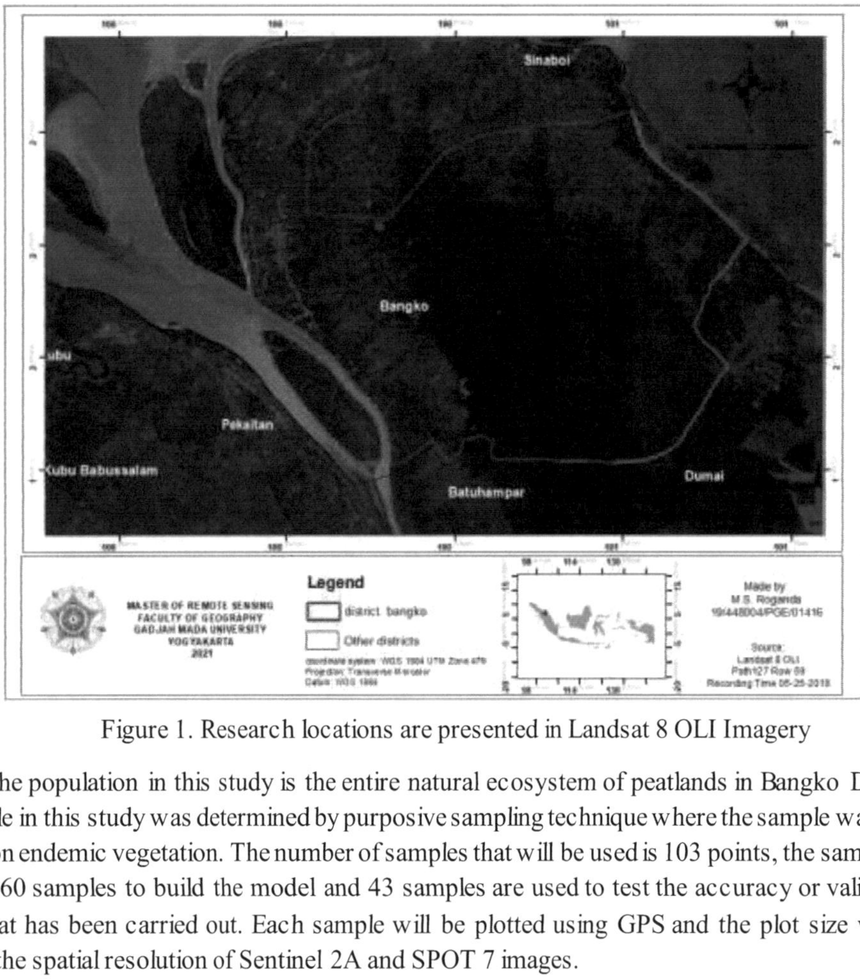

Figure 1. Research locations are presented in Landsat 8 OLI Imagery

The population in this study is the entire natural ecosystem of peatlands in Bangko District, while the sample in this study was determined by purposive sampling technique where the sample was determined directly on endemic vegetation. The number of samples that will be used is 103 points, the sample is divided into two 60 samples to build the model and 43 samples are used to test the accuracy or validation of the model that has been carried out. Each sample will be plotted using GPS and the plot size will take into account the spatial resolution of Sentinel 2A and SPOT 7 images.

The data collection and analysis tools used in this research are AMD Ryzen 5 Laptop Set, ENVI 4.5, ArcGis 10.2, eCognition Developer 64, QGis, Garmin GPS 64s, Microsoft Excel, Microsoft Word and Printer. While the materials used in this research are SPOT 7 and Sentinel 2A images. Materials in the form of SPOT 7 and Sentinel 2A images were corrected for geometry and radiometry. Furthermore, the object-oriented classification process is carried out starting with the image segmentation step, this eCognition software has several types of segmentation algorithms, they are *Chessboard Segmentation, Quadtree Based Segmentation, Multiresolution Segmentation Contrast Split Segmentation, Spectral Difference Segmentation dan Multi-Theshold Segmentation.*

The segmentation algorithm used in this study is Multiresolution Segmentation while the classification process is carried out using the Rule-based classification algorithm. Classification because in this algorithm researchers can enter the characteristics of endemic vegetation as a rule in recognizing endemic vegetation in the natural ecosystem of peatlands. Image classification is based on endemic (Spanned) vegetation as a classification key. A total of 60 sample points were used to run the model. The ruleset classification algorithm as a classification method will be used for the two images used, namely Sentinel 2A and SPOT 7 based on pixels as the unit of analysis that has been applied. Furthermore, the accuracy test is carried out using the Kappa coefficient method, which is to compare the classification results on each image with the actual conditions in the field. To be clearer, it can be seen in the following research flow chart at figure 2:

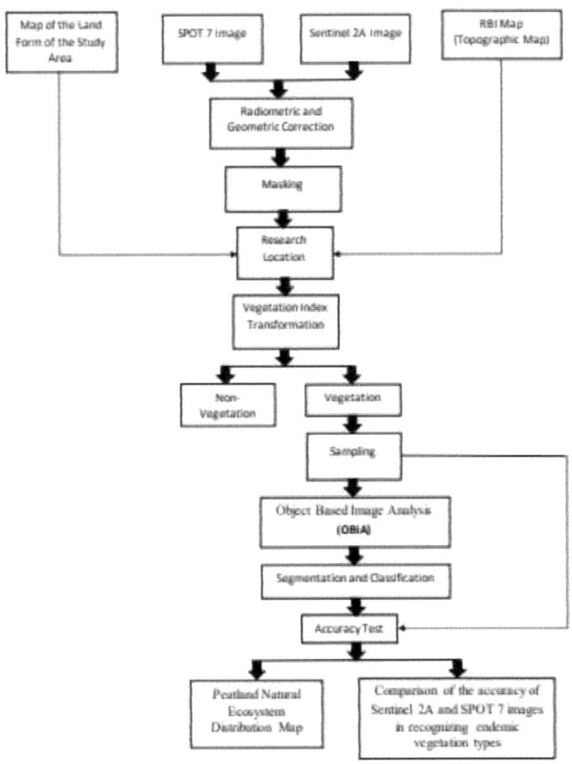

Figure 2. Research Flowchart

4. Results and Discussion

4.1 Characteristics of terentang (endemic) vegetation as ruleset OBIA

In general, OBIA can be divided into 2 processes, namely the segmentation process and the classification process. There are several classification methods using OBIA, namely Field-base Classification, Simple-Sample-based classification, Multistage, Sample based classification, and Rule-based classification. In this study, the Rule-base classification method was used. This method is used because in determining the natural ecosystem of peatlands requires various criteria. One of the difficulties in conducting classification based on rules is in the process of making the rules itself. Classification based on rules is actually an attempt to transfer the logics of interpreting images from the brain/mind and human experience into operations on a computer. This classification strategy will be very successful if these factors are met (Widayani, 2018):

a. Operators/analysts have high knowledge and experience in terms of image interpretation and the object of study.
b. Operator/analyst has the ability to write the rules of interpretation in his mind into computer language.

In determining the criteria/characteristics of the terentang vegetation that will be used as a ruleset in determining the natural ecosystem of peatlands, it will highlight the features that exist in the field. There are two features that will be used as the ruleset, namely Greenness Level and Geometry Area.

a. Greenness level, terentang vegetation has a different level of green from other vegetation. To see the level of greenery, a vegetation index will be used with Sentinel 2A imagery using the NDVI algorithm and SPOT 7 Green index. The greenish value used will be adjusted to the highest value of the 60 samples used.
b. Geometry Area, terentang vegetation has the characteristics of living in clusters/clusters. This stretched life characteristic can certainly be used as a ruleset. The Geometry Area value used will be adjusted to the highest value of the 60 samples used.

4.2 OBIA

4.2.1 Segmentation

Before arriving at the segmentation process to identify endemic vegetation (Spanning), a hierarchical class is needed so that the logic built is clearer and more structured. The hierarchical class that was built can be seen in the following figure 3:

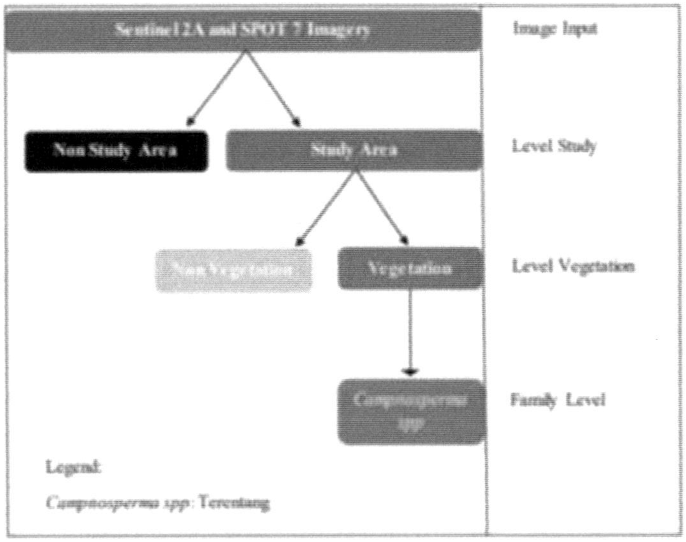

Figure 3. Class Hierarchy

The arrangement of hierarchical classes makes it easier and at the same time reminds back to the initial process to the segmentation and classification process. Based on Figure 3, there are at least 3 stages that have been passed. The first stage is in the form of determining the study area, the second stage is the separation of vegetation and non-vegetation and the third stage is mapping of stretched endemic vegetation as the key to detecting the natural ecosystem of peatlands. To arrive at the mapping of endemic vegetation as the key to detecting the natural ecosystem of peatlands, it is necessary to carry out the next stage, namely segmentation and classification.

Image segmentation is carried out with the aim of grouping pixels that have the same structure, so that each individual structure has a separate area or can be interpreted as a process of grouping neighboring pixels into areas (segments) based on similar criteria. Image segmentation produces "objects", namely groups of pixels which then become the unit of classification analysis. The algorithm used to perform segmentation is Multiresolution Segmentation (MRS). MRS developed by Baatz and Schaepe (2000) is the most widely used segmentation algorithm. This region growing-based segmentation process is run based on five parameters, namely Scale, Color, Shape, Smoothness and Compactness. These parameters can be combined to get various object results, so that they can be adjusted to the operator's wishes regarding the creation of homogeneous objects at the desired resolution.

In this study, the scale parameters used were 10, shape 0.3 and compactness 0.1. The scale parameter is an abstract value to determine the amount of object heterogeneity that is allowed in one object. At the same scale value, heterogeneous appearances will produce smaller object sizes than homogeneous appearances. The scale parameter 10 was chosen because what will be detected is individual vegetation, so it needs very detailed detection. The shape parameter value is given a value of 0.3. A high shape value will cause segmentation to be more emphasized on texture, while emphasis on texture does not always produce the desired image object (Frandi, 2015). The compactness parameter is used to separate compact objects from non-compact objects that have relatively low differences in spectral values. The greater the value of this parameter, the resulting object will have a more compact shape. This value is a deviation from the given ideal compact form. The segmentation results can be seen in Figure 4.

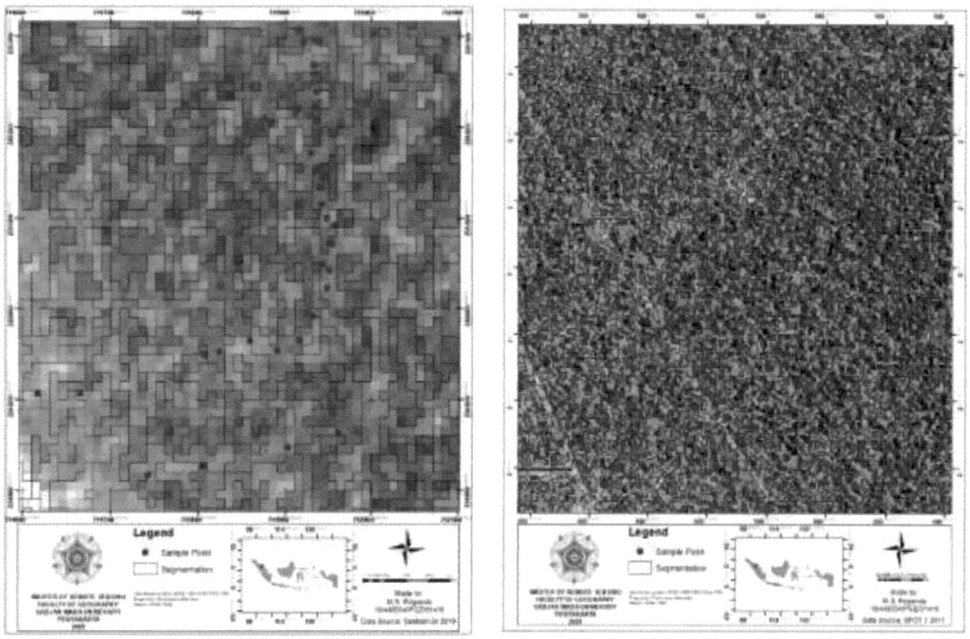

Sentinel 2A SPOT 7

Figure 4. Sentinel 2A and SPOT 7 Image Segmentation Results

Table 1. Total of Sentinel 2A and SPOT 7 Image Segmentation Results

No	Image	Total of Segmentation (KM^2)
1	Sentinel 2A	243.694
2	SPOT 7	4.191.639

Source: Results of data processing with ecognition 2021.

The number of segmentations generated on the Sentinel 2A image is 243,694 while the SPOT 7 image is 4,191,639 segmentations. The spatial resolution of the image affects the number of segmentations produced, but apart from the difference in the spatial resolution of the image, the difference in the number of segmentations is also influenced by the difference in the natural ecosystem of the peatlands in the two images, where the Sentinel 2A image (247,469 KM2) and the SPOT 7 image (447,080). The substantial number of segmentations in this study illustrates the increasingly detailed pixel grouping which then greatly influences the classification results.

4.2.2 Classification

There are several classification methods using OBIA, namely: *Field-base Classification, Simple-Sample-based classification, Multistage, Sample based classification, Rule-based classification*. In this study, the Rule-base classification method was used. This method is used because in figuring out the natural ecosystem area of peatlands it uses endemic vegetation characteristics *(Terentang)* as the classification key. One of the difficulties in conducting classification based on rules is making the rules. This study applies different rules or regulations between Sentinel 2A and SPOT 7 images, the reason for the difference in the application of these rules is due to the different spatial resolutions of the two images. Different spatial resolutions will affect the detail of the segmentation. Images that have high spatial resolution will produce smaller segments. The determination of the threshold value that is used as the Rulset is generated from the smallest value of the vegetation index and the largest value of the geometry area in the 60 samples used. For Sentinel 2A images, the following rulesets are applied in table 2:

Table 2. Sentinel Ruleset 2A for classification of extended endemic vegetation

No	Ruleset	Score
1	NDVI	>0,73
2	Geometry area	<50 Pixel

Source: Research Result

NDVI and Geometry Area are used as a ruleset to identify spanned endemic vegetation as the key in determining the natural ecosystem area of peatlands in Sentinel 2A imagery. These two rulesets adapt the appearance of stretched vegetation in the field, using NDVI to distinguish green spots, while Geometry Area is used to distinguish the area of spanned endemic vegetation groupings. The level of greenery and the size of the area are set according to the field sample that will be used to run the ruleset. The results of the NDVI and Geometry Area ruleset on the Sentinel 2A image can be seen in Figure 5 below:

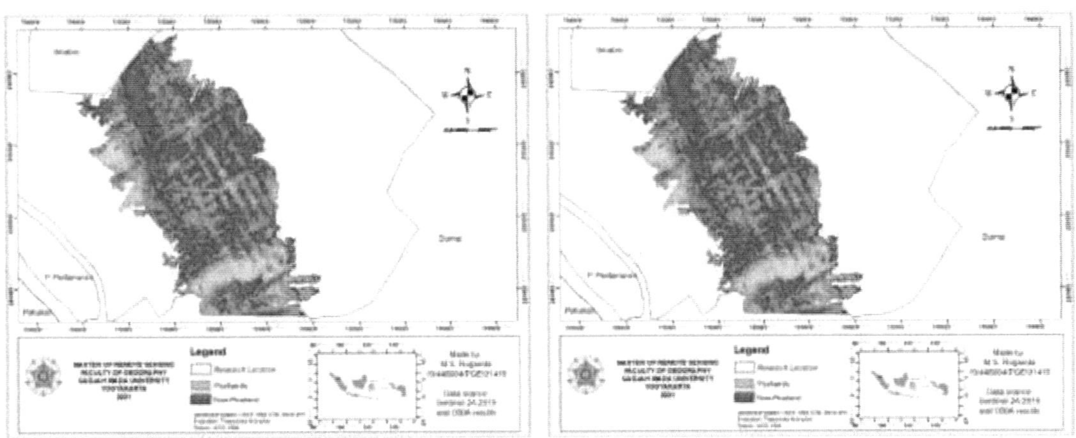

NDVI Geometry Area

Figure 5. Comparison of the results of the NDVI to Geometry Area ruleset step

There has been a change in the natural ecosystem of peatlands (in green) from the process of applying the NDVI ruleset to the Geometry Area Ruleset. the area of the natural ecosystem of peatlands in the application of the NDVI ruleset is 105,447 KM^2 after proceeding to the Geometry Area Ruleset it becomes 94.994 KM^2. This change in area occurred because there was a segment that was eliminated from the NDVI research which had an area of more than 50 pixels. NDVI and Geometry Area are used as a ruleset to find natural ecosystem areas of peatlands in Sentinel 2A image. These two rulesets adapt the appearance of stretched vegetation in the field, using NDVI to distinguish green spots, while Geometry Area is used to distinguish the area of the *Terentang* endemic vegetation grouping. The level of greenery and the size of the area are set according to the field sample that will be used to run the ruleset. The results of the NDVI ruleset and the Geometry Area of the Sentinel 2A image can be seen in Figure 6 below:

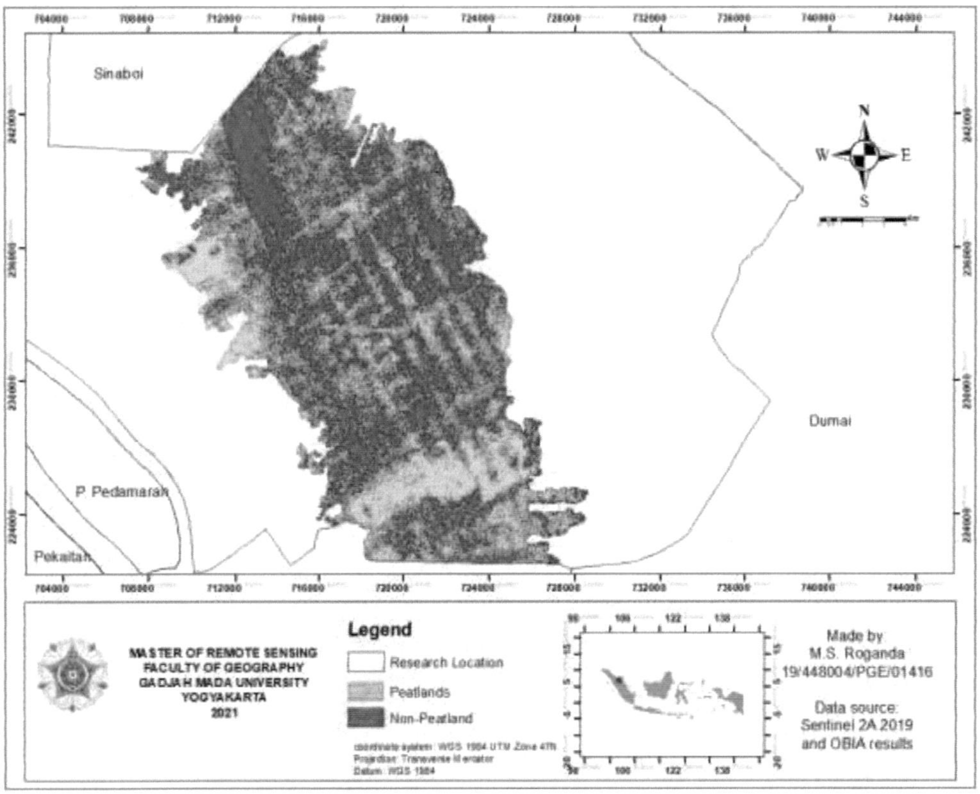

Figure 6. Map of Distribution of Natural Ecosystem Areas of Peatlands in Bangko District, Rokan Hilir Regency with Sentinel 2A image

Based on the application of the ruleset in the eCognition software, the distribution of natural ecosystem areas of peatlands is generated, which is then classified into two classes. The red color is non-peat area while the green color is the natural ecosystem area of peatland. The area of each class can be seen in table 3 below:

Table 3. The area of the natural ecosystem of peatlands using Sentinel 2A image

No	Classification	Area (KM^2)
1	Non-Peatland	152,475
2	Peatland	94,994
Total		247,469

Source: Research Result

Similar to the ruleset in the Sentinel 2A image, the SPOT 7 image also uses 2 rulesets. if the Sentinel 2A image uses NDVI to distinguish the green dots, then the SPOT 7 image uses the Green Index to distinguish the green dots, this is because the SPOT 7 image does not have a NIR band, the Green Index uses the same formula as NDVI, only the NIR band is replaced with a green band, while the Geometry Area ruleset is still being used. Meanwhile, the composition of the values for each ruleset is certainly different due to differences in the spatial resolution of the image, while for the determination of the threshold the value is the same as the Sentinel 2A image which is produced from the smallest value of the vegetation index and the largest value of the geometry area in the 60 samples used. For the SPOT 7 image, the ruleset as shown in table 4 is applied:

Table 4. SPOT 7 Ruleset for classification of stretched endemic vegetation

No	Ruleset	Score
1	Green Index	>0,25
2	Geometry area	<30 pixel

Source: Research Result

After running the ruleset with the Green Index to distinguish the greenness of the endemic vegetation with a value of > 0.25 and Geometry area to distinguish the area of the stretched endemic vegetation grouping with a value of < 30 pixels. The level of greenery and the size of the area are set according to the field sample that will be used to run the ruleset. The field samples that will be used are 60 samples, the same as the Sentinel 2A image. The results of the ruleset of the Green Index and Geometry Area of SPOT 7 images can be seen in Figure 7 below:

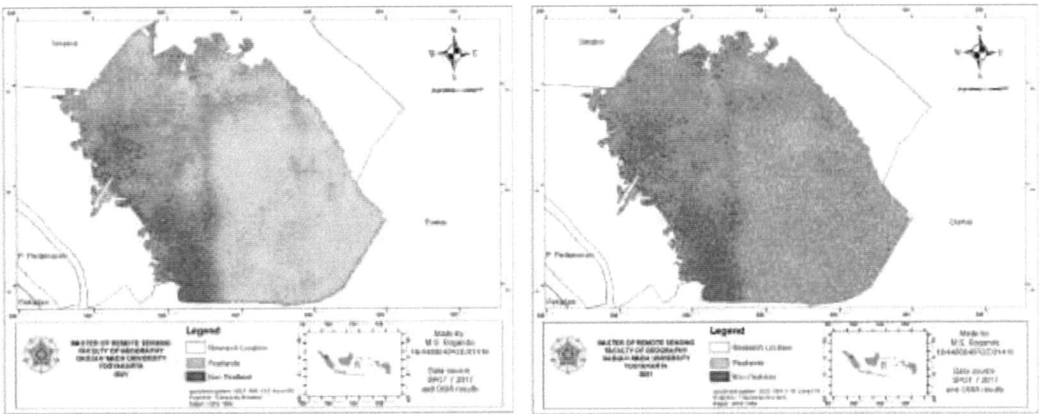

Green Index **Geometry Area**

Figure 7. Perbandingan hasil *ruleset* tahap *Green Indeks* sampai *Geometry Area*

There has been a change in the natural ecosystem of peatlands (in green) from the process of applying the Green Index ruleset to the Geometry Area Ruleset. the area of the natural ecosystem of peatlands in the application of the Green Index ruleset covering an area of 315,104 KM^2 after proceeding to the Geometry Area ruleset it becomes 281,443 KM^2. The total difference in the natural ecosystem of peatlands from the total area of the Green Index ruleset and Geometry Area ruleset is 33,661 KM^2. This change in area occurred because there were segments that were eliminated from the Green Index ruleset which had an area of more than <30 pixels.

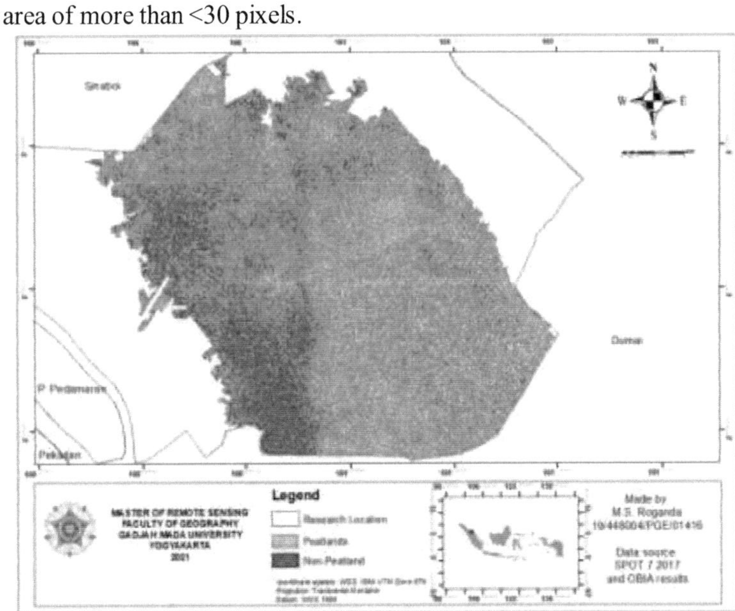

Figure 8. Map of Distribution of Natural Ecosystem Areas of Peatlands in Bangko District, Rokan Hilir Regency with SPOT 7 image.

Based on the application of the ruleset in the eCognition software, the distribution of the Natural Ecosystem Area of Peat on the SPOT 7 image is generated, which is then classified into two classes. The red color is non-peat area, while the green color is the natural ecosystem area of peatland. The area of each class can be seen in table 5 below:

Table 5. The area of the natural ecosystem of peatlands using SPOT 7 image

No	Classifications	Area (KM^2)
1	Non-Peatland	165,637
2	Peatland	281,443
Total		447,080

Source: Research Result

4.2.3 Accuracy test

The steps of testing the accuracy of the OBIA classification are carried out using the Kappa coefficient method, which is to compare the classification results on each image with the actual conditions in the field. The Kappa coefficient is based on the consistency of the assessment by considering all aspects, namely producer's accuracy/omission error and user's accuracy/commission error obtained from the error matrix or confusion matrix.

Table 6. Confusion Matrix for Sentinel 2A image

Interpretation Result (Sample for accuracy test)/User Accuracy (User's Accuracy / Commission Error)	Reference Data (the sample used for the ruleset)/ Maker's accuracy (producer's accuracy)			
	Class	**Peatland**	**Non-peatland**	**Total**
	Peatland	35	8	43

Source: Research Result

Through the error matrix in the table above, Kappa accuracy can be obtained mathematically in the following way.

$$\text{Peatland} = \frac{35}{43} \times 100\% = 81{,}395\%$$

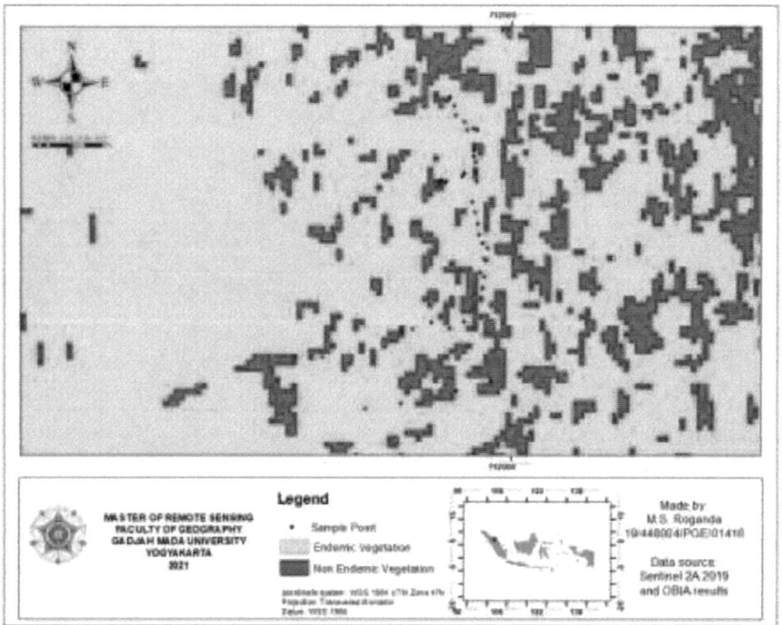

Figure 9. Distribution of accuracy test sample points on endemic vegetation in Kec. Bangko Kab. Rokan Hilir with Sentinel 2A image.

Tabel 7. Confusion Matrix for SPOT 7 image

Interpretation Results (Sample for accuracy test)/User Accuracy (User's Accuracy / Commission Error)	**Reference Data (sample used for ruleset) / Maker's Accuracy (producer's accuracy)**			
	Class	**Peatland**	**Non-Peatland**	**Total**
	Peatland	38	5	43

Source: Research Result

Through the error matrix in the table above, Kappa accuracy can be obtained mathematically in the following way

Peatland $= \frac{38}{43} \times 100\% = 88{,}372\%$

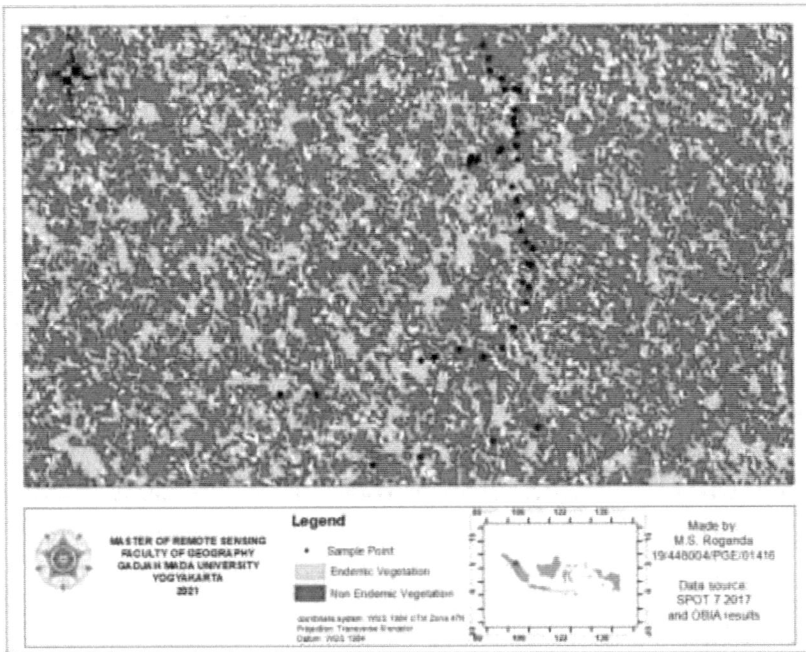

Figure 10. Distribution of accuracy test sample points on endemic vegetation in Kec. Bangko Kab. Rokan Hilir with SPOT 7 image.

5. Conclusion and Suggestions

5.1 Conclusion

a. The natural ecosystem of peatlands through endemic vegetation objects using the OBIA method produces a different distribution from the two images (Sentinel 2A and SPOT 7), Sentinel 2A produces a more dominant distribution on the eastern, northern and western edges, while in the middle it tends to be less, the total area The natural ecosystem of peatland in Sentinel 2A image is 94,994 KM2. In the SPOT 7 image, the distribution of natural ecosystems of peatlands is more dominant in the eastern part while the western part tends to be small, the total distribution of natural ecosystems of peatlands in the SPOT 7 image is 281,443 KM2.

b. The accuracy rate of Sentinel 2A image is 81.395% and SPOT 7 image are 88.372% in recognizing endemic vegetation types in natural ecosystems of peatlands using the OBIA method. The SPOT 7 image has a higher accuracy rate than the Sentinel 2A image, the difference in accuracy is due to the different spatial resolutions of the two images.

5.2 Suggestions

a. The number of samples used to run the model and test accuracy is more than this research, besides that it is necessary to pay attention to the distribution of samples including the distance between samples.

b. The making of the ruleset must be thorough and adopt the factual and real appearance of the characteristics of the object to be used, because the preparation of the ruleset is the key to research using the OBIA method.

c. This research is still far from perfect, so it needs further research using the OBIA method or other methods to be a comparison that will give birth to innovations in mapping the natural ecosystem of peatlands.

References

Akhbar, P. 2017. Identification of Peatlands Using Landsat 8 Oli Tirs Satellite Imagery Based on Geographic Information Systems (Sig) Case Study of Tebing Tinggi Island. Sciencedirect. Vol:462.1-8

Baatz, M. and Schäpe, A. (2000). "Multiresolution segmentation: an optimization approach for high quality multi-scale image segmentation". In: XII Angewandte Geographische Informationsverarbeitung, Wichmann-Verlag, Heidelberg, 2000.

Danoedoro, P. 2012. Introduction to Digital Remote Sensing. Yogyakarta: ANDI Publisher.

Jusuf, M.D. 2017. Peatland Vegetation Species Model Using Hymap Image and Machine Learning. Earth and Environmental Science 98.1-6.

Kononova, M.M. 1968. Transformation of organic matter and their relation to soil fertility. Sov. Soil. Sci. 8:1047-1056.

Lapan. 2018. Remote Sensing Data and Technology Center. 2019.

Lillesand, T.M. & Kiefer, R.W. 1990. Remote Sensing and Image Interpretation (Dulbahri Translation, et al). Yogyakarta: Gadjah Mada University Press.

Stiawan, Y. 2016. Spatial and Environmental Distribution Characteristics of the Sumatran Peat Area Using Multi-Temporal MODIS 250 M. Sciencedirect Data. 33. 117-127.

Suharyadi. 2014. Application of Object-Based Image Analysis (OBIA) for Detection of Land Use Change Using ALOS AVNIR-2 Imagery. 131-138.

Veljanovski, T., Kanjir, U., & Ostir, K. (2011). Object Based Image Analysis of Remote Sensing Data. Geodetski Vestnik, 55/4. http://geodetskivestnik.com/55/4/gv55- 4_665-688.pdf

Widayani, P. (2018). Application of Object Based Image Analysis for Early Identification of Slums Using Worldview-2 Satellite Imagery. Geography Magazine Indonesia Vol. 32, No.2.

IOP Conf. Series: Earth and Environmental Science 1047 (2022) 012018 doi:10.1088/1755-1315/1047/1/012018

Optimizing Image Scale and Resolution to Obtain Lineament Related Structures using ALOS PALSAR Dual Orbit Data

Fadhil Muhammad Akrom, Asep Saepuloh*

Geothermal Master Program, Faculty of Mining and Petroleum Engineering, Bandung Institute of Technology (ITB)
Faculty of Earth Sciences and Technology, Bandung Institute of Technology (ITB)

Jl. Ganesha No. 10, Bandung, West Java, Indonesia 40312

*Corresponding email: saepuloh@itb.ac.id

Abstract. Geological structures identification using conventional optical remote sensing encountered some problems due to atmospheric and surface conditions. Therefore, we exploited the Synthetic Aperture Radar (SAR) in Ascending and Descending orbits to characterize lineaments related to geological structures. We have analyzed two images from the Phased Array type L-band Synthetic Aperture Radar (PALSAR) onboard the Advanced Land Observing Satellite (ALOS). The adjustment of their resolution in the multilooking process was applied to obtain the best understanding between detected lineament and reference fault. The cross-polarized data of ALOS PALSAR level SLC were multilooked with variation of resolution from 15 to 120 m. Obtaining the optimum detected lineament-related faults, which applied the Yamaguchi criterion to extract the lineament visually in the multilooked images. The Kamojang-Guntur Geothermal Fields in West Java were selected as a study area because there is a challenge to confirm geological structure in the area with high alteration and high weathering conditions. According to the proposed criterion, we have obtained those images with resolution of 75 to 90 from the analysis images can represent surface geological structure conditions, and the correlation with existing published geological map. The field verifications have been confirmed that the geological structure indication from images has been proved with the existence of slickenside and morphological conditions such as drainage of rivers, ridges, and valleys. Based on lineaments, stress regimes, and field verifications, it was concluded that Yamaguchi criterion analysis with image resolution 75 to 90 m can describe the presence of regional structures and interpreted structures.

1. Introduction

Java Island in Indonesia located in the subduction zone between the Eurasian and Australian Plate was identified by active volcanoes with high-potential prospects of geothermal energy, especially in the south area of Java Island. Geological structures, such as fault and fracture, may create openings inside the rock to allow geothermal fluids to flow to the surface and create a permeable zone. Those structures can be observed as lineaments in satellite images. In this study, the satellite images were transformed into eight new images with various spatial resolutions from Kamojang-Guntur Geothermal Field as shown in Figure 1. These images were later compared to each other in finding which resolution that best represents faults in the area.

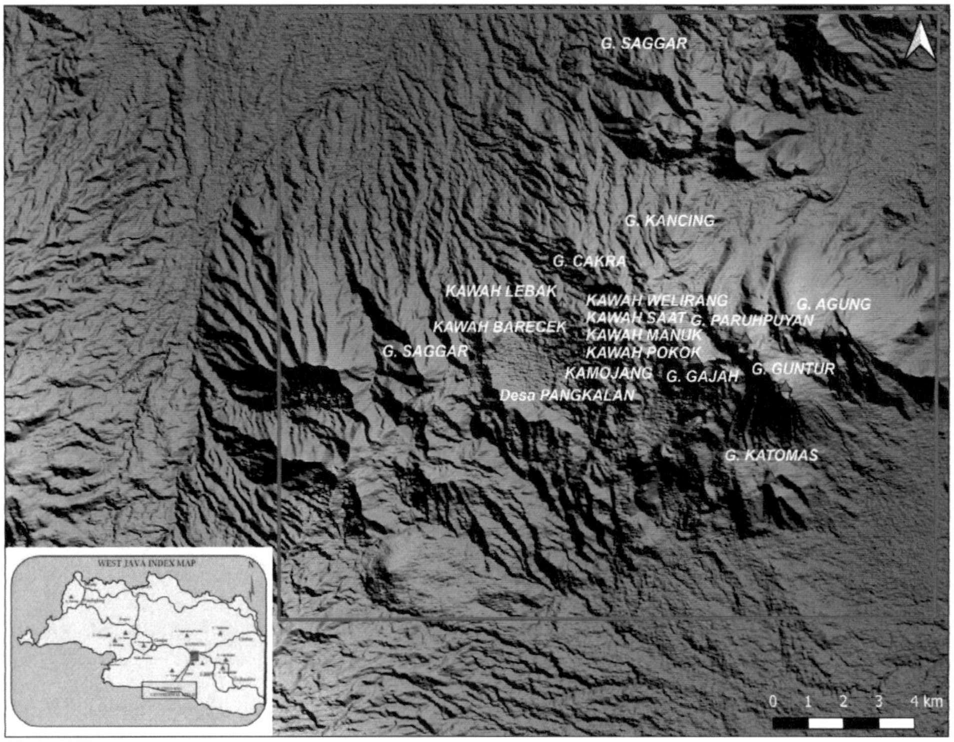

Figure 1. Study area presented by red rectangular overlaid on the Digital Elevation Model (DEM) of Kamojang-Guntur Geothermal Fields.

Geologically the area is covered by eight formations (from older to younger): Pangkalan Formation (Qgpk), Pangkalan-Kendang Formation (Qko), Malabar-Tilu Formation (Qmt), Mandalawangi-Mandalagiri Formation (Qmm), Masigit formation (Qym), Young Volcanics (Qypu), Guntur Lava (Qhg) and Lake Deposits (Qd).

Three important faults in these volcanic areas are Kendang Fault, Masigit Fault, and Guntur Fault, where the orientation both is NE-SW. The orientation of faults is conformity by the tectonic regime in the region where the direction of subduction of Australian Plate is north-south as shown in Figure 2.

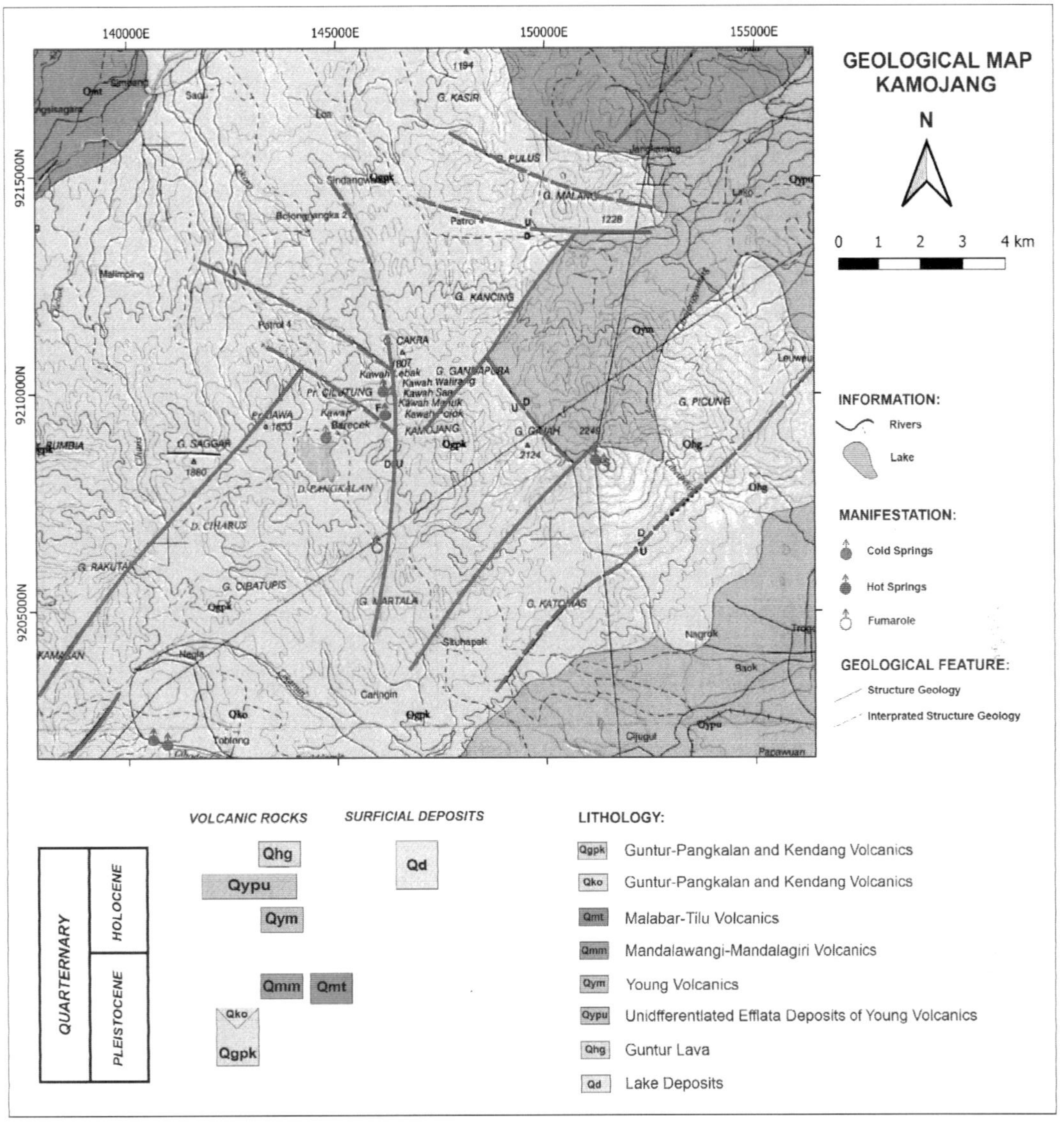

Figure 2. Geological map of Kamojang-Guntur Geothermal Fields showing the distribution of rock formation and faults in study area by Alzwar et al. in [1]

2. Data Collection and Criterion

The research methodology in this study is illustrated by Figure 3.

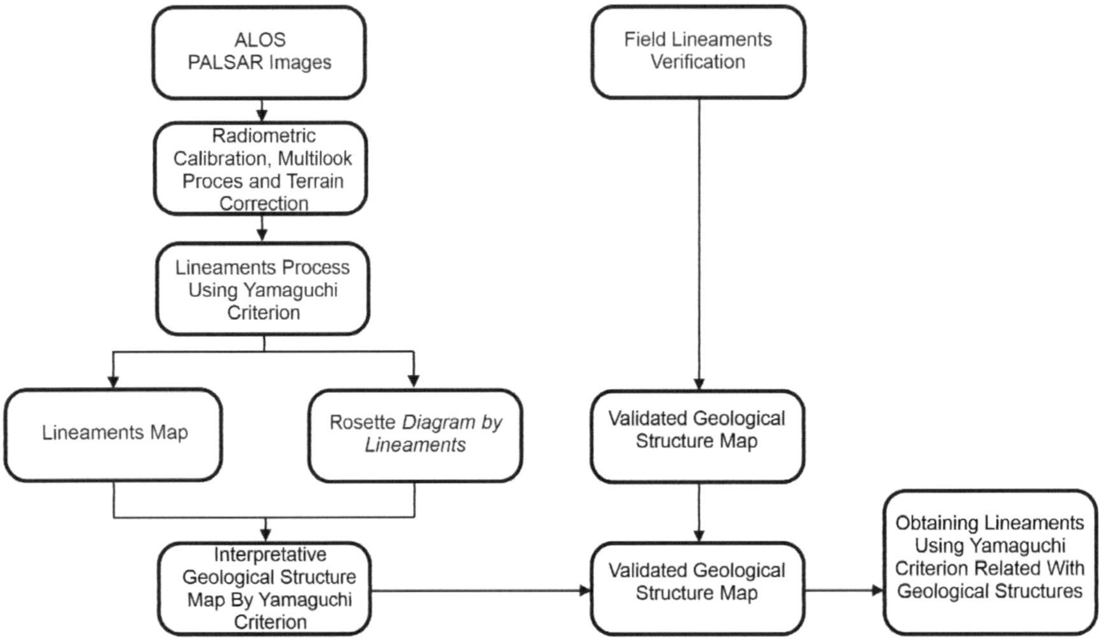

Figure 3. The research methodology in this study

2.1. Data collection and specification

Saepuloh et al. in [2] explained that the SAR back-scattering intensity data represents the power return to the satellite as a function of several surface parameters such as roughness, gradient slope, magnetic, and dielectric property of the ground. The parameters are presented by the geomorphologic and structural feature (GSF) in the SAR back-scattering intensity images. According to the GSF, the ridges, valleys, hills, and rivers are the crucial components for detecting linear features.

These are two orbits of ALOS PALSAR data used in this study; ascending and descending. Both data have been orthorectified in level 1.5. The ascending orbit indicates that the satellite moves from South to North with looking direction to East. Meanwhile, the descending orbit indicates that the satellite moves from North to South by looking directly to the West based on Saepuloh et al. [3]. Detail about data acquisition is listed in Table 1.

Table 1 Detail data used in this study

No	Data type	Data ID	Orbit	Date
1	Single polarization	ALPSRP145807040-L1.1	Ascending	Oct 19, 2008
2	Single polarization	ALPSRP073233760-L1.1	Descending	June 10, 2007
3	DEMNAS	DEMNAS_1308-41_v1.0	-	March 12, 2021

2.2. Image Processing and resolution adjustment

The raw data of ALOS PALSAR were transformed into various spatial resolutions of images and quantified by multi-look processing with eight look number and spatial resolution. After that process, we continue with terrain correction. In which process DEM 1 arc second data were used. In this study, we analyzed the spatial resolution size or spatial resolution of 15 m in an image.

We use the Yamaguchi criterion to obtain more details about the structural features within lineaments in the study area based on Yamaguchi et al. in [4]. The criterion is selected based on image scale (Sk) and spatial resolution (Rs) as follows:

$$Sk = \frac{0.1\,mm}{Rs} \quad (1)$$

The optimal scale for manual or visual lineaments identification needs to be determined because there is relationship between the image scale (Sk) and the number of lineaments. The larger scale, the more lineaments that can be identified. On the other hand, the smaller the scale, the fewer lineaments can be identified. Therefore, it is necessary to adjust the optimal scale to determine the correct screen magnification in the lineament identification process. This is done based on the Yamaguchi criterion which involves the Rs.

3. Results and Discussions

Lineament identification was applied based on spatial resolution of the eight different look numbers as shown in Table 2. These lineaments were identified visually on the calculated image scale from image resolution following Yamaguchi criterion. Accordingly, we can compare the identified lineaments with the existing regional lineaments in the Kamojang-Guntur Geothermal Field that is shown in Figure 4.

Table 2 Detail data used in this study

No	Number of range look	Number of azimuth look	Resolution (m)	Scale
1	1	5	15	1:150000
2	2	9	30	1:300000
3	3	14	45	1:450000
4	4	19	60	1:600000
5	5	24	75	1:750000
6	6	28	90	1:900000
7	7	33	105	1:1050000
8	8	38	120	1:1200000

3.1. Lineament Derived SAR Images

Following Yamaguchi criterion, look number 1×5 produced image resolution 15 m is suitable for lineament identification for image scale 1:150,000. Using this image scale, we could identify and calculated the major lineament direction is N55°E (Figure 5). For look number 2×9 (resolution 30 m), the image scale 1:300,000 produced major lineament direction also to the N55°E. On the other hand, for look number 3×14 (resolution 45 m), the image scale 1:450,000 produced major lineament direction N60°E. For look number 4×19 (resolution of 60 m), the image scale 1:600,000 produced major lineament direction N35°E. For look number 5×24 (resolution 75 m), the image scale 1:750,000 produced major and minor lineament directions N50°E and N135°E, respectively. For look number 6×28 (resolution of 90 m), the image scale 1:900,000 produced two major lineament directions N45°E and N125°E. Those lineament directions were then interpreted as a local geological structure in the study area. For look number 7×33 (resolution 105 m), the image scale 1:1,050,000 produced major lineament direction N45°E. Then, for look number 8×38 (resolution 120 m), the image scale 1: 1,200,000 produced major lineament direction N55ºE direction. According to the explained results, the image resolutions 75 to 90 m produced more complete lineament direction than others (Figure 5).

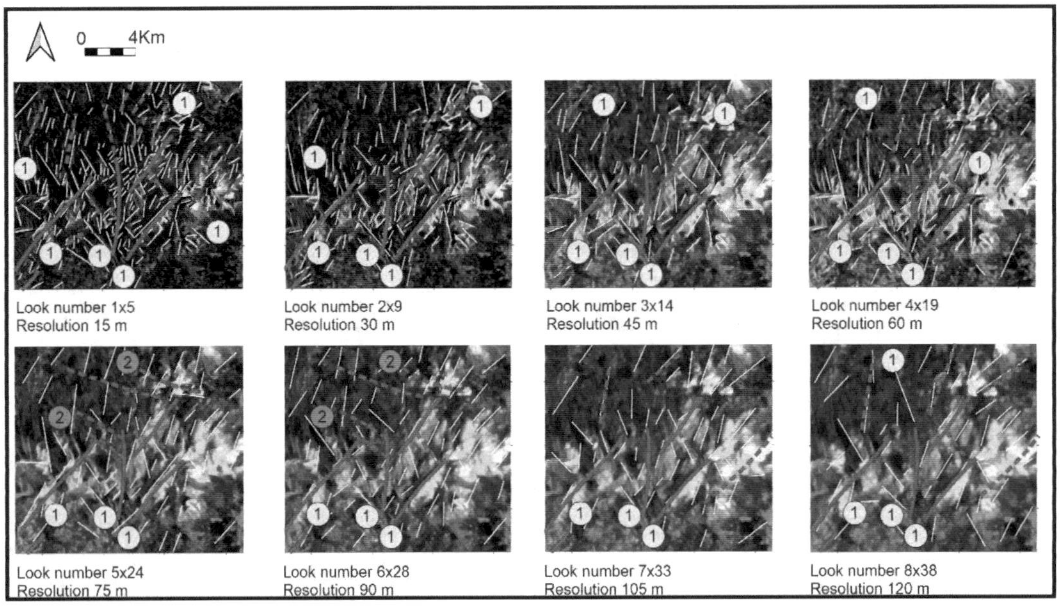

Figure 4. Detected lineaments following Yamaguchi criterion from different image resolution presented by white lines. The red lines are identified fault line and confirmed by geological map, the red dashed lines are interpreted local faults, and the circles with number are main lineament direction following rose diagram in Figure 5.

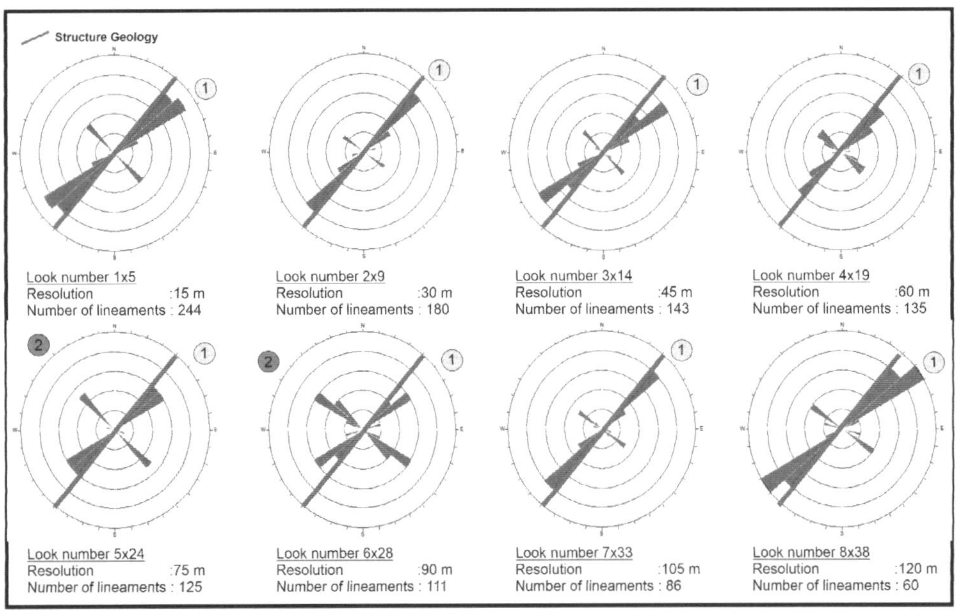

Figure 5. Rose diagram of detected lineaments showing the major lineaments directions pointed by 1, minor lineaments by 2, and regional faults by green line for each image resolution.

3.2. Field Lineament Verification

Based on the field verification results, we collected six data points and observed the indications of the existing of geological structures in study area (Figure 6). The six observation points are explained as follows.

Point 1. At this location, we found that the interpreted fault corresponds well to the straight line of the ridge with direction N312°E.

Point 2. The interpreted fault at this location has good agreement with fault evidence of slickenside. The slickenside has trend N212°E, plunge 14°, and rake 20°. Based on that measured data, the type of fault is strike-slip with same direction as the reference geological structure in the NE-SW region.

Point 3. At this location, the circular feature has been observed. We measured that the north rim of the circular feature trend is N30°E. This trend follows the geomorphology and structural features in the ALOS PALSAR image.

Point 4. The observation point is located at the southern part of the circular feature observed at Point 3. At this point, the measured rim trend is N340°E.

Point 5. At this location, a slickenside as fault indication have been observed and measured with trend N24°E, plunge 14°, and rakes 27°. This measured fault evidence is concordance to the fault interpretation from Yamaguchi criterion lineaments with resolution 75-90 m.

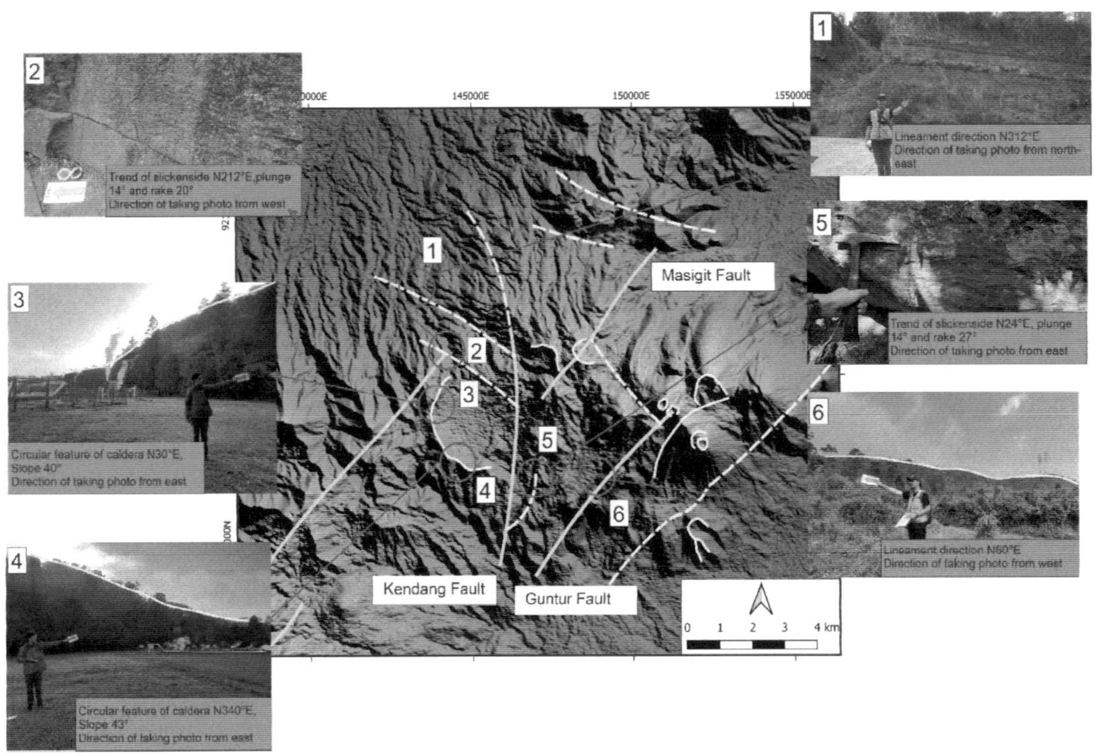

Figure 6. Field photographs confirmed lineament-related structures in the most optimum scale and resolution of ALOS PALSAR images.

Point 6. At this location, we have observed the lineament of the ridge with trend N60°E. This ridge is manifestation of Guntur fault.

Based on the analysis from the eight selected look numbers and field verification, the obtained major lineaments directions N30°E-N60°E concordance to direction 1 (Figure 4). Meanwhile, the direction 2 has not been observed clearly due to limited access, but we have found the ridge as represent of the geomorphology and structural features with direction N312°E. The verified two fault directions located in the same quadrant with major geological structure at NE – SW and NW-SE. The agreement of interpreted faults from extracted lineaments and fault evidence at field represents the correctness proposed criterion.

4. Conclusions

Eight look numbers and spatial resolutions were processed by multilook process from two orbits ALOS PALSAR data: ascending and descending orbits. Each spatial resolution associated with specific parameter and produced different lineaments. The most appropriate lineaments related to resolution were characterized based on lineaments direction, number of lineaments, rose diagram analysis, geological structure map, and field verification. Following this study, we have analyzed that look numbers from 5×24 to 6×24 producing spatial resolution from 75 to 90 m provide the best agreement of the lineament related structures using Yamaguchi criterion and field verification as well as geological map. The extracted lineaments presented the regional fault system and new local faults in the Kamojang-Guntur Geothermal Field. The interpreted faults inferred from distribution lineaments of the 75 to 90 m spatial resolutions coincidence with the fault evidence at field. The selected resolution provides visually the most optimal lineaments related to faults in the ALOS PALSAR images. The circular feature as well as NE-SW and NW-SE faults were identified well in the Kamojang-Guntur Geothermal Field. This interpretation can be used as a first consideration prior to detailed geological field mapping.

Acknowledgments

Authors sincerely thank Badan Informasi Geospasial for kindly providing the Digital Elevation Model Nasional (DEMNAS) and JAXA for providing ALOS PALSAR images for this study. This study was supported by LPPM ITB under research scheme "Riset Unggulan ITB 2021".

References

[1] Alzwar, M., Akbar, N., and Bachri, S. (1992): Geological map of Garut and Pemeungpeuk sheet, Java Directorate of Geology, Bandung
[2] Saepuloh, A., Haeruddin, H., Heriawan, M. N., Kubo, T., Koike, K., and Malik, D. (2018): Application of lineament density extracted from dual orbit of synthetic aperture radar (SAR) images to detecting fluids paths in the Wayang Windu geothermal field (West Java, Indonesia), Geothermics, 72, pp. 145–155.
[3] Saepuloh, A. (2021): Prinsip dan Aplikasi Pengindraan Jauh Geologi Gunung Api Pendekatan Fisis dan Observasi Geologi Lapangan, ITB Press, Bandung.
[4] Yamaguchi, Y., (1985): Image-scale and look-direction effects on the detectability of lineaments in radar images. Remote Sensing of Environment, 17, pp. 117–127.

IOP Conf. Series: Earth and Environmental Science 1047 (2022) 012019 doi:10.1088/1755-1315/1047/1/012019

Seismic Noise Analysis in InaTEWS Earthquake Station Network *(Case Study: Flores Earthquake 7.4, 14 December 2021)*

Arif Rachman Hakim[1,2*], Adhi Harmoko Saputro[1], Supriyanto Rohadi[2], Mohamad Taufik Gunawan[2], Ricko Kardoso[2]

[1] Department of Physics, Faculty of Mathematics and Natural Sciences, Universitas Indonesia, Depok, Indonesia

[2] Meteorological, Climatology and Geophysics Agency of Indonesia

*email: arif.rachman04@ui.ac.id

Abstract. Meteorological, Climatology and Geophysics Agency (BMKG) is the agency responsible for providing information on earthquake parameters in Indonesia. BMKG has installed 411 seismometers network spread across Indonesia, this network is known as the Indonesia Tsunami Early Warning System (InaTEWS) network. The dense distribution of the seismometer is expected to be able to provide accurate earthquake parameter results. However, the accuracy of earthquake parameters does not only depend on many stations but also on the quality of the data recorded by the seismometer. In this study, an analysis of 15 data quality was carried out at the nearest station that recorded the Flores earthquake on December 14, 2021. Analysis of data quality using the power spectral density (PSD) and probability density functions (PDFs) methods, where good data quality can be seen in the spectrum of the seismograph recordings that are between the Peterson's new high noise model (NHNM) and new low noise model (NLNM).

1. Introduction

Meteorological, Climatology and Geophysics Agency (BMKG) has installed 411 seismometers throughout Indonesia (Figure 1), with the aim of closing the earthquake observation network, so that accurate results are obtained. Accurate results depend not only on the number of stations but also on the quality of the data recorded by the seismometer. One thing that affects the quality of the seismometer recording data is the noise level at each station. Noise sources are divided into 3 characters, namely cultural noise sources or sources originating from human activities, both wind, water, and geological sources, and the third is a micro seismic source[1]. BMKG has installed a broadband seismometer that has a frequency range of 0.01 HZ – 50 Hz, wherein in that frequency range there is also seismic [2]. To determine the performance and quality of earthquake observation stations, noise level analysis was carried out using the power spectral density (PSD) and probability density functions (PDFs) methods, where the calculated frequency was compared with Peterson's new high noise model (NHNM) and new low noise model (NLNM) [3]. In this study, data quality analysis was carried out at 15 stations in the Flores Sea area that recorded a tsunami earthquake with a magnitude of 7.4 on December 14, 2021, to determine the quality of data recording at these stations.

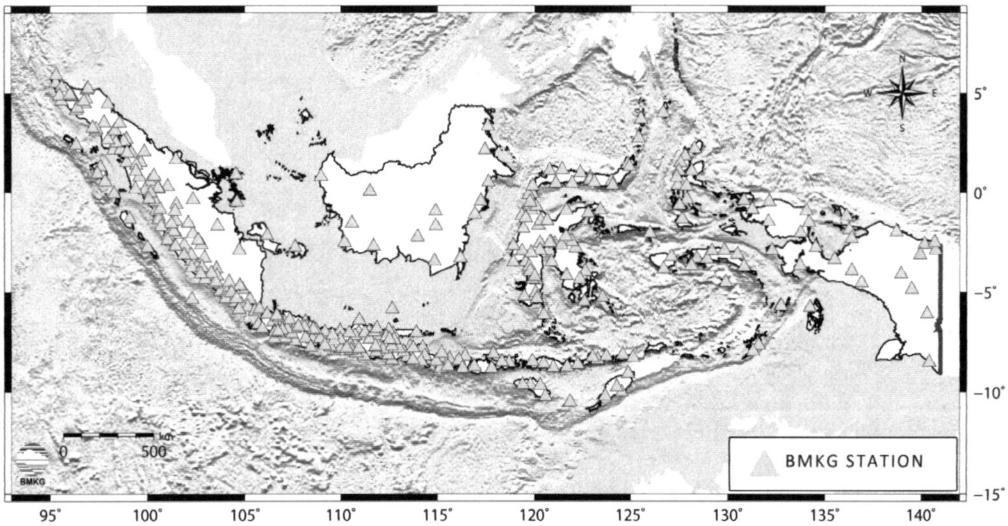

Figure 1. Map of BMKG earthquake stations network in Indonesia with yellow triangle symbols

2. Data and Method

The data used in this study is broadband seismometer recording from earthquake stations, located around the 7.4 M Flores earthquake in the Flores sea area (Figure 2), where data were obtained as many as 15 earthquake observation stations (Table 1) from the BMKG webdc3 repository, with a seismometer recording time span of about 30 days, when an earthquake occurs, and after shock recording (15-16 December 2021) using the vertical (Z) component.

Table 1. 15 Station InaTEWS - BMKG and Location

	Station	Latitude	Longitude
1	BBSI	-5.49	122.56
2	BKSI	-5.32	120.12
3	BSSI	-6.14	120.49
4	EDFI	-8.75	121.69
5	KJCM	-5.56	119.80
6	LBFI	-8.48	119.89
7	LRTI	-8.28	123.00
8	MKS	-5.22	119.47
9	MMRI	-8.69	122.24
10	OMBFM	-8.69	119.82
11	PSJCM	-7.06	120.62
12	RNFM	-8.43	121.03
13	WFTFM	-8.52	122.71
14	WLFM	-8.53	123.53
15	WEFM	-8.52	122.71

In this study, Obspy (library python) [4] is a tool used to analyse noise levels with probabilistic power spectral density (PPSD) class, to calculate PSD and PDF whose calculations are based on routines used by McNamara [1], to determine the PSD of an earthquake station uses the waveform and the instrument response files. The hour-long continuous time series data segments are processed by removing mean, long period trend which reduces the number of operation and long period contamination. Tapering is performed using a 10% sine function to smooth and minimize the discontinuity between the beginning and end of the time series and Fast Fourier Transform (FFT) algorithm[5] applied. Segments are then averaged to provide a PSD and raw frequency distributions constructed by gathering the individual PSDs for each channel. Next, the raw frequencies are binned in octave intervals and distribution bins are normalized by the total number of PSDs to construct the PDF results, for the direct comparison with New Low Noise Model (NLNM) and New High Noise Model (NHNM), the instrument transfer function is deconvolved with the individual time segments [6]. Figure 3 is example PDF result can be used by network operators as a reference in analysing noise levels[1].

Figure 2. Map view of selected area in this study in Flores Sea, green triangle is earthquake station used in this study; red star is epicentre of Flores earthquake 7.4 M

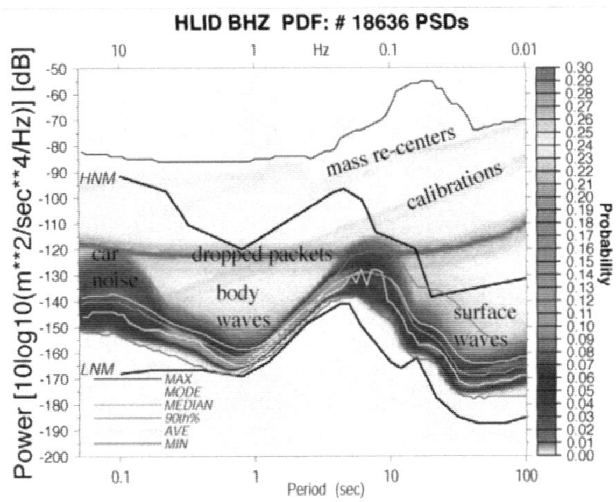

Figure 3. Example PDF for station HLID BHZ, constructed using 18,636 PSDs. Recording system transients and earthquakes are observable in the PDF, can be used for guidance [1]

3. Result and Discussion

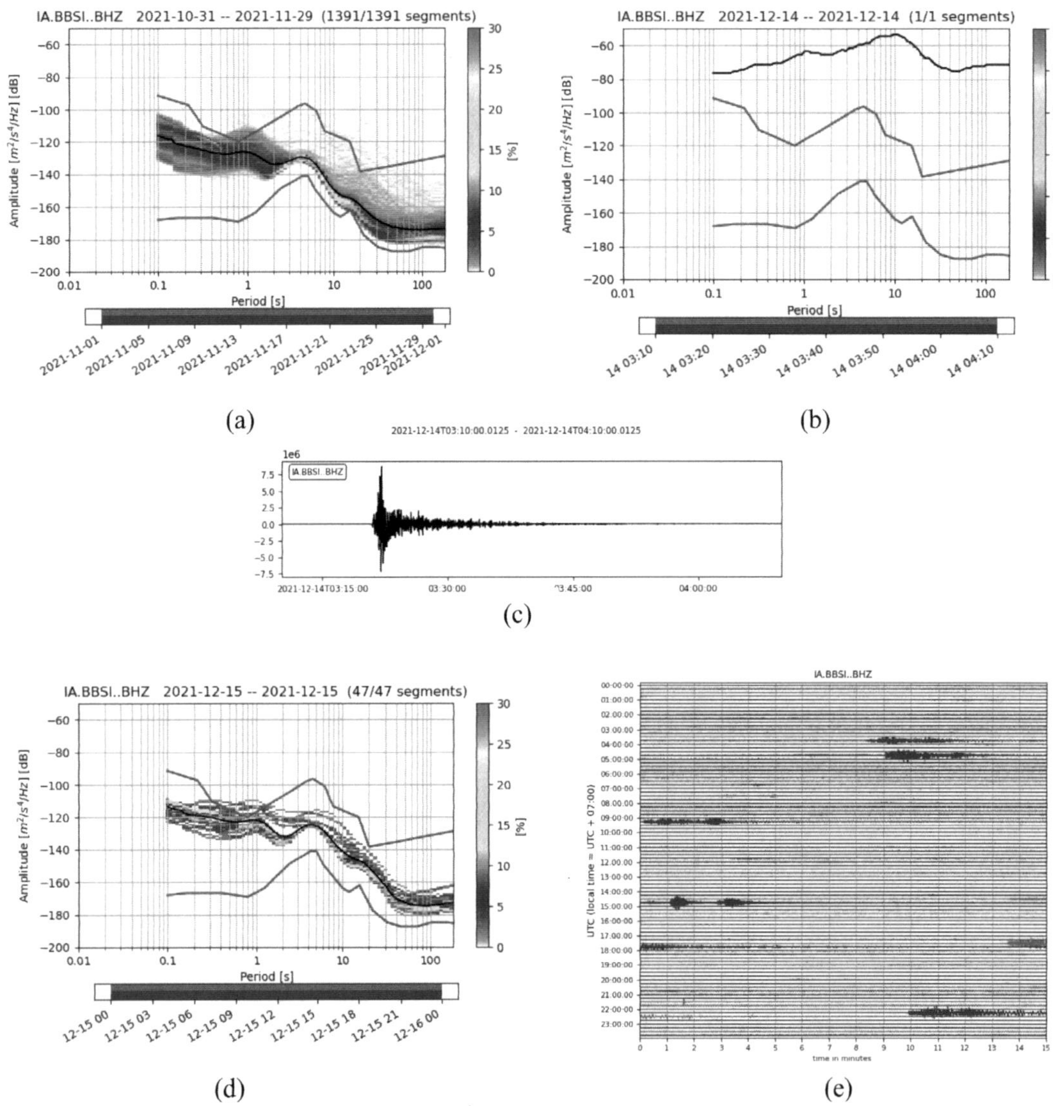

Figure 4. Result for BBSI station (Bau Bau, Buton - Sulawesi). (a) PPSD 30 days (b) PPSD when earthquake occurs (c) waveform mainshock (d) PPSD aftershock 7.4 (e) waveform aftershock

Figure 4.a is the result of PPSD for BBSI station with a data length of about 30 days, the signal spectrum is mostly between the noise of the NHNM and NLNM models, but in the 1s period the spectrum is above the NHNM model around 5%, this indicates human activity or cultural noise, because the earthquake station is in the airport area. While for the period between 1s to 100s the spectrum is in the noise model so it can be said that this station is quite good at recording earthquakes, it can be seen for PPSD (Figure.4.b) when there was a Flores earthquake with a magnitude of 7.4 signal spectrum away from the noise model[7] and the waveform recording (Figure.4.c) was quite clear recording the P and S phases. While the results of the PPSD aftershocks (Figure 4.d) with a data length of one day can be seen in the low period, the recording of small earthquakes is still very clear (Figure 4.e) but not too much, it is possible that the relationship between magnitude and station distance[8] influences earthquake detection, in this case the BBSI station is about 237 km from the epicentre and the effect of noise in the 1s period.

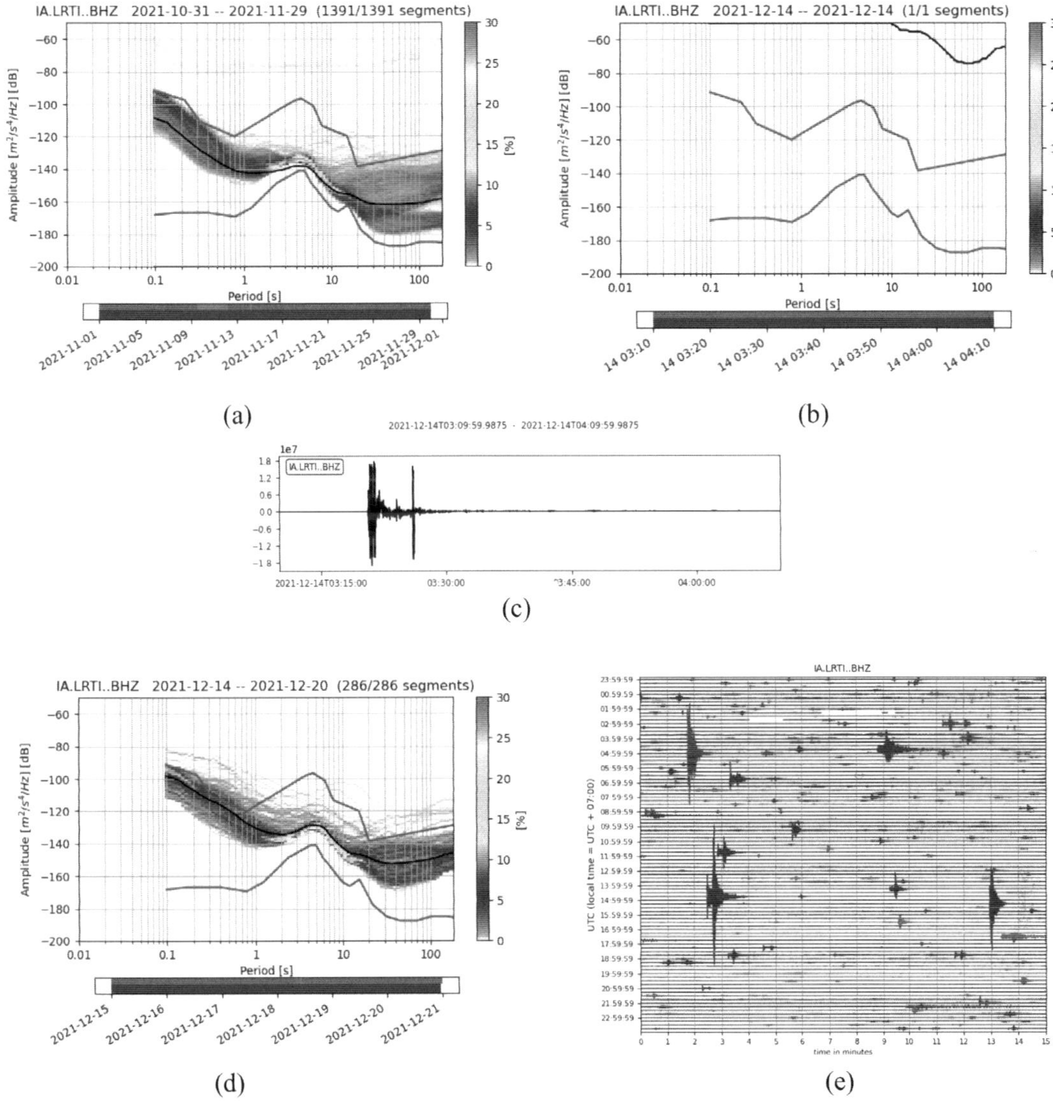

Figure 5. Result for LRTI station (Larantuka NTT). (a) PPSD 30 days (b) PPSD when earthquake occurs (c) waveform mainshock (d) PPSD aftershock 7.4 (e) waveform aftershock

For the LRTI station (Larantuka, NTT) with 115 km distance from epicentre Figure 5. a show that all spectrums are in the NHNM and NLNM noise models, while at the time of the earthquake, the PPSD (b) is far from the noise model, and the earthquake recording is quite clear (c). While the results of the PPSD (Figure 5. d) aftershock with a data length of one day can be seen in the low period the recording of small earthquakes is still very clear (Figure 5. e) and quite a lot of earthquakes with small magnitudes are recorded.

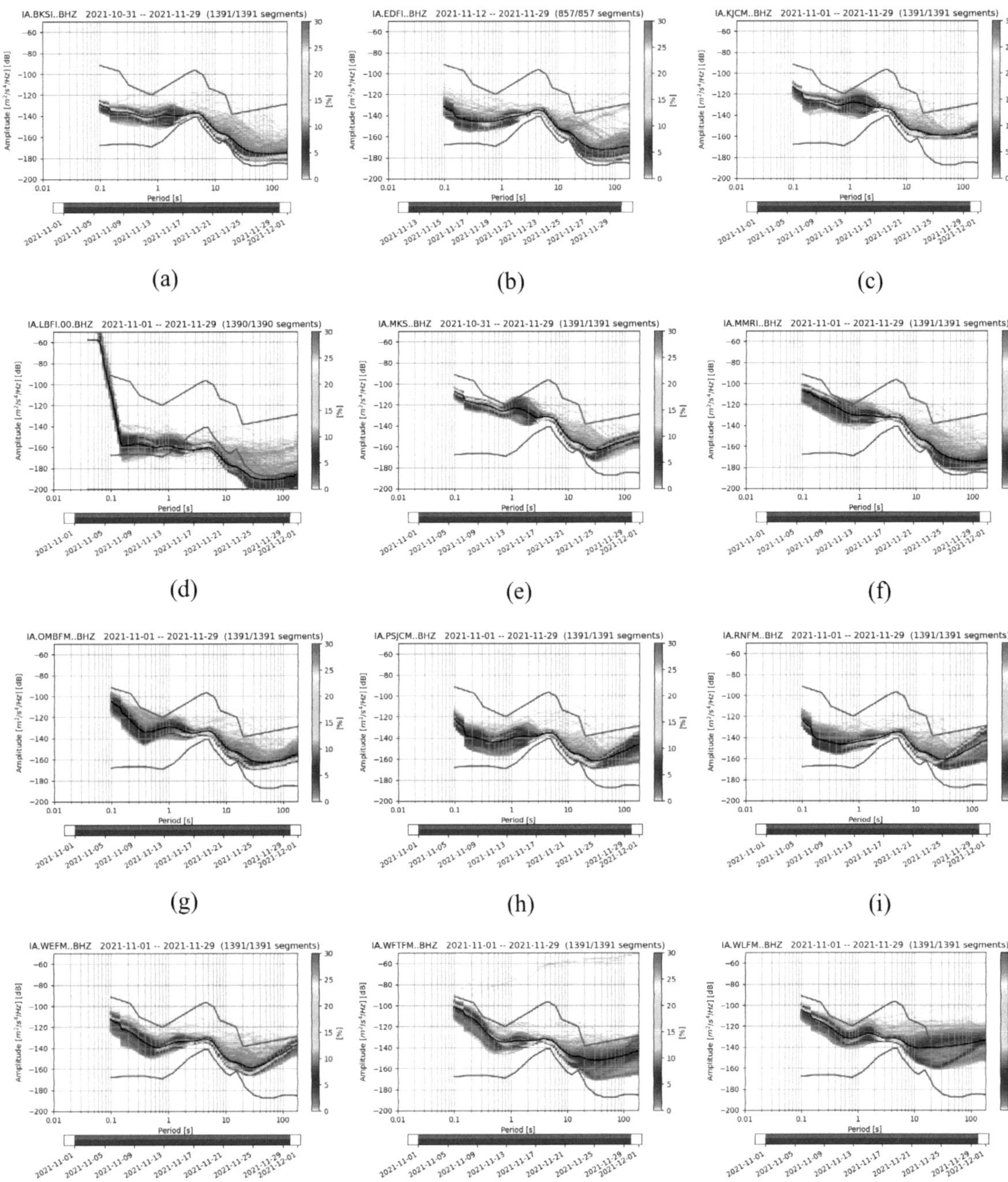

Figure 6. Probability power spectral density (PPSD) for 12 stations. (a) BKSI, Bulukumba Sulawesi (b) EDFI, Ende Flores NTT (c) KCJM, Kelara Sulawesi Selatan (d) LBFI, Labuan Bajo NTB (e) MKS, Makasar Sulawesi (f) MMRI, Maumere Flores NTT (g) OMBFM, Komodo Manggarai Barat NTB (h) PSJCM, Selayar Sulawesi (i) RNFM, Riung Ngada NTT (j) WEFM, Wewaria ENDE NTT (k) WFTFM, Walunggitang Flores NTT (l) WLFM, Wulondoni Lembata NTT

From the other 12 stations (Figure 6) there is one LBFI (Figure 6.d) station whose PPSD results are under the NLNM model, this is due to a station metadata error, it is recommended that the data for LBFI station is not used in further analysis. While for the other 11 stations the signal spectrum at each station is in the NHNM and NLNM noise models, it can be said that this station can be used for earthquake analysis.

4. Conclusion

The probabilistic power spectral density (PPSD) results at several stations show that the spectrum in the low period (0.1 - 1) is still within the noise limits of the NHNM and NLNM models, this indicates that the sensor placement is appropriate and far from disturbing cultural noise. For the mainshock recording, the P and S phases are very clearly visible, and for the aftershock, in the low period a small earthquake (magnitude 3 – 5) is still well recorded, and some stations do not record it is possible due to the distance of the station from the earthquake source. From the results of the PPSD for all stations, it can be concluded that the noise level at each station is still in the noise model so that the station is considered suitable for use in earthquake analysis, and one LBFI station which is not recommended for use in earthquake analysis due to metadata errors so that the PPSD results are below the NHNM noise model.

Acknowledgments

We would like to thank the funding support from the Meteorology, Climatology and Geophysics Agency (BMKG).

References

[1] D. Mcnamara and R. Buland, "Ambient Noise Levels in the Continental United States," *Bulletin of the Seismological Society of America*, vol. 94, pp. 1517–1527, Dec. 2004, doi: 10.1785/012003001.

[2] P. Bormann and E. Bergman, "The new IASPEI Manual of Seismological Observatory Practice," *Seismological Research Letters*, vol. 71, pp. 510–518, Sep. 2000, doi: 10.1785/gssrl.71.5.510.

[3] J. R. Peterson and U. S. G. Survey, "Observations and modeling of seismic background noise," 1993. doi: 10.3133/ofr93322.

[4] L. Krischer *et al.*, "ObsPy: a bridge for seismology into the scientific Python ecosystem," *Computational Science & Discovery*, vol. 8, no. 1, p. 14003, May 2015, doi: 10.1088/1749-4699/8/1/014003.

[5] J. Cooley and J. Tukey, "An Algorithm for the Machine Calculation of Complex Fourier Series," *Mathematics of Computation*, vol. 19, no. 90, pp. 297–301, 1965.

[6] D. Mcnamara, C. Hutt, L. Gee, H. Benz, and R. Buland, "A Method to Establish Seismic Noise Baselines for Automated Station Assessment," *Seismological Research Letters - SEISMOL RES LETT*, vol. 80, pp. 628–637, Feb. 2009, doi: 10.1785/gssrl.80.4.628.

[7] A. Errington, "Sensor placement for microseismic event location," 2006.

[8] D. E. McNamara, C. Hillebrandt-Andrade, J.-M. Saurel, V. Huerfano-Moreno, and L. Lynch, "Quantifying 10 years of improved earthquake-monitoring performance in the Caribbean region," *Seismological Research Letters*, vol. 87, no. 1, 2015, doi: 10.1785/0220150095.

IOP Conf. Series: Earth and Environmental Science 1047 (2022) 012020 doi:10.1088/1755-1315/1047/1/012020

Analysis of Exploration Wells on Aquifer in Kaliwungu Formation and Regional Planning Recommendations

Achmad Darul[1,2] Rizki Kurniawan[2] Dasapta Erwin Irawan[1], Dimas Maulana Wibowo

[1]Fakultas Ilmu dan Teknologi Kebumian, Institut Teknologi Bandung, Jl. Ganesha 10 Bandung 40132, Indonesia
[2]Program studi Teknik Geologi, Institut Teknologi Sumatera, Jl. Terusan Ryacudu, Lampung 35365

E-mail: darul.rochman@gl.itera.ac.id

Abstract. Infrastructure development in the northern coast of Java continues to take place. The study and support of subsurface geological data and the value of hydrogeological parameters is one of the important parameters to support sustainable development. The existence of well exploration reportedly can support researchers for model validation and calibration. We carried out One-Dimensional and Two-Dimensional geoelectrical acquisition along with drilling in the Kaliwungu Formation which was lined with the Halang Formation. Ten points of one dimensional geoelectrical and two lines of geoelectrical acquisition VES method (Vertical Electrical Sounding), with a depth range of 100m Schlumberger and Wenner-Schlumberger configurations for profiling have reflected the study area dominated by aquitard (silty clay) with resistivity values < 20 ohm.m. The result of VES was then followed up by analysing the value of the well logging resistivity which was compared with the lithology of the cutting results in the 108m deep drill hole. Value of hydrogeological properties was carried out by the pumping test, were pumping test analysed using pumping cooper – Jacob single well software, and the result of hydraulic conductivity value was 6×10^{-9}m/s, and the transmittivity value was 5.6×10^{-7} m^2/s. Based on the Farvolden Method, the safe yield value was found to be comparable to the value according to Johson which is 3-8%.

1. Introduction

Infrastructure development absolutely must begin with making careful planning, which is in accordance with various technical criteria, because the infrastructure will be used for a long period. Studies are needed to assess long-term risks such as the risk of land subsidence [1], and management of water resources [2,3]. One of the parameters that must be reviewed is subsurface analysis, including geological conditions and hydrogeological parameters [4]. This research includes one and two-dimensional geoelectrical acquisitions followed by drilling to a depth of 100 meters, to determine subsurface lithology and hydrogeological parameters, which is expected to support sustainable development planning in Arjawinangun Regency-Cirebon and other areas on the northern coast of Java in the Kaliwungu formation [5].

2. Regional Geology

The regional geology of the research area is compared with the Map of Regional Geology-Arjawinangun sheet (Figure 1). Based on the map, this area is located in the Alluvial Deposit bounded by the Kaliwangu Formation (Pk) and Undegradable Volcanic Products (Qyu). The three units are described as follows:

Qa (Alluvial Deposits) are young alluvial deposits consisting of river alluvial deposits, swamp deposits, coastal alluvial deposits, their distribution is in the lowlands in Cirebon Regency and in river valleys that flow in a relatively south-north direction [6]. Pk (Kaliwangu Formation), in this formation many mollusc fossils were found. Stratigraphically in the study area, the Kaliwangu Formation was deposited unconformity with the Upper Halang Formation and the Cinambo Formation. This formation was deposited in a shallow marine environment (Litoral) with a thickness of approximately 600 meters [7]. Qyu (Undecomposed Volcanic Products) It consists of breccia, lava, which is andesitic basalt, tuffaceous sand, and lapilli. This unit was formed as a product of the eruption of Mount Ceremai [8]. It is estimated that this unit has a thickness of about 100 meters. Based on the hydrogeological map of the West Java regional sheet, the area survey is extensive, moderately productive aquifers. aquifers of moderate transmissivity, water table or piezometric head of groundwater near or above land surface. Well yields are generally less than 5 l/sec.

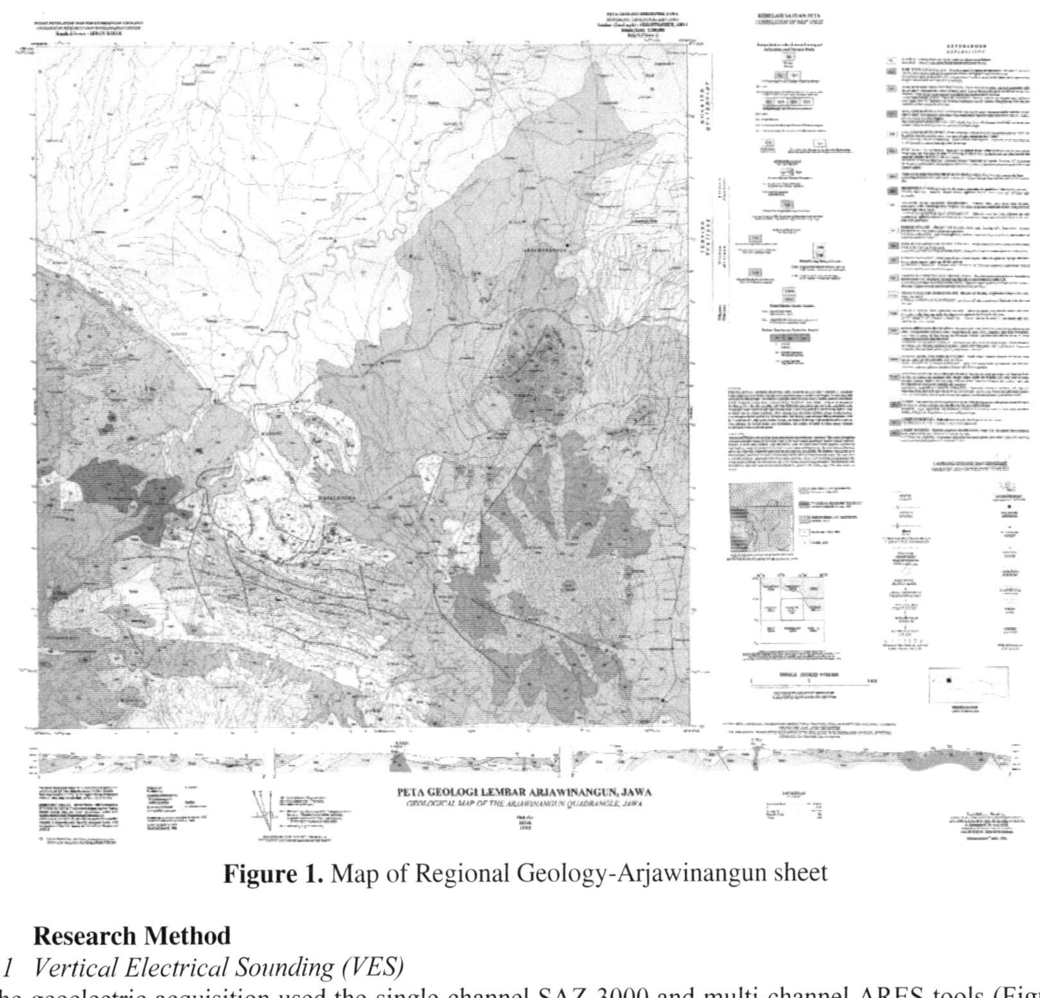

Figure 1. Map of Regional Geology-Arjawinangun sheet

3. Research Method

3.1 Vertical Electrical Sounding (VES)

The geoelectric acquisition used the single-channel SAZ-3000 and multi-channel ARES tools (Figure 2). The method used is VES (Vertical Electrical Sounding) Schlumberger and Wenner-Schlumberger configurations for profiling. Geoelectrical acquisition at the Arjawinangun-Cirebon area, carried out to determine subsurface conditions. The approach is to measure the amount of rock type resistance below the ground surface. Vertical electrical sounding aims to determine variations in the arrangement of rock layers below the ground vertically, by providing an electric current into the ground and the value of the resistivity is measured. The value of the resistivity of rock that is measured directly in the field is the value of apparent resistivity (apparent resistivity). Therefore, the value of resistivity in the field must be calculated and analyzed to get the actual value of resistivity by matching the field curve with the standard and auxiliary curves. The method used in geoelectrical measurement is using the Schlumberger (Figure 3), configuration using the following calculation formula:

$$\rho a = \prod \left(\frac{b^2}{2} - \frac{a}{4}\right)\left(\frac{V}{I}\right)$$

Information :

ρa	:	Apparent resistivity value (ohm meter)
V	:	Potential Differences (mili Volt)
I	:	Current (miliAmper)
b	:	Half current electrode distance (meter)
a	:	Current electrode distance (meter)

Requirements : AB/2 > MN/2

Figure 2. Geoelectrical Equipment SAZ-3000 and ARES

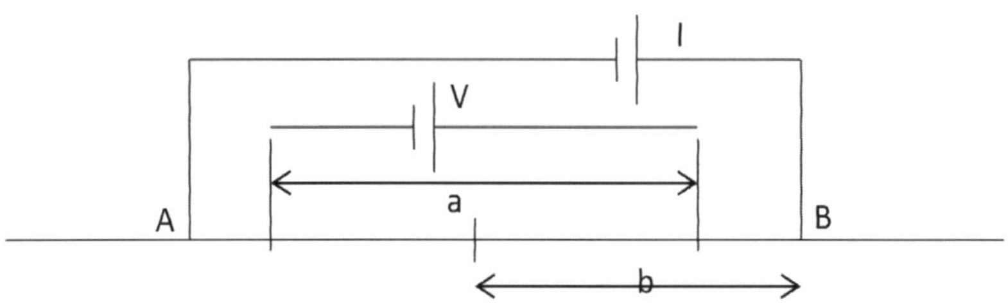

Figure 3. Schematic of a geoelectric survey using the Schlumberger method

Furthermore, in processing and calculating field data to get the actual resistivity value along with the depth, the matching curve method is used. For this survey, the maximum AB/2 length is 500 meters, and the maximum MN/2 length is 25 meters. In field measurements, the reading of the instrument is carried out 4 times (stack) and if necessary, it is carried out up to 16 stacks so that the most stable measurement value is obtained. The distribution of VES and two-dimensional geoelectrical acquisitions in the study area is shown in the following figure (see Figure 4).

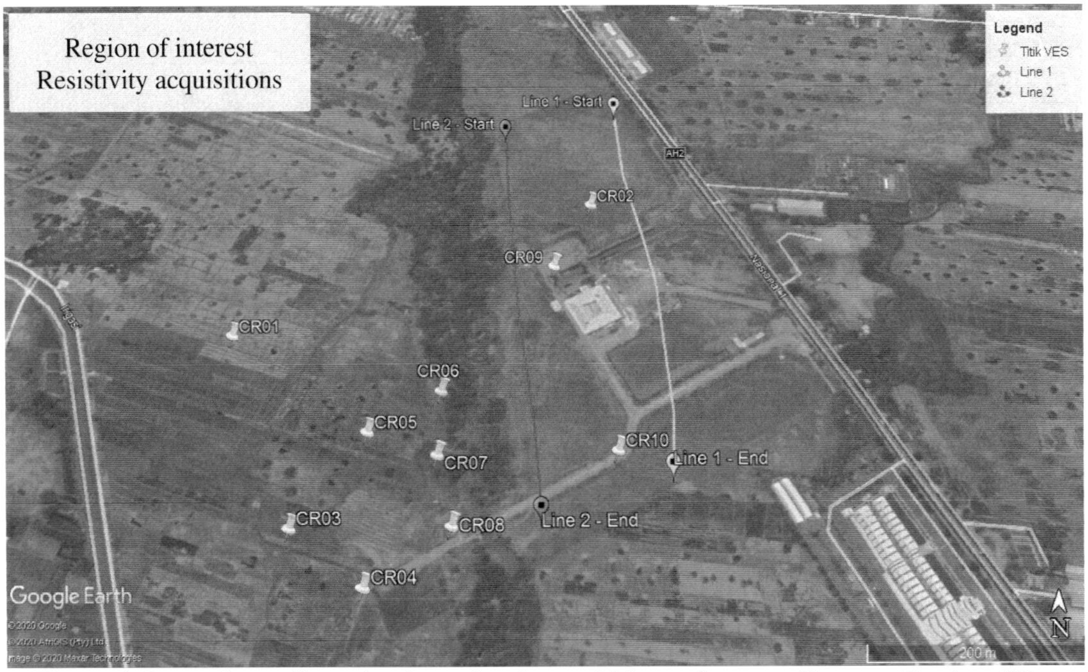

Figure 4. One and two-dimensional geoelectric acquisition at the area of interest. Caption: CR codes 1 to 10 for the one-dimensional geoelectrical acquisition symbol, line-1 and line-2 codes for the two-dimensional geoelectrical acquisition symbol.

3.2 Exploration Drill Investigation

To get better and more accurate results, the subsurface analysis was followed up with an exploration drilling, using a Jarco drilling machine 300 (Figure 5), with a hole diameter of 61 mm and a depth of 108 m. Cutting analysis of lithology (Figure 6) and results of the drilling cutting are arranged in the drill log of the work, arranged according to the construction well. Water level meter is used to determine the groundwater level and a submersible pump for hydrogeological property analysis [9]. The drilling system used is rotary drilling and pull-down pressure accompanied by circulation of drilling mud (mud flush) into the drill hole. Well logging analysis is carried out in the drill hole. Well logging is used to determine the position layers of lithology in the borehole based on the value of the resistivity of each layer of lithology in the in-situ drilled well. Well logging is also used as a comparison of measurements using other methods, such as Gamma Ray. Basically, the well logging method is the same as the surface resistivity investigation, the difference is the arrangement of current and potential electrodes, where in the borehole cross-section, the electrode arrangement is in an open well (without casing). An important stage before well logging is pilot hole activities to find out the lithology in detail. pilot holes are carried out with a small drill bit, for example with a wing bit drill bit to a depth exceeding the planned construction depth. The systematic stages of exploration well drilling include:

- Pilot hole
- Electrical Well Logging
- Cleaning and Reaming Hole
- Well Construction and, Screen
- Depositing Gravel Pads (Gravel Pack)
- Well Development
- Grouting
- Pumping Test
- Finishing

Figure 5. Jacro drill machine 300

Figure 6. Cutting analysis

To determine the value of the hydrogeological property, a pumping test was carried out in the borehole to obtain the flow rate value (the ability of the aquifer to fill exploration wells) using the equation Q = A.v where A is the cross-sectional area of flow and v is its average velocity. Then the flow rate value as input to the Cooper -Jacob single well pumping software to determine the hydrogeological property, the in situ hydrogeological property value was also compared with secondary data referring to Domenico and Schwatz [10] and the results of cutting analysis data (see table 1).

Table 1. Unconsolidated sedimentary materials

Material	Hydraulic Conductivity (m/sec)
Gravel	$3x10^{-4}$ to $3x10^{-2}$
Coarse sand	$9x10^{-7}$ to $6x10^{-3}$
Medium sand	$9x10^{-7}$ to $5x10^{-4}$
Fine Sand	$2x10^{-7}$ to $2x10^{-4}$
Silt, loess	$1x10^{-9}$ to $2x10^{-5}$
Till	$1x10^{-12}$ to $2x10^{-6}$
Clay	$1x10^{-11}$ to $4.7x10^{-9}$
Unweathered marine clay	$8x10^{-13}$ to $2x10^{-9}$

Meanwhile, the safe yield value (groundwater debit that can be utilized) refers to the Farvolden Method value [11], the formula to get the safe yield is as follows:

$$Q_{20} = \frac{4\pi T(H_A/8)}{2.30} S_f = 0.683 T H_A S_f$$

Information :

Q_{20} : 20-year safe pumping rate for the well (m^3/s)
T : Transmissivity (m^2/s)
S_f : Safety factor = 0.7 (no units)
H_A : Available head (m)
Q : Pumping rate used during the pumping test (m^3/s)

For comparison, the value of the in-situ lithology yield measurement, was also compared with that according to Johnson 1967 [12], as listed in the following table, (see table 2).

Table 2. Specific-yield value

Geological Formation	Specific Yield (%)
Clay	3.0
Sandy Clay/Silt	8.0
Very find sand	20.0
Fine sand	23.0
Fine medium sand	25.0
Medium sand	28.0
Medium coarse sand	27.5
Coarse sand	27.0
Gravel fine	25.0
Gravel medium	24.0
Gravel coarse	23.0

4. Results and Discussion

The results of the observations found that the survey area was filled with alluvial deposits that covered the fragmented claystone – siltstone units that were un-conformity, and no outcrops were found on the surface. The dominant fragments found at the research site are lithic andesite, rounded, and gravel in size. not only andesite fragments but also other types of fragments which is quartz and mollusc (Turritella sp.). The thickness of the clay lithology at the study site, and based on its lithological characteristics, indicates that the research location is the Kaliwungu Formation.

Measurement of the physical parameters of groundwater in several dug wells in the survey area, it was found that the pH, Total Dissolved Solids (TDS) and Electro Conductivity (EC) data had a range of pH values of 7.1 – 7.8, TDS 170 – 600, and EC 340 – 1120 µS. Based on the pH data, the groundwater in the study area is classified as neutral, while referring to the EC value and TDS value it classifies it as freshwater (Tikhomirov, 2016. Freeze and Cherry, 1979).

In the geoelectric survey, there are 10 sounding points and 2 profiling lines that can represent the situation and conditions at the location. From the sounding data obtained, it is then processed into an earth model, each geoelectric measurement point is interpreted to determine the type and thickness of the material based on resistivity values which are also based on the geological mapping. After that, interpolation is carried out to form 3D cross-sections and volumes.

The lithology resistivity values in 10 one-dimensional geoelectrical acquisitions (see figure 7 and figure 8), and 2 two-dimensional geoelectrical acquisitions (figure 9), only show two groups which are top-soil, and then dominated by a low resistivity of less than 20 Ohm.m which reflects silty clay lithology (green colour). The presence of silty clay is thought to be from shallow to 100 m. Silty clay is aquitard, this indicates that the study area is the Kaliwungu formation and based on the results of well logging the resistivity value is very low so that it confirms the lithology at the study site.

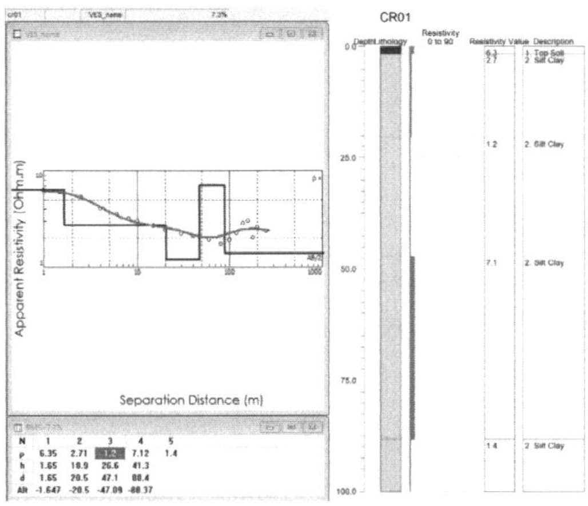

Figure 7. Result of CR01, one dimensional geoelectric acquisition at the area of interest

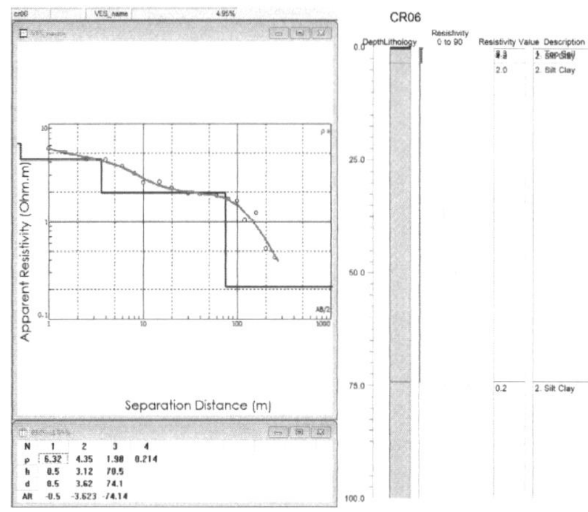

Figure 8. Result of CR06, one dimensional geoelectric acquisition at the area of interest.

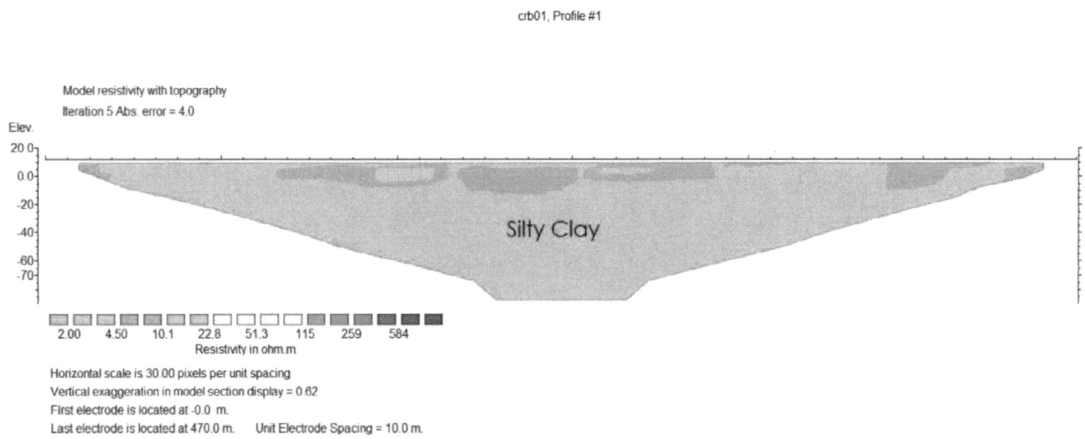

Figure 9. Results of line-1, two-dimensional geoelectrical acquisition.

Not only geoelectric acquisition accompanied by a study of geological conditions and physical properties of groundwater, but also to support development planning recommendations followed up by drilling of exploration wells. Based on the drilling procedure, the results of the drilling cuttings arranged in the drill log are known that from the surface to a depth of 2.00 mm it is top-soil, brown, soft, cemented, plastic, mafic material. at a depth of 3.00m - 09.00m is soil, there are fumic fragments and mafic material < 1 mm. at a depth of 09.00-33.00m is clay and sandy fragments are found. while at a depth of 33.00m - 103.00m it is claystone, siltstone, has sandy fragments, calcareous and tuffaceous materials, calcite material > 1 mm, black glass material < 1mm.

Based on lithological data in boreholes, research on hydrogeological property values was followed up with pumping-test on exploration wells, flow rate measurements obtained an average value of 0.003 lt/s, which is then used in the Pumping Cooper – Jacob single well software to determine the properties hydrogeology, from these data it was found that the value of the hydraulic conductivity = 6 x 10^{-9} m/s, and transmissivity value = 5.6 x 10^{-7} m^2/s (Figure 10). Considering that the data taken is very minimal, to confirm that the calculated data is relevant, secondary data which refers to the table of unconsolidated sedimentary materials values is also taken and matched with cutting data, it is found that the lithology is between silt – clay (1x10^{-11} m/s – 2 x 10^{-5} m^2/s). While the results of the calculation to get the safe yield value based on hydrogeological properties that have been measured in situ, it was found that the value at the research location was very small/close to zero (~ 0) m3/s. The measurement results are comparable based on the table of specific yield values according to the lithology at the research site, that the clay and silt values range from 3-8%.

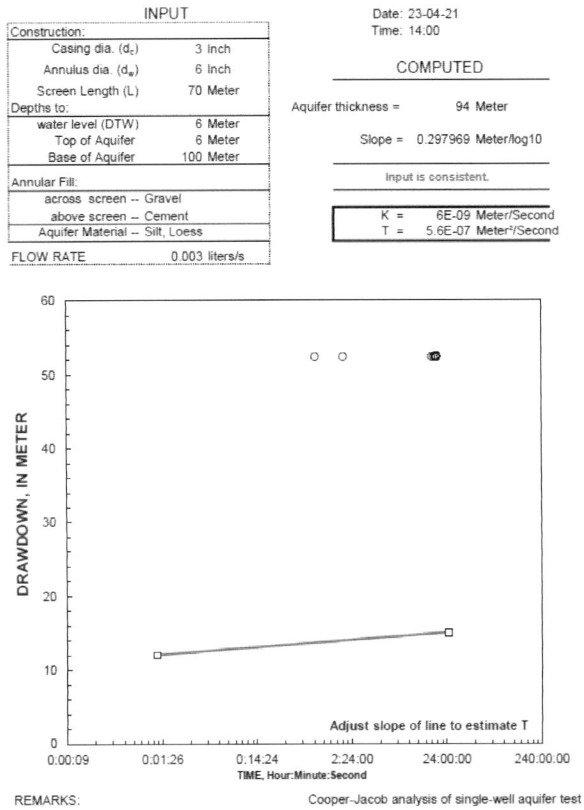

Figure 10. Result of the pumping test, Pumping Cooper – Jacob single well.

Based on this research, the condition of the research area which is dominated by silty clay provides information that the research area is dominated by aquitards, which is not a productive area of groundwater and it is recommended that water resource management consider water filters, rain harvesting. Recommendations for planning infrastructure development, subsurface lithology should be considered based on this analysis.

5. Conclusion

The results of the study describe the condition of the subsurface rock layers vertically and the in situ hydrogeological property measurements. The resulting conclusions are as follows:

- Geoelectrical acquisitions have been carried out at the research site as many as 10 VES measurement points with Schlumberger configuration, and 2 profiling paths using the Wenner-Schlumberger method.
- Based on the results of Geoelectrical acquisitions, the types of resistivity groups at the research site are divided into 2: top-soil and silty clay (low, < 20 Ohm.m).
- Based on cutting analysis from drill data, the lithology of the location area is dominated by clay. This indicates that the study area is the Kaliwungu formation.
- Measurement of hydrogeological property by pumping test found that the value of hydraulic conductivity = 6×10^{-9} m/s, and the value of transmitivity = 5.6×10^{-7} m^2/s. Based on references from the literature, the hydraulic conductivity values range from 1×10^{-11} m/s – 2×10^{-5} m^2/s. Meanwhile, the safe yield of groundwater is close to zero m^3/s and based on references it is around 3 – 8% or very small.

- That the subsurface data is limited, while the infrastructure development continues to take place, will potentially cause unprecedented problems, like land subsidence.
- The data currently available is not sufficient to set up a space utilization model, but subsurface lithology should be considered based on this analysis for infrastructure development planning. At the research location, groundwater resources are suggested as a complement after various other efforts to fulfil water resources such as water filters, rain harvesting.

Acknowledgements
We wish to express our gratitude to Iqbal Fauzi Aditama cs., the team of geoscience exploration that lends their hands and energy in the field survey.

References
[1] Sarah, D., & Soebowo, E. (2018). Land subsidence threats and its management in the North Coast of Java. In *IOP Conference Series: Earth and Environmental Science* (Vol. 118, No. 1, p. 012042). IOP Publishing. https://doi.org/10.1088/1755-1315/118/1/012042
[2] Irawan, D. E., Darul, A., Sumadi, H., Kuntoro, A. A., Argo, T. A., & Nurhayati, Y. (2017). Beban ganda pengelolaan air tanah di kabupaten/kota pasca pembatalan UU No 7/2004 tentang Sumber Daya Air: Ilustrasi dari Kota Bandung. *Jurnal Ilmu Sosial dan Humaniora*, *6*(1), 83-96. http://dx.doi.org/10.23887/jish-undiksha.v6i1.9720
[3] Irawan, D. E., Sulistyawati, E., Midori, A. A., Faisal, B., Darul, A., & Agustin, A. (2018). Surface and Groundwater Interactions: Cikapundung Bandung, Kanal Banjir Timur Semarang and Cisadane Tangerang. In *IOP Conference Series: Materials Science and Engineering* (Vol. 349, No. 1, p. 012013). IOP Publishing. https://doi.org/10.1088/1757-899X/349/1/012013
[4] Darul, A., Irawan, D. E., Trilaksono, N. J., Pratama, A., & Fitria, U. R. (2016). Conceptual model of groundwater and river water interactions in Cikapundung riverbank, Bandung, West Java. In *IOP Conference Series: Earth and Environmental Science* (Vol. 29, No. 1, p. 012026). IOP Publishing. https://doi.org/10.1088/1755-1315/29/1/012026
[5] Sarah, D., Syahbana, A. J., Lubis, R. F., & Mulyono, A. (2011). Modelling of land subsidence along Tanah Mas-Pelabuhan section Semarang city using finite element method. *Riset geologi dan pertambangan*, *21*(2), 105-119. http://dx.doi.org/10.14203/risetgeotam2011.v21.50
[6] Djuri, H. M. D. (1973). Peta Geologi Lembar Arjawinangun, Jawa Skala 1: 1000.000. *Direktorat Geologi Bandung*.
[7] Suhada, D.I., Triono, U., dan Simatupang, D.P., 2007. Inventarisasi Kandungan Gas Daerah Tanjung Medar Kabupaten Sumedang, Provinsi Jawa Barat.
[8] Irawan, D. E., Puradimaja, D. J., Notosiswoyo, S., & Soemintadiredja, P. (2009). Hydrogeochemistry of volcanic hydrogeology based on cluster analysis of Mount Ciremai, West Java, Indonesia. *Journal of hydrology*, *376*(1-2), 221-234. https://doi.org/10.1016/j.jhydrol.2009.07.033
[9] Alvando, P. V., Darul, A., & Irawan, D. E. (2017). Pemodelan Dampak Pemasangan Inclined Drain Hole Pada Wall PIT PT XXX Terhadap Penurunan Muka Airtanah. https://doi.org/10.31227/osf.io/ym925
[10] Domenico, P. A., & Schwartz, F. W. (1998). *Physical and chemical hydrogeology* (Vol. 506). New York: Wiley.
[11] Farvolden, R. N. (1961). *Groundwater resources Pembina area, Alberta* (No. 4). Research Council of Alberta.
[12] Johnson, A. I. (1967). *Specific yield: compilation of specific yields for various materials* (No. 1662). US Government Printing Office.

IOP Conf. Series: Earth and Environmental Science 1047 (2022) 012021 doi:10.1088/1755-1315/1047/1/012021

Monitoring CO and O_3 concentration that caused climate change periodically using Google Earth Engine (study case: Java Island)

N M Ikram[1], L Afifah[1], B S Arthatia[1], S J Wicaksono[1], M Maharani[1], Ediyanto[1], T Ihsanudin[1], D Apriyanti[1]

[1]Geomatics Engineering, Universitas Pembangunan Nasional "Veteran" Yogyakarta

nurmuhammadikram@gmail.com

Abstract. Climate change is the change of temperature and weather pattern in long terms caused by the accumulation of greenhouse emission in the atmosphere. Those emissions act like a blanket covering earth, trapping heat and causing the increase of earth's temperature. The emissions of greenhouse gases causing climate change are CO_2, O_3, N_2O, CH_3, et cetera. The existence of these gases have to be observed periodically for monitoring and controlling purposes. These days, remote sensing is capable of doing that. Sentinel-5P imagery exists as an imagery that can observe gas concentration within the earth's atmosphere. This research is aimed to observe CO and O_3 gas, which cause climate change, using Sentinel-5P satellite imagery using Google Earth Engine (GEE) periodically. GEE is a platform allowing its users to access and process data freely. The result of this research shows that there is a change in CO and O_3 concentration throughout January to November 2021. The highest concentration for CO was in October, and the lowest was in July. Meanwhile, the highest concentration for O_3 was in October, and the lowest was in June. Based on the research done, Sentinel-5P satellite imagery succeeded in monitoring CO and O_3 concentration periodically every month.

1. Introduction

The main cause of climate change is the increase of greenhouse gases (CO_2, CH_4, etc.) within the atmosphere. In 2000, the Ministry of Industry recorded as much as 1,720 ton of CO_2 equivalent emission (Ministry of Industry, 2010), and if there were no act taken to halt its increase, there would be 2,950 ton of CO_2 emission. The increase of CO_2 emission would cause elevation of temperature and global warming after some time, which would cause climate change. According to the Ministry of Industry in 2021, nationally, the biggest patron of greenhouse gas is the forestry sector, followed by transportation, industry, commercial, and domestic sector. In the industrial sector, the main source of greenhouse gases emission came from energy used, followed by processing technology and waste produced from the industry.

The existence of the industrial sector cannot be separated from society development, where production will always develop to fulfil human needs. Naturally, industrial processes would create waste, and how such waste is treated shall impact its surrounding environment.

There are several factors that allow the fact that an increase of certain greenhouse gas emissions would badly impact human life. Efforts are needed to monitor and control the existence of such gases.

Therefore, a periodical monitoring on such gases was done. Google Earth Engine or GEE is a cloud computing media capable of storing and processing big data with scales up to petabyte for analysis and decision making as needed by the user. GEE can help its users to create and run special algorithms and computation or fast calculations as the user needs. Other than that, GEE has an abundance of open source datasets that are renewed and developed every day. These datasets can be used for free for analysis purposes and are connected straight to cloud computing engines. One of the datasets available in GEE is Sentinel-5P recordings.

Sentinel-5P or Sentinel-5 Precursor's mission is to monitor earth's atmosphere. Sentinel-5P carries TROPOMI or Tropospheric Monitoring Instrument that can calculate the atmosphere's concentration with high spatial-temporal resolution. Data gained from the calculation can be used to know air quality, ozone and ultraviolet radiation, and climate monitoring and forecasting.

Processing done in GEE by writers using Sentinel-5P data allowed them to see changes of CO and O_3 concentration in Java Island periodically across 2021. The result of the monitoring then was used to see whether there is a correlation between CO and O_3 concentration with climate change. Beside that, writer was also capable to see the average emission, area with highest gas concentration, and the correlation between both gases.

2. Methods

2.1. Data

The research was done in Java Island, which is located at 105° 12' 44" BT - 114 36' 48" and 5° 54' 54" – 8 46' 48" LS. Java is an island that has the greatest number of industries in Indonesia, especially in Eastern and Western Java. This was according to Wahyuni et al (2021) who said that Eastern Java is an island with one of the greatest number of industries in Indonesia. Meanwhile, Western Java was also claimed to have the greatest industry in Indonesia (Ismaningsih, 2019). According to Nebath et al (2014), the existence of industrial complexes caused an increase in dangerous gas production, including CO and CO_2.

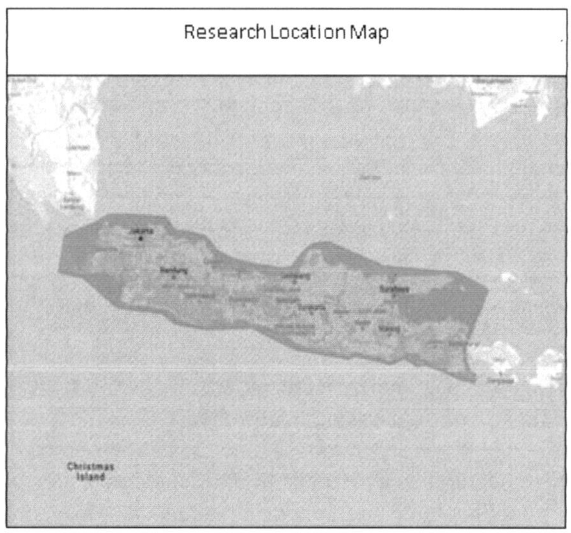

Figure 1. *Research location map in Java Island, Indonesia*

Data used in this research is Sentinel-5P's dataset available in GEE's database. Data taken was Near Real Time (NRTI) data of CO and O_3 in column number density or mol/m2. Dataset taken ranged from January through November 2021, using a one month interval. Data then processed using GEE from its access link https://engine.googleearth.com

2.2. Methods

Generally, methods performed in this research contained three steps: preparation, processing, and analysis. Complete procedure can be seen in Picture 2.

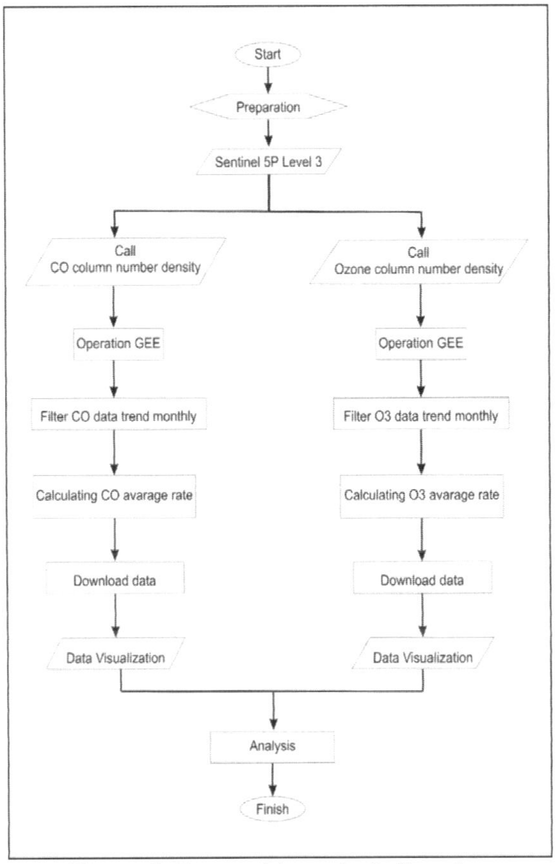

Figure 2. *Research's methods flowchart*

Preparation is the first step done to prepare next steps so they can run smoothly. This step is started by studying literature regarding things connected to research that was about to be conducted. Afterwards, preparation and information seeking about satellite imagery that was about to be used. In GEE, Sentinel-5P imagery offers several types of data, be it Offline or NRTI, for O_3, CO, NO_2, CH_4, SO_2, HCHO, and UVAI. Each type of data available has different recordings based on the band used for each recording. In this research, the data type processed was NRTI for CO and O_3. Bands available in the GEE dataset are shown in table 1 and 2.

Table 1. *NRTI CO bands in Google Earth Engine*

Name	Units	Min	Max	Description
O3_column_number_density	Mol/m^2	0.0047*	0.272*	Total atmospheric column of O_3 between the surface and the top of atmosfer, calculated with the DOAS algorithm.
O3_column_number_density_amf	Mol/m^2	1.92*	6.83*	Weighted mean of cloudy and clear air mass factor (amf) weighted by intensity-weighted cloud fraction
O3_slant_column_number_density	Mol/m^2	0.014*	1.402*	O_3 ring correctes slant column number density
O3_effective_temprature	K	-5962*	936*	Ozone cross section effective temperature
Cloud_fraction	fraction	0*	1*	Effective cloud fraction. See the Sentinel 5P L2 Input/Output Data Definition Spec, p.220.
Sensor_azimuth_angel	degrees	-180*	180*	Azimuth angele of the satellite at the ground pixel locagtion (WGS84); angle measured East-of-North.
Sensor_zenith_angel	degrees	0.098*	66.44*	Zenith angle of the satellite at the ground pixel location (WGS84); angle measured away from the vertical.
Solar_azimuth_angle	degrees	-180*	180*	Azimth angle of the Sun at the ground pixel location (WGS84); angle measures East-of-North.
Solar_zenith_angel	Degrees	8*	80*	Zentih angle of the ssatllite at ground pixel location (WGS84); angle measures away from vertical.
*estimate min or max value				

Table 2. NRTI O3 bands in Google Earth Engine

Name	Units	Min	Max	Description
CO_column_number_density	mol/m²	-279*	4.64*	Vertically integrated CO column density.
H2O_column_number_density	mol/m²	-465360*	3.45844e+07*	Water vapor column.
Cloud_height	m	-8341*	5000*	Scattering layer height.
Sensor_altitude	m	828542*	856078*	Altitude of the satellite with respect to the geodetic sub-satellite point (WGS84)
Sensor_azimuth_angle	degree	-180*	180*	Azimuth angle of the satellite at the ground pixel location (WGS84); angle measured East-of-North.
Sensor_zenith_angle	degree	1*	66*	Zenith angle of the satellite at the ground pixel location (WGS84); angle measured away from the vertical.
Solar_azimuth_angle	degree	-180*	180*	Azimuth angle of the sun at the ground pixel location (WGS84); angle measured East-of-North.
Solar_zenith_angle	degree	9*	80*	Zenith angle of the satellite at the ground pixel location (WGS84); angle measured away from the vertical.
*estimate min or max value				

The next step is processing the data. This step was done to process Sentinel-5P data in GEE to gain CO and O_3 extraction results. The processing began with entering some particular program scripts. This was done twice, for CO and for O_3. CO information extraction script used in GEE can be seen in figure 3. Meanwhile, the script used to extract O_3 information can be seen in figure 4. Then, information gained on both are used to find correlation between CO and O_3 with the creation of CO_2, which is a gas causing global warming.

```
//DATASET CITRA SELAMA TAHUN 2021
var Data1 = ee.ImageCollection('COPERNICUS/S5P/NRTI/L3_CO')
  .select('CO_column_number_density')
  .filterDate('2021-01-01','2021-01-31')
  .filterBounds(jawa)
  .mean();

var Mean_Data1 = Data1.reduceRegion({
  reducer: ee.Reducer.mean(),
  geometry: jawa,
  scale: 300,
})

var Data2 = ee.ImageCollection('COPERNICUS/S5P/NRTI/L3_CO')
  .select('CO_column_number_density')
  .filterDate('2021-02-01','2021-02-28')
  .filterBounds(jawa)
  .mean();

 var Mean_Data2 = Data2.reduceRegion({
  reducer: ee.Reducer.mean(),
  geometry: jawa,
  scale: 300,
})
```

Figure 3. *Part of JavaScript syntax used for CO extraction in Google Earth Engine*

```
//DATASET CITRA SELAMA TAHUN 2021
var Data1 = ee.ImageCollection('COPERNICUS/S5P/OFFL/L3_O3')
  .select('O3_column_number_density')
  .filterDate('2021-01-01','2021-01-31')
  .filterBounds(jawa)
  .mean();

var Mean_Data1 = Data1.reduceRegion({
  reducer: ee.Reducer.mean(),
  geometry: jawa,
  scale: 300,
})

var Data2 = ee.ImageCollection('COPERNICUS/S5P/OFFL/L3_O3')
  .select('O3_column_number_density')
  .filterDate('2021-02-01','2021-02-28')
  .filterBounds(jawa)
  .mean();

 var Mean_Data2 = Data2.reduceRegion({
  reducer: ee.Reducer.mean(),
  geometry: jawa,
  scale: 300,
})
```

Figure 4. *Part of JavaScript syntax used for O_3 extraction in GEE*

These algorithms were run to gain information regarding CO and O_3's column number density, which could reveal CO and O_3's concentration within the atmosphere. Extraction was done for imageries from January to November 2021 (table 3).

Table 3. *Data extraction period*

No	Variable Name	Month
1.	Data1	1 – 31 January 2021
2.	Data2	1 – 28 February 2021
3.	Data3	1 – 31 March 2021
4.	Data4	1 – 30 April 2021
5.	Data5	1 – 31 May 2021
6.	Data6	1 - 30 June 2021
7.	Data7	1 – 31 July 2021
8.	Data8	1 – 31 August 2021
9.	Data9	1 – 30 September 2021
10.	Data10	1 – 31 October 2021
11.	Data11	1 – 30 November 2021
12.	Data12	1 – 31 December 2021

Analysis stage is the last stage to test or describe extracted information gained. Analysis used in this research was statistical analysis. Analysis was done for CO and O_3 information extracted in the data processing stage. The information analyzed was pixel value recorded by CO and O_3 column number density. The change of column number density in each data was observed through a graphic to see changes of CO and O_3 gas concentration in the atmosphere within a time period, shown in table 3. Meanwhile, analysis was also done by observing CO and O_3 concentration distribution visualization through GEE within a time period. This visualization was done by interpolating extracted CO and O_3 data which allowed writers to observe them from GEE. Afterwards, an analysis on CO and O_3 correlation with the creation of CO_2 was conducted. Elements were analyzed based on their possibility to react while the other was in the atmosphere.

3. Result and Discussion

Table 4. *Sentinel-5P imagery data processing result*

Month	CO Concentration (mol/m^2)	O3 Concentration (mol/m^2)
Jan	0.026697671	0.112402
Feb	0.0296675	0.114266
Mar	0.028063732	0.116793
Apr	0.028508367	0.117013
May	0.029871632	0.115901
Jun	0.028290447	0.113132
Jul	0.025689859	0.113405
Aug	0.027920354	0.114845
Sept	0.028943896	0.116299
Oct	0.031130811	0.117417
Nov	0.027103315	0.1161

From processing Sentinel-5P imagery data, writers can produce data regarding the spread of CO and O_3 in Java's atmosphere.

Monitoring done throughout 2021 showed that Java Island donated about 0.029 mol/m^2 of CO gas and 0.115 mol/m^2 of O_3 gas throughout January to November 2021. Monitoring also showed that within those 11 months, the concentration of greenhouse gases emission differed from each month.

O_3 concentration gained from GEE calculations showed that the highest average was in March, May, and October. Meanwhile, the lowest average concentration was in January and November. The highest average CO concentration was in February, May, and October. Meanwhile, the lowest average CO concentration was in July.

The rise of COVID-19 cases in Kudus as one of an industry area in Java impacted widely in industrial production, making it decrease as much as 20% (CNBC Indonesia, 2021). As seen in Picture 5, CO gas concentration dipped from May to July. In May, CO gas concentration rose up to 4%, while O_3 gas concentration decreased as much as 0.1%. In June, CO gas concentration decreased all the way to 5% and in July decreased as much as 10% as well. The anomaly from the increase of CO concentration in the atmosphere happened from August to October, where in August the increase was 8%, 3% in September, and 7% in October. This anomaly most likely happened because the level of PPKM in Java-Bali Islands slowly loosened. Though, in November, the decrease of the gas concentration was drastic, impacted by PPKM Java-Bali that went all the way to November 29, 2021, by 12%. Industrial activity

IOP Conf. Series: Earth and Environmental Science 1047 (2022) 012021 doi:10.1088/1755-1315/1047/1/012021

was hindered once more, where every activity had to fulfill health protocols, such as fully-vaccinated workers and PCR for long-distance travels (CNBC Indonesia, 2021).

Changes and anomalies that occured when observing CO and O_3 gases using Sentinel 5P might be caused by reactions that occured in the atmosphere. This reaction could occur when CO gas collected in the atmosphere reacted with O_3 in the atmosphere. O_2 is the second largest gas component in the atmosphere. This could cause a greater possibility of a reaction between CO and O_2. The reaction between them would produce O_3 and CO_2. Sentinel-5P could monitor CO and O_3 levels in the troposphere, so, the presence of CO_2 could also be monitored indirectly through correlation analysis between CO and O_3. The reaction between elements C and O in the atmosphere is shown in equation 1 (Ambarsari, 2010).

Tropospheric Chemitry

$$CO + 2O_2 = CO_2 + O_3 \quad (1)$$

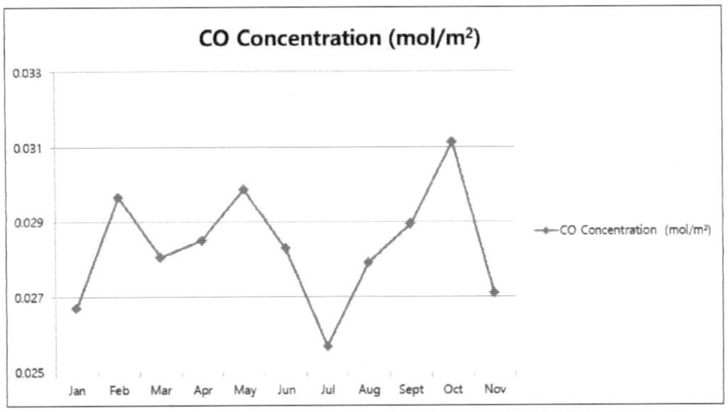

Figure 5. *CO concentration processed from Sentinel-5P data in GEE, from January to November 2021*

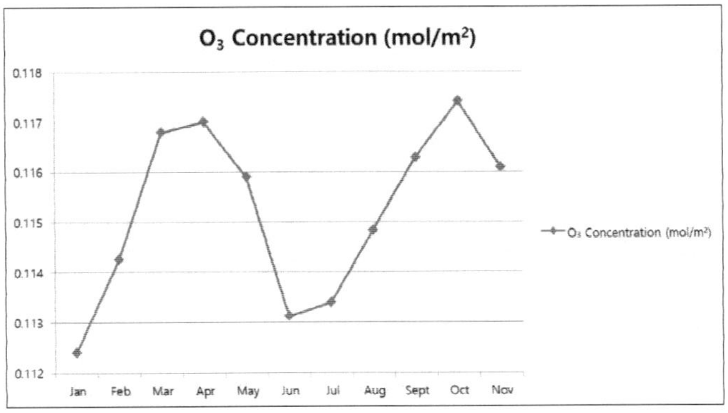

Figure 6. *O3 concentration processed from Sentinel-5P data in GEE, from January to November 2021*

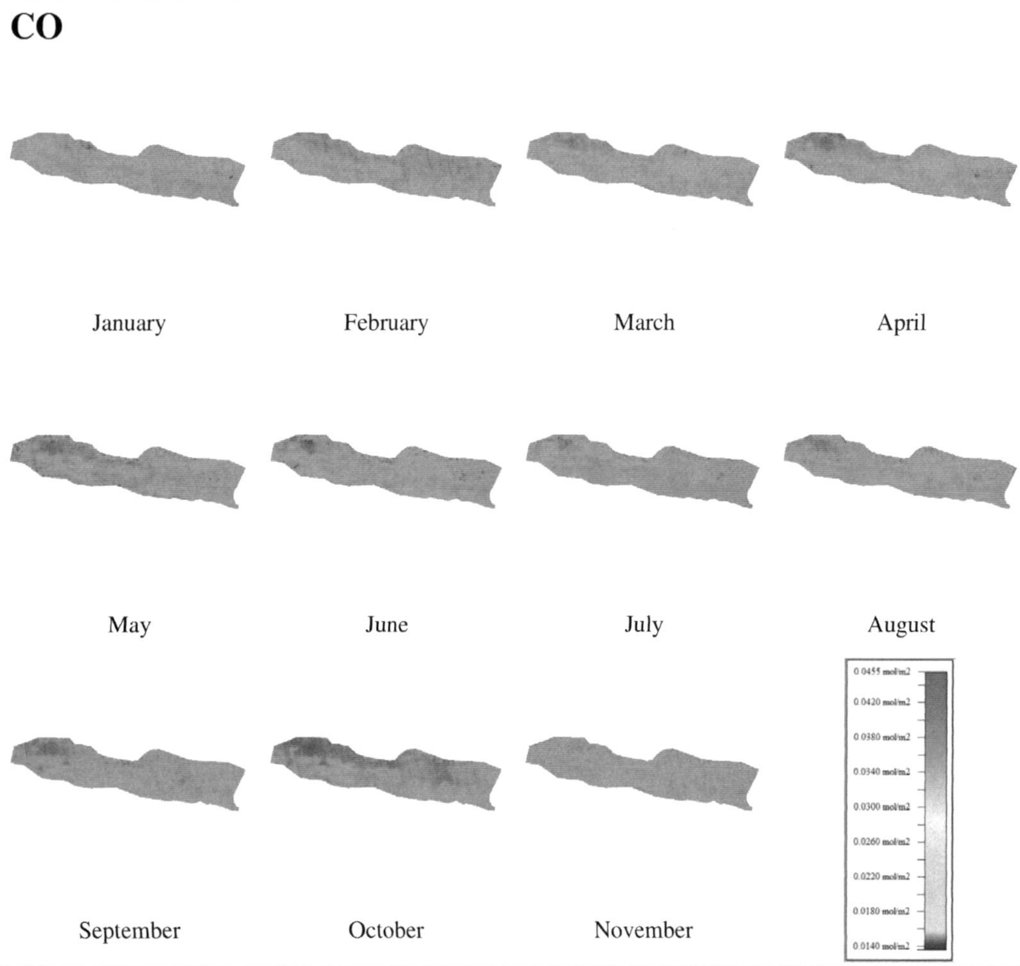

Figure 7*. CO gas dissimilarity in Java Island, from January to November 2021*

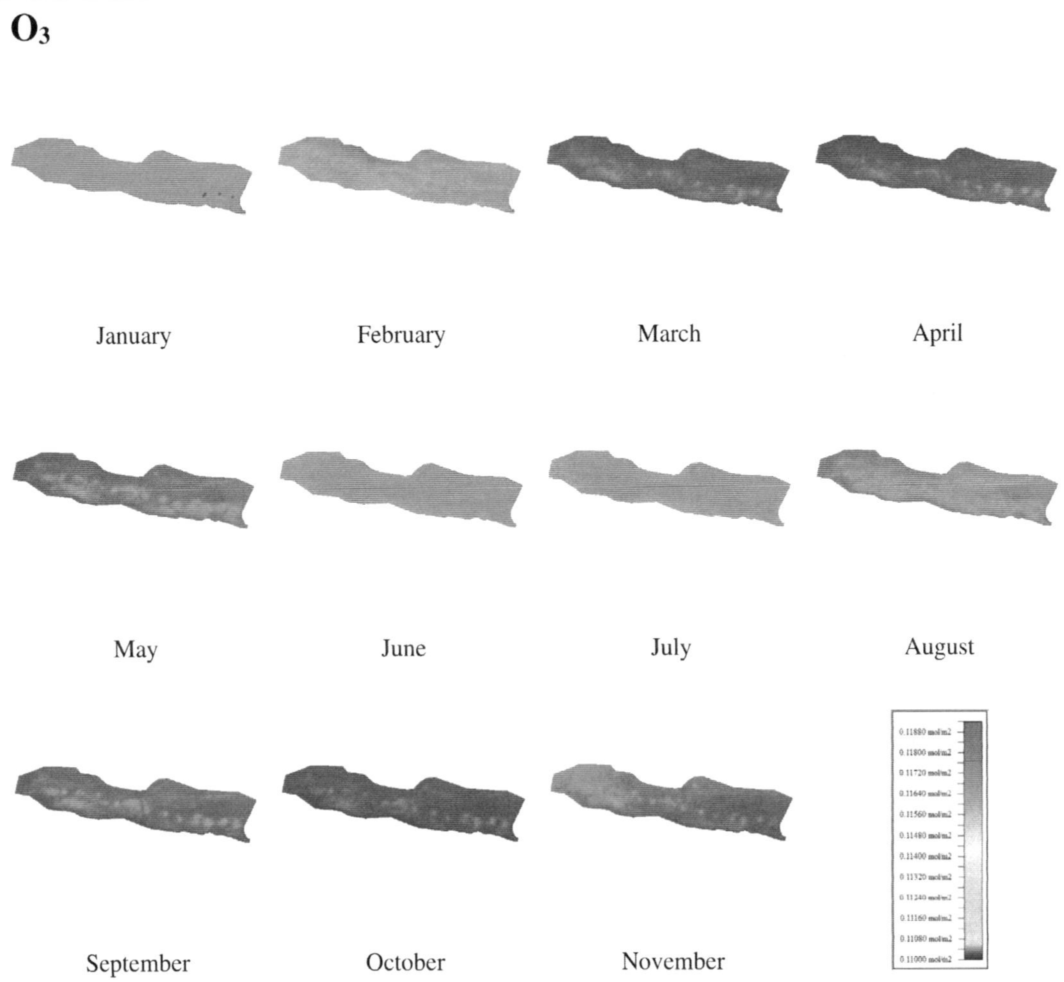

Figure 8. *O_3 gas dissimilarity in Java Island, from January to November 2021*

4. Conclusion

Sentinel-5P satellite imagery using GEE showed change in CO and O_3 concentration in earth's atmosphere from January to November 2021. This was due to the application of PSBB which had an impact on society's overall activity. Extraction gained from research showed that the highest concentration CO was in February, May, and October, and the lowest was in July. Meanwhile, the highest O_3 gas concentrations were in March, May, and October, and the lowest concentrations were in January and November. CO and O_3 concentration within the troposphere would contribute to C and O molecules. From the analysis result, CO and O_3 have correlation in creating CO_2, which is one of the gases causing the greenhouse effect. Due to this, CO and O_3 monitoring indirectly can be used to monitor CO_2 concentration within the troposphere.

References

[1] Ahnaf, J. (2020). Pemanfaatan Citra Sentinel-5P untuk Pembuatan Peta Perbandingan Konsentrasi NO2 Selama Pandemi COVID-19 di DKI Jakarta Menggunakan Google Earth Engine. Doctoral dissertation, Universitas Gadjah Mada.

[2] Ambarsari, N., Komala, N., & Budiyono, A. (2010). Pengaruh karbon monoksida terhadap ozon permukaan. *Widyariset,* 13(3), 59-64.

[3] Anggrainia, T. S., Artaningha, F., Sihotanga, E., Dimara, A., & Saktia, A. (2020). Variasi Emisi Gas Nitrogen Dioksida saat Pembatasan Sosial Berskala Besar di Provinsi Jawa Barat dari Pengolahan Data Satelit Sentinel-5P. *Jurnal Penginderaan Jauh Indonesia* Agustus, 2.

[4] Faisal, M., Prakoso, K. A., Sanjaya, H., & Darminto, M. R. (2021, September). Spatio-Temporal Analysis of Air Pollutants Changes during the COVID-19 Using Sentinel-5P in Google Earth Engine (Case Study: Java Island). *IEEE Asia-Pacific Conference on Geoscience, Electronics, and Remote Sensing Technology (AGERS)*, 102-108.

[5] Gorelick, N., Hancher, M., Dixon, M., Ilyushchenko, S., Thau, D., & Moore, R. (2017). Google Earth Engine: Planetary-scale geospatial analysis for everyone. *Remote sensing of Environment*, 202, 18-27.

[6] Indriyaningtyas, S., Hasandy, L. R., & Dewantoro, B. E. B. (2021). Dinamika Konsentrasi Emisi Gas Karbon Monoksida (CO) selama Periode PSBB Menggunakan Komputasi Berbasis Cloud pada Google Earth Engine. *Majalah Ilmiah Globe*, 35-42.

[7] Ismaningsih, N. K. (2019). Analisis Pengaruh Investasi dan Tenaga Kerja terhadap Nilai Output Industri Manufaktur di Jawa Barat Tahun 2010-2015. Skripsi, Universitas Pasundan. http://repository.unpas.ac.id/43834/

[8] Julianto, F. D., Fathurohman, C. R., Rahmawati, S. D., & Ihsanudin, T. (2021). Shoreline Changes after the Sunda Strait Tsunami on the Coast of Pandeglang Regency, Banten. *International Journal of Remote Sensing and Earth Sciences (IJReSES)*, 17(2), 201-208.

[9] Nebath, E., Pang, D., & Wuwung, J.O. (2014). Rancang Bangun Alat Pengukur Gas Berbahaya CO dan CO2 di Lingkungan Industri. *E-Journal Teknik Elektro dan Komputer*. ISSN: 2301-8402.

[10] Purba, L. S. L., & Harefa, N. (2019). Pengaruh Kandungan Oksigen Udara Sekolah terhadap Konsentrasi Belajar Siswa SMAN 9 Jakarta Timur.

[11] Wahyuni, F.E., Riyanto, W.H., & Sulistyono, S.W. (2021). Analisis Pengaruh Tingkat Pendidikan, Upah Minimum Provinsi Dan Jumlah Industri Terhadap Kesempatan Kerja di Provinsi Jawa Timur Tahun 1995-2020. *Jurnal Ilmu Ekonomi (JIE),* 5(3): 551-562.

[12] Zheng, Z., Yang, Z., Wu, Z., & Marinello, F. (2019). Spatial variation of NO2 and its impact factors in China: An application of Sentinel-5P products. *Remote Sensing*, 11(16), 1939.

IOP Conf. Series: Earth and Environmental Science 1047 (2022) 012022 doi:10.1088/1755-1315/1047/1/012022

The Spatial and Temporal Impacts of Mixed Layer Depth (MLD) variations in West Sumatera Coastal Area

Putri Adia Utari[1*], Okta Fajar Saputra[2], Hilda Lestiana[2]

[1] Research Center for Deep-Sea, Research Organisation of Earth Sciences, BRIN, Jl. Syaranamual, Guru-guru, Poka, Ambon, 97233, Indonesia
[2] Research Center for Geotechnology, Research Organisation of Earth Sciences, BRIN, Komplek LIPI, Jl Sangkuriang Bandung 40135, Indonesia.

* Email: putriadiautari91@gmail.com

Abstract. This paper investigates sea surface height to geoid (SSHG) and mixed layer depth (MLD) variations in the West Sumatera Coastal Area dominated Indian Ocean Dipole (IOD) evolutions using the data of MLD and (SSHG) from 2004 to 2017. The MLD variations in West Sumatera Coastal Area are explicitly investigated through spatial and temporal variations Empirical Orthogonal Functions (EOF). In the previous research, MLD variations in these extensive ocean-atmospheric circulations mainly focused on seasonal variations. The first results of EOF mode indicate 12% of the variability and show MLD variations in West Sumatera Coastal area during boreal fall (September – November). This result indicates that seasonal variation of MLD is slightly weak to the IOD event caused by ENSO signals from atmospheric teleconnection. On the other situation, the seasonal variations of SSHG is tightly related to the Indian Ocean Dipole (IOD) evolutions. The EOF results of SSHG shows 30% variabilities and is stronger than the MLD. The weak (dominant) pattern of these MLD and SSHG variations affects a shoaling layer over the west Sumatra coastal area associated with the negative Indian Ocean during 2010. We suggest that this MLD and SSHG product could be used as a reference for future studies and predict fish productivity.

1. Introduction

The mixed layer depth (MLD) is highly prominent on the surface layer to the depth of vertical mixing which is identified by almost constant temperature, salinity, and density with depth [1]–[3]. The homogeneous parameters are mostly induced by atmospheric fluxes such as wind stress, heat, and freshwater exchange that contribute to MLD variability [4]–[7]. A shoaling (deepening) of the MLD relates to the changes of Sea Surface Height to Geoid (SSHG) that responds to a pole in the west Sumatera coastal area. This location is considered to be a part of negative (positive) Indian Ocean dipole events that generate downwelling (upwelling) that drive deep (shallow) MLD.

The signature of positive (negative) IOD is defined by negative (positive) SST anomalies along the Sumatra and Java coast and a positive (negative) SST over the tropical central-western Indian Ocean[8]–[11]. The evolution of negative IOD events during 2010 crucially pointed out a rich variability of SSHG and MLD over the west Sumatera coastal area. The evolution of negative IOD events during 2010 was the huge warming events after the extreme negative IOD events in 1998. The positive SST anomalies in 2010 was more than 1°C associated with the negative IOD appeared in southeastern tropical Indian Ocean (SETIO). The warm SST anomalies observed more than 1°C from August to October 2010 in the SETIO which took place with a La Niña event in the Pacific.

Numerous investigations have observed the connection between MLD and SSHG to the climate variability (seasonal, interannual, and annual). However, this research confirmed that the MLD and SSHG variability was influenced by negative IOD events during 2010 – 2011 using EOF analysis in the west Sumatra coastal area.

2. Data And Methods

We retrieved data from Argo Mixed Layers from *http://mixedlayer.ucsd.edu/* in this research. The product of MLD merges closely 2,450,000 Argo profiles (through March 2021) with a hybrid method. The common appearance of each profile is demonstrated by this hybrid algorithm, determined for physical properties in the profile, and counted up the sill and gradient MLDs to congest the possibility of MLD values. It subsequently investigates the patterns in the suite to select a final MLD estimate. We use monthly mixed layer depth (MLD) to observe the MLD variability from January 1st 2012 to December 1st 2017. The difference of potential density from the surface are calculated by getting the MLD value, equivalent to a 0.8 C decrease in SST (Kara et al., 2000). This standard reckons to temperature and salinity stratification.

The Global Ocean Data Assimilation System (GODAS) provides Sea Surface Height to Geoid (SSHG) and can be retrieved on *https://psl.noaa.gov/data/gridded/data.godas.html.* The assimilation method is the same 3DVAR scheme as used in RA6, but it has been modified to assimilate salinity. The standard GODAS has been utilized for a reanalysis starting from 1979 to the present. In this reanalysis, GODAS assimilates temperature. This product is utilized for this research from January 1st 2012 to December 1st 2017, with a horizontal resolution of 1°× 1°. The dipole mode index (DMI) data were obtained from Ocean Observations Panel for Climate (OOPC) that could be downloaded on *http://stateoftheocean.osmc.noaa.gov*. The DMI product was used in this research from January 1th 2004 to December 1th 2017. The Dipole Mode Index (DMI) is an index that was signified by the east-west temperature gradient across the tropical Indian Ocean. The DMI data is calculated as the difference between the Western Tropical Indian Ocean (WTIO) and southeastern tropical Indian Ocean (SETIO) [9]. The red box presented the location of WTIO in box A and the location of SETIO in box B, as revealed in **Figure 1**.

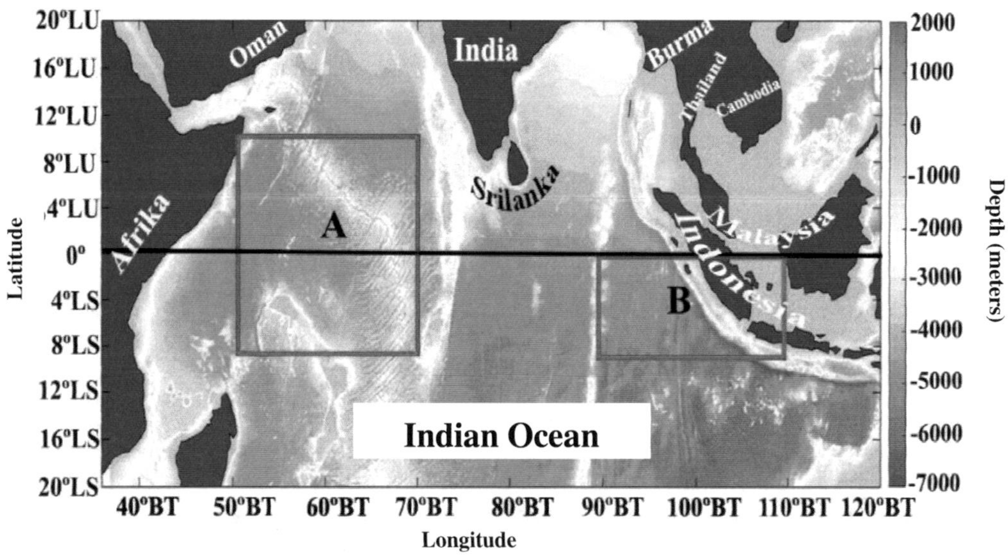

Figure 1. The study area for determining Dipole Mode Index (DMI).

Data sets of MLD, SSHG, and DMI were constructed using their mean climatologies and anomalies. Using these three data from computed anomaly results were evaluated by Empirical Orthogonal

IOP Conf. Series: Earth and Environmental Science 1047 (2022) 012022 doi:10.1088/1755-1315/1047/1/012022

Function (EOF) analysis which uncovers those relationships in space. It then distributes the same time variability and combines them into a single spatial pattern. Integrating a common time variability is based on the data for each of the same months from January to December. This pattern is found by regression across space so that a precise relationship between neighboring sites is found. Ultimately, the patterns are found that show these covarying locations. If one location shows hot and other location shows slightly cooler.

3. Results and Discussions

3.1. *The spatial dynamic of MLD and SSHG anomalies*

The hovmuller diagram shows the MLD anomalies with averaged in 0.5–7.5°S in **Figure 2a.** The pattern of MLD anomalies shows a significant pole at the east of 90°E, during the negative IOD events in 2010 from boreal fall (September – November) to boreal winter (December – February). The thickness is about 20 m that exhibits MLD shoaling along the western coastal area at this period. At the same time, the MLD pattern propagates with the SSHG that acts the same pattern at the 70°E (**Figure 2a**). On other results using EOF results, the first mode of MLD variations confirms that a shoaling variance in the west Sumatra coastal area is about 12% and 8% for second mode of MLD variations (**Figure 2b-c**). They illustrate that the pattern has slightly weaker IOD signals than SSHG variations. These weaken signal of MLD variations using EOF analysis corresponded with atmospheric teleconnection which represented by the strong ENSO signal [12]–[15]. The strength of ENSO signal can weaken or reinforce the DMI evolution signal which was represented by Nino 3 Index shown in **Figure 4a-b**. ENSO exerts a strong influence on the global climate. This shift in convection alters the surface heat fluxes over the tropical oceans by changing the atmospheric circulation remotely via atmospheric wave adjustments [16].

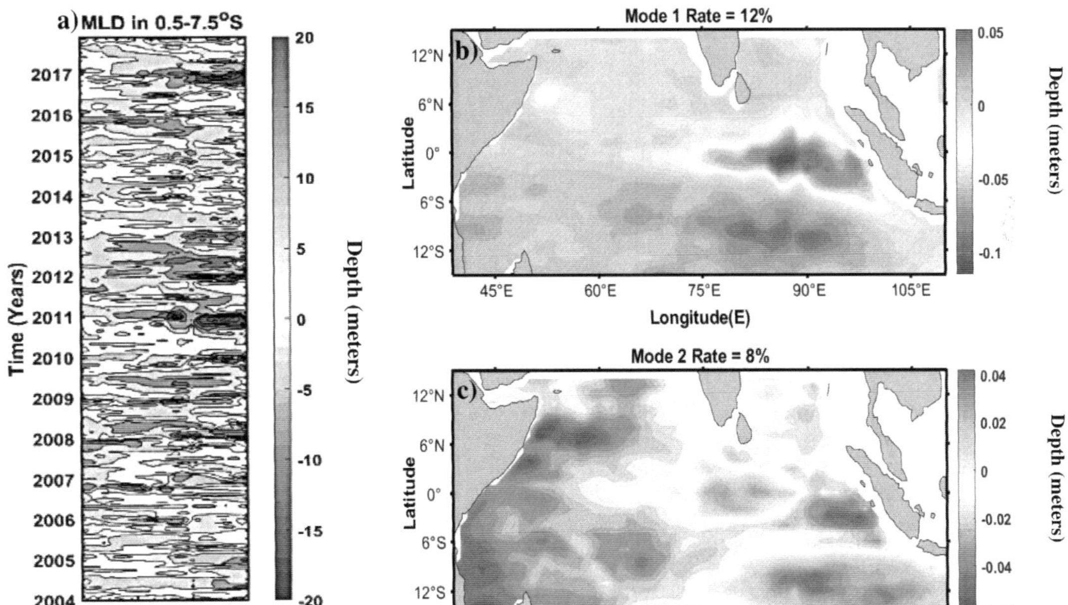

Figure 2. Temporal variability (Hovmuller) sections of anomalies: (a) MLD (shading, meters (m)) showed in left side, (b) first and (c) second mode of MLD variations on right side by using EOF analysis. The research location was averaged in 0.5–7.5°S.

The other hovmuller diagram of SSHG anomalies in the averaged in 0.5–7.5°S figures out on **Figure 3a**. The SSHG anomalies shows a significant pole at the central of 70°E, during the negative IOD events in 2010 from boreal fall (September – November) to boreal winter (December – February). The depth of SSHG variability is about 0.15 m that presents SSHG shoaling along the central of 70°E area at this period. At the same time, the first mode of SSHG displays 30% corresponded the negative IOD events (**Figure 3b**). The second mode of SSHG has small variability about 12%. The SSHG result by using EOF analysis indicates that a shoaling variance in the west Sumatra coastal area clearly. They illustrate that the pattern has the same phase of the corresponding negative IOD event. A shoaling MLD pattern is more significant in the west Sumatra coastal area – the central equatorial Indian Ocean. The different situation over the SETIO region occur. SSHG pattern also demonstrates the same pattern with MLD pattern but different variance for 30% (**Figure 3b**). The variance of SSHG anomalies is about 0.1 m, coinciding with a shoaling trend over the western Java coastal area and West Sumatera. Additionally, it seems that the SSHG response to negative IOD forcing is relatively stronger than MLD, as revealed in **Figure 3b-c**. Nevertheless, despite the differences in both MLD and SSHG patterns, the overall patterns are similar and associated with negative IOD.

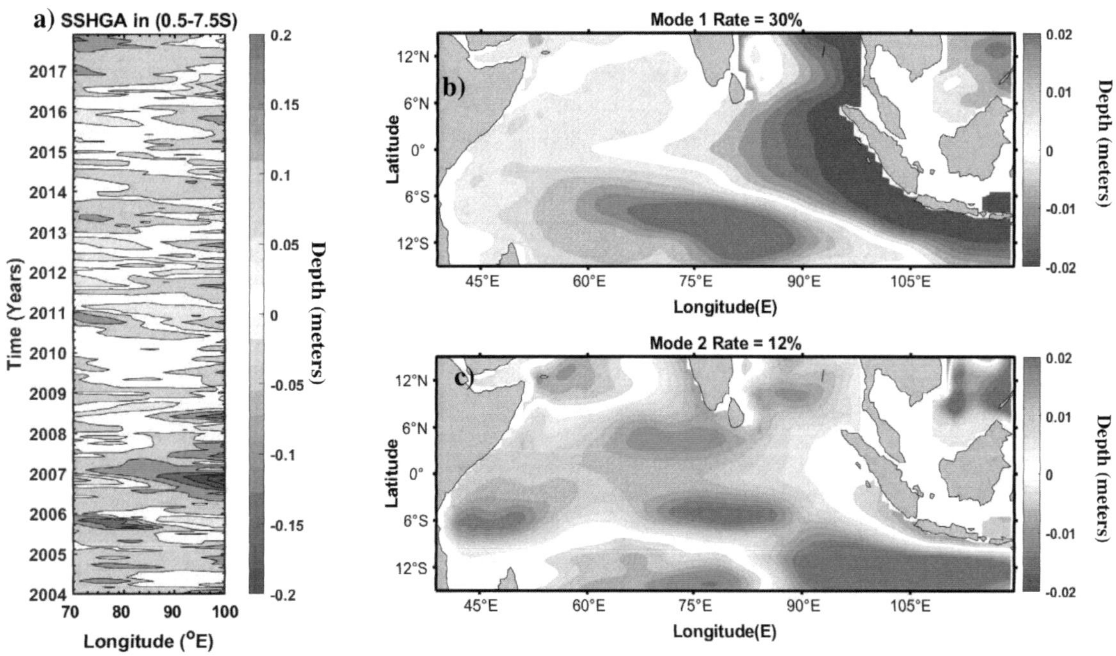

Figure 3. Temporal variability (Hovmuller) sections of anomalies: (a) SSHG (shading, meters(m)) showed in left side, (b) first and (c) second mode of MLD variations on right side by using EOF analysis. The research location was averaged in 0.5–7.5°S.

3.2. The temporal results of MLD and SSHG anomalies

Refer to the previous research, the strong negative SSHG and MLD variability occured from late 2010 to early 2011 [17]. This research was conducted using EOF analysis that the black lines show pc (principal component) results, red lines show DMI evolution and the blue lines show nino3 signals for ENSO evolution (**Figure 4a-b**). For the first mode of MLD, DMI Index (red lines) show a negative anomaly from late 2010 to early 2011, which is about 1°C showed in green circles (**Figure 4a**). The ENSO strength has weaker signal than DMI signals. These pattern demonstrate that the ENSO weaken the DMI signal, it is due to the spatial result of MLD variability in the first mode becoming low variance.[12], [14], [18] Furthermore, this pattern confirmed that the negative IOD occurred in the western Java coastal area and the west Sumatera coastal area during this period. The different MLD patterns for second mode co-occurs with the weak negative IOD which is circled in the black line in **Figure 4b**.

In 2011, a significant negative of SSHG also revealed -2°C for the first mode, which is consistent with the analysis of negative IOD events in **Figure 4c**. Note that the first mode of SSHG variability is much higher than the second mode of SSHG variability in **Figure 4d**. These results confirm that the negative IOD also influences the SSHG anomaly using the EOF method. The west Sumatra coastal area is induced by strong wind anomalies, which could also force the upwelling (or downwelling) Rossby waves. It then contributes to a shoaling MLD and SSHG. This enables MLD/SSHG to complement the triggering mechanism that allows negative (positive) IOD events to occur along the coast of West Sumatra.

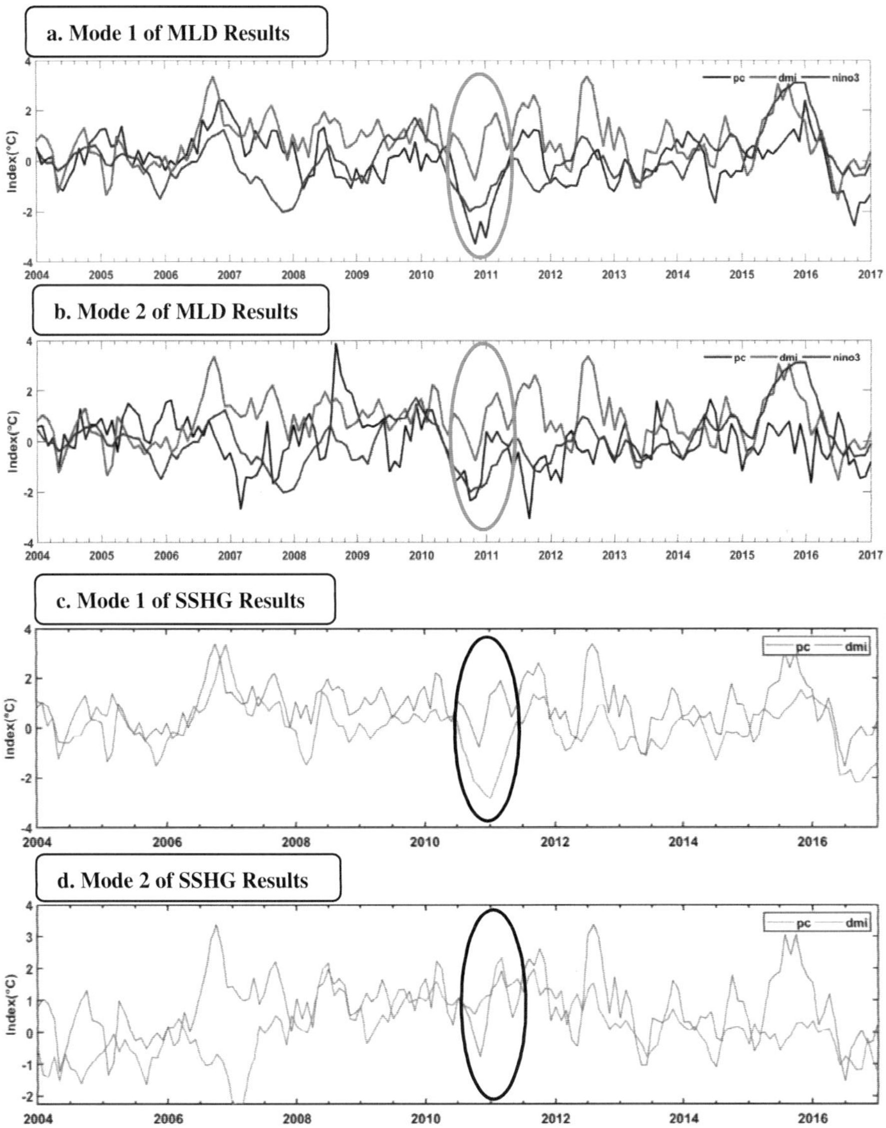

Figure 4. Time series of: (a) mode 1 of MLD variations for DMI Index (red lines, °C), PC1 results (black lines, °C), and Niño Index (blue lines, °C) (b) mode 2 of MLD variations with the same component **Fig. 4a**. (c) mode 1 of SSHG variations (line, °C), (d)) mode 2 of SSHG variations (line, °C). The research location was averaged

in 0.5–7.5°S using EOF analysis. The orange lines on **Figure 4c-d** show principal component and blue lines show dipole mode index (DMI).

4. Conclusion

This research provides a detailed description of the SSS variations in the south western Tropical Indian Ocean (SWTIO) which is based on Argo products and GODAS profiles to observe the 2010 negative IOD event and the fundamental dynamics of SSHG and MLD. From the spatial and temporal impacts, we can explain that the negative IOD 2010 significantly impacts MLD and SSHG in the western Sumatra Indian Ocean until 2011. Firstly, a significant shoaling of MLD is also identified over West Sumatera Indian Ocean. This basin-wide tendency originates on WTIO during boreal fall and weakens to the tropical Indian Ocean (TIO) during boreal winter. Secondly, EOF analysis unveils that the shoaling MLD and SSHG in the Western Sumatera Indian Ocean primarily affected the negative IOD events during 2010. The involvement of the negative IOD events to the western Sumatra Indian Ocean are insignificant, as revealed in **Figure 3b**. Thirdly, SSHG analysis shows the same pattern with MLD that is primarily involved the negative IOD during 2010 – 2011. MLD and SSHG results show that they significantly contribute to the western Sumatera coastal area, and the surface heat flux plays a damping effect. The deeper (shallow) MLD and SSHG variations in SETIO may support greater basin-wide warming than cooling.

Acknowledgments

The author thanks to the supporting fund in this research by LIPI's COREMAP-CTI 2021–2022 (17/A/DK/2021). The author also thanks to anonymous reviewers for beneficial suggestions and commentaries that could develop the perspective discussions on the dynamic of MLD and SSHG along the tropical Indian Ocean (TIO).

References

[1] M. G. Keerthi, M. Lengaigne, J. Vialard, C. de Boyer Montégut, and P. M. Muraleedharan, "Interannual variability of the Tropical Indian Ocean mixed layer depth," *Clim. Dyn.*, vol. 40, no. 3–4, pp. 743–759, Feb. 2013, doi: 10.1007/s00382-012-1295-2.

[2] N. K. Vissa, A. N. V. Satyanarayana, and B. Prasad Kumar, "Comparison of mixed layer depth and barrier layer thickness for the Indian Ocean using two different climatologies," *Int. J. Climatol.*, vol. 33, no. 13, pp. 2855–2870, 2013, doi: 10.1002/joc.3635.

[3] J. Yu, B. Gan, Z. Jing, and L. Wu, "Winter Extreme Mixed Layer Depth South of the Kuroshio Extension," *J. Clim.*, vol. 33, no. 24, pp. 10419–10436, Dec. 2020, doi: 10.1175/JCLI-D-20-0119.1.

[4] A. Kumar and R. Bhatla, "Modeling the mixed layer depth in Southern Ocean using high resolution regional coupled ocean sea ice model," *Model. Earth Syst. Environ.*, no. 0123456789, 2021, doi: 10.1007/s40808-021-01321-2.

[5] E. Panassa, C. Völker, D. Wolf-Gladrow, and J. Hauck, "Drivers of Interannual Variability of Summer Mixed Layer Depth in the Southern Ocean Between 2002 and 2011," *J. Geophys. Res. Ocean.*, vol. 123, no. 8, pp. 5077–5090, 2018, doi: 10.1029/2018JC013901.

[6] A. Yamamoto, H. Tatebe, and M. Nonaka, "On the emergence of the Atlantic multidecadal SST signal: A key role of the mixed layer depth variability driven by North Atlantic oscillation," *J. Clim.*, vol. 33, no. 9, pp. 3511–3531, 2020, doi: 10.1175/JCLI-D-19-0283.1.

[7] S. H. Lim, C. J. Jang, I. S. Oh, and J. J. Park, "Climatology of the mixed layer depth in the East/Japan Sea," *J. Mar. Syst.*, vol. 96–97, pp. 1–14, 2012, doi: 10.1016/j.jmarsys.2012.01.003.

[8] B. Lu *et al.*, "An extreme negative Indian Ocean Dipole event in 2016: dynamics and predictability," *Clim. Dyn.*, vol. 51, no. 1–2, pp. 89–100, 2018, doi: 10.1007/s00382-017-3908-2.

[9] N. H. Saji, P. N. Goswami, P. N. Vinayachandran, and T. Yamagata, "Saji,N.A et al,.dipole mode

in the tropical indain ocean," *Nature*, vol. 401, no. September, pp. 360–363, 1999.

[10] T. Izumo *et al.*, "Influence of the state of the Indian Ocean Dipole on the following years El Nĩo," *Nat. Geosci.*, vol. 3, no. 3, pp. 168–172, 2010, doi: 10.1038/ngeo760.

[11] C. Ihara, Y. Kushnir, and M. A. Cane, "Warming trend of the Indian Ocean SST and Indian Ocean dipole from 1880 to 2004," *J. Clim.*, vol. 21, no. 10, pp. 2035–2046, 2008, doi: 10.1175/2007JCLI1945.1.

[12] J. A. Carton, S. A. Grodsky, and H. Liu, "Variability of the oceanic mixed layer, 1960-2004," *J. Clim.*, vol. 21, no. 5, pp. 1029–1047, 2008, doi: 10.1175/2007JCLI1798.1.

[13] S. A. Klein, B. J. Soden, and N. C. Lau, "Remote sea surface temperature variations during ENSO: Evidence for a tropical atmospheric bridge," *J. Clim.*, vol. 12, no. 4, pp. 917–932, 1999, doi: 10.1175/1520-0442(1999)012<0917:RSSTVD>2.0.CO;2.

[14] A. Santoso, A. Sen Gupta, and M. H. England, "Genesis of Indian Ocean mixed layer temperature anomalies: A heat budget analysis," *J. Clim.*, vol. 23, no. 20, pp. 5375–5403, 2010, doi: 10.1175/2010JCLI3072.1.

[15] Y. Du, T. Qu, G. Meyers, Y. Masumoto, and H. Sasaki, "Seasonal heat budget in the mixed layer of the southeastern tropical Indian Ocean in a high-resolution ocean general circulation model," *J. Geophys. Res. C Ocean.*, vol. 110, no. 4, pp. 1–15, 2005, doi: 10.1029/2004JC002845.

[16] S. Chakravorty, J. S. Chowdary, and C. Gnanaseelan, "Epochal changes in the seasonal evolution of tropical Indian Ocean warming associated with El Niño," *Clim. Dyn.*, vol. 42, no. 3–4, pp. 805–822, 2014, doi: 10.1007/s00382-013-1666-3.

[17] Q. Sun *et al.*, "Evolution of Sea Surface Salinity Anomalies in the Southwestern Tropical Indian Ocean During 2010–2011 Influenced by a Negative IOD Event," *J. Geophys. Res. Ocean.*, vol. 124, no. 5, pp. 3428–3445, 2019, doi: 10.1029/2018JC014580.

[18] S. A. Rao, S. K. Behera, Y. Masumoto, and T. Yamagata, "Interannual subsurface variability in the tropical Indian Ocean with a special emphasis on the Indian Ocean Dipole," *Deep. Res. Part II Top. Stud. Oceanogr.*, vol. 49, no. 7–8, pp. 1549–1572, 2002, doi: 10.1016/S0967-0645(01)00158-8.

Love Wave Dispersion Analysis from Ambient Seismic Noise in Banyumas Basin, Central Java, Indonesia

A H Kusuma[1], Zulfakriza[2], A D Nugraha[2], S Rosalia[2], A Setiawan[3], Marjiono[3]

[1]Geophysical Engineering Study Program, Faculty of Mining and Petroleum Engineering, Bandung Institute of Technology, Bandung, Indonesia.
[2]Global Geophysics Research Group, Faculty Mining and Petroleum Engineering, Bandung Institute of Technology, Bandung, Indonesia
[3]Center for Geological Survey, Geological Agency of Indonesia, the Ministry of Energy and Mineral Resources, Bandung, Indonesia

E-mail: alfankusuma46@gmail.com

Abstract. The Banyumas Basin is one of a sedimentary basins in the southern coast of Central Java that has many manifestations of the presence of hydrocarbons in the form of oil and gas seeps found on the surface. Its presence is covered by a thick subvulcanic sedimentary layer. A conventional seismic active method which is usually used to delineate subsurface structures in an hydrocarbon area, however it can be challenging to imaged the structures beneath surface in areas covered by thick sub-volcanic sediments. Another method is needed to investigate the subsurface image in a thick sub-volcanic area, such as Banyumas basin. We cross correlated of the ambient seismic noise at transversal component to retrieve the love wave empirical Green's function. We used 40 broad band seismic stations for 60 days recording from Center for Geological Survey - Indonesian Geological Agency. The daily cross-correlated data then stacked to obtain the Love waves Green's function. The Green's function is clearly seen in 1-10 s period band. Then, the obtained Green's function will be analyzed to get the Love waves group velocity.

Keywords: Ambient noise tomography (ANT), Love Waves, Sub-volcanic area

1. Introduction

Central Java is one of the islands in Indonesia which has complex geological structure as a result by the movement of convergent plates. Tectonics on Java Island is a result by the interactions between the Eurasian Plate with the India-Australia [2]. The Eurasian Plate has moved to the southeast since the Oligocene, while the India-Australia Plate has moved north and subducted under the island systems of Sumatra and Java [15]. This tectonic activity causes the development of fault areas and basins, which are important components of the petroleum system.

The Banyumas Basin is a sedimentary basin that is located in the central South Java Fore Arc Basin, where the morphology of the coastline in Central Java is indented due to transtension tectonic processes, namely the presence of 2 (two) pairs shear faults, namely the Muria-Kebumen Fault with a North-East-South-West trend and the Pamanukan-Cilacap Fault with a North-West-Southeast trend [4] . This tectonic activity has caused the development of faults and basins in the area, which are important for oil and gas plays. This area, however, is covered by a thick layer of subvolcanic sediments [3][5][9], which causes issues in oil and gas exploration because conventional seismic active methods, which are typically used to describe subsurface structures in hydrocarbon areas, have low sensitivity when used in areas covered by thick subvulcanic sediments [14][17].

In cases in areas with low seismicity, one of the methods being developed is the ambient noise tomography (ANT) method [16] . The ANT method is a method that utilizes ambient seismic noise as data containing information that can be used to describe the dispersion of surface wave properties in the soil layer by reconstructing the surface wave portion of the Green function and the measured dispersion curve. Consequently, one can use the same methods as conventional surface wave tomography for ambient noise measurements. This technique has been used extensively for the study of continental or regional crust and upper mantle. Several studies have been conducted regarding the ANT method [7][8][19], some of them have used ANT in delineation and mapping of basins [10][11][12]. ANT has also been applied in Indonesia to determine S-wave velocity structures in Bali [6], Western Java [13] 18], Lombok [14], and Central Java [22]. Based on those study, the ANT method is believed to be able to describe the structure of the basin quite well. In this study, we will describe the preliminary result of the dispersion curves to generate the Love waves group velocity, and give general structure analysis beneath Banyumas Basin.

2. Data And Method

2.1 Data

In this study, we used Ambient Noise Data for Love Waves obtained from the Bandung Geological Survey Center. We used seismic data recorded at 40 portable 3-C broadband borehole seismic stations from the Geological Survey Center (Badan Geologi) (shown in Figure 1) spread between Banjar and Cilacap. The data used in this study is love wave recording data, which was taken for 2 months (recording period June and July 2018).

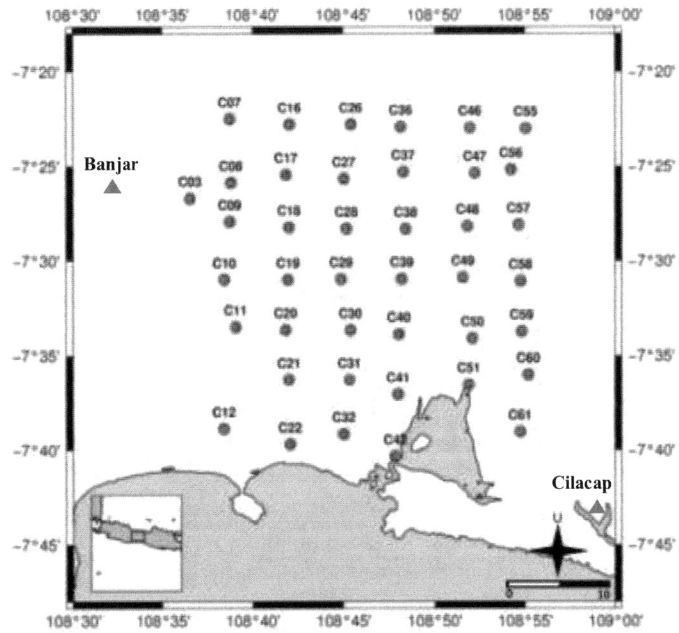

Figure 1. Stations distribution in the research area. Red point is the stasions used in this study.

2.2 Ambient Noise Tomography

Ambient Noise Tomography (ANT) is a method for obtaining tomography of the subsurface wave velocity of the earth by utilizing noise data from the environment (ambient noise). Ambient noise is a natural seismic wave that has a low amplitude originating from meteorological activities such as ocean waves, wind, and wave sources originating from human activities [15]. The ANT method does not require an active source or an earthquake source, making this method quite popular, especially for research in areas with low seismicity. The main principle of ANT is to utilize seismic interferometry methods to obtain surface wave velocity. Ambient seismic noise interferometry refers to the process of cross-correlation between two sets of ambient seismic noise data that is recorded by two receivers in an acoustic diffusion wavefield to produce EGF at the receiver position as if one of the stations acts as the source.
The cross-correlation process between two stations, X_A and X_B [11] can be written as:

$$G(X_A, X_B, t) = \int_{-\infty}^{\infty} V_1(X_A, \tau)\, V_2(X_B, t+\tau)\, d\tau \qquad (1)$$

Where $V_1(X_A, t)$ and $V_2(X_B, t)$ represent the seismic data recorded at stasions A and B, t is the time variable, and τ is lag time. Using equation (1), we can increase the signal to noise ratio to determine the earth response between 2 pairs of stations.

2.3 Method

Then we used the common procedure [1] to obtain the Empirical Green's Function (EGF) for each pair station. First we did data preparation. At this stage, ambient noise data recorded every 10 minutes is combined with waveforms into daily data, then daily data conversion was carried out in miniseed format for each station to be converted into SAC format, resampled data from 100 Hz to 10 Hz, de-trending, de-meaning , spectral whitening, time-domain normalization. We applied 1-10 s multiple period bandpass filter in the single data preparation process. After a series of these processes have been carried out, then the component rotation process was carried out to obtain the transverse transverse direction component, and cross-correlation was carried out with 10 s lag time. After that, we calculated the cross-correlation using daily data and combined the results of the correlation for each station and the results were in the form of EGFs. The EGFs generated between pairs of stations provide information about the travel time of surface waves and the dispersion between stations. The next step was to extract group velocity from EGFs by picking the dispersion curve.

3. Result and Discussion

The Green Empirical Function (EGF) obtained from cross-submissions all stasiun pairs shows a clear "V" shaped envelope in the "cross chorelogram plot" (Figure 2), it's mean that that the the bandpass with 1-10 s suitable for all stasiun for generating signal with high resolution (based on signal to noise ratio).

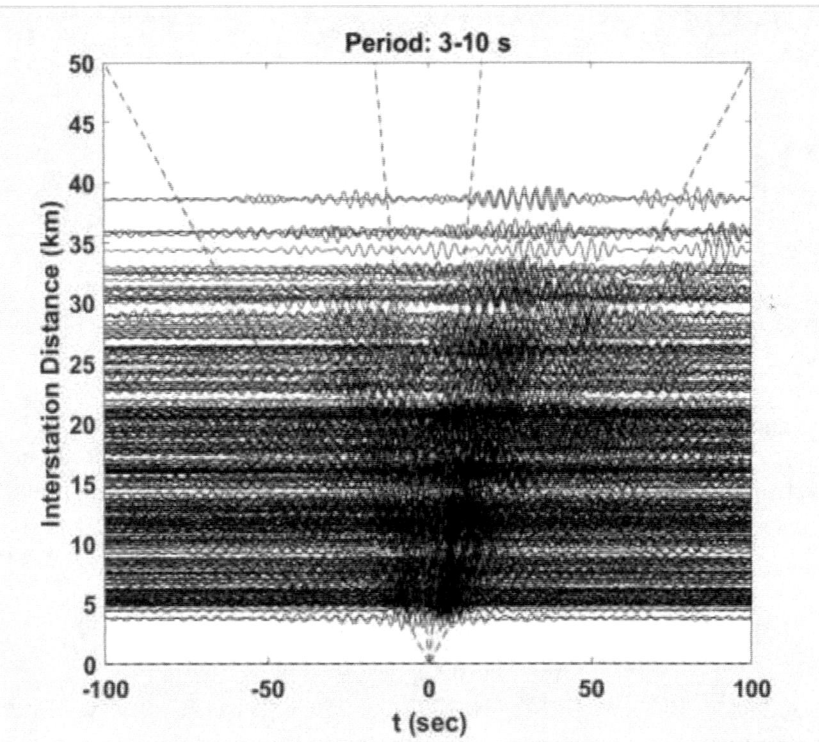

Figure 2. The 1-10 s bandpass filtered Empirical Green's Function (EGF) of all stasiun pairs. The black line are the EGF from all stasiun pairs, and the red line are helpline to show Vshape from the EGF.

We used he method that developed by Yao [20] [21] to extract the dispersion curve from each cross-correlation product (figure 3). Here, we obtained the dispersion curve of all station pairs (figure 3). The graph of all station pairs dispersion curve showed that the group velocity ranged from 1 to 2.5 km/s with period band of 1 to 10 s. The lower period dispersion curve indicated the shallow subsurface structure which might produce higher resolution than the deeper structure due to its denser curve. Lower group velocity could correlate with sediment layer, while the higher group velocity could show the denser rock structure beneath the research area which need further study to understand the subsurface structure better.

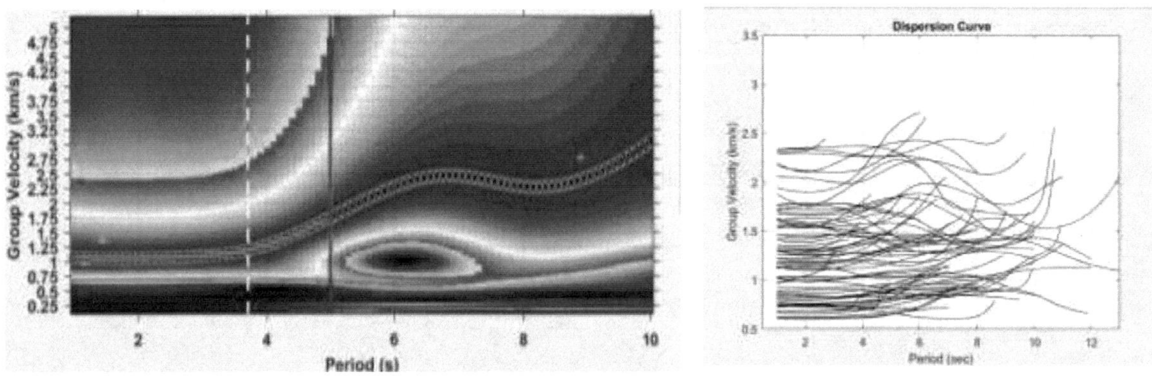

Figure 3. Dispersion curve picking of C03-C07 station pair (left) with red points are selected dispersion curve for this station pair, yellow and red line indicated the wavelength is equal to 2 and 1, consecutively. Selected dispersion curve of all station pairs with value ranges from 1 to 2.5 km/s (right).

4. Conclusion

From this preliminary study, we obtained Love waves group velocity of all station pairs from the cross-correlation product of the transverse component ambient noise data in western part of Java. The EGF is spotted at period band 1-50 s and showed clear "V" shaped pattern. The group velocity ranged from 1-2.5 km/s at period band 1-10 s. Low period group velocity indicated sediment layer, meanwhile the high period correlated with the denser structure beneath the research area. For future research, the results of this group velocity can be used to conduct further research for imaging the subsurface shear velocity of the Banyumas basin.

5. Refrences

[1] Bensen G D, Ritzwoller M H, Barmin M P, Levshin A L, Lin F, Moschetti M P, Shapiro N M, and Yang Y, (2007) *Processing seismic ambient noise data to obtain reliable broad-band surface wave dispersion measurements* Geophys J. Int. **169** 1239–1260.

[2] Hall R, (2012) *Late Jurassic–Cenozoic reconstructions of the Indonesian region and the Indian Ocean Tectonophysics* **570-571** 1-41.

[3] Hidayat H, (2020) *Interpretasi Struktur Geologi Bawah Permukaan Berdasarkan Updating Data Gaya Berat Cekungan Banyumas Jawa Tengah* Jurnal Geologi dan Sumber Daya Mineral **21** 111-118.

[4] Fahlevi M R, Syavitri D, and Herdiansyah F, (2021) *Analisis Lingkungan Pengendapan Daerah Karangpucung Kabupaten Banyumas, Journal Of Geoscience Engineering and Energy* **2** 2715-5358.

[5] Junursyah G M L, (2019), *Survei Geomagnet Untuk Mendeliniasi Cekungan Sedimen yang Tertutupi Batuan Vulkanik di Daerah banyumas dan Sekitarnya dalam Eksplorasi Hidrokarbon* February.

[6] Martha A A, Cummins P, Saygin E, and Widiyantoro S, (2017) *Imaging of upper crustal structure beneath East Java-Bali Indonesia with ambient noise tomography* Geoscience letters **4(14)**, 1-12.

[7] Nicolson H, Curtis A, Baptie B, and Galetti E (2012) *Seismic interferometry and ambient noise tomography in the British Isles* Proceedings of the Geologists' Association 123:74-86.

[8] Porritt R W, Miller M S, O'Driscoll J, Harris C W, Roosmawati N, and Costa L T, (2016) *Continent-arc Collision in the Banda Arc imaged by ambient noise tomography* Earth and Planetary Science Letters **449** 246-258.

[9] Purwasatriya E B, Surjono S S, and Amijaya D H, (2019) New paradigm to understanding turbidite sediment in Banyumas basin New Paradigm to Understanding Turbidite Sediment in Banyumas Basin. AIP Conference Proceedings 1–9.

[10] Arroucau P, Rawlinsosn N, Sambridge M, (2010) *New insight into Cainozoic sedimentary basins and Palaeozoic suture zones in southeast Australia from ambient noise surface wave tomography* Geophysical Reserch Letter, VOL. 37,

[11] Saygin E, Cummins P R, Cipta A, Hawkins R, Pandhu R, Murjaya J, Irsyam M, Widiyantoro S, and Kennett B L N, (2016) *Imaging Arcitecture of the Jakarta Basin Indonesia with transdimensional inversion of seismic noise* Geophysical Journal International **204** 918–931.

[12] Schippkus S, Zigone D, Bokelmann G, and Group W, (2018) *Ambient-noise tomography of the wider Vienna Basin region* 102–117.

[13] Rosalia S, (2020) *Crustal Structure Imaging Beneath Western Java Using Seismic Ambient Noise Tomography* Dissertation of Doctoral Program Institut Teknologi Bandung.

[14] Setiawan A, Zulfakriza Z, Nugraha S D, Rosalia S, Priyono A, Widyantoro S, Sahara D P, Marjiyono M, Januar H S, Lelono A B, Asep K, Hidayat (2021) *Delineation of Sedimentary Basin Structure bneath the Banyumas Basin, Central Java, Indonesia, Using Ambien Noise Tomography*, Geoscience Letters.

[15] Simons W J F, Socquet A, Vigny C, Ambrosius B A C, Abu S H, Promthong C, Subarya C, Sarsito D A, Mathussen S, Morgan P, and Spakman W (2007) *A decade of GPS in Southeast Asia: Resolving Sundaland motion and boundaries* J Geophys Res **112** B06420.

[16] Wapenaar K, Neut J, Van D R E, and Draganov, D (2011). *Seismic interferometry by crosscorrelation and by multidimensional deconvolution a systematic comparison*. 1335–1364.

[17] Warnana D D, and Syaifudin F, (2017) *Pengolahan data seismik pada daerah batuan vulkanik*: Jurnal Geosaintek

[18] Wuryani S D, Yudistira T, and Widiyantoro S, (2019) *Surface wave tomography using seismic ambient noise data for subsurface imaging beneath Bandung basin, West Java and its surrounding* In: IOP Conference Series. Earth and Environtmental Science

[19] Yuan S, Shangxu W, Ming Ma, Yongzhen Ji, and L D, and Abstract—In. (2017) *Sparse Bayesian Learning-Based Time-Variant Deconvolution* Ieee Transactions On Geoscience And Remote Sensing 1–13.

[20] Yao H. van D H, and de Hoop M V, 2006 *Surface-wave array tomography in SE Tibet from ambient seismic noise and two-station analysis* : I - Phase velocity maps. Geophys.J. Int **166(2)**.

[21] Yao, H (2015) Manual for EGF Analysis Time Frequensi Dispersion Software University of Science and Technology of China

[22] Zulfakriza Z, Saygin E, Cummins P R, Widiyantoro S, Nugraha A D, Lühr B G, and Bodin T,

(2014) *Upper crustal structure of central Java, Indonesia, from transdimensional seismic ambient noise tomography. Geophysical Journal International* **197** 630–635.

Warming trend during Medieval Climate Anomaly in Tropical Maritim Indonesia: Records from *Porites* Corals

S.Y. Cahyarini[1], M. Hendrizan[1,2], D.A. Utami[1]

[1] Paleoclimate and Paleoenvironment Research Group, Research Center for Geotechnology, Research Organisation of Earth Sciences- BRIN
[2]Doctoral student of Earth Sciences Study Program, Faculty of Earth Sciences and Technology-ITB

E-mail: sriy004@lipi.go.id

Abstract. A period that lasted from 900-1300 AD has been known as the Medieval Climate Anomaly. This period has been indicated as the warming period of the earth's temperature. However, this warming phenomenon is still the subject of debate today, whether global or regional warming. Several studies concluded that the Medieval warm period is a global phenomenon and an important warm period, although external forcing is mainly similar to the present day. Several climate archives from Indonesia i.e. lake sediment, marine sediment and speleothem show a warming trend during this period. In this study, the seasonal warming trend during the Medieval climate is resolved from *Porites* coral from Lampung Bay, Indonesia. However, the seasonal temperature magnitude during the Medieval period is lower than today. This study confirms the coral medieval climate records from Mentawai islands. The result suggests that a warming trend occurred in Indonesia during the Medieval climate anomaly.

1. Introduction

The Medieval climate is variously defined from A.D. 800–1250[1], from A.D. 900-1300[2], from the 5th to 13th century [3]. The climate during Medieval is interpreted as warmer than today's climate, but if so, then those opposed to global warming used to claim that fossil fuel is not the cause of warming today [4]. The temperature variability of Medieval climate and its magnitude is 50% lower than today, recorded in coral from the Sunda Strait [2]. The concept of the medieval warm period was first launched in 1965 by Lamb [5], which found based on paleoclimate data around 1100 to 1200 A.D., when the European climate was wet summer and winter severity. However, the magnitude of a global or even hemispheric Medieval temperature anomaly is never attempted to estimate. The temperature estimation from Lamb's studies only pertains to western Europe.

There is still a lack of Medieval climate records at the maritime tropic. The published records which covered the Medieval period are from [6] and [7] from the Marine Sediment archive, from Speleothem archive [8], from Lake Sediment archives [9]. However, those available Medieval records are still in low resolution i.e. hundreds-thousands years data resolution. The high-resolution climate records which covered Medieval periods are from coral archives from [10] and [2], which provide monthly resolution climate data.

In this study, we focus on the Medieval period of 900 A.D. to 1300 A.D. We concentrate on the reanalysing coral proxy from [10] and [2] and investigate SST trend during the period of 900A.D. -1300

A.D recorded in coral from Indonesia and whether the other proxy temperature from other climate archives from Indonesia shows a similar trend.

2. Methods

Published proxies from different natural archives from different location across Indonesia (Figure 1) is used in this study. We use available coral proxy temperature published data set from Lampung Bay-Sunda Strait [2], from Mentawai islands [10]. Mg/Ca foraminifera content in marine sediment from Makassar Strait [6][7] is also used for temperature proxy data. The precipitation proxy data i.e δWax content in the lake sediment [9] and speleothem [8] were also used in this study. Those proxy datasets are reanalysed to extract the past temperature and precipitation variability and trend. Table 1 shows a reference list of the proxy data description used in this study. Proxies time series data is then selected for a period of 900-1300 as our focus in this study. Due to different data lengths and intervals, then the data was interpolated to get the same length and interval of the date.

Temperature proxies from modern coral used in this study are from Timor Island [12], Seribu islands [13], and Enggano islands [2] (Figure 2). Statistical analysis of time series data is implemented on the proxies. Advanced statistical analysis is processed using Caliza software application from climate risk [14].

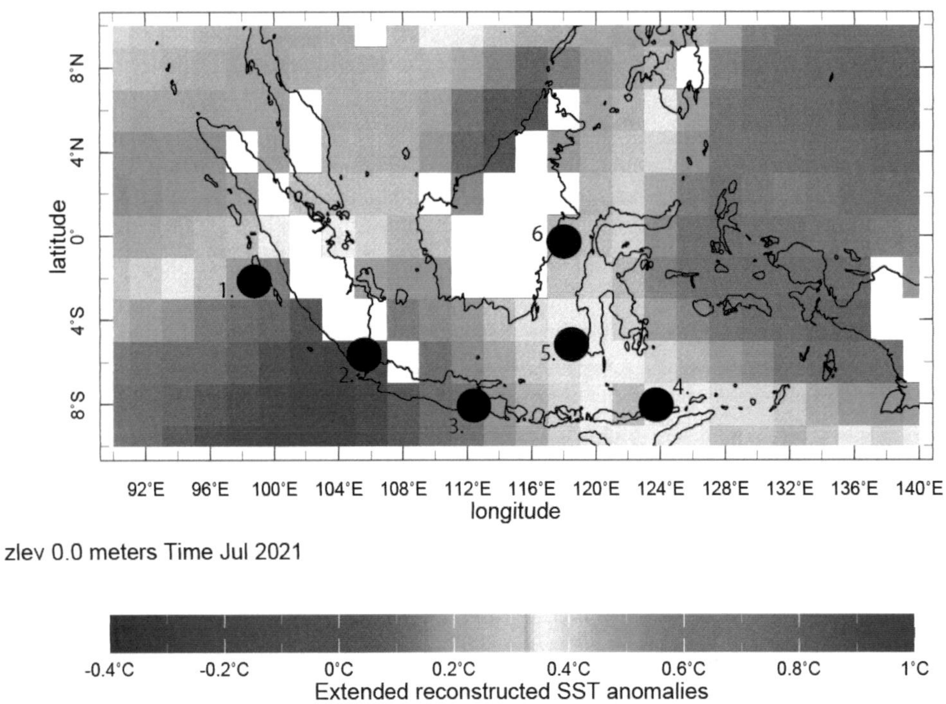

Figure 1. Map of Sea surface temperature anomaly (SSTa) during boreal summer. The SST dataset from the ERSST v.5. Map is generated from the https://iridl.ldeo.columbia.edu/SOURCES/.NOAA/.NCDC/.ERSST/.version5/.anom/. Black dot is location of published geochemical proxies derived from different natural archives i.e. 1, 2: coral; 3: lake sediment; 4: speleothem; 5,6: marine sediment (see text for discussion).

IOP Conf. Series: Earth and Environmental Science 1047 (2022) 012024 doi:10.1088/1755-1315/1047/1/012024

Table 1. Reference list of the proxy data covering Medieval period used in this study

Location	Coordinate	Proxy	Archives	Dated Age	Dating Method	covered-MCA age	Season	resolution	Reference
South west Sulawesi-Makassar strait	3°53'S 119o27'E; 5°12'S 117o29'E	Mg/Ca; d18O	Marine Sediment		14C	905-1305AD	SST; rainfall	10yr/ sample	[6]
Makassar strait	3.566°S, 119.383°E	dDwax	Marine Sediment	1310yrBP- (12.7–11.5 ka) BP	14C	~897-1298AD	Precipitation	100yr/sample	[11]
Liang Luar cave, Flores	8° 320 S, 120° 260 E;	d18O	Speleothem	6.5 ±0.05 kyrBP; 12.6±0.12KyrBP	U-Th	~902-~1300AD	Precipitation	8yr resolution/1mm sample increment	[8]
Lake Lading East Java Indonesia	8°0.5290S, 113°18.750E,	dDwax	Lake Sediment	850CE-1986CE	14C;210Pb	~904-1303AD	Precipitation	8.5yr/sample	[9]
Mentawai island-West Sumatra	2.37°S to 3.18°S 99.74°E to 100.52°E	d18O	Coral	710 cal yr BP (1240 CE) to -69 cal yr BP (2019 CE)	U-Th		SST	monthly	[10]
Lampung Bay Sunda Strait	105.578 E, 5.749 S	Sr/Ca	Coral	1100-1140	U-Th		SST	monthly	[2]

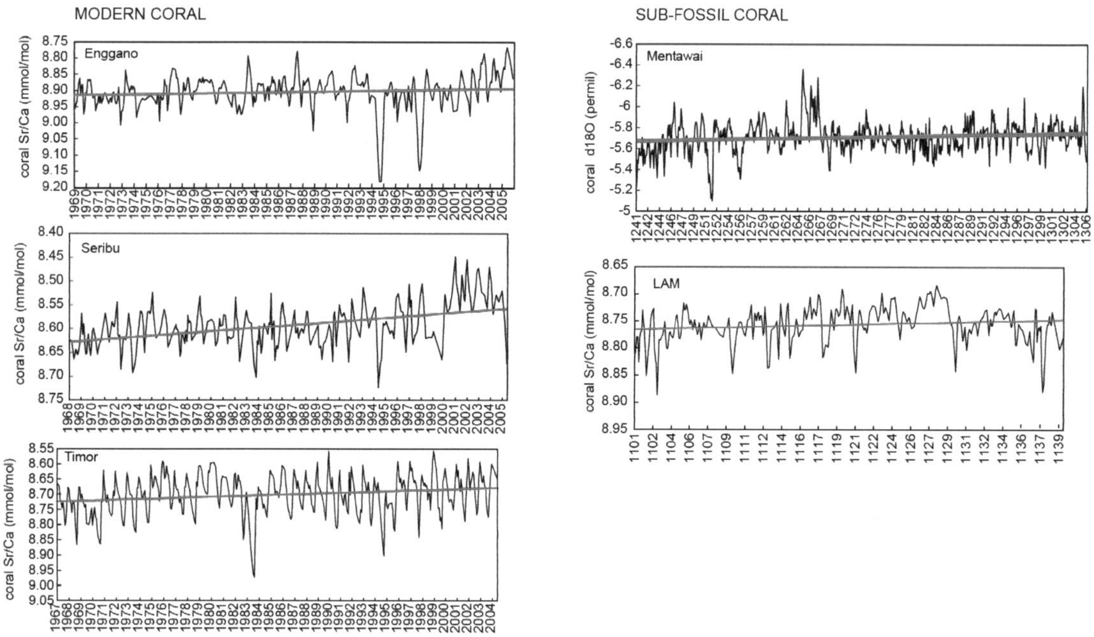

Figure 2. Temperature proxy records from modern coral (Enggano [2], Timor [12], Seribu [13]) and sub fossil coral (Mentawai[10], Lampung Bay-Sunda Strait [2]) covering Medieval period. The red line is linear trend. The proxy data is in monthly resolution representing seasonal variation of temperature.

3. Results and Discussions

At present, Indonesian climate variability is influenced by the monsoon, interannual climate phenomena from the Pacific and Indian Ocean, and Indonesian Throughflow (ITF) and South China Sea Throughflow (SCS). The North West (NW) monsoon commonly brings much rainfall; meanwhile South East (SE) monsoon brings less rainfall which is dry season over Indonesia. The fresher water from SCS flows further to the eastern part of Indonesia during the SE monsoon. Oppositely during the NW monsoon, the more saline water from the ITF exit passage Makassar Strait and also Banda Sea region will flow further to the west [15][16]. In the western part of Indonesia, in the Sunda Strait throughflow the intraseasonal variability is strong and the strongest energy occurring in the boreal spring, which can reverse the Sunda Strait throughflow [17]. This intraseasonal variability is related to local zonal wind and the sea level gradient along the strait [17]. In the Sunda Strait and surrounding region the fluctuation of the SST and SSS is strongly influenced by the monsoon [17][18]. The zonal wind during SE monsoon is the important factor to trigger and propagate the upwelling in the southern Java which cause the low SST. During the Northwest (NW) monsoon, the water mass from Karimata Strait meets with the water mass from the Indian Oceans that was transported into the Java Sea through the Sunda Strait and moves eastward [18] which cause the salinity in the Java Sea increased due to the inflow of higher salinity from the Indian Ocean and the lower Sea Surface Temperature (SST) in the region. Reversely during SE monsoon, the current flows westward and turns northward through the Karimata Strait. The water mass is transported out from the Java Sea to the Indian Ocean[18] which cause lower salinity higher SST in the Java Sea. In annual mean scale the water mass transport from Java Sea to Indian oceans. The modern coral from the Sunda Strait [2] and Mentawai [10] records SST variability in this region, which shows clearly seasonal variability of SST. The SST trend shows increasing seasonal temperature of about 0.5 to 1.5°C recorded from the modern coral from Enggano [2], Seribu [13] and Timor waters [12] (Figure 2).

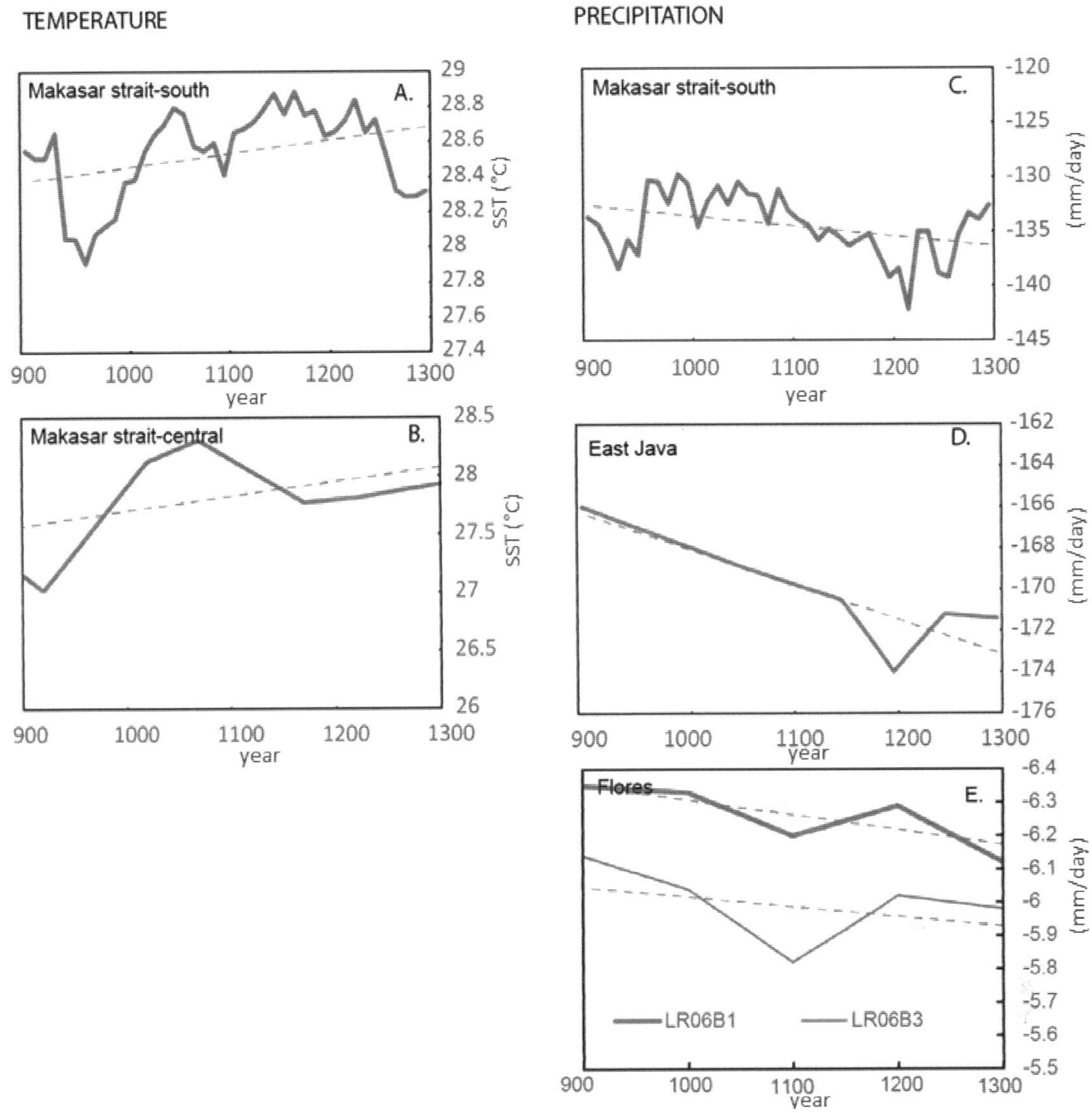

Figure 3. Sea surface temperature (A, B) and Precipitation (C, D, E) in interdecadal resolution derived from natural archives. SST extracted from marine sediment (A) Makassar Strait-south and (B) Makassar Strait-central. Precipitation data is extracted from (C) Marine and (D) Lake Sediment, (E) Speleothem. The data is selected for the period of MCA 900-1300.

Medieval period has been known as warm climate, this also shown in several proxy records from Indonesia. We select period of 900-1300 for each temperature -derived proxy from several different archives i.e. marine sediment from [6] [7] and coral from [10][2] to complement each other for the time scale and data resolution. Coral Sr/Ca changes is about 0.05 mmol/mol for increasing temperature of 1°C [19][20] and coral $\delta^{18}O$ changes of 0.22‰ for increasing temperature of 1°C [21]. In this study modern coral Sr/Ca increase of ±0.025 mmol/mol (equal with ±0.5°C in terms of SST anomalies assuming a Sr/Ca-temperature relationship of -0.05 to -0.06 mmol/mol per 1°C, as estimated via the modern coral Sr/Ca-SST calibrations [6]) recorded by Timor and Enggano islands corals and 0.075 mmol/mol (equal with 1.5°C) recorded by Seribu islands coral. The Medieval sub-fossil coral $\delta^{18}O$ increase of 0.072 ‰ (equal with 0.33°C) recorded from Mentawai coral and 0.02 mmo/mol (equal with

0.4°C) recorded from coral Sr/Ca Sunda Strait. In summary, the seasonal scale Medieval temperature increasing trend (0.33-0.40°C) is lower than the today increasing temperature trend (0.5 -1.5°C) which recorded in coral (Figure 2).

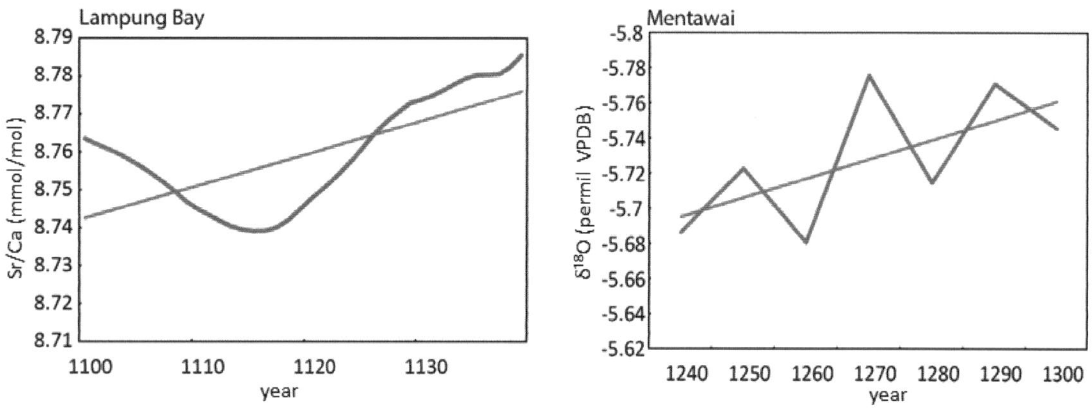

Figure 4 Interdecadal variation of Sr/Ca content from sub-fossil coral from Lampung Bay and coral $\delta^{18}O$ from Mentawai. Lower coral Sr/Ca or $\delta^{18}O$ represent higher SST variation (see text for discussion). The red line is linear trend.

We interpolated the proxy temperature data extracted from marine sediment from Makassar Strait (i.e data source from [6][7] to get the same time interval of 50yrs. Interdecadal variation of temperature derived from marine sediment shows increasing temperature trend of 1-1.5°C during the Medieval period (Figure 3). The decadal variation recorded in coral is obtained by 10yrs running mean of the proxy, the records show increasing decadal temperature trend at Mentawai (period of 1240-1300) of 0.4°C (for coral $\delta^{18}O$ changes of 0.22 ‰/°C) and at Sunda Strait (period of 1100-1140) the increasing temperature trend of 0.6°C (for coral Sr/Ca changes of 0.05 mmol/mol/°C) (Figure 4). Precipitation derived from lake sediment shows declining precipitation trend during the medieval at tropical maritime (Figure 3).

During Medieval climate period intensities of ENSO increase [22] recorded in coral proxies from the South China Seas, this is also shown in coral from Lampung Bay [2] and Mentawai [10] where in the late Medieval period the interannual climate event is clearly shown SST anomaly [2]. The increasing of interannual climate phenomena correlated with the warming trend of SST during Medieval period. The weak of Indian monsoon [23] may play key role also to generate warmer and wetter climate during Medieval.

Declining precipitation at the maritim tropical during Medieval could be due to the southward migration of the ITCZ. At present time the southward migration of ITCZ caused by northern hemisphere summer insolation cause declining rainfall during summer Indian monsoon, these changes also coincide with the reversal of Indian Ocean SST, which witnessed as high ENSO variability [24]. Coral oxygen-isotope records in the Indian Ocean indicate a linkage between Indian Summer Monsoon rainfall and Indian Ocean Dipole (IOD) [25] during the last 60–70 years.

4. Conclusions

Temperature shows increasing trend during Medieval recorded from the proxy temperatures, however today's temperature increases larger than during Medieval period. At present time seasonal SST trend increase of about 1 to 1.5°C and of about 0.5°C at the Medieval period. At the tropical maritim Indonesia, Medieval climate regionally experiences warm and wet in the seasonal and decadal-interdecadal scale as well.

5. References

[1]. Rein B, Luckge A, and Sirocko F. 2004 *Geophys. Res. Lett* vol. 31 L17211 doi:10.1029/2004GL020161
[2]. Cahyarini, S.Y., Pfeiffer, M., Reuning, L, Liebetrau V, Dulo WC, Hideko T, Anwar IP, Utami DA, Garbe-Schoenberg D, Hendrizan M, Eisenhauer A. 2021 *Sci Rep* vol. 11 14952 https://doi.org/10.1038/s41598-021-94465-1
[3]. Britannica, The Editors of Encyclopaedia. "Middle Ages". *Encyclopedia Britannica*, Invalid Date, https://www.britannica.com/event/Middle-Ages. Accessed 25 December 2021
[4]. Bradley RS, Hughes MK, Diaz HF 2004 *Science* vol. 302 www.sciencemag.org
[5]. Lamb HH 1965 *Palaeogr. Palaeoclimatol. Palaeoecol*. 1,13
[6]. Oppo DW, Rosenthal Y and Linsley BK 2009 *Nature* 460 doi:10.1038/nature08233
[7]. Hendrizan M, Kuhnt W and Holbourn A 2017 *Paleoceanography* 32 1054–1069 https://doi.org/10.1002/2016PA003030
[8]. Griffiths ML, Drysdale RN, Gagan MK, Zhao J-X, Ayliffe LK, Hellstrom JC, Hantoro WS, Frisia S, Feng Y-x , Cartwright I, Pierre E St., Fischer M. J. and Suwargadi B.W 2013 *Nature GeoScience*. 2 597-664 doi:10.1038/ngeo605
[9]. Konecky BL, Russell JM, Rodysill JR, Vuille M, Bijaksana S, and Huang Y 2013 *Geophys. Res. Lett.* 40 386– 391 doi:10.1029/2012GL054331
[10]. Abram NJ, Wright NM, Ellis B, Dixon BC, Wurtzel JB, England MH, Ummenhofer CC, Philibosian B, Cahyarini SY, Yu TL, Shen CC, Cheng H, Edwards RL & Heslop D 2020 *Nature* https://doi.org/10.1038/s41586-020-2084-4
[11]. Tierney JE, Oppo DW, Rosenthal Y, Russell JM, and Linsley BK. 2010 Paleoceanography 25 PA1102 doi:10.1029/2009PA001871
[12]. Cahyarini SY, Pfeiffer M, Nurhati IS, Aldrian E, Dullo WC and Hetzinger S. 2014 *J. Geophys. Res. Oceans* 119 4593-4604 DOI: 10.1002/2013JC009594
[13]. Cahyarini, SY *et al.* 2016 *Mar. Pollut. Bull.*, doi:10.1016/j.marpolbul.2016.04.052
[14]. Caliza^{TM} Version 03. www. climate-risk-analysis.com December 2014
[15]. Gordon AL, Susanto DR and Vranes R 2003 *Nature* 425 824-828.
[16]. Sprintall J *et. al.* 2019 *Front. Mar. Sci.*, https://doi.org/10.3389/fmars.2019.00257
[17]. Shujiang Li *et al*. 2018 *J. Oceanography* https://doi.org/10.1007/s10872-018-0476-y
[18]. Putri, M.R. 2005 Dissertation Fachbereich Geowissenschaften der Universitaet at Hamburg
[19]. Cahyarini SY, Pfeiffer M, Timm O, Dullo WC, and Garbe-Schoenberg D. 2008 *Geochim. et Cosmo. Acta* 72/12 2841-2853 DOI: 10.1016/j.gca.2008.04.005.
[20]. Cahyarini SY, Pfeiffer M, and Dullo WC 2009 98 31-40 DOI: 10.1007/s00531-008-0323-2
[21]. Gagan MK, Ayliffe LK, Hopley D, Cali JA, Mortimer GE, Chappel J, McCulloch MT, and Head MJ. 1998 *Science* 279 1014-1018
[22]. Jiang L, Yu K , Tao S , Wang S, Han T, and Jiang W 2021 *Paleoceanography and Paleoclimatology* 36 e2020PA004173. https://doi. org/10.1029/2020PA004173
[23]. Loo YY, Billa L, Singh A 2015 *Geoscience Frontiers* 6 817-823
[24]. Reddy P, Gandhi N, Krishnan R 2021 *Quarternary International* https://doi.org/10.1016/j.quaint.2021.06.018
[25]. Saji NH, Goswami BN, Vinayachandran PN and Yamagata T 1999 *Nature* 401 360-363.

Acknowledgement
Acknowledge to the AvH Digital Cooperation fellowship program to SYC ref 3.4-IDN/1158893. This manuscript contributes to Nusantara grant entitled Coastal Evolution in Indonesia: Holocene-Future Interplay Between Coral Reefs and Mangrove during Sea level Changes, SK No 3558d/IT1.B07.1/KP/2021. This manuscript also contributes to pendanaan riset RP Kebencanaan OR IPK BRIN TA 2022 to DAU. Main contributor author SYC, contributor author member MH, DAU.

Characterization Of Sedimentary Process In An Estuarine Environment Based On Grain Size Trend Analysis And Geochemical Trace Element In The West Flood Canal River, Semarang

[1]S N Devienna; [2,3]S H Nugroho; [1]A Kurniasih; [2,3]P S Putra; [4]Lestari; [4]S P Awibowo; [4]M T Kaisupy; [2]Amar; [2]Jakah

[1]Department of Geological Engineering, Universitas Diponegoro, Semarang
[2]Research Center for Geotechnology, National Research and Innovation Agency (BRIN), Bandung
[3]Department of Geological Engineering, Institut Teknologi Bandung, Bandung
[4]Research Center for Oceanography, National Research and Innovation Agency (BRIN), Jakarta
Corresponding authors : syafanatasyadevienna@gmail.com (SND); sept005@brin.go.id (SHN); anis.kurniasih@live.undip.ac.id (AK)

Abstract. The West Flood Canal River is one of the major rivers that flows the central of Semarang Bay. The west flood canal river is located in the middle of a densely populated residential and used as city drainage in the Semarang. Unfortunately, the upper part of the river has been swallowed up. This river has unique sedimentation dynamics, so it becomes a suitable research object to determine the relationship between river sedimentation processes and the concentration of trace elements contained in the sediment. A core sample was taken at the mouth of the west canal flood river for research purposes. The analysis included grain-size trend analysis (GSTA) and geochemical trace elements on 40 sub-samples of sediment core slices within 1 cm intervals. Afterward, the GSTA results were grouped by clustering to obtain lithofacies. Each lithofacies has geochemical characteristics that indicate changes in environmental conditions over a period of time. The GSTA showed that the sediment core is dominated by coarse silt to fine silt. The cluster showed there are at least 5 lithofacies, which have different sedimentary characteristics. In general, the character of the sediment which is dominated by coarser grain size has a higher concentration of Fe, Ca, Mn, Ti, and lower levels of K. On the other hand, the sediment group which has a sedimentary character dominated by fine grain size has a lower content of Fe, Ca, Mn, Ti, and higher K levels.

Keywords: sediment characteristics, trace element, grain size, geochemical, west flood canal river, Semarang.

1. Introduction

The research location is located around the West Flood Canal Estuary, West Semarang sub-district, Central Java. The West Flood Canal River is a river that has the potential as a medium for carrying waste, including heavy metal waste, which is disposed of by industrial activities including the textile industry, metal and machinery industry, chemical industry, and ceramic industry [1]. The estuary of the West Flood Canal is a transitional area between land and sea that is most affected by pollution. This estuary is characterized by an area that has a quite high turbidity due to river water input and sediment resuspension. Heavy metal research in Muara West Semarang Flood Canal River has done a lot especially concentration of dissolved heavy metals and total heavy metal in sediment [2].

This study uses grain-size trend analysis (GSTA) and geochemical analysis approaches. GSTA is commonly used to distinguish physical and textural characters which serve to determine the transport and deposition of sediments [3]. Several studies have been carried out, especially regarding sediment characteristics and geochemical content [4]. Different environmental conditions can affect sediment characteristics and geochemical content. This river has unique sedimentation dynamics, so it becomes a suitable research object to determine the relationship between river sedimentation processes and the concentration of trace elements contained in the sediment..

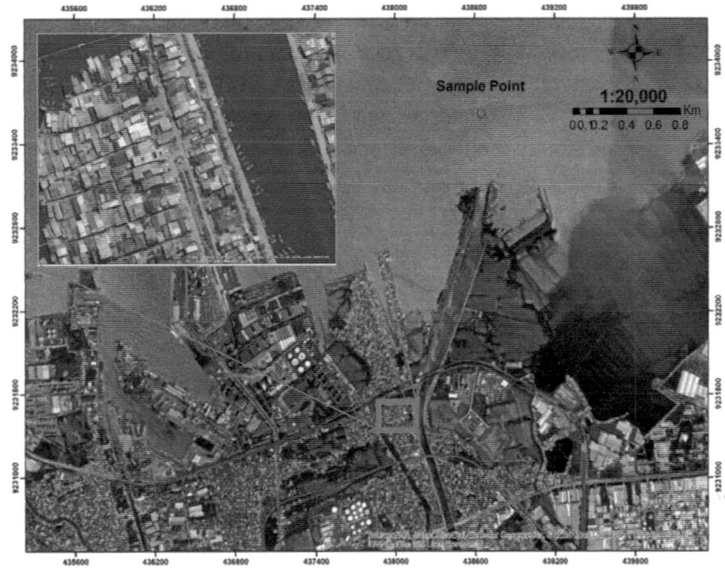

Figure. 1. Research location sampling point. The red box is a zoom-in of residential and industrial areas. The red dot is the sample point.

2. Geography dan Climate

The city of Semarang is located between Mount Ungaran (elevation 2000 m) to the south and the Java Sea to the north, with a geographical position between 6° 50' -7 °10' south latitude and 109° 50' -110° 35' east longitude. The city of Semarang is bordered by Kendal Regency in the West and Demak Regency in the East. The area of Semarang is 37,333.83 ha with 99 sub-districts and 117 sub-districts. The population of the city of Semarang in 1991 was 1,154,536 inhabitants. In addition, industrial locations in Semarang are not located in one particular area but are widely distributed in the east, west, north, and southeast of the city.

The local climate is influenced by the Northwest Monsoon which occurs between October-May (rainy season) and the Southeastern Monsoon which occurs between May-October (dry season). Monthly rainfall in the watershed during 1958 - 1990 ranged from 1,474 - 3,181 mm in the West Flood Canal. The average amount of rainfall ranges from a low of 70 mm in August and a high of 365 mm in January. The average monthly wind flow ranges from 4.5 to 5.8 km/hour and the temperature ranges from 26.5-28.3^{o} C, the highest is 33.3 C, the lowest is 23.2 C [5].

Based on the regional geological map sheet Magelang – Semarang [6], the research location is located in the alluvial unit is alluvial deposits of beaches, rivers, and lakes. The beach deposits consist of clay, silt, and sand with a thickness of 50 m or more. River and lake deposits consist of gravel, sand,

IOP Conf. Series: Earth and Environmental Science 1047 (2022) 012025 doi:10.1088/1755-1315/1047/1/012025

and silt with a thickness of 1-3 m. The gravels are composed of andesite, limestone, and a little sandstone.

3. Data and Method

The research material used is A sediment core was taken from the West Flood Canal, Semarang (110°23'37" E and 6° 56' 44' S) as part of the LIPI's Coremap expedition on March 2021. The thickness of the sediment sample taken is ± 40 cm from the surface. The data used are 40 sub-sample of sediment core slices, 40 sub-sample of grain size analysis, and 40 sub-sample of geochemical analysis.

A sediment core was cut at 1 cm intervals to produce 40 sediment samples. 40 The samples were measured the chemical content and grain size of the sediment. Geochemical analysis used the Thermo NITON XL3t 500 by Thermo Fisher Scientific. Meanwhile, sediment grain size analysis used Mastersizer 2000 by Malvern Panalytical. The results of sediment grain-size analysis were clustered based on lithofacies using PAST4.03 software.

4. Results

The results of the calculation of grain size, there are 4 parameters namely mean, sorting, skewness, and kurtosis are presented in the form of the table below.

Table 1. The result of grain size parameters of sediment

CODE	MEAN (μm)	SORTING (μm)	SKEWNESS (μm)	KURTOSIS (μm)	SAND	MUD
BKB 1	13.05	6.052	0.266	0.624	34.0%	66.0%
BKB 2	14.03	5.400	0.123	0.669	30.0%	70.0%
BKB 3	15.37	5.476	-0.002	0.680	30.9%	69.1%
BKB 4	4.415	2.595	-0.012	0.938	0%	100.0%
BKB 5	4.056	2.402	-0.001	0.961	0%	100.0%
BKB 6	4.743	2.957	0.083	0.995	1.2%	98.8%
BKB 7	5.050	3.065	0.095	0.992	1.5%	98.5%
BKB 8	4.930	2.787	0.054	0.934	0%	100.0%
BKB 9	4.417	2.643	0.030	0.928	0%	100.0%
BKB 10	4.449	2.646	0.053	0.974	0%	100.0%
BKB 11	4.728	2.854	0.056	0.974	0.4%	99.6%
BKB 12	12.99	4.763	-0.003	0.708	20.5%	79.5%
BKB 13	7.327	4.105	0.147	0.897	7.5%	92.5%
BKB 14	9.534	4.428	0.149	0.773	12.6%	87.4%
BKB 15	14.33	4.757	-0.007	0.723	22.6%	77.4%
BKB 16	18.58	5.423	-0.150	0.692	32.8%	67.2%
BKB 17	14.77	5.709	0.106	0.695	28.0%	72.0%
BKB 18	12.35	4.345	0.100	0.899	15.1%	84.9%
BKB 19	20.04	5.769	-0.012	0.709	33.9%	66.1%
BKB 20	25.66	5.501	-0.318	0.706	41.1%	58.9%
BKB 21	45.43	3.912	-0.592	1.491	61.3%	38.7%
BKB 22	23.33	5.189	-0.373	0.691	39.3%	60.7%
BKB 23	14.51	4.642	-0.015	0.706	23.5%	76.5%
BKB 24	8.111	4.010	0.128	0.935	8.6%	91.4%
BKB 25	6.377	3.704	0.147	1.025	6.2%	93.8%
BKB 26	10.13	4.594	0.166	0.823	14.9%	85.1%
BKB 27	10.64	4.525	0.117	0.782	14.9%	85.1%
BKB 28	11.22	4.852	0.118	0.806	17.1%	82.9%
BKB 29	7.508	3.871	0.104	0.927	6.7%	93.3%
BKB 30	7.373	3.750	0.079	0.917	5.6%	94.4%
BKB 31	11.29	4.502	0.110	0.831	15.5%	84.5%
BKB 32	4.564	2.657	0.007	0.920	0%	100.0%
BKB 33	4.446	2.736	0.021	0.922	0%	100.0%
BKB 34	4.569	2.801	0.028	0.936	0%	100.0%
BKB 35	4.594	2.763	0.006	0.935	0%	100.0%
BKB 36	7.236	4.236	0.182	1.106	8.8%	91.2%
BKB 37	13.02	6.773	0.253	0.857	23.5%	76.5%
BKB 38	6.075	3.520	0.122	0.979	4.4%	95.6%
BKB 39	5.045	2.992	0.040	0.969	1.4%	98.6%
BKB 40	6.549	3.257	-0.020	0.892	1.6%	98.4%

IOP Conf. Series: Earth and Environmental Science 1047 (2022) 012025 doi:10.1088/1755-1315/1047/1/012025

The results of the measurement of the grain size of the sediment are presented in the form of a graph (figure 2). The results of the analysis are used to determine the grain size characteristics at the research site. The difference in grain size characteristics of sediment at a certain depth is grouped using PAST4.03 software which is presented in the form of clusters (figure 3).

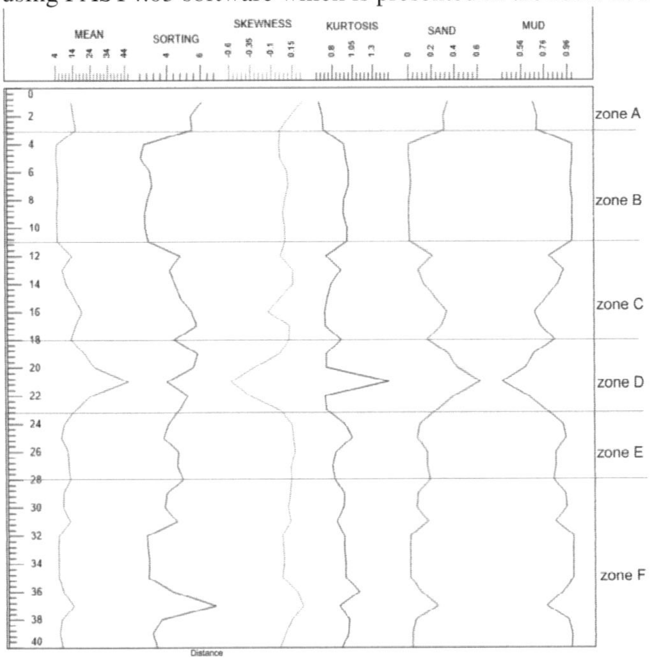

Figure. 2. Result of grain-size trend analysis using software Strater 5.

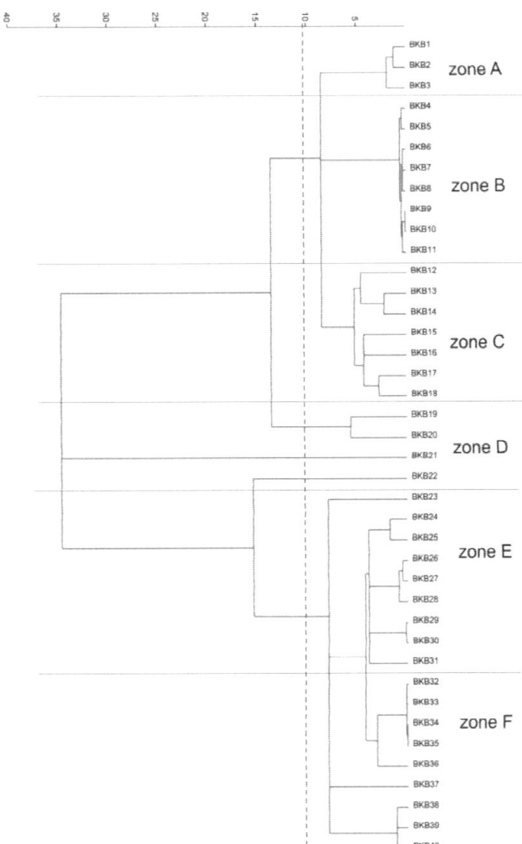

Figure. 3. Result of grain-size clustering using software PAST4.03.

The parameters used to determine the grain size characteristics of the sediments in the study area are mean and sorting. Mean value shows the average grain-size of the sediment meanwhile sorting value represents sediment sortation. Based on the graph, zone A has a mean value belonging to the silt medium with a value of around 13.05 – 15.37 µm. Zone B has an average grain-size of fine silt with values around 4056 – 5.05 µm. Zone C is medium silt with values around 9.534 – 18.58 µm. Zone D is coarse silt of about 20.04 – 25.66 µm. Zone E is medium silt with values around 8.111 – 14.51 µm and Zone F is fine silt with values around 4,446 – 7,236 µm.

Based on the sorting value, zone A is categorized as very poorly sorted, which means that in zone A the degree of uniformity of the grain size is very poor. Zone B is classified as poorly sorted, meanwhile zone C, D, and E are categorized as very poorly sorted. Zone F is included in poorly sorted.

The results of measuring the content of geochemical elements are presented in table (table 2.) and graphical form (figure 4.). The results of the analysis are used to determine the geochemical minor and major elements contained in the sediment.

Table 2. geochemical elements of sediment sample

CODE	Zr	Sr	Rb	Zn	Fe	Mn	Ti	Ca	K
BKB 1	133.31	147.68	29.12	52.66	31343.52	521.44	3536.19	15257.26	5467.41
BKB 2	122.72	150.32	26.41	43.43	25652.9	454.83	3332.57	13335.7	6048.84
BKB 3	123.42	167.78	26.24	40.9	29600.19	516.46	3461.51	15569.92	5359.91
BKB 4	132.59	166.76	23.24	45.31	32652.86	550.29	3597.32	15682.32	5711.36
BKB 5	127.19	134.93	29.44	53.73	26218.24	556.88	2994.58	13356.95	5414.39
BKB 6	109.93	132.53	28.17	58.27	23285.26	416.26	2909.53	14506.36	6746
BKB 7	106.66	116.63	27.45	41.47	22092.78	460.06	3264.82	12558.32	6543.84
BKB 8	80.41	93.41	25.54	42.05	19010.44	508.52	2966.56	8909.98	5531.81
BKB 9	102.61	89.77	35.67	62.91	22591.11	486.23	1729.96	6155.34	3547.12
BKB 10	110.31	138.55	29.65	47.63	23962.73	630.92	2801.29	13145.03	5412.51
BKB 11	126.8	152.3	31.47	54.13	33757.18	826	2515.65	11977.31	4580.25
BKB 12	106.83	126.08	27.03	53.65	24941.16	565.76	2890.08	12938.33	5418.1
BKB 13	119.95	133.14	24.65	46.34	23233.21	394.28	3105.2	14242.26	6566.16
BKB 14	109.14	125.92	26.73	47.68	24188.25	566.81	3142.81	14031.28	6461.3
BKB 15	114.2	132	27.71	50.86	24581.01	452.94	3142.87	14825.94	6246.43
BKB 16	102.74	164.21	31.77	57.63	26787.34	443.13	2972.58	14626.43	5873.42
BKB 17	108.31	174.01	29.37	80.98	31452.65	669.2	3071.04	14169.39	5229.41
BKB 18	101.91	169.27	30.55	59.37	29510.77	540.1	3086.68	15219.31	6176.84
BKB 19	96.97	152.32	24.48	52.63	25653.56	587.1	2956.62	13746.26	5196.34
BKB 20	89.29	131.92	22.6	48.98	24097.54	390.62	2518.65	13035.32	4784.5
BKB 21	209.84	203.28	23.17	66.65	43265.54	791.86	4577.61	19184.03	4917.49
BKB 22	241.41	199.56	19.86	58.85	42421.21	851	4530.88	20031.7	5051.87
BKB 23	240.67	220.89	27.67	63.16	50054.46	867.01	4805.55	20951.75	4008.75
BKB 24	134.92	186.56	27.61	49.38	30779.64	681.13	3709.3	20208.83	5685.37
BKB 25	178.86	200.9	26.77	57.32	34551.14	648.43	3975.54	22147.06	5956.5
BKB 26	138.98	175.11	25.36	44.44	29416.63	658.44	3158.12	16891.96	5262.31
BKB 27	98.66	157.99	29.59	47.38	25692.32	514	3338.57	16409.35	6765.64
BKB 28	96.47	156.7	21.74	43.76	24570.43	454.98	2874.89	16076.12	5889.19
BKB 29	95.49	146.16	25.57	46.52	26718.57	426.74	3070.6	16976.52	6379.23
BKB 30	101.04	151.49	29.18	56.12	26887.72	427.75	2874.95	14445.52	5773.16
BKB 31	99.47	181.95	28.13	60.76	30222.61	477.62	3407.74	16357.29	5382.64
BKB 32	79.2	86.46	32.95	43.61	19360.56	411.72	2024.84	7966.32	4285.52
BKB 33	87.19	105.25	27.76	34.23	22292.77	555.54	3296.33	13250.54	7572.14
BKB 34	88.1	98.7	33.12	49.97	21739.64	530.97	2797.01	9701.27	6317.83
BKB 35	87.38	100.07	25.11	30.04	19981.6	397.14	2647.28	11542.32	5094.61
BKB 36	76.22	100.37	27.45	51.5	20050.04	370.57	2451.82	9874.03	4891.62
BKB 37	83.23	139.7	28.61	38.32	25555.54	628.7	2439.48	12266.82	5052.78
BKB 38	121.5	176.86	31.23	87.98	38117.89	669.39	2779.13	12970.15	5511.29
BKB 39	95.68	123.72	29.6	33.05	22551.42	419.96	3319.47	14664.3	5973.98
BKB 40	81.72	75.28	27.11	58.11	21840.14	737.04	2316.47	6055.35	3848.54

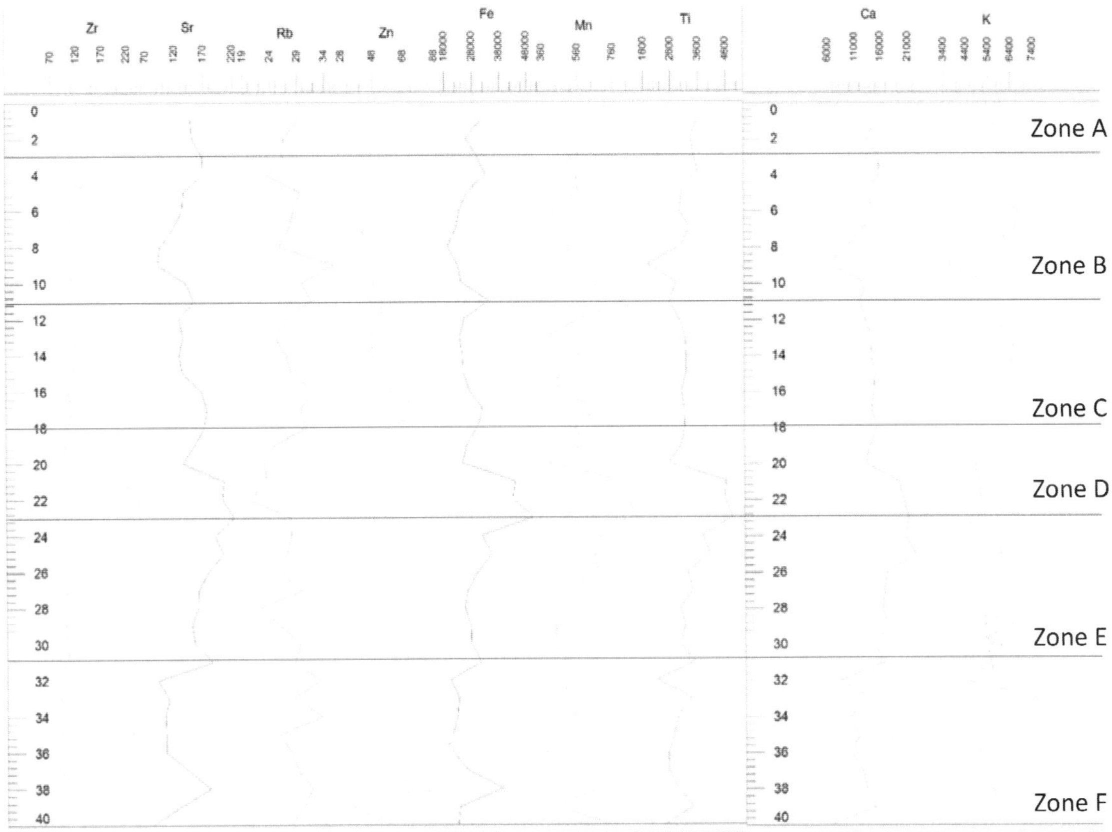

Figure 4. the trend of geochemical analysis.

Based on the trend of the geochemical analysis (figure 3.), the geochemical elements are contained in the sediment consist of the elements Zr, Sr, Rb, Zn, Fe, Mn, Ti, Ca, and K. The elemental content of Zr is around 76.22 – 241.41 ppm, the content of Sr is around 93.41 – 220.89 ppm, the content of Rb is around 19.86 – 35.67 ppm, the content of Zn is around 30.04 – 87.98 ppm, the content of Ca is around 6055.35 – 22147.06 ppm, the content of Fe is around 19010.44 – 50054.46 ppm, the content of Ti is around 1729.96 – 4805.55 ppm, the content of K is around 3547.12 – 6765.64 ppm and the content of Mn is about 370.57 – 687.01 ppm. Zone A, B, C, E, and F have low Fe, Mn, Ti, Ca levels, and high K levels. Meanwhile, zone D has high Fe, Mn, Ti, Ca levels and low K levels.

The difference in grain size and trace elements can be influenced by several factors such as transportation mechanism, river currents, and the origin of the sediment supply [4,6,7]. Various grain size tends to be coarser showing a sudden change in transportation mechanism, possibly due to fluctuations in water level conditions [9]. The supply of sediment carried by river currents mostly comes from land rather than the sea. This can be proven by the trace elements contained in the sediment, namely the high values of Fe and K. This can be indicated that there has been an overflow of river water carrying sediment from the mainland due to flooding in the past few years. The supply of sediment from the mainland is carried away by the river. Zone D has coarse grain size and is defined as the result of erosion by waves which then transported and deposited. River currents entering seawater experienced a slowdown. In the zone that has a fine grain size are transported with the suspension process and as the magnitude sediment particles then become saltation and also traction. The dominance of the traction transport type, saltation, and suspension are also different for each research area sample. The difference may be caused by variations in grain size as well as flow strength deposition [10].

High Ca elements are indicated by the reworked sedimentary rocks containing high Ca deposited [11]. Fe and Mn had frequently been used as an indicator of natural changes in the heavy metal carrying capacity of the sediment and its concentration has sometimes been related to the abundance of metal-reactive compounds supposedly and not significantly affected by human actions [12]. Meanwhile, high Mn and Fe are indicated by Fe input from rivers and estuaries originating from industrial waste, residential waste, and deposited in sediments [11]. This is supported by the research area that is close to residential and industrial areas. Pollutants that enter the river's mouth will be scattered and undergo a deposition process. Sediment absorption of metal particles is more dominant because metals tend to bind to iron hydroxide, manganese hydroxide, and organic matter. Naturally, heavy metals (Fe) that enter the marine environment include rivers and erosion caused by waves. Meanwhile, sediments containing Mn are in a reduced state where Mn is dissolved and buried in the sediment.

5. Conclusion

Sediment characteristics in the studied area are dominated by coarse silt - fine silt. Traces elements are contained in the sediment which is dominated by the content of Ca, Mn, and Fe. Factors that affect sediment characteristics are sediment sources, the velocity of river currents, and erosion. Meanwhile, the high content of Ca, Mn and Fe elements are the results of reworked rock containing high Ca and waste pollution from industry and settlements.

6. References

[1] Bappedal Propinsi Jawa Tengah 2002 Laporan program kali bersih tahun 2002. *Laporan Penelitian*. Semarang: Pemerintah Propinsi Jawa Tengah. Hal 13 – 14.

[2] Maslukah L. 2007. Konsentrasi logam berat (Pb, Cd, Cu, Zn) terlarut dalam seston dan dalam sedimen di estuari Banjir Kanal Barat, Semarang . J. Aquatik. **2**(1):1-4.

[3] Nugroho S H, Putra P S and Pamungkas A D G 2020 Karakteristik Sedimen Holosen–Pleistosen Berdasarkan Ukuran Butir Pada Sedimen Inti Ew17-08, Tepi Barat laut Paparan Sunda. *Jurnal Geosaintek*, 6(2), 107-116.

[4] Nugroho S H and Putra, P.S., 2017 Spatial Distribution of Grain Size and Depositional Process in Tidal Area along Waikelo Beach, Sumba. *Marine Georesources and Geotechnology* **36**(3): 299-307. DOI: 10.1080/1064119X.2017.1312649.

[5] TICA Study Team 1994 Interim Report on Flood Control in BanjirKanal Barat and Babon Rivers, Semarang, Indonesia p.3-41.

[6] Thanden,dkk 1996 Regional Geological Map Sheet Magelang – Semarang.

[7] Environmental Impact Management Agency (BAPPEDAL) 1994 PROKASIH (Clean Water Program) *: Evaluating The Last Four Years and Looking to the Future,* Jakarta, Indonesia.

[8] Angusamy N G and Rajamanickam V 2006 Depositional environment of sediments along the Southern Coast of Tamil Nadu, India. *Oceanologia* **48**: 87-102.

[9] Nugroho S H and Putra P S 2019 Characteristics of Grain Size Distribution on Beaches Sediment of Sumba Island, Nusa Tenggara Timur Based on Grainsize and Geochemical Data. *Jurnal Geologi dan Sumberdaya Mineral*, ***20***(3), 153-164.

[10] Hukubun C, Nurdrajat D J S and Prabowo, H 2020 Rekonstruksi Lingkungan Pengendapan Berdasarkan Analisis Granulometri Sedimen Core No 7 Pada Perairan Laut Arafura, Maluku. *Geoscience Journal*, **4**(2), 163-172.

[11] Takarina, N 1996 Heavy Metal Contents in Surficial Sediment of Banjir Kanal Barat and Babon Rivers, Semarang, Central Java, Indonesia. Doctoral dissertation (McMaster University).

[12] Barreiro R , Real C, Carballeria A 1994. Heavy metal horizontal distribution in surface sediments from a small estuary (Pontedeume, Spain). *Sci. Total Environ.,* **154**: 87 -100.

7. Acknowledgment

The study was partially funded by LIPI's Coremap-CTI 2021-2022 (grant number 17/A/DK/2021) who received by L. and SHN. This study was also partially funded by Laboratory of Energy Resources, Geological Engineering Department Diponegoro University. We thank the Deputy of Earth Sciences, Research Center for Oceanography of BRIN, Research Center for Geotechnology of BRIN. We are also very grateful to two anonymous reviewers for their useful critical reviews. SND, SHN, AK designed the research concept, prepared and wrote the manuscript. SND contributed to data processing. PSP and L contributed by providing suggestions and input, as well as review-edited the manuscript. SPAW, MTK prepared the samples and tools in the cruise and A and J in the laboratory. L, SHN, SPAW, and MTK joined the cruise and obtained the samples.

The Relationship of Lithology with Landslide Occurrences in Banyumanik and Tembalang Districts, Semarang City

D Trisnawati [1], Najib [1*], A S Hidayatillah[1]

[1]Geological Enginering, Diponegoro University, Semarang, 50275, Indonesia.

Corresponding author: najib@ft.undip.ac.id

Abstract. Movement of soil and rock masses down the slope or known as landslide is a natural activity that can pose a danger to humans. Generally, landslide can be influenced by the slope, slope materials (rock/soil), and rainfall. Morphologically, the study area, which is located in the borders of Tembalang and Banyumanik Districts, Semarang City, is a hill that is quite steep. It has a potential to experience landslides. Landslide occurences at the study area do not always occur on steep slopes. Therefore, the aim of this study is to determine the relationship between the slope materials and the occurrences of landslide. Data was obtained through mapping to determine geological conditions and enlisting landslide occurrences at the study area. Furthermore, the correlation between the number of occurrences and the type of slope materials at the location of the ground motion was carried out using information value method. Based on geological mapping, 5 types of lithology were obtained, namely: alluvium, breccia, sandstone, siltstone, and claystone. Out of 10 landslides that were found, 6 of them occurred in claystone lithology. We interpret that landslide occurrences in study area are relating to the lithology claystone with information value 0.416. The claystone, a part of the Kerek Formation, physically were easily degraded. In addition, this lithology mostly found in open slopes, where the process of erosion by water is more intense.
Keyword : Landslide, Lithology, Claystone, Kerek Formation, IVM

1. Introduction

Semarang city cannot be separated from geological disasters. Based on records from the Regional Disaster Management Agency (BPBD) of Semarang City, geological disasters that often occur in Semarang City are tidal floods, floods, and landslides [1]. The graph of the trend of disaster events during the period of 2012 to 2020 that can be seen in **Figure 1.** Its shows that landslides are a more frequent disaster than other geological disasters. Therefore, landslides are still an interesting object to be studied in Semarang City. Several studies on landslides in Semarang City with various topics have been carried out. Among them were carried out in the period between 2017 to 2021 by Hani'ah et al. [2], Yulianto et al. [3] [4], Supriyadi et al. [5], Tjahjono et al. [6], Khumaedi et al. [7], Ansrory et al. [8], Rahman et al. [9], Ardinugroho and Handayani [10], Astuti et al. [11], and Hidayatillah et al. [12]. From this research, no one has discussed the relationship between lithology and landslides.

Landslide occurs because it is influenced by various factors but some published references can be used to determine the cause of it. According to Nurdin and Kubota [13], each causative factor of the occurrence of landslides have different values in the arrangement of the landslides susceptibility map. In general, landslides is influenced by slope, slope materials, rainfall, and land use.

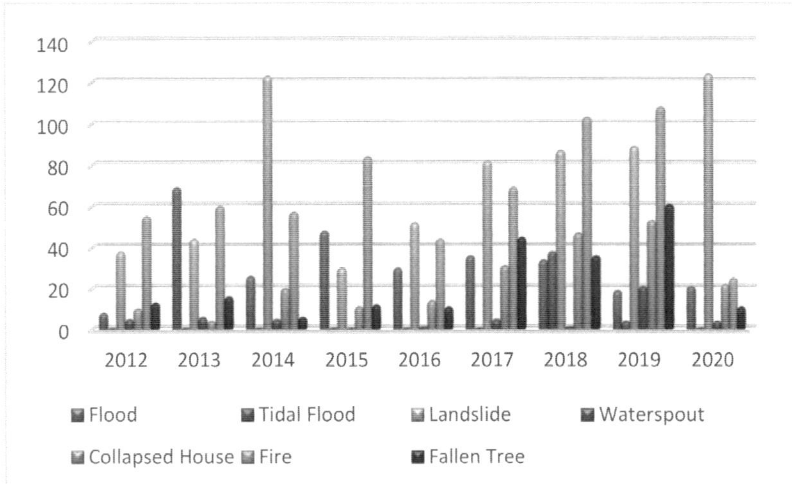

Figure 1. Graph of the trend of disaster events in Semarang City during 2012 - 2020 period [1].

Slope materials, soil or rocks, is one of the most important factors in the study of landslides. Each region has a different types of rock and soil. It will produce different values in the arrangement of the landslide susceptibility map [13]. Several studies on the influence of lithology on the landslides occurrence have been carried out by several researchers including Safaei et al. [14], Henriques et al. [15], Peruccacci et al. [16], Bhandari and Dhakal [17], and Su et al. [18]. Based on Geological Map of Semarang, study area is composed of the Kerek Formation, Kalibeng Formation, Kaligetas Formation, and Quaternary Deposits [12][19]. This indicates the possibility of fnding lithological variations in the study area. The objective of this study was to determine the relationship between lithology and soil movement that occurred in the study area.

2. Method

2.1. Study Area

The study area (**Figure 2**) is located between Tembalang District and Banyumanik District. Based on the Geological Map Sheet Semarang [19] the study area consists of 4 formations, namely the Kerek Formation, Kalibeng Formation, Kaligetas Formation, and Quaternary Deposits. The Kerek Formation is the oldest formation in the study area, above it lays the Kalibeng Formation, followed by the Kaligetas Formation, and the youngest is the Quaternary Sediment. The rock formation is estimated to be traversed by a normal fault trending west-east. Several studies have stated that there is a relation between landslides and the presence of geological structures [4][12][20][21][22]. The presence of geological structures in the study area has effects on the landslides occurrence [12]. From the SPOT image (**Figure 2**), the land use of the location is dominated by settlements. The existence of settlements near the slopes creates a potential of disaster if landslides occur. Topographically, the study area is 48 to 180 meters above sea level, with a height difference of 100 meters.

2.2. Mapping

Mapping was imposed to obtain lithology and landslide data. When the lithological properties become important as internal factor concerning erosion activity, the lithological map has crucial importance for erosion mapping studies [15]. For this reason, a detailed lithological map is needed to obtain the suitability of the data in the research. The inventory of landslide occurrences was carried out by the Semarang City BPBD only for landslides that occur around settlements. Therefore, it is necessary to gather data on the occurrence of landslides in the study area.

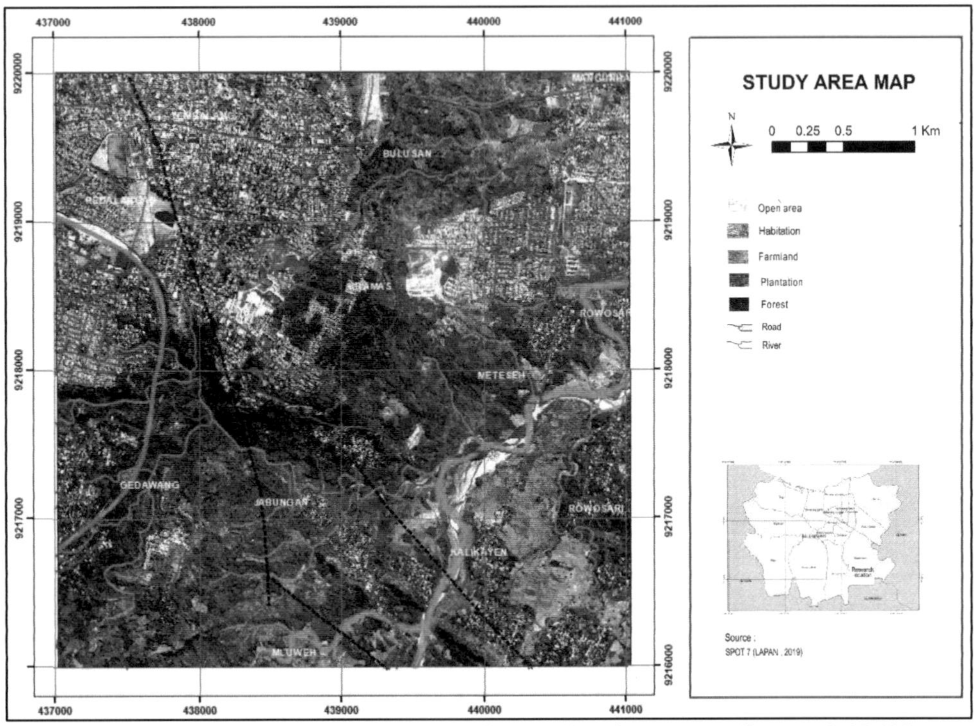

Figure 2. Visual Appearance of study area from SPOT 7 image space(Source : LAPAN)

2.3. Information Value Method (IVM)

The IVM method is one of the bivariate statistical methods used to model the influence of environmental factors on the occurrence of landslides. IVM is a statistical method of conditional analysis in determining the statistical relationship between the occurrence of landslides and its causal factors [14]. The information value (IV) of each variable x is given by Yin and Yang (1988)[14][22] :

$$IVx = log\frac{Tx/Nx}{T/N}$$

Where,

Tx : number of terrain units with landslides and presence of variable

Nx : number of terrain units with variable x

T : total number of terrain units with landslides

N : total number of terrain units

3. Result and Discussion

3.1. Lithological Units Of The Study Area

Based on the mapping of the study area, several lithological units were obtained. Lithological units are determined based on lithological characteristics that can be seen in the study area. Based on the order of formation, lithological units are presented from oldest to youngest in the description below :

a. Claystone Unit

Light gray to dark gray with clay grain size or <1/256 mm on the Wentworth scale [23]. The grains are rounded, well sorted, sealed with carbonated cement. The lithology was found in fresh to moderately weathered (**Figure 3.a**). These units are part of the Kerek Formation which was formed during the Miocene.

b. Siltstone Unit

It has a light brown to dark gray color with silts grain size or 1/16-1/256 mm [23]. The grains are rounded, well sorted, sealed with carbonated cement. The lithology was found in fresh to moderately weathered (**Figure 3.b**). These units are part of the Kalibeng Formation which was formed in the Late Miocene - Late Pliocene.

c. Sandstone Unit

Yellowish-brown to light brown with a grain size of coarse (1/2 - 1 mm) to very fine sand (1/8 - 1/16 mm) [23]. The grains are rounded, poorly sorted, closely packed with carbonate cement. The lithology was found in fresh to moderately weathered **(Figure 3.c)**. Similar to siltstone lithological units, these units are also part of the Kalibeng Formation.

d. Volcanic Breccia Unit

This unit is a volcanic product was found in a moderate to fresh weathered state, light brown to dark in color with a rock fragment of 2-5 cm. It has a matrix of andesite, subangular-subrounded rock fragments, poorly sorted, openly packed with non-carbonated cement (**Figure 3.d**). This lithological unit belongs to the Kaligetas Formation which is Pleistocene in age.

e. Alluvium Unit

This deposit is the result of an exogenic process that works at the study area and is a quartenary age deposit. Light to dark gray with fragments of 2-20 cm in size, with a matrix of fine to very fine sands. Subangular-subrounded fragments, poorly sorted, open-packed with non-carbonated cement (**Figure 3.e**).

Figure 3. Lithological unit of the study area (a) Claystone Unit, (b) Siltstone Unit, (c) Sandstone Unit, (d) Volcanic Breccia Unit, and (e) Alluvium Unit.

3.2. Landslide Occurrences

Based on the results of the mapping conducted at the study area, 10 locations of landslides were found. The distribution of landslide points can be seen in **Figure 4**. It is mostly developed on the south side of the study area. Landslides occur at the study area ensue on the slopes of more than 20°, which is morphologically typical of land with a steep to a very steep slope. This morphology is in the center of the study area extend from north to south.

There are two types of landslides found in the study area which are rotational and fall. The rotational type of landslide is a movement that occurs due to the movement down the slope with the materials can

IOP Conf. Series: Earth and Environmental Science 1047 (2022) 012026 doi:10.1088/1755-1315/1047/1/012026

be soil or rock. It can be identified from the presence of a clear collapse plane at the contact between the landslide mass and the underlying lithology. Fall-type landslides can occur in soil or rock materials, especially those on the side of cliffs with little or no displacement field [22]. Data on the occurrence of landslides that have been collected from the research location are presented in **TABLE 1**.

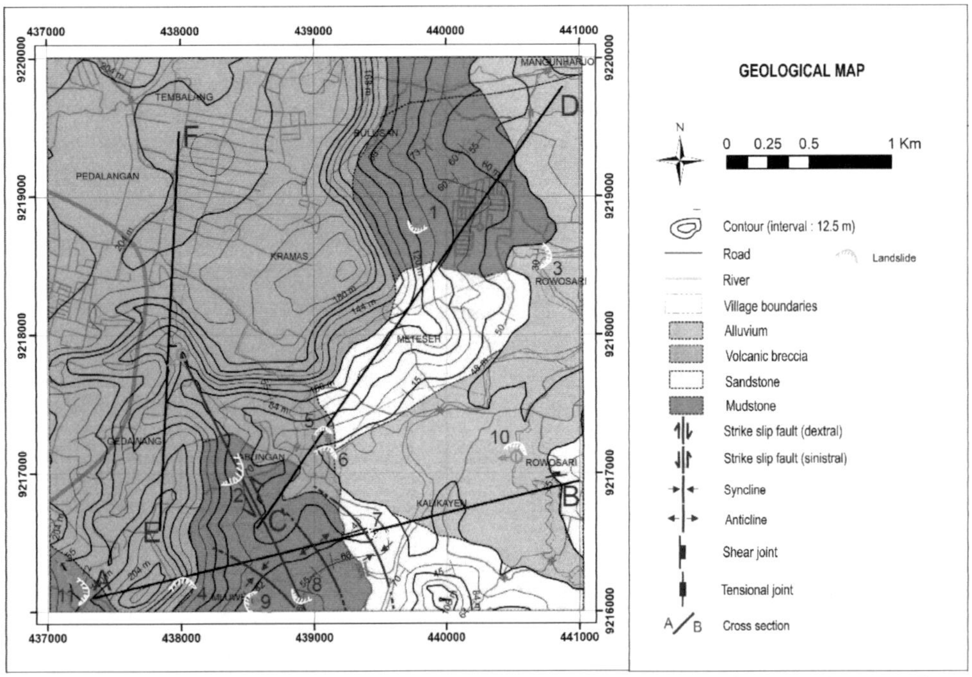

Figure 4. The map shows the distribution of landslide occurrence in the study area [12].

TABLE 1. Data of landslides occurred in the study area, modified from Hidayatillah et al [12]

No.	Landslide Type	Location	Direction of Landslide	Dimension	Slope	Lithology
1	Rotational	Meteseh	N 30° E	4 x 10 m	30°	Claystone
2	Rotational	Jabungan	N 245° E	17 x 12 m	70°	Claystone
3	Fall	Rowosari	N 300° E	3 x 10 m	35°	Claystone
4	Rotational	Gedawang	N 196° E	5 x 20 m	60°	Volcanic Breccia
5&6	Rotational	Jabungan	N 170° E	7 x 20 m	50°	Volcanic Breccia
7	Rotational	Jabungan	N 35° E	7 x 12 m	45°	Sandstone
8	Fall	Mluweh	N 40° E	4 x 80 m	40°	Claystone
9	Fall	Mluweh	N 138° E	10 x 120 m	60°	Claystone
10	Fall	Rowosari	N 186° E	4 x 43 m	80°	Sandstone
11	Fall	Gedawang	N 280° E	3 x 40 m	20°	Claystone

3.2.1. Rotational Landslide

The rotational type of landslide occurred at the point of occurrence 1, 2, 4, 5, 6, and 7. In several landslide locations, this type appeared to have a circular shape on the landslide crown as shown in **Figure 5**. Generally, the circular shape found in the study area occurred at the slopes of claystone and sandstone and are associated with cliffs that have a slope of more than 30°. The contact between the landslide mass and the lithology of the slope is visible.

Figure 5. Rotational type landslides at the study site occur in claystone and sandstone. Occurred at the lithology of the slope composition in the form of claystone intercalation sandstone at (a) point 1 and (c) point 2, while (b) occurred at point 7 with the lithology of the slope composition in the form of sandstone intercalation claystone.

The type of rotational landslide occurs in volcanic breccias (points 5 and 6) at the study area, may occur due to several causes, such as volcanic breccia which lays on above claystone lithology; exogenic processes that worked on the slope; and loading on the slope (e.g construction, water infiltration, human activity, etc) can cause landslides. The location of the landslides (**Figure 6**) is on the river border which is 100 meters from the river body. Landslide at points of 5 and 6 are a type of rotational landslide with more than 1 landslide crowns.

3.2.2. Fall Landslide

Fall landslide is often found on river slopes. Fluviatile processes that works on riverbanks result in scouring the bottom of the slope. Therefore, due to the loss of slope material at the bottom, the top of the slope does not have support so it collapse. Several landslides at the study site can be seen in **Figure 7**, which generally occur, in the cut-off slope morphologies.

Figure 6. Rotational landslide at point 5 (top) and point 6 (right) with the lithology of volcanic breccia.

Figure 7. Falling landslides in the study area: (a) point 3, (b) point 8, (c) point 9, (d) point 10, and (e) point 11.

3.3. Relation Between Lithology and Landslide

In the geological cycle, lithology is a product of geological processes associated with the asthenosphere and troposphere. Processes related to the asthenosphere will affect the basic properties of the lithology itself, including the color, composition, and engineering properties of the rock (i.e unit weight, compressive strength, shear stregth, RQD, etc) . On the other hand, the surface process, which is the interaction between the troposphere and the lithosphere will directly affect humans being. In some literature, lithology is one of the factors that cause landslides and is a parameter in arranging vulnerability and landslide susceptibility maps [2],[3],[4],[14][15],[16],[17].

In the study area, based on the mapping results, there are 5 lithological units that have different characteristics. Of the five lithologies, landslides occurred only in 3 types of lithology : claystone, sandstone, and volcanic Breccia. Claystone is the oldest lithology at the study site, it's part of the Kerek Formation [12]. There are 6 points of landslide occurrence in the claystone from 10 events after inventoried, at the point of occurrence 1, 2, 3, 8, 9, and 11. In the sandstone, there were 2 landslide events, points 7 and 10, while in the volcanic breccia there are 2 landslide events, points 4 and 5.

Based on the IV calculation as shown in **TABLE 2**, the total value of IV is 0.694. This value indicates that there is a relation between landslides occurrence and lithology. Furthermore, the largest IV value is indicated in claystone with a value of 0.416. Therefore, claystone has a very important role in the occurrence of landslides at the study site. A total of 6 out of 10 landslides occurred in claystone (**TABLE 1**).

TABLE 2. IVM calculation based on the type of landslide in the study area

Variable	RL	FL	Grand total	RL	FL	Population	IV
Claystone	2	4	6	40%	80%	60%	0.416
Sandstone	1	1	2	20%	20%	20%	0.139
Volcanic Breccia	2	0	2	40%	0%	20%	0.139
Grand total	5	5	10	100%	100%	100%	0.694

Note = RL: Rotational Landslide; FL: Fall Landslide

Claystone in the study area which is part of the Kerek Formation composed of Illite, Haolisite, and Montmorillonite. Based on Diansisti [24], in the weathered forms of claystone the presence of clay minerals has a higher ratio to non-clay minerals. The combination of rainfall, slope, geological structure, and lithology that occurred in the past will affect the rock weathering process. Rocks that already have a certain mineral composition that can be converted into clay minerals, will produce a weathered rock with more of a clay mineral composition. The presence of clay mineral in the rock reduces the strength of the rock [25].

Folds and faults in the study area indicates tectonic processes that worked on the formation of the study area in the past. During the formation, these structures leave traces on the claystone, such as cracks, known as joint structures. The joint structure is a discontinuous plane in the rock that becomes the path for the initial geomorphic process to work. Furthermore, previous research has also proven that there is a link between the existence of structures and landslides [12].

4. Conclusion

Landslide is a common geological process that occurs on the earth's surface. This process is influenced by several factors including topography, lithology, climate, and land use. As many as 6 out of 10 landslides in the study area occur in claystone. This is 60% of the population of landslide in the study area. Based on the calculation of IV, a value of 0.416 is obtained, which means that there is a relations between landslides and claystone in the study area compared to other lithologies.

Acknowledgments
This research is fully supported by the Faculty of Engineering, Diponegoro University based on contract number 2496/STGL02/UN7.5.3.2/PP/2020. The authors also thank the National Institute of Aeronautics and Space of Indonesia (LAPAN) for the satellite imagery support. Also to all contributors and assistants that made this research possible.

References

[1] BPBD Kota Semarang, "Data Bencana," http://bpbd.semarangkota.go.id/pages/data-bencana (diakses pada 10 September 2021)

[2] Hani'ah., H. S. Firdaus, and A. L. Nugraha, "Analysis of Environmental Vulnerability in The Landslide Areas (Case Study: Semarang Regency)," *IOP Conference Series: Earth and Environmental Science*. Vol. 98. No. 1. IOP Publishing, 2017. [Online]. Available: https://iopscience.iop.org/article/10.1088/1755-1315/98/1/012013/meta [Accessed on December, 2021].

[3] T. Yulianto, R. Gernowo, and S. Widada, "Landslide Potency At Trangkil, Gunung Pati Based On The Groundwater Flow Pattern And The Value Of Safety Factor," *Journal of Physics: Conference Series, , 1025(1)* doi:10.1088/1742-6596/1025/1/012028, 2018. [Online]. Available: https://iopscience.iop.org/article/10.1088/1742-6596/1025/1/012028/meta [Accessed on December, 2021].

[4] T. Yulianto, S. Suripin, and H. Purnaweni, "Zoning Landslide Vulnerable Area According To Geological Structure, Slopes, And Landuse Parameters In Trangkil Sukorejo Gunungpati Semarang City's Residental Area," *Journal of Physics: Conference Series (Vol. 1217, No. 1, p. 012029).* IOP Publishing, 2019. [Online]. Available: https://iopscience.iop.org/article/10.1088/1742-6596/1217/1/012029/meta [Accessed on December, 2021].

[5] Supriyadi., R. H. Hidayatullah, R. Kusumawardani, and T. C. Upomo, "Landslide potency based on microseismic data-case study in Trangkil area Gunungpati, Semarang," *Journal of Physics: Conference Series* (Vol. 1170, No. 1, p. 012073). IOP Publishing, 2019. [Online]. Available: https://iopscience.iop.org/article/10.1088/1742-6596/1170/1/012073/meta [Accessed on December, 2021].

[6] H. Tjahjono, Suripin, and Kismartini, "Spatial analysis on landslide disaster risk in the semarang city, central java, indonesia," *Disaster Advances, 12*(7), 49-58, 2019.

[7] Khumaedi, N. M. D. Putra, and A. Z. Fikriyah, "Identification of landslide potential in gajahmungkur village semarang using ground shear strain analysis," *Journal of Physics: Conference Series, 1170*(1) doi:10.1088/1742-6596/1170/1/012072, 2019. [Online]. Available: https://iopscience.iop.org/article/10.1088/1742-6596/1170/1/012072/meta [Accessed on December, 2021].

[8] Ansosry, H. A. Rahman, & F.R. Ramadhan, " Landslide Mitigation of Banjir Kanal Semarang with Grouting Method", *Journal of Physics: Conference Series,* 1387 (1), 2019. [Online]. Available: https://iopscience.iop.org/article/10.1088/1742-6596/1387/1/012099/meta [Accessed on December, 2021].

[9] B. Rahman, R. Fimana, & U. C. Sari, "The Analysis of Landslide Vulnerability in Settlement of Citizenship Associaton VI, Sukorejo Village, Semarang", *IOP Conference Series: Earth and Environmental Science*. Vol. 328. No. 1. IOP Publishing, 2019. [Online]. Available: https://iopscience.iop.org/article/10.1088/1755-1315/328/1/012035/meta [Accessed on December, 2021].

[10] N. S. Ardinugroho, & W. Handayani, " Landslide Community Resilience: An Examination of Six Neighborhoods in Sukorejo, Semarang" *IOP Conference Series: Earth and Environmental Science*. Vol. 447. No. 1. IOP Publishing, 2020. [Online]. Available: https://iopscience.iop.org/article/10.1088/1755-1315/447/1/012015/meta [Accessed on December, 2021].

[11] B. Astuti, I. Suryaningsih, A. Rusilowati, Khumaedi, & H. H. Kusuma, " Science Literacy Profile of Student on Landslide Disaster Mitigation in Semarang City", *Journal of Physics: Conference Series,* 1918 (1), 2021. [Online]. Available https://iopscience.iop.org/article/10.1088/1742-6596/1918/2/022017/meta [Accessed on December, 2021].

[12] A. S. Hidayatillah, Najib, D. Trisnawati, U. C. Nugroho, R. F. Hidayat, F. Febyani, B. T. Widodo, G. Yogiswara, & A. Ilma, " Geological Structure Influence on Landslide Occurences : Tembalang and Banyumanik sub-districts, Semarang City", *AIP Conference Proceedings*. Vol. 2363, no. 1, 2021. [Online]. Available : https://aip.scitation.org/doi/abs/10.1063/5.0061436. [Accessed on December, 2021]

[13] P. F. Nurdin & T. Kubota, " GIS-based Landslide Susceptibility Assessment And Factor Effect Analysis By Certainty Factor In Upstream of Jeneberang River, Indonesia", Journal of Geomatics and Planning, Vol. 5, No. 1, pp. 75 – 90, 2018.

[14] M.Safaei, H. Omar, B. K. Huat, & Z. B. M. Yousof, "Relationship between Lithology Factor and Landslide Occurence Based on Information Value (IV) and Frequency Ratio (FR) Approaches - Case Study in North of Iran," *Electronic Journal of Geotechnical Engineering*, Vol.17,pp.79-90,2012.

[15] C. Henriques, J. L. Zezere, & F. Marques, "The role of the lithological setting on the landslide pattern and distribution", *Engineering Geology*, Vol. 189, pp.17-31, 2015.

[16] S. Peruccacci, M. T. Brunetti, S. Luciani, C. Vennari, & F. Guzzeti, " Lithological and Seasonal Control on Rainfall Thresholds for The Possible Initation of Landslides in Central Italy", *Geomorphology*, Vol. 139-140, pp. 79-90, 2012.

[17] B. P. Bhandari & S. Dhakal, " Lithological Control on Landslide in the Babai Khola Wathersed, Siwaliks Zone of Nepal", *American Journal of Earth Sciences*, Vol. 5, no.3, pp 54-64, 2018.

[18] M. Su, K. Cheng, Y. Liu, Y. Xue, P. Wang, K. Zhang, & C. Li, " Combining Geophysical Methods, Drilling, and Monitoring Techniques to Investigate Carbonaceous Shale Landslide Along A Railway Line : A Case Study on Jiheng Railway, China*", Bulletin of Engineering Geology and Enviroment*, Vol 80, pp. 7493-7506, 2021. https;//doi.org/10.1007/s10064-021-02365-5.

[19] R. E. Thanden, H. Sumadirdja, P. W. Richards, K. Sutisna, and T. C. Amin, "*Geological Map of The Magelang and Semarang Sheets, Jawa, 2nd Edition*," Geological Research and Development Centre, 1996.

[20] S. Lee, U. Chwae, K. Min, Landslide susceptibility mapping by correlation between topography and geological structure: the Janghung area, Korea, *Geomorphology* Vol 46. pp 149-162, 2002.

[21] Y. Yuskar, D. B. E. Putra, A. Suryadi, T. Choanji, C. Cahyaningsih,"Structural Geology Analysis In A Disaster-Prone Of Slope Failure, Merangin Village, Kuok District, Kampar Regency, Riau Province," *Journal of Geoscience, Engineering, Environment, and Technology,* vol.2, no. 4, pp249-254, 2017.

[22] E. M. Lee & D. K. C. Jones, " *Landslide Risk Assessment*", Thomas Telford Publishing, London, 2004, ISBN : 0727731718.

[23] Wentworth, C.K. 1922. A Scale of Grade and Class Terms for Clastic Sediments. *Journal of Geology, Vol. XXX*: 377-392.

[24] A. Diansisti, " Hubungan Antara Komposisi Mineral, Sifat Fisik, Slaking, dan Kuat Geser Pada Batulempung Formasi, Kerek Studi Kasus : Daerah Susukan, Ungaran Timur, Kabupaten Semarang Jawa Tengah", 2015. Retrieved from https://digilib.itb.ac.id/index.php/gdl/view/19893

[25] A. D. Regmi, K. Yoshida, M. R. Dhital, & K. Devkota, " Effect of Rock Weathering, Clay Mineralogy, and Geological Structures in the Formation of Large Landslide, A Case Study from Dumre Besesi Landslide, Lesser Himalaya Nepal", *Landslides*, Vol. 10, pp. 1- 13, 2013.

Estimation of B-Value Variation as Earthquake Precursor in Java Region with Maximum Likelihood Method

D Arubi [1*], Zulfakriza[2,3], S Rosalia[2,3], D P Sahara[2,3], and N T Puspito[2,3]
[1] Undergraduate Program of Geophysical Engineering Program, Faculty of Mining and Petroleum Engineering, Institut Teknologi Bandung, Indonesia
[2] Global Geophysics Research Group, Faculty of Mining and Petroleum Engineering, Institut Teknologi Bandung, Indonesia
[3] Center for Earthquake Science and Technology, Institut Teknologi Bandung, Indonesia

*corresponding author's e-mail: dinar.kuliah@gmail.com

Abstract. Java Island is one of the areas with a population density of 1,171 people/km^2, which is relatively high compared to other regions in Indonesia. The tectonic setting of the Java region is affected by the interaction of tectonic plate subduction between the Indo-Australian Plate and the Eurasian Plate which moves at the rate of 67 mm/year. In addition, the presence of several active faults on Java Island also contributes to the tectonic setting. The implication of this setting is a relatively high level of potential earthquake events and some of them are destructive earthquakes. The estimation of b-value variations by using Gutenberg-Richter Equation plays an important role in estimating the occurrence of earthquakes. The aims of this study are to determine the role of spatial and temporal variations of b-value as earthquake precursors on Java Island. In this research, earthquakes analysis was carried out in the form of spatial and temporal variations of b-value from earthquake catalog data in 1900-2016 for the Java Island region. Based on the estimation of earthquake recurrence time for magnitude 6.5, 7, and 7.5, and also the analysis of spatial variation of b-value shows the areas with the potential for significant earthquakes in the west and eastern part of Java region. In addition, our result also obtains the existence of a seismic gap zone in the southern part of Central Java.

1. Introduction

The tectonic setting of the Java region is affected by the interaction of tectonic plate subduction between the Indo-Australian Plate and the Eurasian Plate in the southern part of Java and Nusa Tenggara region. The presence of plate subduction and active faults causes many earthquakes that occur around Java Island. The total area of Java Island is ±129,442 km^2 with a population reaching more than 150 million people. With an area of only 6.74% of Indonesia's territory, Java Island has the highest population density in Indonesia [1], this makes the occurrence of earthquakes with large magnitudes and potentially damaging earthquakes will greatly impact the population.

As an effort to mitigate earthquakes in Java, it is necessary to map the area based on the potential of earthquakes occurrence and the possibility of their recurrence time. B-value as one of the parameters of the intensity of the earthquake is used to show the gradient of the relationship between earthquake frequency and magnitude from a linear equation. Variations in b-value depend on the type of stress and tectonic characteristics of an area [2].

This study will analyze b-value variations and create earthquake recurrence maps. By conducting b-value research and mapping the distribution of earthquakes in the Java Island region, it is expected to be part of mitigation efforts and reduce the risks due to earthquakes in the future.

2. Data and Method

This study uses earthquake catalog data from The National Centre for Earthquake Studies (PusGeN) that occurred in Indonesia from October 7th, 1900 to August 23rd, 2016. This study focuses on earthquakes that happened around Java region as shown in Figure 1.

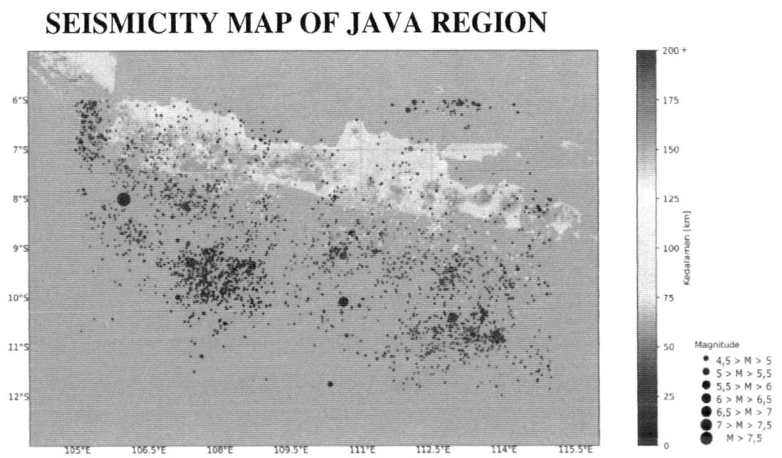

Figure 1. Seismicity map of the study region. The colour gradation from blue to red indicates depth and the size of the dot indicates an increase in magnitude.

Calculation of b-value from earthquake catalog done by using the maximum likelihood method with the help of ZMAP software. B-value calculation is carried out using the maximum likelihood method. The maximum likelihood method proposed by Aki-Utsu (1965) estimates the b-value parameter with uniform earthquake weighting,

$$b = \frac{\log e}{\bar{M} - M_{min}}$$

the a-value can be determined using the following equation.

$$a = \log N(M \geq M_{min}) + \log (b\ ln10) + M_{min}b$$

where $\bar{M}$ is the average magnitude and M_{min} is the minimum magnitude from the earthquake catalog data used. The standard deviation can be calculated using the following equation.

$$\delta b = 2.30b^2 \sqrt{\sum_{i=1}^{n} (M_i - \bar{M})^2 / n(n-1)}$$

In this study, the b-value calculation was carried out for earthquake events in the Java region and earthquakes originating from the Java megathrust zone. Earthquake catalog data is then declustered using Gardner-Knopoff method. Estimation of the magnitude of completeness (Mc) is carried out using the maximum curvature method [3, 4].

Calculation of the spatial variation of b-value is done by dividing the research area into a grid of 0.1°×0.1°. Processing of spatial variation of b-value is carried out using the Fixed Mc method. An earthquake density map in the study area was also made to see the distribution of earthquakes and the possibility of seismic gaps.

Calculation of b-value temporal variation is carried out using earthquake data without going through the declustering process, using the sliding time window method [5]. The results of the spatial and temporal variation of the b-value parameter are also mapped into an earthquake recurrence time map at a certain magnitude. In this study, earthquake recurrence time maps were made for earthquakes in the megathrust zone with a magnitude of M≥6.5; M≥7.0; and M≥7.5. The recurrence time of earthquakes is calculated based on the seismicity index or the cumulative frequency of earthquakes per year. The steps taken in processing the research data are shown in Figure 2.

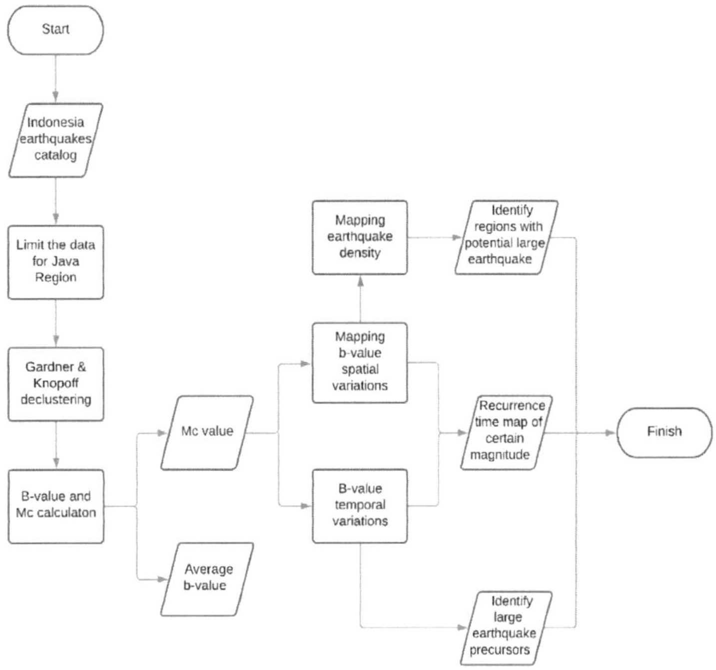

Figure 2. Flowchart of data processing.

3. Result and Discussion

The distribution of the magnitude and frequency curves of earthquakes shows an inverse relationship. The calculated b-value parameter shows a different value for the Java Island as a whole is 1.36±0.14 and the b-value for the megathrust zone is 1.19±0.2. The calculation result of b-value in the megathrust zone is relatively smaller, it can be interpreted that the stress that occurs is influenced by deeper earthquake events [1].

The spatial variation of b-value in the Java Island region and the megathrust zone of Java are shown in Figure 3. Both results show a fairly low distribution of b-value marked in blue in the west and east regions, while the central region has a larger b-value anomaly. The b-value anomaly in the central part of Java is caused by the confluence of structural patterns trending NW-SE with NE-SW [6]. A high b-value parameter can also be influenced by the heterogeneity of the rocks that make up an area [7].

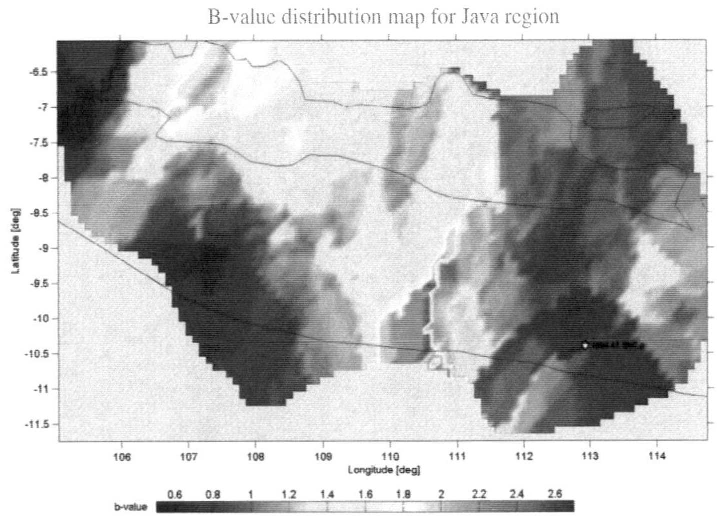

(a)

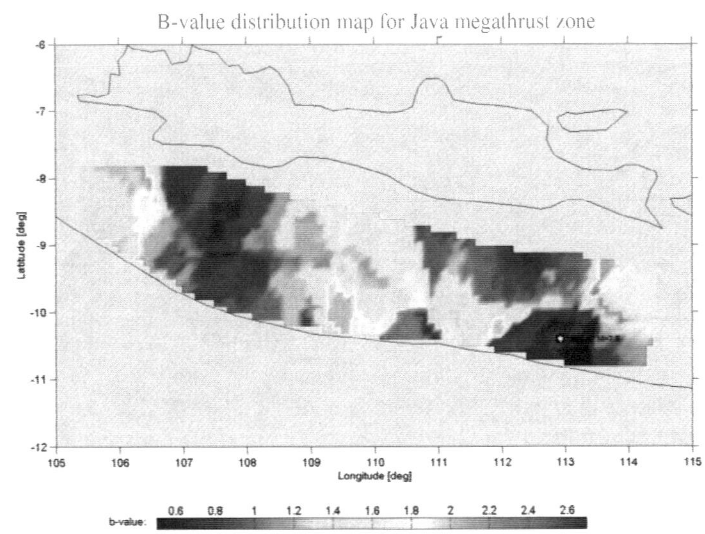

(b)

Figure 3. B-value spatial variation of (a) Java region, and (b) Java megathrust zone. The yellow star indicates mainshock with M > 7.

The earthquakes density map of the Java Island region and the megathrust zone of Java shows that there is a seismic gap zone in the central region (Figure 4). Areas with a higher frequency of small-magnitude earthquakes have less chance of major earthquakes occurring in the future.

IOP Conf. Series: Earth and Environmental Science 1047 (2022) 012027 doi:10.1088/1755-1315/1047/1/012027

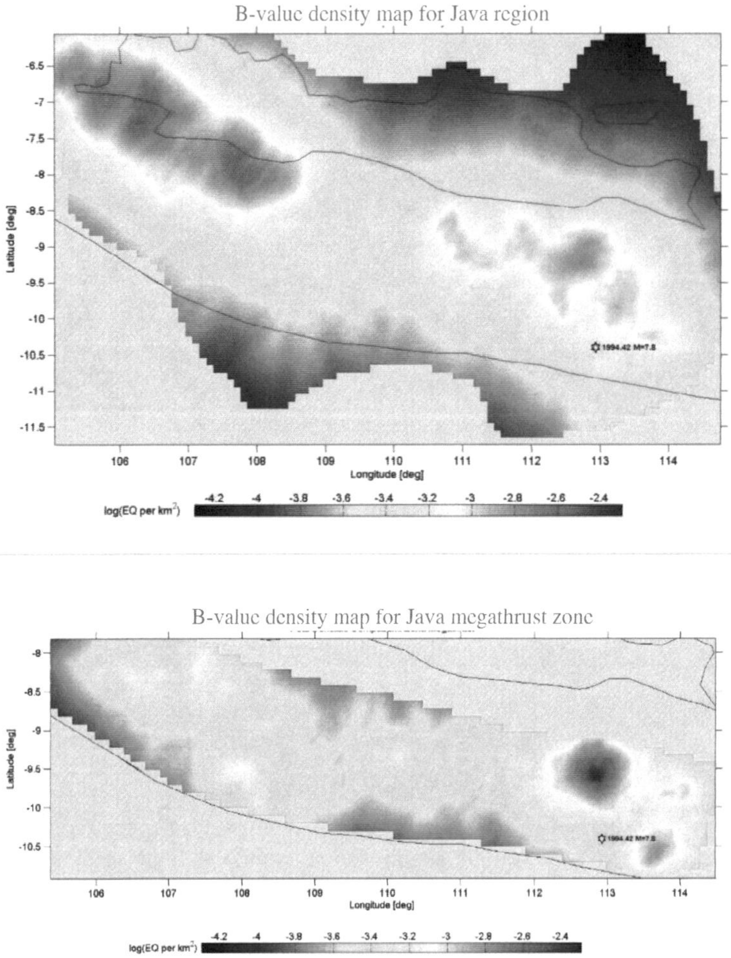

(a)

(b)

Figure 4. Earthquake density map of (a) Java region, and (b) Java megathrust zone. The yellow star indicates mainshock with M > 7.

The temporal variation shows the b-value pattern in a certain time range and its relation to an earthquake event. Temporal variation of b-value is carried out without data declustering [8] so that the pattern of the overall earthquake occurrence can be seen. The results of the temporal variation of b-value for the entire Java region (Figure 5) and the megathrust zone (Figure 6) showed a significant decrease that occurred due to the occurrence of the same large earthquake, namely the Banyuwangi (1994) earthquake with magnitude 7.2 and the Pangandaran (2006) earthquake with magnitude 7.7.

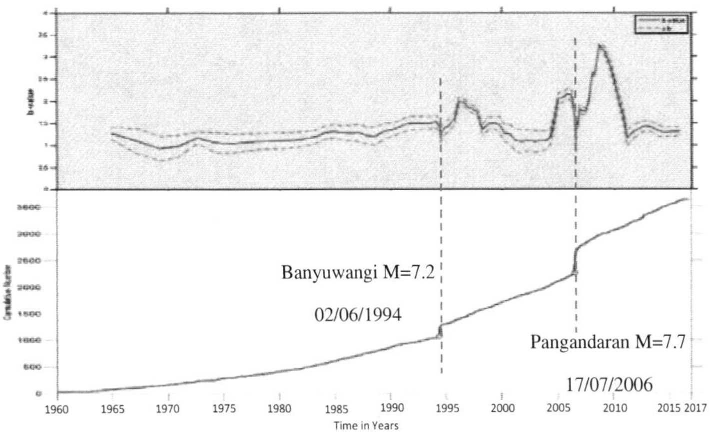

Figure 5. B-value temporal variation of Java region, the yellow star indicates mainshock with M > 7 and the red line indicates the relation of mainshock with b-value declination.

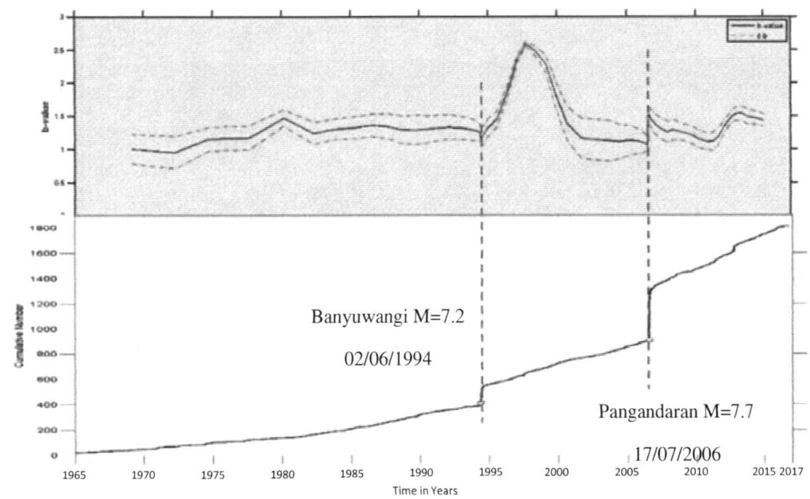

Figure 6. B-value temporal variation of Java megathrust zone, the yellow star indicates mainshock with M > 7 and the red line indicates the relation of mainshock with b-value declination.

The possibility of recurrence of large earthquakes was observed based on the distribution of the mainshock earthquake in the megathrust zone of Java. The analysis was carried out for earthquakes with a magnitude greater than 6.5; 7; and 7.5. The earthquake recurrence time map has a shorter time distribution in the west and east regions, while for the middle region the recurrence time tends to be lower (Figure 7). From the earthquake recurrence time map, it can be seen that the western and eastern regions of Java Island need to get more attention in earthquake mitigation.

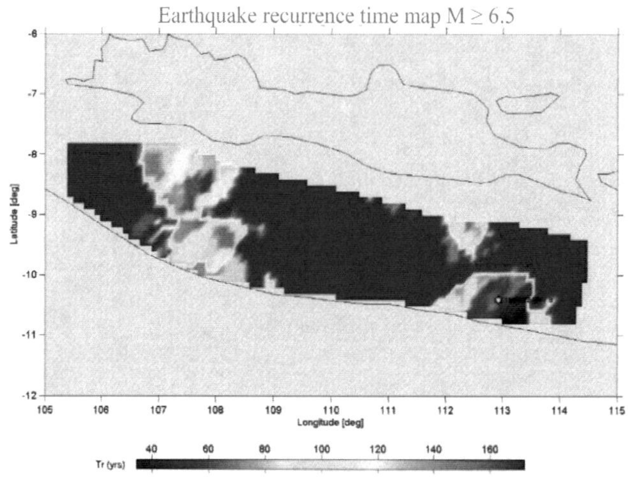

(a)

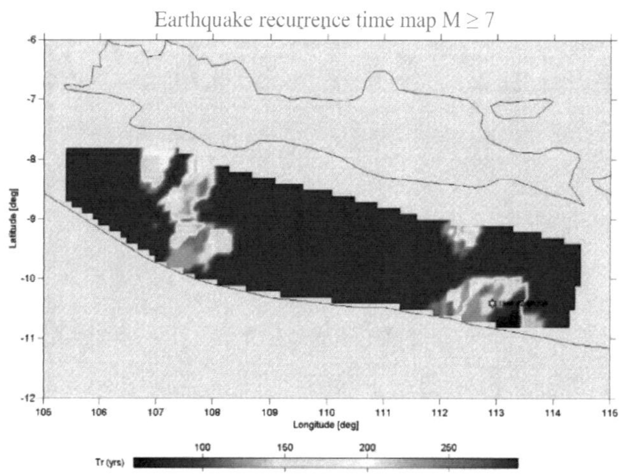

(b)

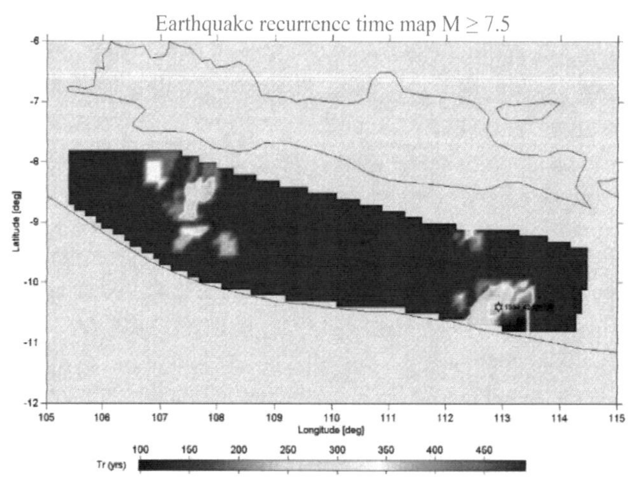

(c)

Figure 7. Earthquake recurrence time map in Java megathrust zone (a) M≥6.5; (b) M≥7.0; and (c) M≥7.5. Yellow star indicates mainshock with M > 7.

4. Conclusion

Based on the results and analysis that has been done, it can be concluded that the average b-value for Java region is 1.36±0.14 and the b-value for Java megathrust zone is 1.19±0.2. The large difference in the b-value of the whole Java region with the b-value of the megathrust zone of Java can be caused by differences in stress.

The spatial variation map shows the b-values on the grids of the study area ranging from 0.6 to 2.6, areas with low b-values are in the western and eastern parts of Java region. Analysis of temporal variation of b-value shows that large earthquakes usually occur preceded by a significant decrease in b-value. Temporal variation of b-value can be observed periodically and can be used as a precursor to a large earthquake [9].

Based on the earthquake recurrence time maps, areas with short recurrence times are around the western and eastern parts of Java Island, so that these areas need more attention for the benefit of mitigation and disaster risk reduction due to earthquakes.

5. Acknowledgement

We are grateful to The National Centre for Earthquake Studies (PusGeN) for the catalog data. The maps and figures were generated with ZMap Software.

References

[1] Badan Pusat Statistik 2020
[2] C R Allen, P Amand, C Richter, and J Nordquist 1965 *Bull. Seismol. Soc. Am.* **55** 753-797
[3] A Mignan and J Woessner 2012 *Community Online Resource for Statistical Seismicity Analysis*
[4] H E A Hafiez 2015 *Arabian Journal of Geosciences* **8(11)** 9315–9323
[5] D Ngadmanto 2010 *Center of Research and Development Meteorological, Climatological, Geophysical Agency (BMKG)*
[6] A H Satyana 2006 *American Association Petroleum Geologists, International Conference and Exhibition* 90061
[7] T Yeken 2016 *Op. Geosc.* **8(1)**
[8] S Rohadi, G Hendra, A R Mezak 2007 *Jurnal Meteorologi dan Geofisika* **8(1)** 42-47
[9] P Nuannin and O Kulhánek 2012 *J. of Earth Science and Engineering* **2** 166-188
[10] C Ghita, M Diaconescu, I A Moldovan, E Oros, and E G Constantinescu 2020 *Romanian Reports in Physics* **72(1)** 704
[11] W Hammilton 1979 *U.S. Geological Survey Professional Paper* 1078
[12] M Irsyam, P R Cummins, M Asrurifak, L Faizal, D H Natawidjaja, S Widiyantoro, … A J Syahbana 2020 *Earthquake Spectra* 1-25
[13] F Kebede and O Kulhánek 1994 *Physics of the Earth and Planetary Interiors* **83(3-4)** 249–264
[14] L Mizrahi, S Nandan, and S Wiemer 2021 *Seismological Research Letters* **92(4)** 2333–2342
[15] Nadiva, Fahira, Ernandi, dan Madlazim 2020 *Jurnal Inovasi Fisika Indonesia* **9(3)** 24-30
[16] K R Newcomb and W R McCann 1987 *J. Geophys. Res.* **92(B1)** 421
[17] P Nuannin 2006 *Digital Comprehensive Summaries of Uppsala Dissertations from the Faculty of Science and Technology* 183
[18] PuSGeN (Pusat Studi Gempa Nasional) 2017 *Pusat Penelitian dan Pengembangan Perumahan dan Permukiman Badan Penelitian dan Pengembangan Kementerian Pekerjaan Umum dan Perumahan Rakyat*
[19] P A Rydelek and I S Sacks 1989) *Nature* **337(6204)** 251–253
[20] D Schorlemmer, S Wiemer and M Wyss 2005 *Nature* **437(7058)** 539–542
[21] Supartoyo, Surono, E T Putranto 2014 *Departemen Energi dan Sumber Daya Mineral, Badan Geologi, Pusat Vulkanologi dan Mitigasi Bencana Geologi*
[22] S Wiemer and M Wyss *J. Geophys. Res.* **102(B7)** 15115–15128
[23] D Yu 2013 *Massachusetts Institute of Technology. Department of Earth, Atmospheric, and Planetary Sciences*

Seismic Hazard Potential in Yogyakarta Based on HVSR Curve Estimation

Zega, B.N.[1], Zulfakriza, Z.[2,3], Rosalia, S.[2,3], Puspito, N.T.[2,3]

[1]Geophysical Engineer Undergraduate Program, Faculty of Mining and Petroleum Engineering, Bandung Institute of Technology, Bandung, Indonesia
[2]Global Geophysics Research Group, Faculty of Mining and Petroleum Engineering, Bandung Institute of Technology, Bandung, Indonesia
[3]Center for Earthquake Science and Technology, Bandung Institute of Technology, Bandung, Indonesia

*E-mail: fttmbril22@gmail.com

Abstract. In 2006, the Special Region of Yogyakarta was shaken by a destructive earthquake. The United States Geological Survey (USGS) recorded that the earthquake had a magnitude of 6.3 M_w at a depth of 12.5 km below the surface that was triggered by strike slip fault activity. The severe damage occurred in Bantul district and Klaten district, where the Klaten area location were far from the earthquake epicenter. This proves that the magnitude and the earthquake source distance are not the only parameters for seismic hazard potential, but rather the presence of local site effects and building conditions. The present study aims to determine the characteristics of the soil and estimate the seismic hazard potential by using HVSR method. Horizontal to Vertical Spectral Ratio (HVSR) is one method that can be used to obtain the subsurface information from single station measurements. Furthermore, the inversion of Rayleigh wave ellipticity curve is used to obtain 1-D of shear wave velocity (Vs). The parameters from HVSR calculation and Rayleigh wave inversion were mapped to understand the subsurface structure beneath the Yogyakarta area. Based on the soil classification to Vs, Yogyakarta area is categorized to SD (Stiff soil/soft soil) and SC (Very dense soil and soft rock). The results of the mapping analysis indicated that Bantul district is an area with the highest potential of seismic hazard.

1. Introduction

The tectonics of Indonesia is located between several tectonic plates, that is Eurasia, Australia, Pacific, and Philippine Sea plates [1]. Consequently, these conditions cause Indonesia to get activity from earthquakes and volcanism frequently, and also has a rough topography. [2]. In the last two decades, several destructive earthquakes have been recorded with significant magnitudes, one of them is the Yogyakarta earthquake in 2006. The earthquake causes 4,772 fatalities, 17,772 injured, and 204,831 buildings destroyed [3].

Damage of a building from seismic waves propagation can be influenced by the soil properties and the building conditions [4]. Microtremor is one method that can be used to find out the soil properties and identify the potential effect caused by earthquake waves propagation [5]. The microtremor data can be analyzed by Horizontal to Vertical Spectral Ratio (HVSR) method to provide several parameters as a reference in earthquake hazard level in an area [6]. Several parameters obtained from the calculation of the HVSR curveare frequency dominant, amplification factor, and Vs^{30}. This research was conducted to analyze the microtremor data of the 15 GeoForschungsZentrum (GFZ) stations network that includes almost all areas of Yogyakarta by using the HVSR method.

2. Geological Setting

As shown in Figure 1, the Yogyakarta area is located of the southern coast of Java and close to the Australian plate boundary which was affected by the uplift of the Southern Mountain and the Kulon Progo mountains during the early Pleistocene [2,7]. As a consequence, there are a series of volcanoes along the island of Java including Mount Merapi, and became an area affected by tectonic activity.

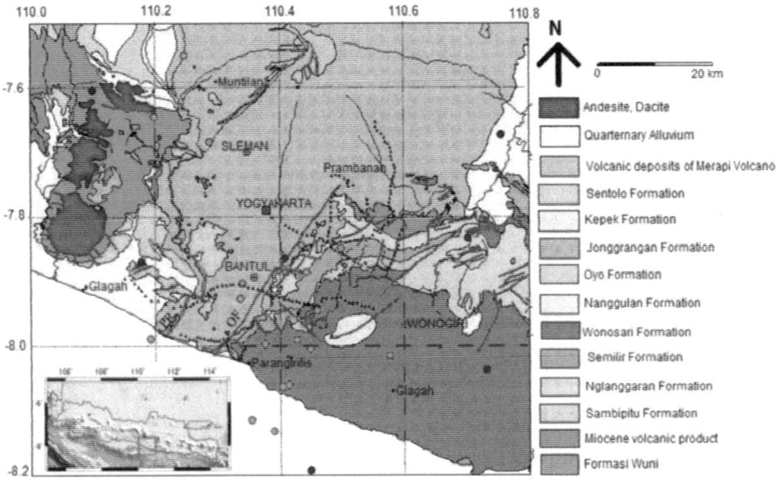

Figure 1. Geological Map of Yogyakarta area. The red line represents the Opak fault line [5] and the blue and green dots represent the location of epicenter [8].

Morphologically, the Yogyakarta area is divided into several major morphological units. First, the Kulonprogo and Gunungkidul regency from west Yogyakarta are mountains formed in the Tertiary. Second, The Sleman regency from north Yogyakarta was formed by sedimentation from the eruptions of Mount Merapi and the soil profile is mainly composed of coarse sand and gravel. And the last, the Kulonprogo regency from south Yogyakarta is lowland area where the soils structure were formed during the Quaternary period. The soil consists of several young and thick deposits and is crossed by an opaque fault line, which have a high seismic hazard potential [9].

3. Data and Method

3.1 Data

The data is used from the GeoForcshungsZentrum (GFZ) network by using 15 single station broad-band seismometers (Figure 2) distributed in the Yogyakarta area after the 2006 earthquake occurred. The stations are recorded full day from June to August 2006. The digital recorder of the seismometer was connected to Global Positioning System (GPS) antenna, making it possible to receive daily coordinates and time. Data recording results in the form of the time series waveform in miniSEED format (Standard Exchange of Earthquake Data). The data from GFZ network in the form of a waveform in the time domain that consist of three main components, such as: vertical components, North-South (NS) horizontal components, and East-Weast (EW) horizontal components with a measurement duration of 1 hour (Figure 3A).

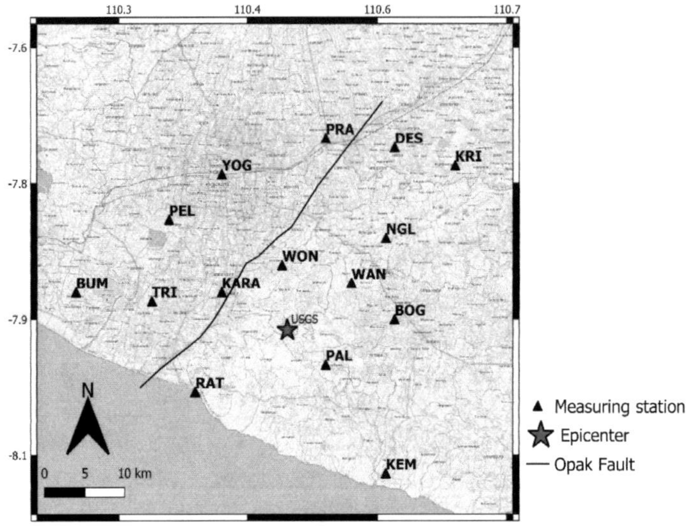

Figure 2. Measuring station distribution in the Yogyakarta area.

3.2 Method

3.2.1 Horizontal to Vertical Spectral Ratio (HVSR)

The fundamental concept of this method is to make a comparison between the spectrum of the horizontal component and the vertical component of a wave [5]. In this study, we used Geopsy software to calculate the spectrum of the measuring station components. Furthermore, filtering of non-stationary signals is carried out by applying the short-time-average through the long-time-average (STA/LTA) algorithm. The filtering needs to be done to reduce noise caused by interference during recording data acquisition. The signal recording from the time domain is converted into frequency domain by the Fast Fourier Transform (FFT) process with a frequency range of 0.5-20 Hz and 60 s timewindow (Figure 3B). The 3 components of the signals can be determined by:

$$A_H = \sqrt{\frac{A_{NS}^2 + A_{EW}^2}{2A_V^2}} \tag{1}$$

where A_{NS}, A_{EW}, and A_V are the components of the Fourier amplitude spectra in the North-South, East-West, and Vertical, respectively.

The output from this processing is an H/V curve that provides the value of dominant frequency and amplification factor. The dominant frequency is estimate as follow:

$$f_0 = \frac{v_s}{4h} \tag{2}$$

and amplification factor is:

$$A_0 = \frac{v_b}{v_s} \tag{3}$$

where the f_0, v_s, h, and v_b are the frequency dominant, shear wave velocity on the surface layer, thickness, and shear wave velocity of basement, respectively [10]

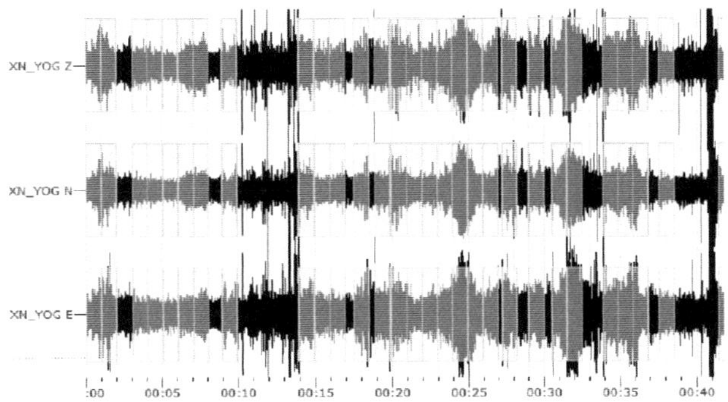

Figure 3. Waveform of YOG station with 3 components that are selected by time window for 1 hour.

3.2.2 Inversion Process

The inversion process is used to transform the measurement data into model parameters. This method assumes that the microtremor data and the H/V curve are dominated by surface waves called the elliptical Rayleigh waves [11]. The ellipticity of Rayleigh waves is inversed to investigate shallow subsurface structure by using DinverPackage from Geopsy in dinver format to obtain the minimal misfit equation. The misfit equation is estimated based on Neighbourhood algorithm as follow:

$$misfit = \sqrt{\frac{1}{N}\sum_{i=1}^{N}\left(\frac{Di-Mi}{\sigma_i}\right)^2} \tag{4}$$

where the N, Di, Mi, and σi are the number of data, inversion data result, observation data result, and standard deviation [12].

According to Herak (2008) there are six parameters that affecting the HVSR curve, such as: compression wave velocity (v_p), shear wave velocity (v_s), thickness (h), density (ρ), and attenuation factors (Qp, Qs [13]. Therefore, to find out the level of influence of the six parameters, an inversion process was performed on HVSR curve. In this study, we define the model parameters initial trial and error by correlating parameters to conditions geology of the research area. This is concluded in order to obtain a comparison between the observation data and the calculated data as small as possible.

Table 1. Model parameters used in the inversion process.

Layers	v_p(m/s)	v_s(m/s)	Depth (m)	Poisson's Ratio	ρ (kg/m^3)
1	100-500	150-500	100		
2	300-1000	150-750	20-100 (Thickness)		
3	600-1500	500-1000	20-100 (Thickness)	0,2-0,5	2000
4	900-2000	500-1250	30-100 (Thickness)		
5	1200-2500	500-1500	30-200 (Thickness)		
6	1500-3000	1000-1750	400 (half space)		

4. Results and Discussion

4.1 H/V Curve

The results of the H/V calculation from all measuring stations obtained three types of H/V curves which are characterized based on the number of the peaks. The three types of curves are: clear peak curve, multiplepeaks curve, and flat curve. These three types of curves are interpreted based on [14]. As shown in Figure 5, the shape of the peak on the H/V curve indicates presence or absence of impedance contrast in the subsurface layer. The clear peak curve (a) indicates the presence of impedance contrast, the multiple peaks curve (b) indicates the presence of impedance contrast with the boundary between new layer and old layer, and the flat curve (c) indicates the absence of impedance contrast in the subsurface layer.

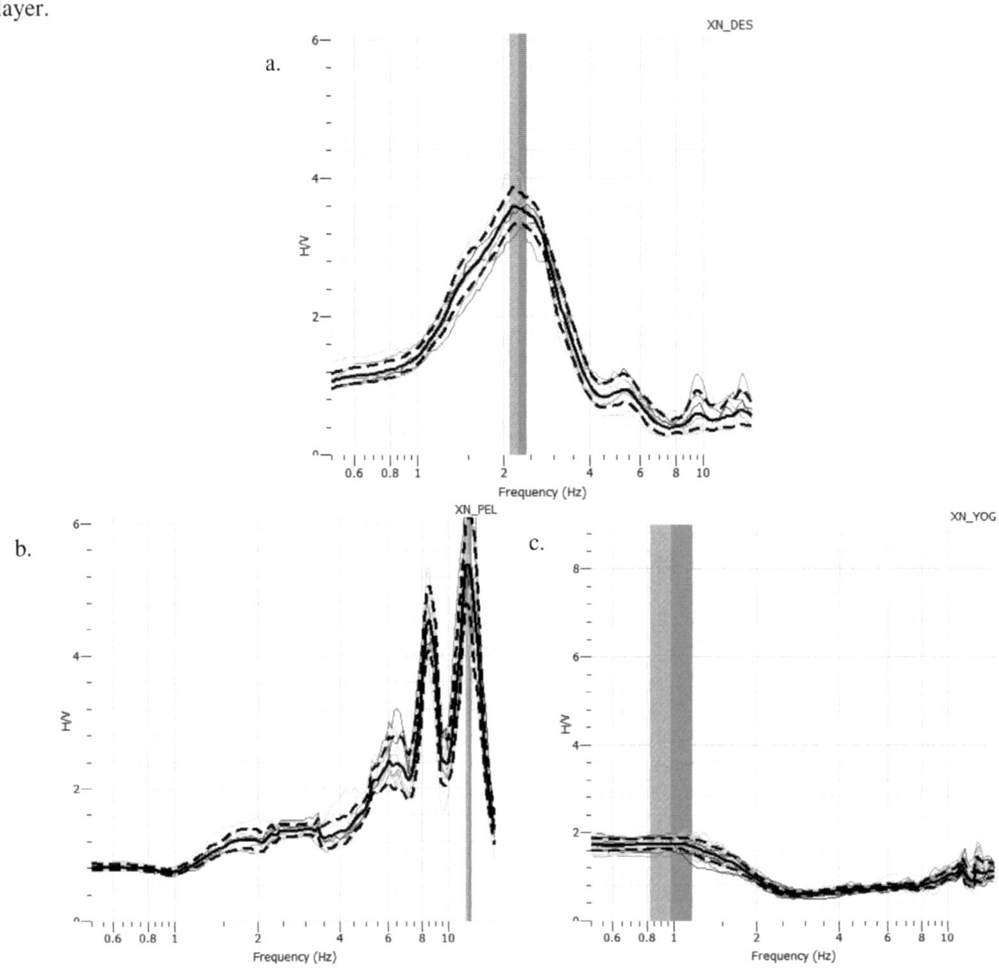

Figure 4. Three types of H/V curves, a. Clear peak curve, b. Multiple peaks curve, c. Flat curve. Every spectrum color shows window selection result, black line shows the mean of the HVSR value, and the black dotted line shows the lower and upper standard deviations.

4.2 Frequency Dominant and Amplification Factor

The result of the H/V calculation have a dominant frequency ranges from 1.05 – 13.7 Hz and amplification factor ranges from 1.17 – 7. Figure 5,6 show the distribution of dominant frequency and amplification factor for 15 measuring station. The dominant frequency is associated with the thickness of the sediment [10]. The areas with low dominant frequency indicate that has relatively deep bedrock depths, on the other hand, the areas that have a high dominant frequency indicate that has relatively thin bedrock depths.

Amplification factor is the ratio of the surface to subsurface layer amplitude [14]. Inother words, the amplification factor value depends on the geometry of the sediment structure. The Yogyakarta area is dominated by amplification factor ranges from 1 – 3.3, but around the PEL station, the value of the amplification factor is more than 6.3. The highest value of the amplification factor is due to PEL station located at Sentolo Formation (Bantul district), which is composed of several different layers [16]. The values of dominant frequency and amplification factor indicate Bantul is a vulnerable area of seismic hazard.

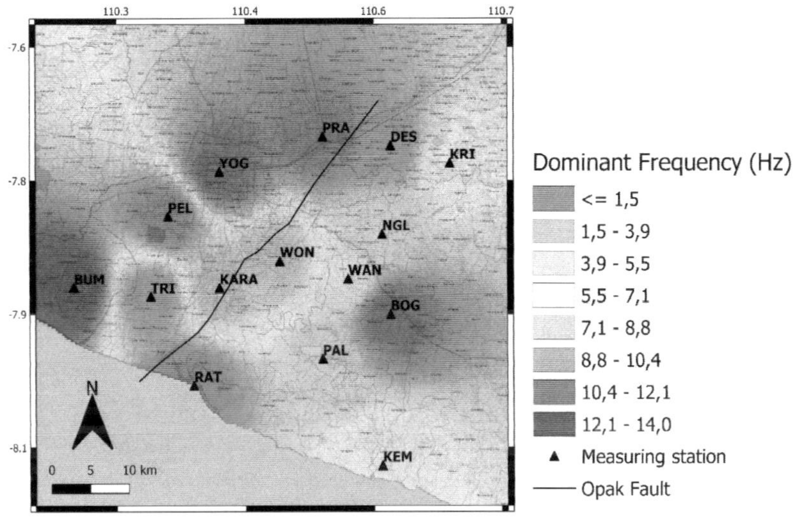

Figure 5. Dominant frequency distribution map of Yogyakarta area

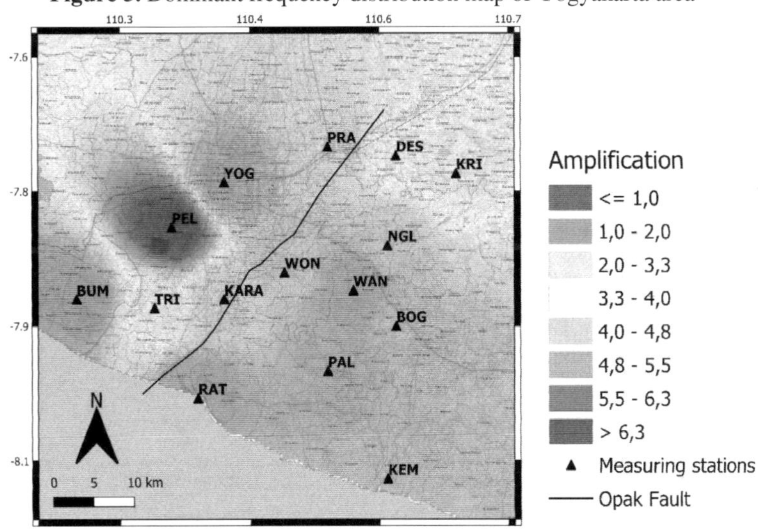

Figure 6. Amplification distribution map of Yogyakarta area

4.3 Ground Profile and Vs^{30}

After the specific parameter model is set, the ellipticity curve and ground profile can be displayed to approach the shear wave velocity in each layer. According to Table 1, the inversion process obtained the estimation model of the ground profile and the ellipticity curve in 1D. The color spectrum of the curve is the result of the estimation model of inversion process for each value of misfit. The red line is the inversion model that has the minimum misfit value. As an inversion concept, the inversion process is worked by iterating the initial model to the best model obtained with the smallest misfit. This result produces a minimum misfit value of 0.38 and maximum misfit value of 0.5 shown by Figure 7.

After the Vs value is obtained from inversion process, then the average value of Vs is calculated at a depth of 30 meters below the ground surface as in equation (2). This result provides the average of Vs^{30} value between 240 – 600 m/s as shown in Figure 8. Based on the value, the Yogyakarta area can be classified according to the type of soil as shown in table 2. Yogyakarta area is categorized into soil profile types SC and SD. However, the value of Vs^{30} is not enough to classify the soil profile types in an area, it is caused the structure of the sediment is also influenced by the value of dominant frequency [10].

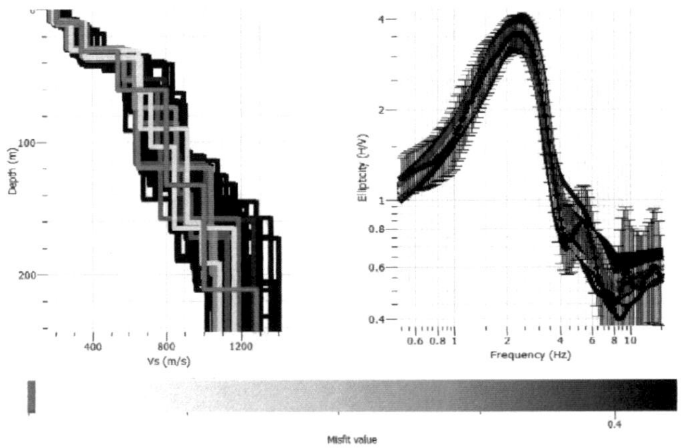

Figure 7. 1D ground profile (left) and ellipticity curve (right)

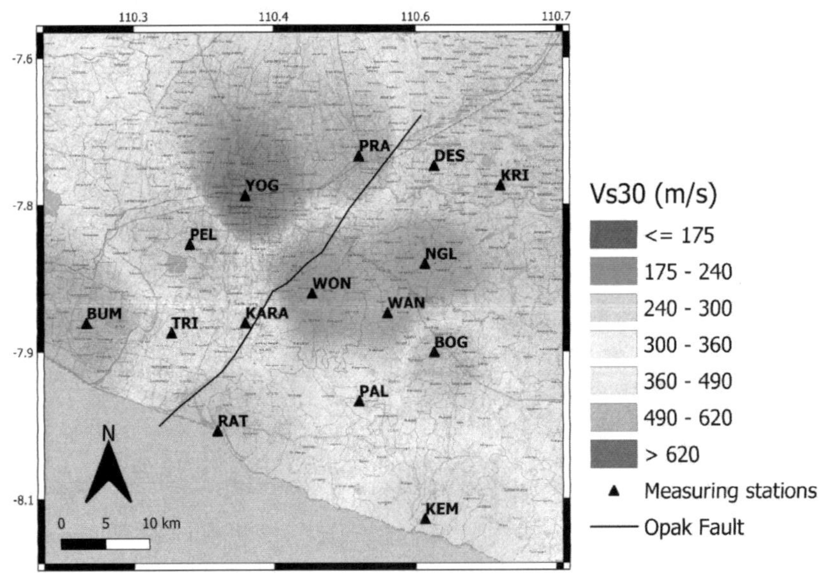

Figure 8. Vs^{30} distribution map of Yogyakarta area

Table 2. Site classification based on Vs^{30} according to SNI 1726:2012

Soil profile type	Description	Vs^{30}	Station
SA	Hard rock	> 1.500 m/s	
SB	Rock	750 – 1.500 m/s	
SC	Very dense soil and soft rock	350 – 750 m/s	DES, KEM, KRI, NGL, PAL, PRA, WAN, BUM, WON
SD	Stiff soil	175 – 350 m/s	BOG, YOG, TRI, RAT PEL, KARA
SE	Soil	< 175 m/s	

5. Conclusion

The result of calculating 3 signal components in each station produced the 3 types of H/V curves based on number of the peaks. The H/V curves obtained the value of dominant frequency and amplification factor. The inversion process from the ellipticity curve of Rayleigh waves produced the value of Vs^{30} that categorized the Yogyakarta area to SD or Stiff soil, and SC or Very dense soil and Soft rock. Those values can indicate the seismic hazard potential in Yogyakarta area. Bantul is an area with a high potential of seismic hazard. This is because the Bantul area is dominated by clear peak and multiple peaks H/V curves and high dominant frequency with the highest amplification factor.

6. References

[1] PUSGEN. 2017. *Peta Sumber dan Bahaya Gempa Indonesia Tahun 2017*. Badan Penelitian dan Pengembangan Kementrian Pekerjaan Umum dan Perumahan Rakyat, ISBN 978-602-5489-01-3.

[2] Hamilton, W. 1979. *Tectonics of the Indonesian region*. U.S. Geol. Surv. Prof. Pap. 1078, U. S. Geol. Surv., Boulder, Colo.

[3] BMKG. 2019. *Katalog Gempa bumi Signifikan dan Merusak 1821-2009*. Sub Bidang Mitigasi Gempa bumi, Jakarta.

[4] Gosar A. 2009. *A microtremor HVSR study of the seismic site effects in the area of the town of Brežice (SE Slovenia)*. Acta Geotech Slov. 2 pp 31–45.

[5] Nakamura, Y. 2008. *On the H/V spectrum. The 14th World Conference on Earthquake Engineering* in Beijing, China.

[6] Nakamura, Y. 1989. *A method for dynamic characteristics of estimation of subsurface using microtremor on the ground. Q. Rep. RTRI,* **30**(1), 25–33.

[7] Rahardjo dkk., 1995, *Peta Geologi Lembar Yogyakarta*. Jawa, Pusat Penelitian dan Pengembangan Geologi, Bandung.

[8] J. Benfield and NEIC. *Significant Earthquake of the World*. USGS webpages, at: http://earthquake.usgs.gov/earthquakes/eqarchives/significant.

[9] Hartantyo, E. 2018. *Comparison of 10% and 2/3 of 2% PE for 50 Years Seismic Hazard at Yogyakarta Special Province (YSP), Indonesia Constructed From the Probabilistic Seismic Hazard Analysis*. IJCIET, Vol. 9, No. 09. pp. 1596-1597.

[10] Nakamura, Y. 2008. *On the H/V spectrum*. Proceedings of the 14th World Conference on Earthquake Engineering (14WCEE), Beijing, China.

[11] Ullah, I. 2017. *Near-Surface Characterization from the H/V Spectral Curves along with the Joint Inversion of the Ellipticity and Dispersion Curves*. SÃO PAULO, hal. 195.

[12] Sambridge, M. 1999. *Geophysical inversion with a neighbourhood algorithm-I. Searching a parameter space*. Geophysical Journal International, 138(2), 479–494.

[13] Herak, M. 2008. *ModelHVSR: a Matlab tool to model Horizontal-to-Vertical Spectral Ratio of Ambient Noise*. Computers and Geoscience, 34, 1514-1526.

[14] SESAME. 2004. *Guidelines for The Implementation Of The H/V Spectral Ratio Technique on Ambient Vibrations*. Europe: SESAME European research project.

[15] Nakamura, Y. 2000. *Clear identification of fundamental idea of Nakamura's technique and its Applications*. Proc. of the 12th World Congresss on Earthquake Engineering, Aucklland,

New Zealand.
[16] Van Bemmelen, R.W. 1949. *The Geology of Indonesia.* Vol. I.A. General Geology. Martinus Nyhoff, The Hague.

Acknowledgment
The research was supported by Global Geophysics Research Group of Bandung Institute of Technology, Bandung, Indonesia.

Numerical Modelling of Wave Attenuation in Pile Breakwater

F.A.R. Abdullah[1], T. Suprijo[1], H. Diastomo[1], G. Napitupulu[2]

[1] Oceanography Research Group, Faculty of Earth Sciences and Technology, Institut Teknologi Bandung, Jalan Ganesha 10, Kota Bandung, Jawa Barat, 40132 Indonesia

[2] Oceanography Study Program, Faculty of Earth Sciences and Technology, Institut Teknologi Bandung, Jalan Ganesha 10, Kota Bandung, Jawa Barat, 40132 Indonesia

email: far.abdullah@itb.ac.id

Abstract. Pile breakwater is an artificial coastal structure consisting of closely spaced piles. Its benefits are similar to other coastal structures by attenuating the energy of the incoming wave. In this paper, the wave attenuation caused by the pile structure is investigated using a numerical model, namely the SWASH model. The model allows for simulating non-hydrostatic phenomena in coastal waters as driven by waves. The model setup is conducted in a rectangular wave flume with paddles to produce monochromatic waves on the left and the wave absorber on the other side. Several scenarios were carried out to determine the effect of variations in wave parameters. To deal with the efficiency of numerically modelled pile breakwater, we also investigate the transmission coefficient. We compare the result of the numerical model with experimental data, which took place in a wave flume at Balai Penelitian dan Pengembangan Pantai, Buleleng, Bali, Indonesia. The validation results demonstrate the capability of the SWASH Model to approximate the incident and transmitted wave height with good accuracy compared to experimental data.

1. Introduction

Coastal abrasion is a crucial problem faced by coastal communities today. In general, this phenomenon is triggered by high sea waves that occur in certain seasons. Ocean waves are a phenomenon of vertical movement of water masses due to the transfer of momentum. Generally, waves in the sea are generated by the wind blowing over the sea level [1]. Waves are generated in the area affected by the wind. The transfer of energy from the wind to the sea surface due to wind gusts causes irregular ripples called seas [2]. The waves then propagate out of the generating area and towards the coast. Characteristics of waves that have come out of the generating area have a more regular waveform called a swell [3]. Waves that enter shallow water will undergo a transformation due to changes in depth. Effects that affect waves as they enter shallow water include shoaling, refraction, diffraction, and breaking waves. Shoaling is a phenomenon where waves experience changes in height, length, and speed of wave propagation. Refraction is the phenomenon of bending the direction of waves due to changes in bathymetry. Diffraction is a change in the direction of propagation of a wave when it hits a structure and propagates towards the back of the structure. Breaking waves occur where the steepness of the waves exceeds the speed of wave propagation so that water will spill towards the front of the wave crest.

The development of coastal protection is necessary to prevent more severe abrasion and nurture the coastal ecosystem. Several solutions protect the coastal area, including nature-based coastal protection and hard engineering structure as a breakwater. The capability of nature-based coastal protection has been investigated in several reports [4, 5, 6]. However, it takes a long time to prepare nature to protect the coast. One alternative solution is to build hard structures in the coastal area. This paper investigated an alternative breakwater structure using a vertical pile structure.

Pile foundation is part of the structure used to receive and transfer (transmit) the load from the superstructure to the supporting soil located at a certain depth. Long piles transmit the load to deeper soil. Pile breakwater is a type of perforated breakwater. This breakwater was developed from perforated wall Silencer Jarlan and Wiegel pile breakwater. This type is an alternative to conventional breakwaters. According to [7], pile breakwaters have several advantages, including being made of low-cost industrial materials such as pipes, easy to install and remove, having an environmental impact compared to conventional rubble-mound breakwaters, and not interfering with beach aesthetics and water quality. However, much research on the behavior of waves around breakwaters, especially those of the pile type, has not been done. Therefore, it is hoped that the wave behavior around the pile-type breakwater will be comprehensively known through this research.

2. Methodology

2.1. Numerical model

SWASH is numerical tool developed by TU Delft. It can be used to simulate unsteady, free-surface, non-hydrostatic, rotational flow in coastal area which is driven by waves and wind forces. The governing equations are described by nonlinear shallow water equations as consist of conservation of mass and momentum [8]. The equations are given by

$$\frac{\partial \zeta}{\partial t} + \frac{\partial hu}{\partial x} + \frac{\partial hv}{\partial y} = 0 \quad (1)$$

$$\frac{\partial u}{\partial t} + u\frac{\partial u}{\partial x} + v\frac{\partial u}{\partial y} + g\frac{\partial \zeta}{\partial x} + \frac{1}{h}\int_{-d}^{\zeta} \frac{\partial q}{\partial x} dz + c_f \frac{u\sqrt{u^2+v^2}}{h} = \frac{1}{h}\left(\frac{\partial h\tau_{xx}}{\partial x} + \frac{\partial h\tau_{xy}}{\partial y}\right) \quad (2)$$

$$\frac{\partial v}{\partial t} + u\frac{\partial v}{\partial x} + v\frac{\partial v}{\partial y} + g\frac{\partial \zeta}{\partial y} + \frac{1}{h}\int_{-d}^{\zeta} \frac{\partial q}{\partial y} dz + c_f \frac{v\sqrt{u^2+v^2}}{h} = \frac{1}{h}\left(\frac{\partial h\tau_{yy}}{\partial x} + \frac{\partial h\tau_{yx}}{\partial y}\right) \quad (3)$$

Where x, y represent horizontal coordinates, t is time, $\zeta(x, y, t)$ is surface elevation measured from undisturbed water level, $d(x, y)$ is still water depth, $h = \zeta + d$ the total water depth, $u(x, y, t)$ and $v(x, y, t)$ are horizontal components of water velocity in x and y directions, respectively. $q(x, y, t)$ is the non-hydrostatic pressure, g is gravitational acceleration, c_f is dimensionless bottom friction, and $\tau_{xx}, \tau_{xy}, \tau_{yy}$, and τ_{yx} are the horizontal turbulent stress terms. In addition, SWASH implements a finite difference method on staggered grid which has second order accuracy [8,9].

2.2. Model setting

The pile breakwaters are modelled as an array of vertical solid cylinders with 100 centimeters height and the diameter of 0.06 meters, and the set-up of the piles are accordance with the experimental set-up (Figure 1). To validate the numerical results, we compare our results with the experimental results which carried out at Balai Penelitian dan Pengembangan Pantai, Buleleng, Bali.

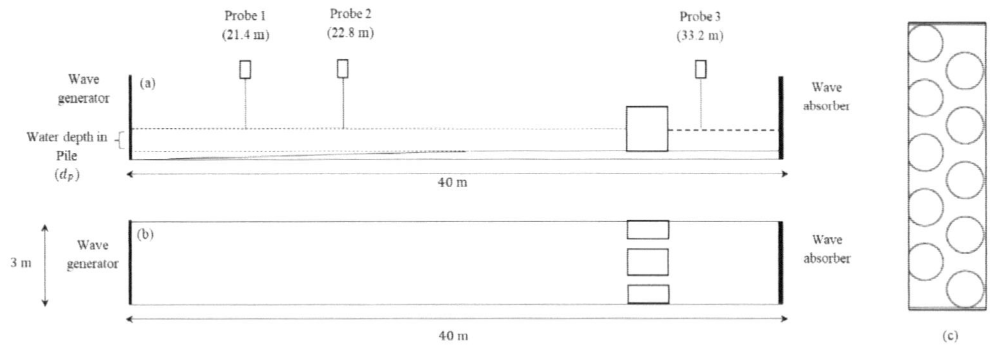

Figure 1. Experimental set-up (a) cross section of the wave flume, (b) top view (c) pile breakwater layer

The physical experiments were conducted in a wave flume of 40 meters and 3 meters wide. The wave flume has a wave generator on the left side channel and a wave absorber on the other side. The three probes were placed along the wave flume to record the wave signal simultaneously. Probe 1, Probe 2, and Probe 3 were located 21.4 meters, 22.8 meters, and 33.2 meters from the wave generator, respectively. Water depth in pile is 0.6 meters. The piles are arranged in staggered with certain gap. We used numerical horizontal grid size of 0.1 meters and with time step of 0.04 seconds. Several model scenarios were performed to investigate the effect of wave parameters on wave energy reduction, it can be seen in Table 1.

Table 1. Model Scenario with Variation of Wave Height and Wave Period

Scenario	*H* (meters)	*T* (seconds)
SN01	0.2	2
SN02	0.2	2.5
SN03	0.2	3
SN04	0.15	2
SN05	0.15	2.5
SN06	0.15	3

3. Result and discussion

In order to test and validate the SWASH model, we compare numerical results and experimental data in several scenarios, see Figure 2. We used Root Mean Squared Error (RMSE) and Correlation Coefficient (CC) to evaluate the model performance.

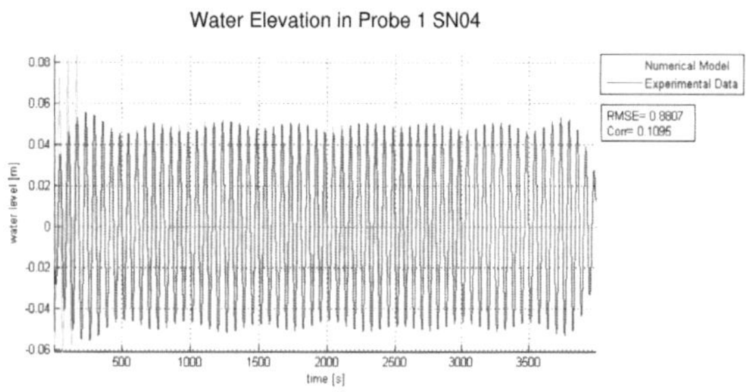

(a)

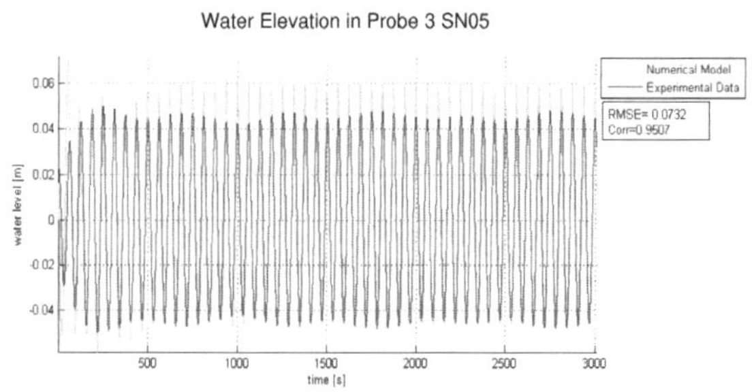

(b)

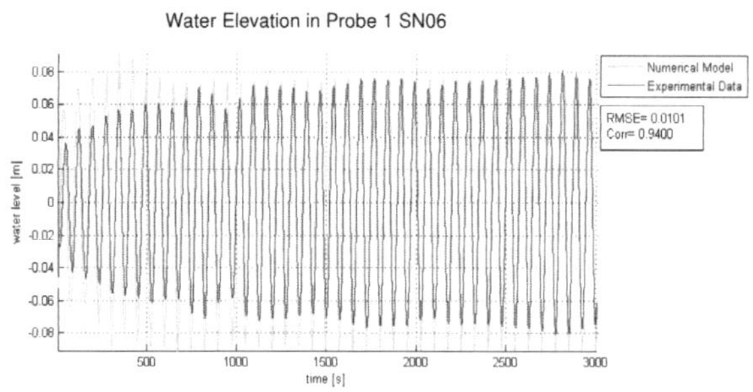

(c)

Figure 2. Comparison of water elevation (numerical result and experimental data) in (a) Probe 1 SN04 (b) Probe 3 SN05, and (c) Probe 1 SN06.

Figure 2 shows the overlayed time series of water elevation from three scenarios of SWASH model results and experimental data. In general, the water level from SWASH model results was able to simulate experimental, indicated by the high correlation and low RMSE values. Even though the SWASH model results tend to overestimate the experimental data in terms of wave amplitude. The error is expected due to the wave reflection phenomenon which is not considered in the numerical model. Table 2 gives detailed information about the correlation and RMSE for each scenario of the numerical model and experimental data.

Table 2. Validation of numerical model with experimental data for all scenarios

Scenario	Probe	RMSE	CC	Scenario	Probe	RMSE	CC
SN01	1	**0.052**	0.891	SN04	1	0.109	0.880
	2	0.073	**0.900**		2	**0.036**	**0.958**
	6	0.187	0.812		6	0.116	0.863
SN02	1	0.211	**0.969**	SN05	1	0.109	0.888
	2	**6.0610e-04**	0.864		2	0.086	**0.989**
	6	0.078	0.955		6	**0.073**	0.950
SN03	1	**0.034**	**0.932**	SN06	1	**0.010**	**0.940**
	2	0.048	0.860		2	0.044	0.901
	6	0.045	0.931		6	0.035	0.923

The result shows that probe 2 in SN05 has the highest value of CC and probe 2 in SN02 has the highest accuracy with accuracy of 6.0610e-04 for RMSE.

3.1. Transmission and reflection coefficient

To investigate the effect of wave parameter variation in wave transmission. We adopted method of Goda and Suzuki [10], to calculate the transmission coefficient. The formula is given by

$$K_t = \frac{H_t}{H_i} \tag{4}$$

Where H_t is transmitted wave height that can be obtained from probe 3 which is located behind the piles and H_i is incident wave height. H_t can be obtained by using zero-up crossing of recorded wave signal in probe 3.

Table 3. Transmission Coefficient for all scenarios

Scenario	K_t (Experimental data)	K_t (Numerical model)	Percentage Error (%)
SN01	0.867	0.909	13.033
SN02	0.813	0.698	14.145
SN03	0.837	0.699	14.576
SN04	0.842	0.898	6.651
SN05	0.218	0.202	6.881
SN06	0.821	0.707	13.764

The comparison between the experiment and the SWASH model result of the transmission coefficient is presented in Table 3. SN05 has the lowest absolute error, 0.056. Meanwhile, the highest absolute error is in SN03 and SN06, which is expected due to the SWASH model limitation on reaching the greater value of the wave period. However, in general, the SWASH model is able to predict transmission coefficient with an acceptable value of percentage error (below 14.57%).

3.2. Influence of wave steepness on wave transmission.

The transmission coefficient (K_t) is the amount of transmitted energy after passing through the piles. Based on the measurement results and calculation, it can be seen the influence of the wave steepness (H_i/gT^2, where T is wave period) to the value of K_t. The value of K_t as a function of H_i/gT^2 are shown in the graphs of Figure 3. Figure 3 shows that the larger value of H_i/gT^2, the greater the value of K_t and vice versa. The value of the transmission coefficient (K_t) in experimental data ranges from 0.8 to 0.85. Meanwhile, in SWASH result K_t varies from 0.7 to 0.9. The values of K_t in both experimental and SWASH result have similar upward trend.

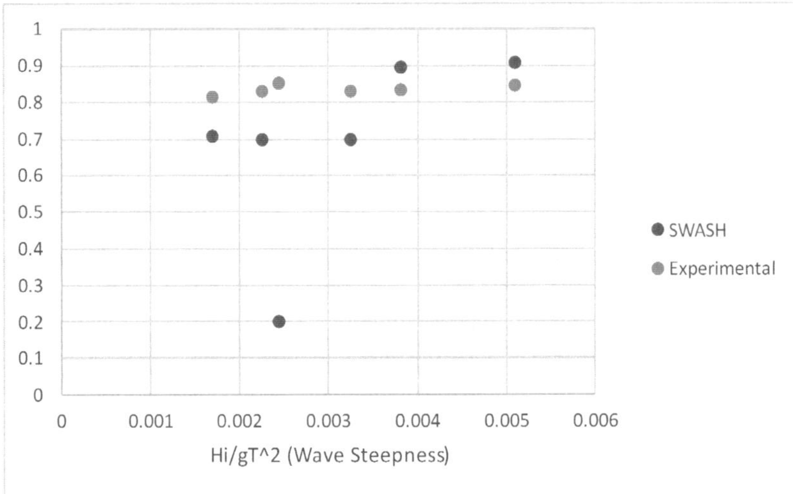

Figure 3. Influence of wave steepness on wave transmission (K_t).

3.3. Influence of wave steepness on wave reflection.
The wave reflection coefficient can be defined as follows:

$$K_r = \frac{H_r}{H_i} \tag{5}$$

Where H_r is reflected wave height and H_i is incident wave height. H_r can be obtained by using zero-up crossing of recorded wave signal in probe 2. The value of K_r as a function of H_i/gT^2 are shown in the graphs of Figure 4. The value of the reflection coefficient (K_r) in experimental data ranges from 0.2 to 0.4. Meanwhile, in SWASH result K_r varies from 0.3 to 0.52.

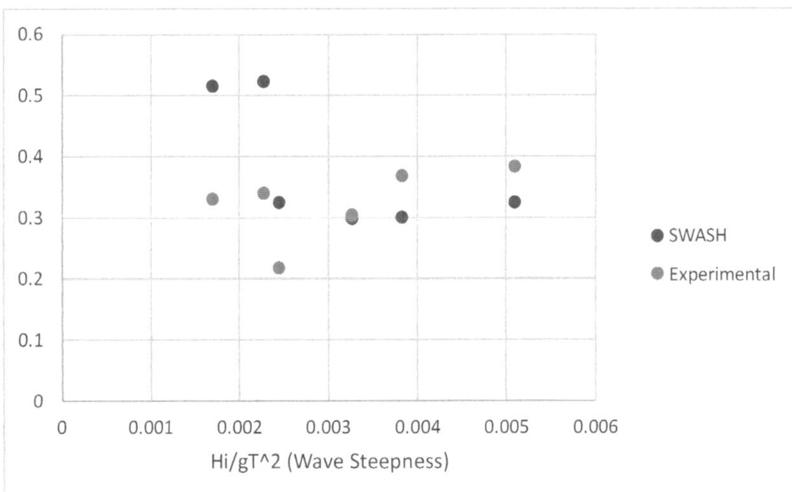

Figure 4. Influence of wave steepness on wave reflection (K_r).

4. Conclusion

The SWASH model has successfully simulated wave attenuation in pile breakwater structures in the present study. The numerical results show good agreement with experimental data carried out in Balai Penelitian dan Pengembangan Pantai, Buleleng, Bali. We also investigate the transmission coefficient to deal with the efficiency of numerically modelled pile breakwater. The SWASH model can predict the transmission coefficient with an average percentage error of 11.5%. The influence of wave steepness with transmission coefficient also shows that the SWASH model has a similar upward trend with the experimental data. Meanwhile, the influence of wave steepness on wave reflection is given different results between the SWASH model and experimental data.

References

[1] Dean, R. G. dan Dalrymple, R. A. 2004. *Coastal Processes with Engineering Applications*. New York: Cambridge University Press

[2] Davidson-Arnott, R. 2010. *An Introduction to Coastal Processes and Geomorphology*. New York: Cambridge University Press

[3] Holthuijsen, L. H. 2007. *Waves in Oceanic and Coastal Waters*. New York: Cambridge University Press

[4] Hu, Z., T. Suzuki, T. Zitman, W. Uittewaal and M. Stive. 2014 "Laboratory study on wave dissipation by vegetation in combined current-wave flow." *Coastal Engineering* 88: 131–142.

[5] Zhang, Y., Y. Yang, K. Yang, X. Tan, X. Sun, B. Leng, C. Zhou and B. Zhu. 2020. "Non-linear Wave Attenuation Quantification Model Improves The Estimation of Wave Attenuation Efficiency of Mangroves." *Estuarine, Coastal and Shelf Science 245.*

[6] Abdullah, F. A. R., S. Husrin and H. Bachtiar. 2019. "Experimental Data Analysis of Wave Attenuation in Mangroves." *Journal of Physics: Conference Series* 1245.

[7] Koftis TH, Prinos PA, Aftias MI. 2012. "Experimental study of a multiple-row pile breakwater." *InProc. of 4th Int. Conf. on the Application of Physical Modelling to Port and Coastal Protection.*

[8] Zijlema, M., Stelling, G., Smit, P., 2011. SWASH: an operational public domain code for simulating wave fields and rapidly varied flows in coastal waters. *Coastal Eng*. 58 (10), 992—1012.

[9] Zijlema, M., Stelling, G.S., 2005. Further experiences with computing non-hydrostatic free-surface flows involving water waves. *Int.J. Numer. Method Fluids* 48 (2), 169—197.

[10] Goda, Y., Suzuki, Y., 1976. Estimation of incident and reflected wave in random wave experiments. *In: Proceedings of 15th Conference on Coastal Engineering, Honolulu, Hawaii, USA*, 828—845.

IOP Conf. Series: Earth and Environmental Science 1047 (2022) 012030 doi:10.1088/1755-1315/1047/1/012030

A review study of landslide analysis in Belimbing road, Pejagoan, Kebumen – Central Java, from hydrogeological aspects

N Mareta[1], S P Saputro[1], S Winduhutomo[1] and R F Lubis[1]

[1] Research Center for Geotechnology, National Research and Innovation Agency (BRIN), Jl. Sangkuriang, Bandung, 40135, Indonesia

E-mail: nandianthea@gmail.com

Abstract. Landslides have occurred on the Belimbing road, Pejagoan, Kebumen – Central Java which has affected 4 houses. The direction of landslide movement is N10°E, with rotational landslide type. The lithology of the landslide location is river deposits that have not been compacted (alluvium). Morphology is plain with steep bordering the Luk Ulo river. The slope bordering the river is 45 – 60° with a shape influenced by lateral erosion of the Luk Ulo river. The result of the analysis of the direction of groundwater flowing around the landslide location showed that there was one peak with the direction of flow towards the north towards Luk Ulo river. When it rains, the infiltrated rainwater will seep and increase of volume of the groundwater. Increasing volume of groundwater will flow in the direction of the flownets and increase the pressure in the direction of the flow of groundwater. The drainage (consisting of five water channels) is located south of the landslide point, namely channel 1 (T1), channel 2 (T2), channel 3 (T3), channel 4 (T4), and channel 5 (T5). There should be a change in the direction of drainage water flow. Changes in the direction of drainage flow can be attempted by closing T4 and only one discharge point, namely at T2.

1. Introduction

Kebumen is one of the regencies in Central Java where landslides frequently occur [1]. Often these events lead to large material losses. Landslide events or mass movements of soil, rock, or combination thereof, usually occur on natural slopes or man-made slopes. Often, this landslide event can be a disaster that causes a lot of harm to property and life. A landslide event will be categorized as a disaster if it occurs in an area inhabited by humans or in place of human activity. So, the aspects of human presence or the influence of human activities are very important in determining whether a landslide (soil, rock, or a mixture of both) is considered a disaster or not.

Karnawati [2] and Guerra et al. [3], explained that the causes of soil/rock movement are divided into controlling factors and triggering factors for soil/rock movements. Controlling factors are factors that make the condition of slope vulnerable or ready to move, including morphological conditions, stratigraphy (rock types and their relationships with other rocks in the vicinity), geological structures, geohydrology, and land use [4]. Movement trigger factors are processes that change of slope from a vulnerable condition or ready to move to a critical condition and finally move. Generally, these processes include rain infiltration, earthquake vibrations, or heavy vehicles/equipment, as well as human activities that result in changes in load or land use on slopes [4-6].

IOP Conf. Series: Earth and Environmental Science 1047 (2022) 012030 doi:10.1088/1755-1315/1047/1/012030

A landslide has occurred beside Belimbing road, damaging 4 houses. The direction of the landslide movement is towards the Luk Ulo River at latitude 7°40'19"S and longitude 109°39'0.6"E. Administratively, the landslide location is in the Pejagoan district, Kebumen regency. Jalan Belimbing is a road that is quite busy for two-wheeled and four-wheeled vehicles. This road is also one of the entrances to housing located in the west. The landslide occurred in July 2020. Initial checks were carried out on Monday 4 January 2021. Landslide management has been handled by the BPBD, Kebumen regency.

1.1. Regional geology and regional hydrogeology landslide site

Based on the Geological Map from [7], the rock in the landslide location is alluvium (Qa), namely river sediment deposits consisting of clay, sand, gravel, and uncompacted gravel, and is the result of deposition of the Luk Ulo river (Figure 1). The alluvium is the youngest rock unit in the Holocene-Recent. Beneath the alluvium is the Halang Formation (Tmph), consisting of calcareous sandstone, marl-claystone, and calcarenite inserts. The Halang Formation is of the Middle Miocene-Pliocene age. The Halang Formation is thought to be a lithological layer under the alluvium or as a basement of the alluvium.

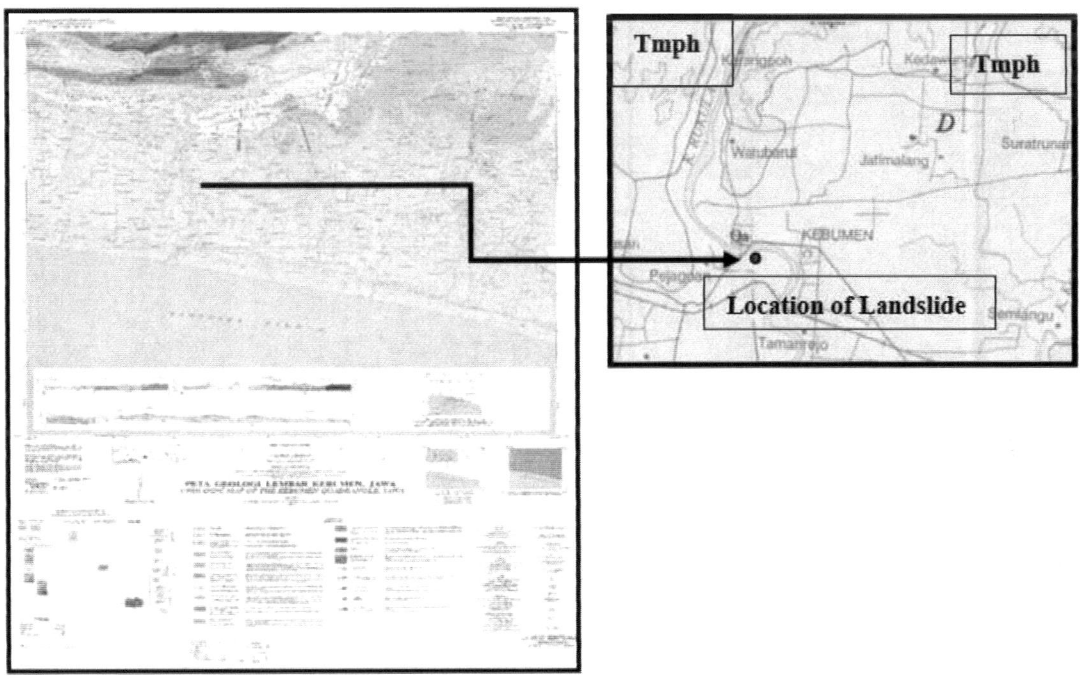

Figure 1. Left, map of geology. Right, inset of landslide location.

Based on the Hydrogeological Map from [8], sheet VI, Pekalongan, scale 1:250.000 compiled by the Directorate of Environmental Geology, the landslide location is included in the medium productivity aquifer with wide distribution, this type of aquifer is an aquifer with medium continuity (Figure 2). Pisometry height or groundwater level above or near the ground level, the well-dug discharge is generally 5 – 10 l/s. The lithology, including alluvium of plains and river deposits, generally, the graduation is medium-high.

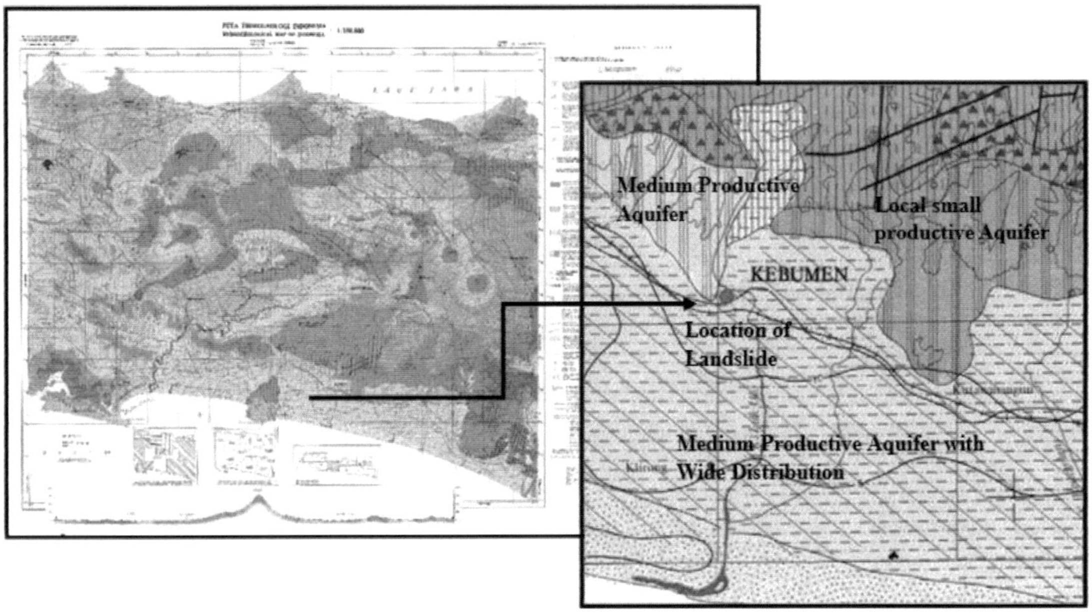

Figure 2. Left, map of hydrogeology. Right, inset of landslide location.

2. Research methods

2.1. Method of collecting data

The data collection methods used are as follows (Figure 3):

- Study of literature/secondary data related to landslides on Belimbing road, Pejagoan district, Kebumen regency, Central Java province. Secondary data includes regional geological conditions, geomorphology, hydrogeological conditions, and the vulnerability of soil movements in the landslide location area. Coordination with related agencies, namely BPBD Kebumen Regency, BIKK LIPI, and other agencies.
- Field surveys to obtain primary data include morphological of landslide site, measurement and plotting well dugs around landslide site to find out the direction of water flow and finally plotting and describing the waterways (drainnage) around the landslide site. Field survey to determine the direction of groundwater flow and drainage conditions of waterways around the landslide location.

2.2. Data analysis method

Primary and secondary data obtained are then processed and analyzed. The analyses process is assisted by using Surfer software. Soil data, morphology, water flow and waterways (drainage) are also overlayed in this analysis process. Analytical data from various factors also compare with landslide elsewhere [4, 6, 9, 10].

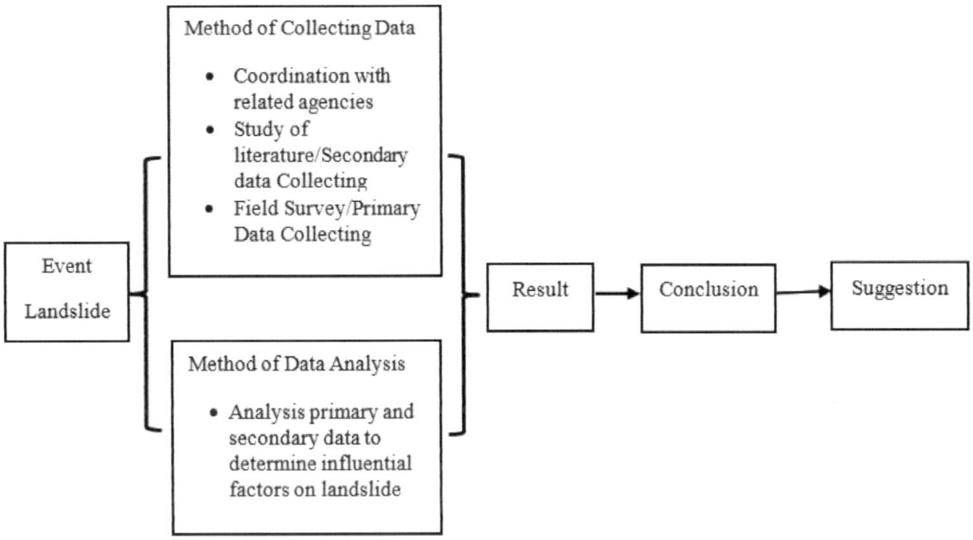

Figure 3. Research flow chart.

3. Results and discussion

3.1. Geology and geomorphology of landslide location

The rocks at the landslide location are various debris materials in the size of clay, sand, and gravel from the Luk Ulo (fluvial) river deposits that have not been compacted and are loose. The dark-colored overburden which is the result of weathering of various uncompacted debris materials with a thickness of 1-2 m.

The geomorphology of the landslide location is plain with steep bordering the Luk Ulo river (Figure 4 and 5). The slope bordering the river is 45-60° with a shape influenced by lateral erosion of the Luk Ulo river. The Luk Ulo river is a mature river, which can be seen from the meander shape which indicates that the lateral erosion is greater than the vertical erosion (Figure 4).

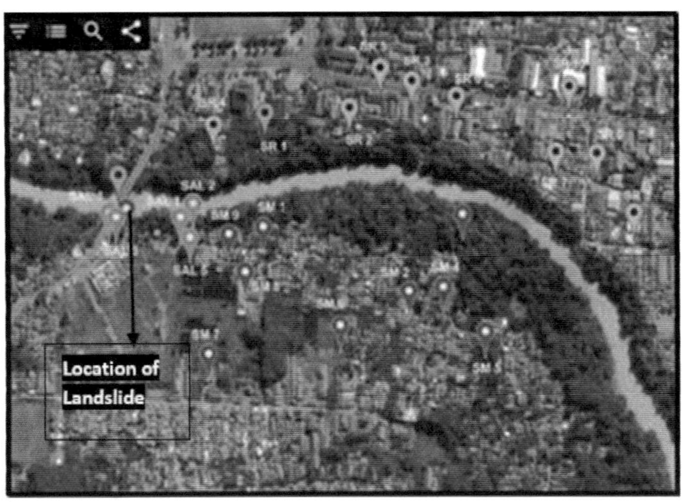

Figure 4. Existing condition landslide location and its surroundings from google map.

Figure 5. Landslide location in Pejagoan village, Pejagoan district. On the left, the photo is taken from the top of the Rainbow Bridge. On the right, the photo is taken from the southeast of the landslide location.

3.2. The direction of groundwater flow and drainage

Groundwater analysis was carried out to determine the flownets (direction of groundwater flow) around the landslide location. The data taken included plotting the location of the well-dug and the depth of the groundwater table. South of the Luk Ulo river there are 9 well-dugs recorded. The groundwater level ranges from 0.26 – 5.48 masl with the lowest depth at STA 6 at 0.26 masl and the highest depth at STA 3 at 5.48 masl. The average depth of the groundwater table is 2.06 masl (Figure 6). Well-dug data are detailed in Table 1.

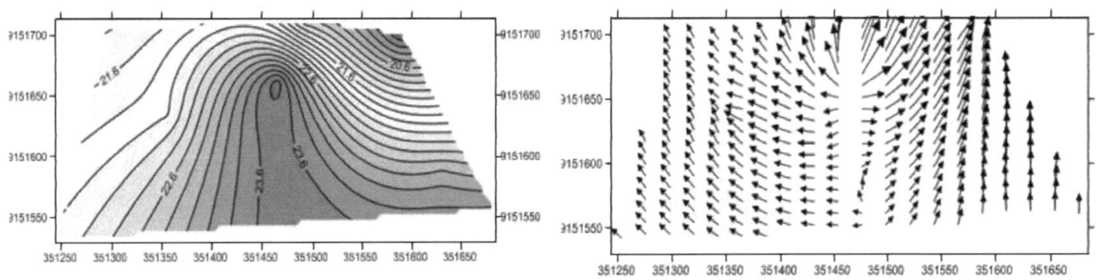

Figure 6. Left, groundwater table depth contour. Right, groundwater flow direction.

The results of the analysis of the direction of groundwater flow around the landslide location showed that there was one peak with the direction of flow towards the north towards the Luk Ulo river. When it rains, the infiltrated rainwater will seep and increase the volume of groundwater [11, 12]. Increasing the volume of groundwater will flow in the direction of the flownets and increase the pressure in the direction of the flow of groundwater (Figure 7).

The water channels (drainage) are located south of the landslide point, namely channel 1 (T1), channel 2 (T2), channel 3 (T3), channel 4 (T4), and channel 5 (T5). The flow direction of the water channel from T5 to T1 goes to T2 and T3. T2 enters the Luk Ulo river. From T4 into T5, and into the Luk Ulo river. From Figure 7 below, it can be seen that T1 is the intersection point of the water channel. The altitude of T1 is 48 meters above sea level, while T3 is 44 meters above sea level. The gradient between T1 to T3 is 7.20. Between T1 to T3, there are two monitoring tubs to ensure the drainage channel is not clogged. The water channel also shows that there are two water discharge points, namely at points T2 and T4. These two discharge points directly into the Luk Ulo river. The location of the landslide is located between the two water discharge points (Figure 8).

Table 1. Well data around landslide location.

ID	X	Y	Z (masl)	Depth of Groundwater Table (m)	Groundwater Level (masl)	Information
STA 1	351294.200	9151705.023	24	-2.62	21.38	
STA 2	351492.190	9151713.580	25	-4.22	20.78	
STA 3	351590.597	9151702.463	25	-5.48	19.52	
STA 4	351630.679	9151584.324	25	-1.93	23.07	
STA 5	351684.805	9151554.022	25	-1.55	23.45	South of River
STA 6	351479.020	9151563.885	24	-0.26	23.74	
STA 7	351243.821	9151529.416	23	-0.95	22.05	
STA 8	351358.371	9151634.274	23	-1	22	
STA 9	351465.843	9151661.139	25	-1.1	23.9	

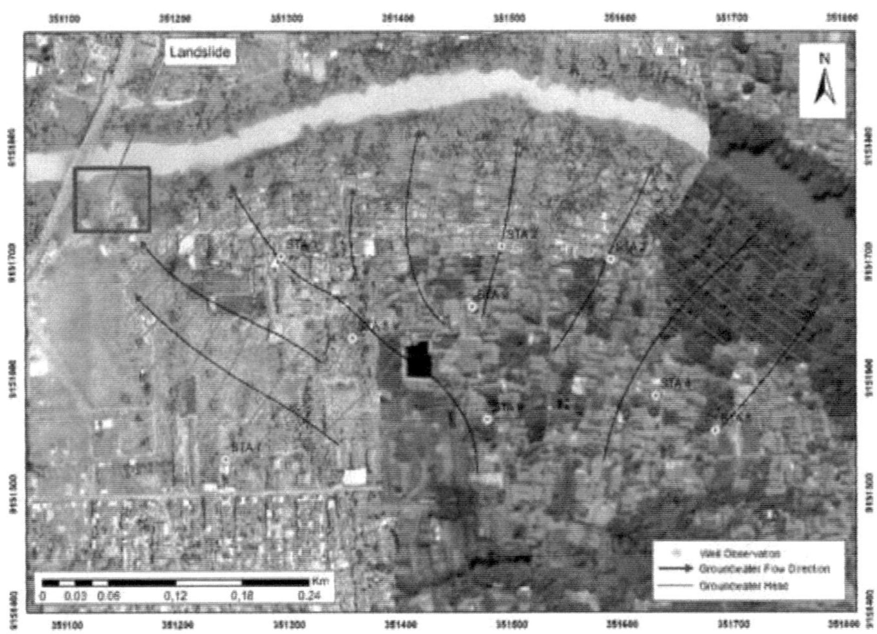

Figure 7. Plotting map of dug well point and groundwater flow direction.

The location, which lies between two discharge points, is a vulnerable area for landslides due to the accumulation of water from the two discharge points and also water from the Luk Ulo river. Seeing and considering the hydrology of waterways around the landslide location, there should be a change in the direction of drainage water flow. Changes in the direction of drainage flow can be attempted by closing T4 and only one discharge point, namely at T2. The direction of drainage flow is attempted from T3 to T1 and then to T2. Spatula (bypass) from T5 to T1 to T2 continues directly to the Luk Ulo river to facilitate the flow direction. Water from the southwest is channeled directly to T3, then to T1, and continues to T2. The dimensions and depth of the channel also need to be widened and deepened so that the flow gradient from T3 or T5 to T1, then to T2 becomes smoother (Figure 9).

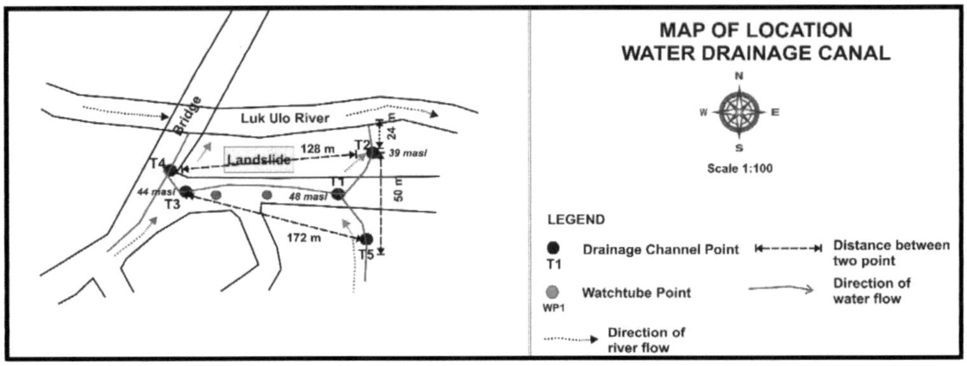

Figure 8. Waterways (drainage) map.

3.3. *Controlling factors and trigger factors*

Controlling factors are factors that make the condition of a slope vulnerable or ready to move, including morphological conditions, stratigraphy (rock types and their relationship with other rocks in the vicinity), geological structures, geohydrology, and land use [4, 6, 13-15]. The landslide control factors at the Belimbing road location are:

1. Morphology, the slope of the landslide is very steep with an angle of more than 60°.
2. Stratigraphy, rock types in the form of river deposits (alluvium) that have not been compacted with a thickness of more than 5 m. Under the alluvium is the Halang Formation consisting of calcareous sandstone, marl claystone, and calcarenite inserts.
3. Hydrogeology, medium productivity aquifer with wide distribution, this type of aquifer is an aquifer with medium continuity. Pisometric height or groundwater level above or near the ground level.
4. Land use, construction of houses, and roads that are quite crowded by vehicles.

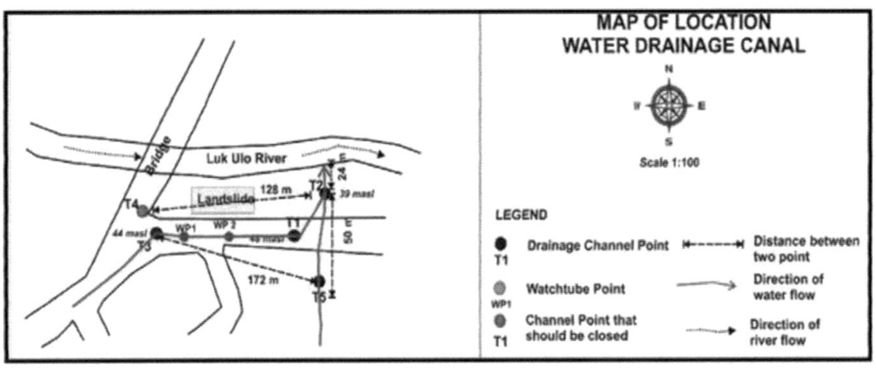

Figure 9. Suggested waterway (drainage) map.

Trigger factors are processes that change a slope from a vulnerable condition or ready to move to a critical condition and finally move. The factors that trigger landslides on the Belimbing road are:

1. Vehicle vibration.
2. Drainage around the landslide location.

4. Conclusions

The rock at the landslide site is in the form of river sediment (alluvium) which consists of various debris in the size of clay, sand, gravel, and uncompacted gravel. The cover soil is weathered with a darker color with a thickness of 1-2 m. The geomorphological condition is in the plains bordering the river which forms a steep slope.

The type of landslide on Belimbing road is a rotation with an area of 1728 m^2 with a movement direction of N10°E. This type of landslide is in the form of debris slide material which continues to become debris flow when mixed with the Luk Ulo river flow.

The direction of groundwater flow around the landslide location is to the north towards the Luk Ulo river. When it rains, the infiltrated rainwater will seep and increase the volume of groundwater. Increasing the volume of groundwater will flow in the direction of the flownets and increase the pressure in the direction of the flow of groundwater.

There are five waterways, two of which are discharge leading to the Luk Ulo river. The location of the landslide is between the two discharges of the water flow.

5. Suggestions

It is advisable to change the drainage flow by doing a spatula (bypass) from T5 to T2, and then towards the river. The channel at T4 is closed or diverted directly to T3 and then to T1 to avoid water buildup under the bridge. If possible, the Belimbing road is moved to the south to widen the distance between the road and the landslide location.

Acknowledgements

The authors would like to thank the Head of Research Center for Geotechnology – BRIN for giving permission for this research and the field team who have assisted in collecting primary data in the field. We also appreciated reviewers and all parties involved in this writing and provided suggestions and corrections.

Contribution Statement

Nandian Mareta is the main contributor of this paper, with support from Sugeng Purwo Saputro, Sueno Winduhutomo, and Rachmat Fajar Lubis.

References

[1] Mareta N, Widiyanto K, Ansori C, Widianto E and Purwoto H 2019 *Prosiding Geodiversity: Seminar Nasional Ilmu Kebumian 2019. Riset Ilmu Kebumian untuk Pengembangan Geopark Nasional* ISBN 978-602-496-202-9 (Paper in Bahasa Indonesia)

[2] Karnawati D 2007 *Dinamika Teknik Sipil* **7** (2) (Paper in Bahasa Indonesia)

[3] Guerra A J T, Fullen M A, Jorge M C O, Bezerra J F R and Shokr M S 2017 *Pedosphere* **27** (1) 27-41

[4] Edwards P J, Williard K W J and Schoonover J E 2015 *Journal of Contemporary Water Research & Education* **154** (1) 3-20

[5] Asdak C 1995 *Gadjah Mada University Press* ISBN: 979-420-737-3 (Book in Bahasa Indonesia)

[6] Boongaling C G K, Faustino-Eslava D V and Lansigan F P 2018 *Land Use Policy* **72** 116-128

[7] Asikin S, Harsolumakso A H, Busono H and Gafoer S 1992 *P3G Bandung*

[8] Effendi A T 1985 *Directorate of Environmental Geology Bandung*

[9] Pareta K and Pareta U 2012 *International Journal of Science and Technology* **1** (2) ISSN 2049-7318

[10] Noorollahi Y, Sadeghi S, Yousefi H and Nohegar A 2018 *International Journal of Hydrology* **2** (2) 137-148

[11] Lari S, Frattini P and Crosta G B 2014 *Engineering Geology* **182** (A) 3-14

[12] Qi W, Xu C and Xu X 2021 *Natural Hazards Research* **1** (3) 103-108

[13] Cullen C A, Al-Suhili R and Khanbilvardi R 2016 *Remote Sensing* **8** (10) 866

[14] Donini M, Napolitano E, Salvati P, Ardizzone F, Bucci F, Fiorucci F, Santangelo M, Cardinali M and Guzzetti F 2017 *Landslides* **14** 1521-1535

[15] Luna B Q, Blahut J, van Asch T, van Westen C and Kappes M 2016 *Geoenvironmental Disasters* **3** 29

Australian–Indonesian monsoon rainfall responses to the northern hemisphere climatic changes prior to the Last Glacial Maximum: an early indication

R D W Ardi[1,2], Aswan[1], K A Maryunani[1], E Yulianto[3], P S Putra[1,3] and S H Nugroho[1,3]

[1] Geological Engineering Study Program, Faculty of Earth Sciences and Technology, Institut Teknologi Bandung, Bandung, Jawa Barat 40132, Indonesia

[2] Agrotechnology Study Program, Faculty of Science and Technology, Universitas Nahdlatul Ulama Purwokerto, Banyumas, Jawa Tengah 53144, Indonesia

[3] Research Center for Geological Disaster, National Research and Innovation Agency, Bandung, Jawa Barat 40135, Indonesia

Email: dwa.ryan@students.itb.ac.id

Abstract. The evidences of Australian-Indonesian monsoon (AIM) rainfall response to the northern hemisphere climatic changes from both marine and terrestrial proxies are well established for the Last Glacial Maximum (LGM) – Deglaciation time interval but not in the previous time interval. Sediment cores from off south Sumba (ST10) and off north Sumba (Sumba strait) (ST14) were analyzed using X-Ray Fluorescence method to obtain elemental proxies. Elemental ratios which reflect terrigenous input (Ti/Ca and K/Ca) are used to infer the AIM rainfall changes since ~42 ka BP. AIM rainfall changes indicate a clear response to the Heinrich Events prior to the LGM (H2, H3, and H4). H2 and H3 are corroborated with the AIM rainfall increases in off south Sumba (~24 – 26 ka BP and ~30 – 31 ka BP) while the AIM rainfall increase (~39 – 41 ka BP) in the Sumba strait indicates a response to H4. The cooler temperature during the Heinrich Events could enhance the Northern Hemisphere (NH) cold surges which eventually pushed the Austral summer Intertropical Convergence Zone (ITCZ) southward in a similar fashion to the Last Deglaciation period (~18 – 11 ka BP). The southward movement of Austral summer ITCZ should increase the research area's exposure time to the Tropical Rain Belt (TRB) during the Australian – Indonesian summer monsoon (AISM) which eventually triggers the AIM rainfall increase. The Sumba strait AIM rainfall unresponsiveness to H2 and H3 could be related to the constant exposure time to the TRB which indicates that the southern limit of Austral summer ITCZ during its northernmost shift didn't reach the latitude of Sumba Strait. Comparison with other proxies from the same site and other rainfall proxies from southern Indonesia and the northern Australia regions is needed to confirm the spatial extend of those responses.

Keywords: Australian-Indonesian monsoon; elemental ratio; paleoclimate; southern Indonesia.

1. Introduction

Monsoonal rainfall are among the important climatic parameters in the Indonesian region as it is affecting the livelihood of millions [1], especially on the densely populated southern Indonesia region. Australian-Indonesian monsoon (AIM), which is part of the southern hemisphere global monsoon system [2,3], is the acting monsoonal system in this region [4]. This monsoon system brought large amount of rainfall during its summer phase, which occurs during October – March and reaches its peak aroung the southern hemisphere (Austral) summer (December – Januari) [4,5].

Previous paleoclimate studies in southern Indonesia and northern Australia region, particularly on the last glacial timescales indicated the fluctuations of AIM rainfall [6–12]. The abrupt northern hemisphere climatic changes related to the ice volume fluctuations are among the driving factors to the AIM rainfall changes, particulary within the Last Deglaciation (~18 – 11 ka BP) timescale [8,13]. For this period, the evidences of AIM rainfall responses to the northern hemisphere climatic changes (i.e. Heinrich Event 1, Bølling-Allerød, and Younger Dryas) are evident in both marine [7,10,11] and terrestrial proxies [12,14]. The northern hemisphere climatic events, especially Heinrich Events (H) were also occurred before the Last Glacial Maximum (LGM) [15,16], however, proxy evidences of their relation to the AIM rainfall are still lacking. This research will utilize geochemical proxies obtained from deep sea sediment cores on off south Sumba and Sumba strait (**Figure 1**.) to confirm the response of AIM rainfall to the northern hemisphere climatic changes up to ~20k years before the LGM (~42 ka BP).

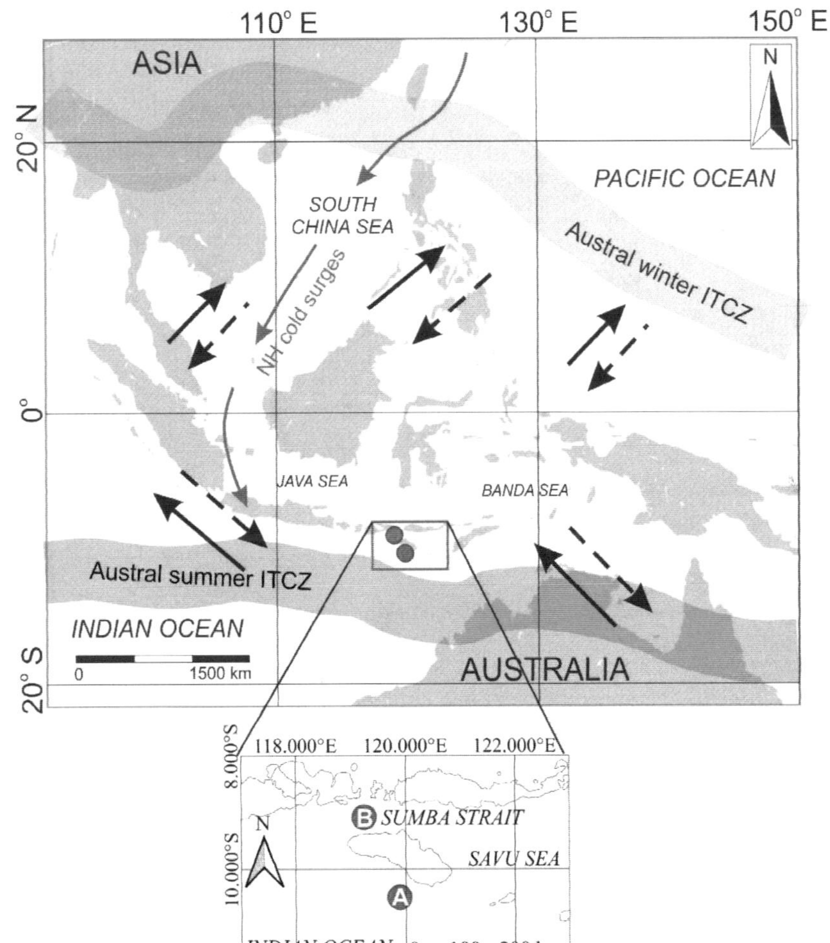

Figure 1. Modern climate of Indonesia region, compiled from previous studies [4,11,17]. Black arrows: dominant wind direction during Australian-Indonesian winter monsoon (AIWM) (solid) and Australian-Indonesian summer monsoon (AISM) (dashed). The red rectangle indicates the Sumba region and two red circles inside point the location of ST10 (A) and ST14 (B).

2. Regional climate

As part of the Indo-Pacific Warm Pool (IPWP), the climatic condition of the Indonesia region is driven by the Asia-Australia regional monsoon system (including AIM), the *El Nino* Southern Oscillation (ENSO), and the Indian Ocean Dipole (IOD) [18]. In southern hemisphere part of the Indonesian archipelago (southern Indonesia) including the Sumba area, the climatic condition is dominated by the AIM indicated by the bi-annual (one peak and one trough) monthly rainfall [4,5]. The wet (dry) Australian-Indonesian summer (winter) monsoon occurs during October – March (April – September) and peaked around the Austral summer (winter), when the Intertropical Convergence (ITCZ) deviates northward (southward) from the equator [4,19] (**Figure 1.**). The spatial range of ITCZ defines Tropical Rain Belt (TRB), the high rainfall zone within the tropical region [20]. The northward (southward) position of ITCZ affects the dominant wind direction which is northwest (southeast) during the Australian-Indonesian summer (winter) monsoon. The southward deviation of ITCZ during the Austral summer is triggered by the northern hemisphere (NH) cold surges that blow southward through the South China Sea [17].

3. Data and Method

Two sediment cores were taken from off south Sumba (ST10) and Sumba strait (ST14) as part of the 2016 Widya Nusantara Expedition (E-WIN) using Baruna Jaya VIII research vessel operated by Research Center for Oceanography, Indonesian Institute of Sciences (LIPI) (now Research Center for Geotechnology, National Research and Innovation Agency/BRIN) (**Table 1**.). Each core was measured using the Thermo NITON XL3t 500, a Field Point X-Ray Fluorescence (FP-XRF) device at 1 cm intervals to obtain the elemental data. A total of 379 data points were obtained from both ST10 (136 data points) and ST14 (243 data points), however, only data points represent the time period before LGM were discussed in this study.

Table 1. Marine sediment cores used in this study.

Core	Location	Coordinate		Depth (m)
		Latitude (°E)	Longitude (°S)	
ST10	Off south Sumba	119.885	10.491	1004
ST14	Sumba strait	119.245	9.079	1283

This study uses elemental ratios to infer the past changes of AIM rainfall to overcome the limitation of single elemental data as paleoclimate proxies [21–23]. The presentation of elemental ratios in logarithmic is also utilized to solve the asymmetrical values caused by numerator/denominator arrangement [24]. The ratio of Titanium/Calcium (Ti/Ca) and Potassium/Calcium (K/Ca) which represent riverine (terrigenous) input are used to infer the past AIM rainfall changes [10,11,25] under the assumption that the rainfall enhancements corroborate with the increase of riverine input to the marine environment [7]. Ti was derived from the rock-forming minerals of the mafic and ultramafic rocks on the surrounding islands [26,27] while K could be the product of micas weathering found from the land outcrops [28]. Ca, which mainly originated from the marine biogenic activity is used as a denominator to remove the biogenic components from the elemental record [26,29,30]. Due to the presence of exposed carbonate formations in Sumba Island [31], the use of Ca as a denominator is needed to be validated. Validation is performed through the linear correlation of Ti vs Ca and K vs Ca (**Figure 2**.). The low values of correlation coefficients (R^2) indicate the overall marine source of Ca and confirm its effectiveness as a denominator for the terrigenous input proxies.

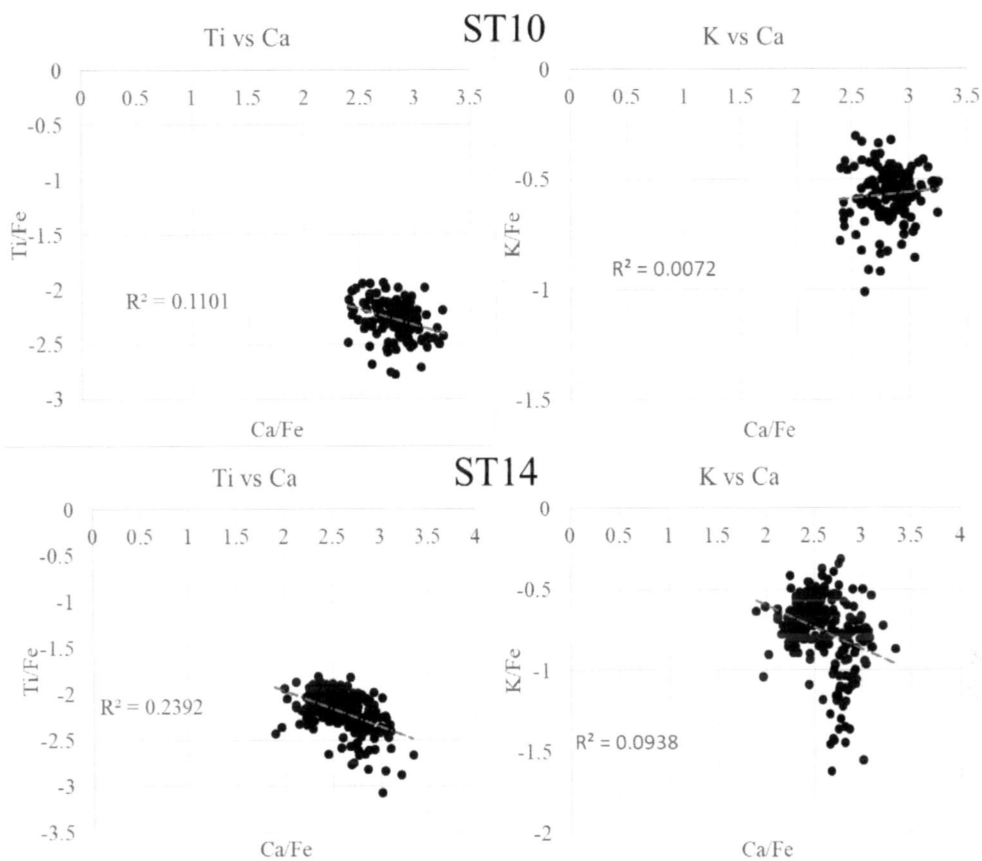

Figure 2. Linear correlation of Ti vs Ca and K vs Ca. Each element normalized by Iron (Fe). The low correlation coefficients confirm that the source of Ca (marine) is different from Ti and K (terrigenous).

Geochronology of ST10 and ST14 cores were based on three Accelerator Mass Spectometry (AMS) ^{14}C measurements (two on ST14 and one on ST10) with the core top sediment was assumed to be aged -66 BP (Present = 1950 Anno Domini/AD, ST10 and ST14 cores were retrieved on 2016) (**Table 2**.). The AMS ^{14}C measurements were conducted at Beta Analytic Inc., USA and the radiocarbon ages were converted to calibrated calendar and radiocarbon ages with High Probability Density Range (HPD) Method [32] using MARINE13 calibration dataset [33] with a reservoir correction (ΔR) of 51 ± 87 years. The age model for each depth interval was calculated using Bayesian approach [34] with rbacon package (version 2.4.2) [35] in R [36] (**Figure 3**.). Afterwards, the mean ages from the age modelling is used to plot the elemental ratio data for further analysis.

Table 2. AMS ^{14}C ages and calibrated 14C ages of ST10 and ST14.

Lab code	Core	Depth intervals	Materials	^{14}C age (BP)	Calibrated ^{14}C age BP
Beta - 572243	ST10	136 – 137 cm	Foraminifera: *Neogloboquadrina* spp.	33280 ± 200	35721 – 34286
Beta - 572244	ST14	242 – 243 cm	Foraminifera: *Neogloboquadrina* spp.	39860 ± 420	43947 – 42510
Beta - 572245	ST14	124 – 125 cm	Foraminifera: *Neogloboquadrina* spp.	14900 ± 40	17899 – 17330

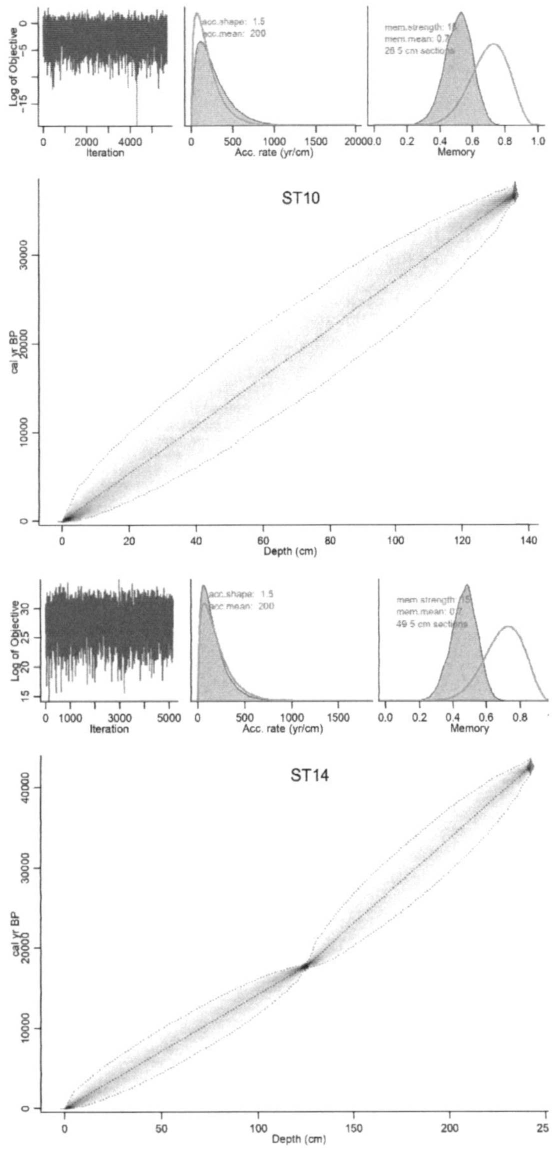

Figure 3. ST10 (upper) and ST14 (lower) age models based on Bayesian approach [34] using rbacon package (version 2.4.2) [35] in R [36]. Upper panels depict the Markov chain Monte Carlo (MCMC) iterations (left panel), the prior (green curves) and posterior (grey histograms) distributions for the accumulation rate (middle panel), and memory (right panel) [35]. Bottom panel shows the calibrated ^{14}C dates (transparent blue) and the age-depth model (darker grays indicate more likely calendar ages, gray stippled lines show 95% confidence intervals, and red lines indicate single 'best' model based on the mean ages) [35].

4. Findings and Discussions

A comparison of the elemental ratio data (Ti/Ca and K/Ca), Greenland ice core (GISP2) oxygen isotope $\delta^{18}O$ [37], and insolation [38] was made to reveal the relation between AIM rainfall and northern hemisphere climatic changes for the time period before LGM (**Figure 4**.). Here the Greeenland ice core $\delta^{18}O$ is used to represent the northern hemisphere climatic changes [39]. The AIM rainfall on ST10 (off south Sumba) indicates enhancement at Heinrich Event 3 (H3, ~30 – 31 ka BP), Heinrich Event 2 (H2, ~24 – 26 ka BP), and LGM (~23 – 18 ka BP) while on the Sumba strait the AIM rainfall enhancement was only detected at Heinrich Event 4 (H4, ~39 – 41 ka BP) and LGM. It is suggested that the the past increases (decreases) in AIM rainfall were analogous to Australian-Indonesian summer (winter) monsoon enhancement due to the higher (lower) rainfall during Australian-Indonesian summer (winter) monsoon in present day Southern Indonesia [7,40]. Based on these assumption, it can be inferred that the rainfall increases detected from the elemental ratio indicate AISM enhancement.

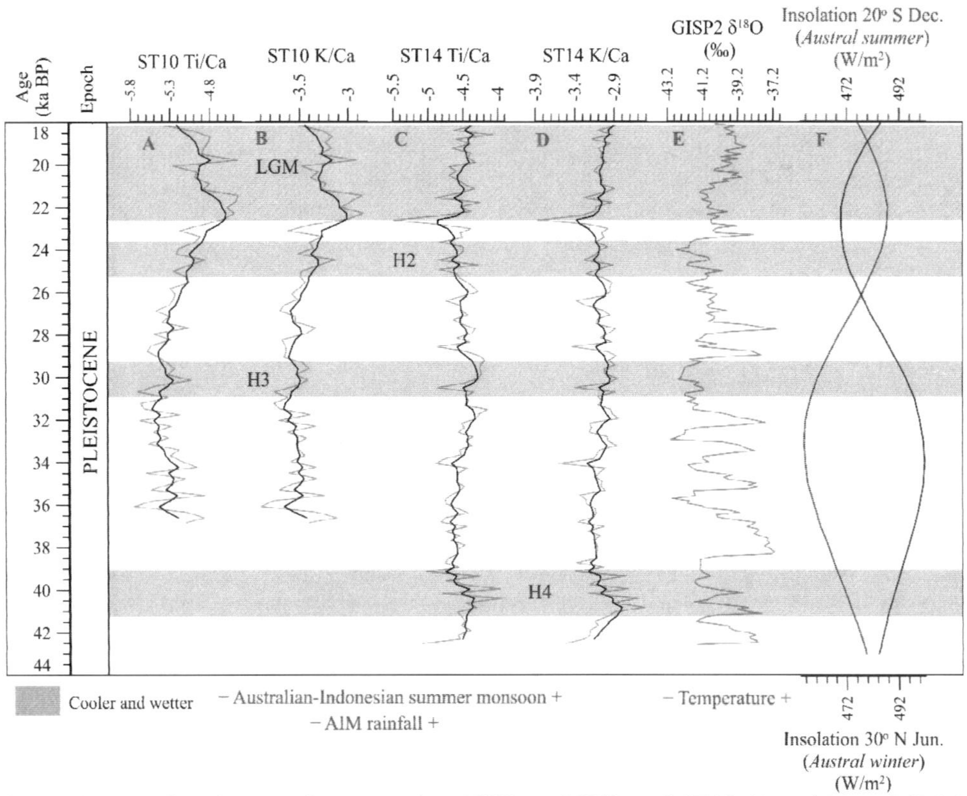

Figure 4. Plot of terrigenous input proxies (Ti/Ca and K/Ca) of ST10 (A and B) and ST14 (C and D) compared to Greenland ice core (GISP2) $\delta^{18}O$ (E) [37] and insolation (Austral summer and Austral winter) (F) [38] at ~42 – 18 ka BP.

Rainfall enhancement at the LGM which is detected on both off south Sumba and Sumba strait was corroborated with the peak increase of Austral summer insolation. This finding supports previous suggestions about the orbital-scale variability of AISM (AIM rainfall) detected from both marine and terrestrial proxies on southern Indonesia region [7,12]. The increase of Austral summer insolation could triggers stronger northwest winds which bring more mositures to the southern Indonesia region [4].

Similar to the more studied cool climatic events during the last deglaciation (YD and H1) [10–12], the cooler temperature during H2, H3, and H4 was most likely enhance the Northern Hemisphere (NH) cold surges which eventually pushed the Austral summer Intertropical Convergence Zone (ITCZ) southward. The southward movement of Austral summer ITCZ should increase the research area's exposure time to TRB during the AISM and eventually induces the AIM rainfall increase. The lack of AIM rainfall response to H2 and H3 in Sumba strait could be related to the constant exposure time to the TRB during the period after H4 and before the LGM (~39 – 23 ka BP) which indicates that the southern limit of Austral summer ITCZ during its northernmost shift didn't reach the latitude of Sumba Strait (**Figure 5.**). This indicates the southward deviation of southern limit of the austral summer ITCZ compared to its position and during the prior (before ~39 ka BP) and later (after ~23 ka BP) periods inferred from previous studies [10,11] To confirm these findings, it is necessary to study other monsoon proxies (e.g. foraminifera abundances, foraminifera oxygen isotop/$\delta^{18}Oc$) and generate more robust age models on ST10 and ST14 cores. In addition, further studies involving both elemental ratios and other monsoon proxies from other locations in southern Indonesia region are needed to confirm the spatial extend of those responses.

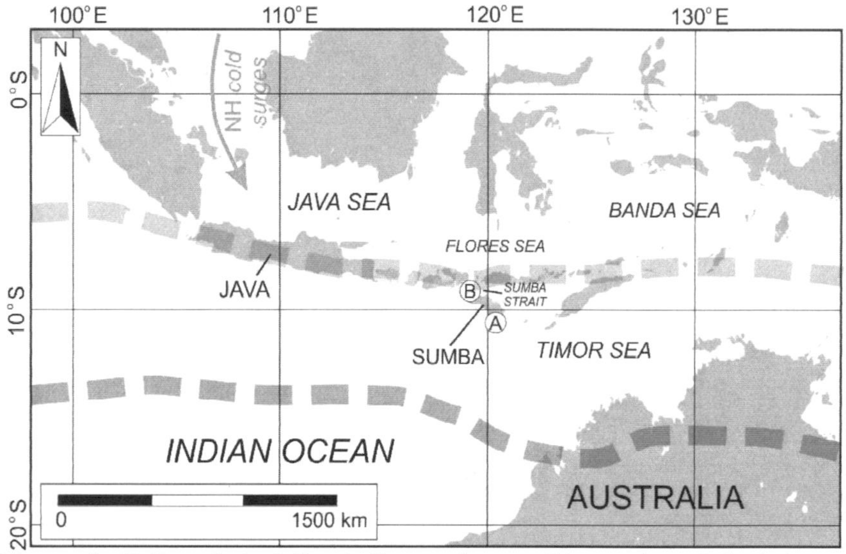

Figure 5. Yellow dashed line indicates the possible southward deviation of southern limit of the austral summer ITCZ during the drier (lower rainfall) periods during ~39 – 23 ka BP compared to its position during the prior (before ~39 ka BP) and later (after ~23 ka BP) periods inferred from previous studies (red dashed line) [10,11]. Blue dashed line indicates southern limit of the austral summer ITCZ during the wetter periods (higher) rainfall periods [10,11]. During ~39 – 23 ka BP, ST14 (B) was positioned northward of the austral summer ITCZ southern limit while ST10 (A) was located southward of it.

5. Conclusion

Elemental ratios (terrigenous input proxies) data from off south Sumba and Sumba strait indicate AIM rainfall increases as responses to the cool northern hemisphere climatic events prior to LGM (H2, H3, and H4). The AIM rainfall increases was related to the southward movement of Austral summer ITCZ due to NH surges enhancement during the cool climatic events. The Sumba strait AIM rainfall unresponsiveness to H2 and H3 could be related to the constant exposure time to the TRB which indicates that the southern limit of Austral summer ITCZ during its northernmost shift didn't reach the latitude of Sumba Strait. Comparison with other proxies from the same site and other rainfall proxies from southern Indonesia and the northern Australia regions in further studies is needed to confirm the spatial extend of those responses.

6. References

[1] Webster P J, Magaña V O, Palmer T N, Shukla J, Tomas R A, Yanai M dan Yasunari T 1998 Monsoons: processes, predictability, and the prospects for prediction *J. Geophys. Res. Ocean.* **103** 14451–510

[2] Wang P X, Wang B, Cheng H, Fasullo J, Guo Z, Kiefer T dan Liu Z 2017 The global monsoon across time scales: Mechanisms and outstanding issues *Earth-Science Rev.* **174** 84–121

[3] Wang P X, Wang B, Cheng H, Fasullo J, Guo Z T, Kiefer T dan Liu Z Y 2014 The global monsoon across timescales: coherent variability of regional monsoons *Clim. Past* **10** 2007–52

[4] Wheeler M C dan McBride J L 2005 Australian-Indonesian monsoon *Intraseasonal Variability in the Atmosphere-Ocean Climate System* (Berlin, Heidelberg: Springer Berlin Heidelberg) hal 125–73

[5] Aldrian E dan Susanto R D 2003 Identification of three dominant rainfall regions within Indonesia and their relationship to sea surface temperature *Int. J. Climatol.* **23** 1435–52

[6] Liu S, Zhang H, Shi X, Chen M, Cao P, Li Z, Troa R A, Zuraida R, Triarso E dan Marfasran H 2020 Reconstruction of monsoon evolution in southernmost Sumatra over the past 35 kyr and its

response to northern hemisphere climate changes *Prog. Earth Planet. Sci.* **7** 30

[7] Mohtadi M, Oppo D W, Steinke S, Stuut J B W, De Pol-Holz R, Hebbeln D dan Lückge A 2011 Glacial to Holocene swings of the Australian-Indonesian monsoon *Nat. Geosci.* **4** 540–4

[8] Denniston R F, Wyrwoll K H, Asmerom Y, Polyak V J, Humphreys W F, Cugley J, Woods D, LaPointe Z, Peota J dan Greaves E 2013 North Atlantic forcing of millennial-scale Indo-Australian monsoon dynamics during the Last Glacial period *Quat. Sci. Rev.* **72** 159–68

[9] Griffiths M L, Drysdale R N, Gagan M K, Zhao J X, Ayliffe L K, Hellstrom J C, Hantoro W S, Frisia S, Feng Y X, Cartwright I, Pierre E S, Fischer M J dan Suwargadi B W 2009 Increasing Australian-Indonesian monsoon rainfall linked to early Holocene sea-level rise *Nat. Geosci.* **2** 636–9

[10] Ardi R D W, Aswan, Maryunani K A, Yulianto E, Putra P S, Nugroho S H dan Istiana 2020 Last Deglaciation—Holocene Australian-Indonesian Monsoon Rainfall Changes Off Southwest Sumba, Indonesia *Atmosphere (Basel).* **11** 932

[11] Kuhnt W, Holbourn A, Xu J, Opdyke B, De Deckker P, Röhl U dan Mudelsee M 2015 Southern Hemisphere control on Australian monsoon variability during the late deglaciation and Holocene *Nat. Commun.* **6** 5916

[12] Ayliffe L K, Gagan M K, Zhao J, Drysdale R N, Hellstrom J C, Hantoro W S, Griffiths M L, Scott-gagan H, Pierre E S, Cowley J A dan Suwargadi B W 2013 Rapid interhemispheric climate links via the Australasian monsoon during the last deglaciation *Nat. Commun.* **4** 1–6

[13] Mohtadi M, Oppo D W, Lückge A, DePol-Holz R, Steinke S, Groeneveld J, Hemme N dan Hebbeln D 2011 Reconstructing the thermal structure of the upper ocean: Insights from planktic foraminifera shell chemistry and alkenones in modern sediments of the tropical eastern Indian Ocean *Paleoceanography* **26** n/a-n/a

[14] Denniston R F, Wyrwoll K, Polyak V J, Brown J R, Asmerom Y, Wanamaker A D, Lapointe Z, Ellerbroek R, Barthelmes M, Cleary D, Cugley J, Woods D dan Humphreys W F 2013 A Stalagmite record of Holocene Indonesian e Australian summer monsoon variability from the Australian tropics *Quat. Sci. Rev.* **78** 155–68

[15] Heinrich H 1988 Origin and Consequences of Cyclic Ice Rafting in the Northeast Atlantic Ocean During the Past 130,000 Years *Quat. Res.* **29** 142–52

[16] Lynch-Stieglitz J, Schmidt M W, Gene Henry L, Curry W B, Skinner L C, Mulitza S, Zhang R dan Chang P 2014 Muted change in Atlantic overturning circulation over some glacial-aged Heinrich events *Nat. Geosci.* **7** 144–50

[17] Suppiah R dan Wu X 1998 Surges, cross-equatorial flows and their links with the Australian summer monsoon circulation and rainfall *Aust. Meteorol. Mag.* **47** 113–30

[18] Abram N J, Gagan M K, Liu Z, Hantoro W S, McCulloch M T dan Suwargadi B W 2007 Seasonal characteristics of the Indian Ocean Dipole during the Holocene epoch *Nature* **445** 299–302

[19] Huang E, Tian J dan Liu J 2015 Dynamics of the Australian-Indonesian monsoon across Termination II: Implications of molecular-biomarker reconstructions from the Timor Sea *Palaeogeogr. Palaeoclimatol. Palaeoecol.* **423** 32–43

[20] Denniston R F, Ummenhofer C C, Wanamaker A D, Lachniet M S, Villarini G, Asmerom Y, Polyak V J, Passaro K J, Cugley J, Woods D dan Humphreys W F 2016 Expansion and Contraction of the Indo-Pacific Tropical Rain Belt over the Last Three Millennia *Sci. Rep.* **6** 34485

[21] Croudace I W, Rindby A dan Rothwell R G 2006 ITRAX: Description and evaluation of a new multi-function X-ray core scanner *Geol. Soc. Spec. Publ.* **267** 51–63

[22] Weltje G J dan Tjallingii R 2008 Calibration of XRF core scanners for quantitative geochemical logging of sediment cores: Theory and application *Earth Planet. Sci. Lett.* **274** 423–38

[23] Dypvik H dan Harris N B 2001 Geochemical facies analysis of fine-grained siliciclastics using Th/U, Zr/Rb and (Zr + Rb)/Sr ratios *Chem. Geol.* **181** 131–46

[24] Aitchison J 1982 The Statistical Analysis of Compositional Data *J. R. Stat. Soc. Ser. B* **44** 139–60

[25] Steinke S, Prange M, Feist C, Groeneveld J dan Mohtadi M 2014 Upwelling variability off

southern Indonesia over the past two millennia *Geophys. Res. Lett.* **41** 7684–93
[26] Kissel C, Laj C, Kienast M, Bolliet T, Holbourn A, Hill P, Kuhnt W dan Braconnot P 2010 Monsoon variability and deep oceanic circulation in the western equatorial Pacific over the last climatic cycle: Insights from sedimentary magnetic properties and sortable silt *Paleoceanography* **25** 1–12
[27] Hemming S R 2007 Terrigenous Sediments *Paleoceanography, Physical and Chemical Proxies* hal 1776 – 1786.
[28] Nesbitt H W, Markovics G dan Price R C 1980 Chemical processes affecting alkalis and alkaline earths during continental weathering *Geochim. Cosmochim. Acta* **44** 1659–66
[29] Hu D, Böning P, Köhler C M, Hillier S, Pressling N, Wan S, Brumsack H J dan Clift P D 2012 Deep sea records of the continental weathering and erosion response to East Asian monsoon intensi fi cation since 14 ka in the South China Sea *Chem. Geol.* **326–327** 1–18
[30] Liu B, Wang Y, Su X dan Zheng H 2013 Elemental geochemistry of northern slope sediments from the South China Sea : Implications for provenance and source area weathering since Early Miocene *Chemie der Erde - Geochemistry* **73** 61–74
[31] Abdullah C I, Rampnoux J P, Bellon H, Maury R C dan Soeria-Atmadja R 2000 The evolution of Sumba Island (Indonesia) revisited in the light of new data on the geochronology and geochemistry of the magmatic rocks *J. Asian Earth Sci.* **18** 533–46
[32] Bronk Ramsey C 2009 Bayesian Analysis of Radiocarbon Dates *Radiocarbon* **51** 337–60
[33] Reimer P J, Edouard Bard B, Alex Bayliss B, Warren Beck B J, Paul Blackwell B G dan Christopher Bronk Ramsey B 2013 Intcal13 and Marine13 Radiocarbon Age Calibration Curves 0–50,000 Years Cal Bp *Radiocarbon* **55** 1869–87
[34] Blaauw M dan Christeny J A 2011 Flexible paleoclimate age-depth models using an autoregressive gamma process *Bayesian Anal.* **6** 457–74
[35] Blaauw M, Christen J A dan Aquino M A 2020 rbacon: Age-Depth Modelling using Bayesian Statistics. *R Packag. version 2.4.2.*
[36] Core Development Team R 2020 A Language and Environment for Statistical Computing *R Found. Stat. Comput.* **2** https://www.R-project.org
[37] Grootes P M dan Stuiver M 1997 Oxygen 18/16 variability in Greenland snow and ice with 10 −3 - to 10 5 -year time resolution *J. Geophys. Res. Ocean.* **102** 26455–70
[38] Berger A L 1978 Long-term variations of daily insolation and Quaternary climatic changes. *J. Atmos. Sci.* **35** 2361–7
[39] Mohtadi M, Prange M, Oppo D W, De Pol-Holz R, Merkel U, Zhang X, Steinke S dan Lückge A 2014 North Atlantic forcing of tropical Indian Ocean climate *Nature* **508** 76–80
[40] Steinke S, Mohtadi M, Prange M, Varma V, Pittauerova D dan Fischer H W 2014 Mid- to Late-Holocene Australian-Indonesian summer monsoon variability *Quat. Sci. Rev.* **93** 142–54

7. Acknowledgments

The study was partially funded by Priority Research and Demand Drive Research Coremap-CTI (Number: SP DIPA- 079.01.1.664156/2020), and we express our gratitude to Dr. Dyah Marganingrum for the permission to participate in this research. This study was also partially funded by the Indonesian Endowment Fund for Education (LPDP) (grant number: 202001110215954). We also thank the Research Center for Oceanography of National Research and Innovation Agency (BRIN), especially Udhi Hermawan, Ph.D. as the chief scientist of E-WIN 2016 for the permission of data usage and administrative assistance and all crews of the Baruna Jaya VIII R.V., especially Singgih Adi Wibowo for technical supports and data collecting. The Research Center for Geological Disaster of BRIN (formerly known as Research Center for Geotechnolog of BRIN) is thanked for the laboratory facilities. RDWA as the main contributor designed the research concept, analyze the data, prepared and wrote the manuscript. A, KAM, and EY as co-contributors supervised the research concept designing and the data analysis process. A, KAM, EY, PSP, and SHN as co-contributors provided suggestions and input, as well as review-edited the manuscript. PSP and SHN joined the E-WIN cruise and obtained the samples.

The significant meaning of biodegradation to the petroleum system of the Seram Basin, eastern Indonesia: evaluation on the basis of physio-chemical composition variation of Seram oils

Barry Majeed Hartono[1], Eddy Ariyono Subroto[1], Asep Heri Patria Kesumajana[1], Roy Andrianto[2], Joko Wahyudiono[3], Galung Malvinas[2], Bagus Priyanto[1]

[1]Geological Engineering, Faculty of Earth Sciences and Technology, Institut Teknologi Bandung, Indonesia
[2]CITIC Seram Energy Limited, Indonesia
[3]Pusat Survei Geologi, Badan Geologi, Indonesia

Corresponding Author Email: barry.majeed@students.itb.ac.id

Abstract. It has been widely accepted since the 1980s that Seram crude oils belong to one family that originated from the Triassic carbonate source rock. Even though these oils share similar nature and origin, they exhibit wide physio-chemical composition variations in Seram Basin. Moreover, this variation is also noted within a similar oil field. Hydrocarbon alteration, especially biodegradation, a process of hydrocarbon destruction by microbial activities, has been attributed to this variation. However, the degree and extension of biodegradation in the Seram Basin are still poorly understood. Therefore, crude oils accumulated in main reservoirs, the Pre-Tertiary Manusela Carbonate and the Pliocene Fufa Clastic reservoirs, retrieved from Lofin, Oseil, Nief, and Bula Fields, are comprehensively studied. In addition, several oil seep samples were also utilized for this study together with crude oil geochemistry data compiled from previous publications. The occurrence of biodegradation is also validated with the geological conditions that include resevoir temperatures and salinities. Based on the geochemical signatures, there is a wide range of biodegradation degrees within the Seram Basin with the biodegradation degree tends to increase and more destructive to Bula Field which is situated in the northern part of the basin. The biodegradation pattern is linked to hydrocarbon generation and migration history. The Neogene deformation in eastern Indonesia is interpreted to be responsible for the different biodegradation process within Seram Basin. This process should be well understood and not be neglected during petroleum exploration, especially in eastern Indonesia. This study shows that such simple process has a deep meaning on the petroleum system history, especially the hydrocarbon preservation in the reservoir.

1. Introduction

One of the important aspects of the petroleum system is hydrocarbon preservation. Commonly, the expelled crude oil undergoes post-expulsion alteration during the migration or after being accumulated in the reservoir such as biodegradation [1, 2], water-washing [3], evaporative fractionation [4], and migration contamination [5]. Biodegradation is a hydrocarbon destruction process by microbial activities that modify the physio-chemical composition of accumulated hydrocarbon in the reservoir [1, 2, 6, 7, 8]. Geochemists must identify this process since the compositional changes of hydrocarbon creates complexity on the geochemical correlation (e.g. Shetland Basin [9]; Western Canadian Sedimentary Basin/WCSB [10]). Moreover, this process significantly affects the commerciality of oil production [2, 7].

Despite the effects of biodegradation on the economic values of oil production, the variation of the biodegradation degree could be utilized to understand the petroleum system of one basin, especially on the hydrocarbon generation and preservation subsystems. This study demonstrates how the different degrees of biodegradation in each field could tell the petroleum system histories with a case from Seram Basin, eastern Indonesia. Seram Basin was chosen since biodegradation is known to extensively affect the hydrocarbon in this area [11, 12, 13]. The understanding regarding the primary controls on crude oil composition such as depositional environment, source of organic matter, and thermal maturity has also been well established [11, 14]. Such information is compulsory prior to comprehensively understand the degree and effect of biodegradation to the Seram oils [2].

2. Biodegradation and Its Effects on Crude Oil

Biodegradation is a microbial process so that there are certain reservoir conditions to accommodate bacterial activities. The reservoir temperature must be below ~160°F or ~75°F as the bacteria still withstand the conditions [1, 2, 6]. Moreover, Palmer et al. [3] suggest that the bacteria will eventually die when the reservoir salinity reached ~100 PSU (practical salinity unit). If the conditions were achieved, biodegradation most likely occurred with the process start to "destroy" the light fraction hydrocarbons (~nC_6-nC_{12}) followed by the heavier fractions (~nC_{20}) [1, 2]. Biodegradation also destroys the normal alkane fraction followed by cycloalkane and aromatic fractions [2, 3]. Consequently, biodegradation tends to lower the API gravity of crude oil (into heavy oil) and increase the oil's viscosity. In terms of major hydrocarbon fractions, the saturate fraction is significantly reduced by biodegradation contrasting with the polar compound (nitrogen, sulphur, and oxygen; NSO) that tends to increase.

For the degree of biodegradation, this study adopts the biodegradation scale from Wenger et al. [2] that divides biodegradation into four levels, respectively, light (1), medium (2), heavy (3), and severe (4-10) biodegradations. The normal alkanes fractions start affected by the light biodegradation and completely destroyed during the heavy biodegradation, followed by isoprenoids (branched alkane) and cycloalkanes that are completely destroyed at the early phase of severe biodegradation (4) [2]. Bicyclic alkanes were destroyed at level 5 biodegradation, while steranes and hopanes were destroyed at level 7 biodegradation. In addition, biodegradation is also responsible for the forming of 25-norhopanes (dimethylated hopane) during level 6-7 biodegradation [1, 2]. Further biodegradation (level 8-9) would destroy the previously formed 25-norhopanes. Diasteranes and aromatic fractions are the most resistant to biodegradation [2, 6].

3. Data and Methods

To understand the biodegradation processes in Seram Basin, 19 crude oil samples were retrieved from main producing oil fields in the basin (figure 1). Specifically, two oil samples are taken from Lofin Field, five samples from Nief Field, seven samples from Oseil Field, and five samples from Bula Field. The origin and nature of these oils have been comprehensively studied by Hartono et al. [14] so that the primary controls on oil composition could be eliminated. Seram oil geochemical data from Price et al. [11] is also utilized in this study. The possibilities of biodegradation to occur are validated with the reservoir conditions that encompass temperature and salinities. Then, the compositional variation of oils is analyzed in a geographic manner. Therefore, the biodegradation trend in Seram Basin could be observed and relate with the regional tectonic events, even further to petroleum system histories of Seram Basin.

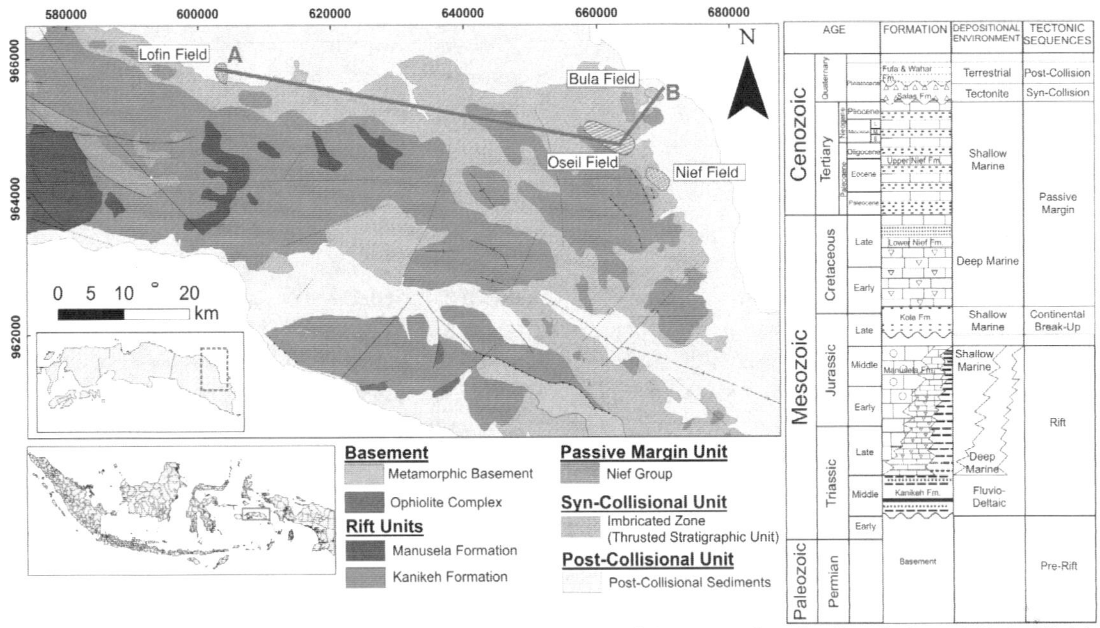

Figure 1. Main producing oil fields in Seram Basin which are Lofin, Oseil, Nief, and Bula Fields (base map based on geological map modified from [15, 16]). The stratigraphy of Seram Basin is complex that is composed by Permian basement, Mesozoic syn-rift sediments, Tertiary drift sediments, and collisional sediments [17].

4. Petroleum Geology of Seram Basin

The petroleum system of Seram Basin is dependent on its complex basin history. Basin filling in Seram Basin could be divided into five periods (figure 1) [17, 18]. The first period is marked by the deposition during the rifting event that is concordant with the initiation of the Gondwana break-up in the Triassic-Jurassic [17]. During this event, fluvio-deltaic Kanikeh Formation, shallow marine carbonate Manusela Formation, and deep marine carbonate Saman-Saman Formation are deposited (figure 1) that controls the generation subsystem of Seram Basin since the stratigraphic intervals are capable to act as source rock [11, 13, 19]. Oil geochemistry study revealed that Seram oils are originated from Pre-Tertiary carbonate intervals [11, 12, 13, 14, 19]. Recent studies suggest that both Manusela and Saman-Saman Formations may act as source rock since the oils have two distinct characteristics, respectively, shallow marine carbonate (Family A) and deep marine carbonate (Family B) origin [14]. Moreover, Manusela Formation acts as major oil-producing Pre-Tertiary carbonate reservoir within the basin.

Continous rifting events resulted in Gondwana to break-up during Late Jurassic to Cretaceous that is marked by hiatus and later followed by the deposition of Kola Shale (figure 1) [17]. Passive margin sediment start deposited in Seram Basin during Early Cretaceous to Late Miocene. At this time, Seram Basin is a starved basin so that the basin filling is dominated by shelf carbonate deposition which is Nief Formation [17]. Both Kola and Nief Formations act as a good seal for the Pre-Tertiary petroleum system.

One important major tectonic event that occurred during Mio-Pliocene is the collision of the Banda Microcontinent with the Australian Plate [17, 20]. The collision event is responsible for the formation of the thrust belt in Seram Basin. Hartono et al. [14] suggest that the thrusting period plays a vital role in the petroleum system of Seram Basin. First, the thrusting gives additional overburden resulting in the Pre-Tertiary carbonate source rocks maturation. Second, the thrusting event creates fracture networks allowing intraformational migration to occur. Last, the thrusting event leaked the previously accumulated oil in Manusela Formation to the post-collisional sediment, Fufa and Wahai Formations, or also as oil seepage via imbricate zone. The oil accumulated in Fufa Formation was then sealed by the intraformational shale creating Late Tertiary petroleum system [17]. The hydrocarbon preservation is changed and followed the latter event.

5. Result and Discussions

5.1 Early Detection for Biodegradation

The evidence of biodegradation in Seram oils is detected based on the relation between percent sulphur and API gravity with a general decreasing trend of sulphur with increasing API values of Seram oils (figure 2). The trend is clearly observed in oils retrieved from Pre-Fufa and Fufa clastic reservoirs, while oils from Manusela carbonate reservoir display a slight decreasing trend. While sulphur-rich oil has been attributed derived from marine carbonate source rock and immature source rock [11, 19], enrichment of sulphur could also be the result of biodegradation. Price et al. [11] pointed out that the sulphur-rich oil in Seram is originated from the nature of the oil, maturation, and biodegradation. However, this study believed that the increasing sulphur with decreasing API values is resulted dominantly from post-expulsion alteration, specifically biodegradation. The interpretation is based on several studies that conclude Seram oils are originated from marine carbonate source rock [11, 12, 13, 14, 19]. If the origin of the oils is similar, the sulphur should be in the same concentration in all samples. Moreover, Seram oils have identical maturity (Rc ~0.7; peak mature; Hartono et al., unpublished data) so that the enrichment of sulphur must not be related to maturity.

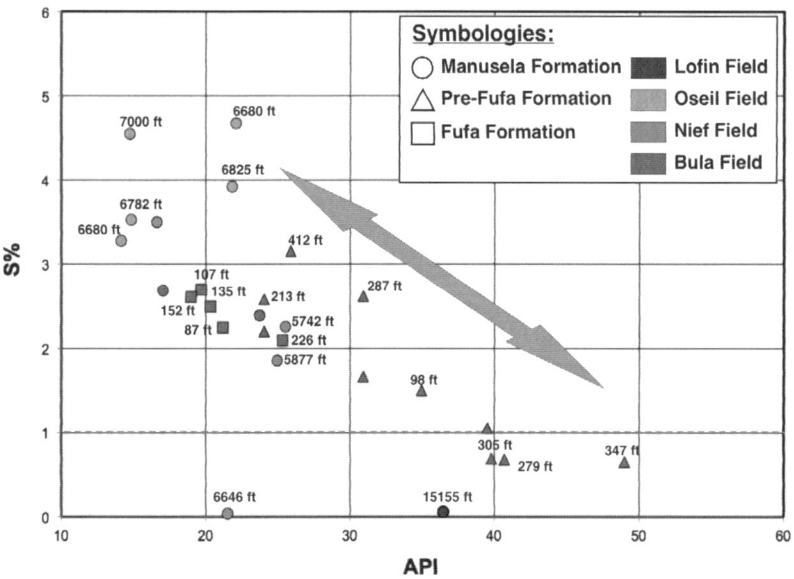

Figure 2. Cross plot of API values and percent sulphur displaying a negative correlation that indicates possible biodegradation.

The possibilities of biodegradation are also observed in the fashion of API-depth, S-depth, and saturate/polar (Sat/NSO)-depth relationship (figure 3). Generally, the API values and Sat/NSO ratio tend to be higher with increasing depth in Seram oils from three main reservoirs in Seram. In contrast, percent sulphur is relatively lower in the deeper reservoir (figure 3). These relationships displayed by Seram oils could directly indicate the occurrence of biodegradation. In the deeper reservoir, the reservoir temperature is relatively higher and most likely have high salinities (released from diagenetic fluids) that inhibit microbial activities [1, 3] so that the API values and the saturate fractions are intact.

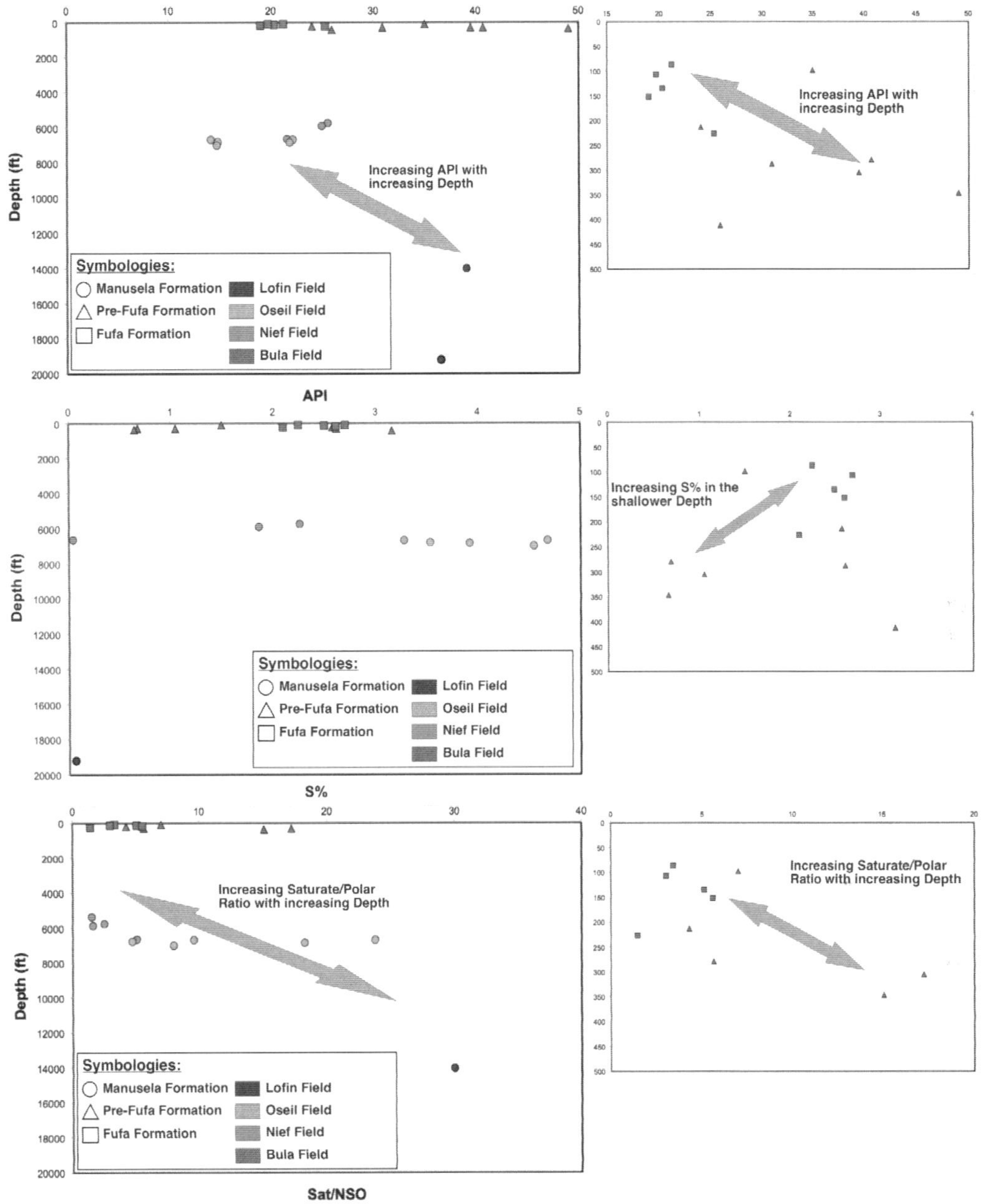

Figure 3. API-depth (above), S-depth (middle), and saturate/polar (Sat/NSO)-depth (below) plots. API and Sat/NSO ratios of oils retrieved from main reservoirs in Seram Basin, Manusela carbonate, and Pre-Fufa and Fufa clastics, are relatively higher, while the S tend to be lower in the deeper reservoir which is possibly associated with biodegradation.

5.2 Seram Oils Physio-Chemical Compositional Variation: A Geographical Manner

The relationship of physio-chemical parameters and depth (figure 3) shows that saturate and NSO fractions reflect the different degrees of biodegradation in this study. However, biodegradation could not only be seen in a single vertical point. Larter et al. [7] and Shuqing et al. [10] demonstrate that spatial physio-chemical variation in WCSB provides essential information regarding biodegradation. Therefore, spatial variation is also observed across the oil fields in this study.

Oils from Lofin Field have the highest API values with an average of 37.75°. However, geographically, the API values significantly decreased to the east and southeast, specifically from the west to the east, 17° in Oseil Field, 28° in Bula Field, and 22.5° in Nief Field, respectively (figure 4). Saturate and NSO fractions that have been considered to be sensitive parameters to biodegradation, display a clear pattern with E-W trend. The saturate fraction is significantly reduced from Lofin Field with the highest saturate fraction (74.59%) in the west to the eastern and southeastern oil field with oils. In addition, the NSO fraction is increased with a similar trend displayed by the saturate fraction with NSO fraction from the oil in Lofin Field reached 1.7% and increased to Bula Field reaching 17% in the relatively east and Nief Field reaching 15.38% in the southeast (figure 4).

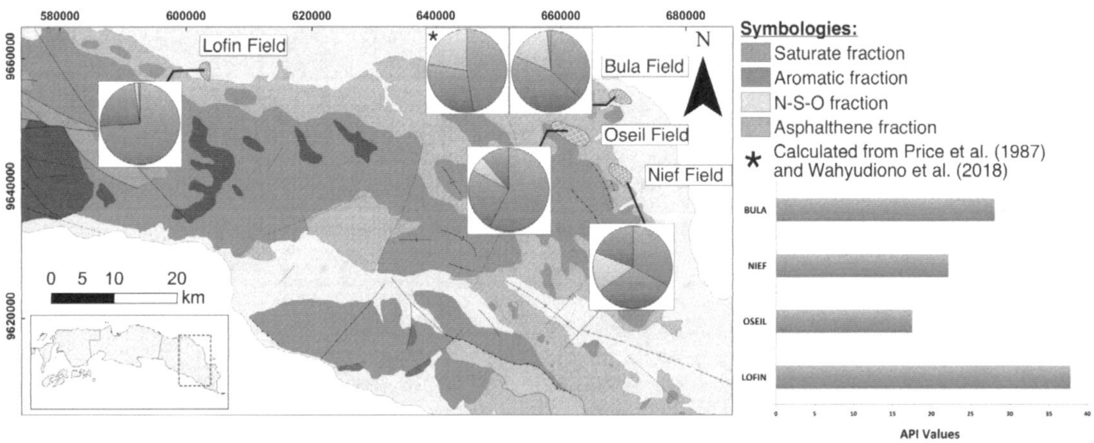

Figure 4. The physio-chemical composition variation across oil fields in Seram, from the west to the east, Lofin, Oseil, Bula, and Nief Fields (base map based on geological map modified from [15, 16]). The saturate fraction tends to decrease, while the NSO fraction tends to increase to the eastern and southeastern oil fields.

Whilst the spatial trend could be explained by the different origin and maturity of oils between the oil [21] retrieved from the east and west oil fields, Hartono et al. [14] suggest that Seram oils display similar origin and maturity. Indeed, oils in the southern part have deep marine carbonate affinities and shallow marine carbonate in the north [14], the differences of saturate and NSO fractions between the east and west oils should not be that significant because deep and shallow marine carbonate still classified as "carbonate source rock" so the expelled crude oils have identical hydrocarbon fractions [11, 22, 23]. In this case, the variation between these oils possibly resulted from hydrocarbon alteration during migration [4, 5] or after accumulated in the reservoir [1, 2, 3, 21]. However, most likely, based on the relationship of several parameters with depth (figure 4) and also API-S (figure 3), the spatial compositional variation between oils across Seram is linked with alteration after oils have been accumulated, specifically biodegradation. Some notes must be taken that biodegradation could occur with other alterations such as water washing [3]. Therefore, with adequate data, the possibilities of other alterations are open.

5.3 Biodegradation in Each Fields

In a geographical manner, it has been understood that biodegradation is possibly more extensive and intensive in the eastern and southeastern oil fields, specifically in Nief and Bula Fields. Oils from Lofin Field are relatively intact from any hydrocarbon alteration. First, oils from L-1 and L-2 wells have high API values, respectively 39° and 36.5°. These oils are also composed of a high saturate fraction that reached 59.98% and 89.2% which is commonly found on unaltered oil [2, 7, 21]. Further examinations from the chromatogram suggest that the n-alkanes of L-2 oil (figure 5) are still intact, especially the nC_6-nC_{12} still visible suggesting no biodegradation or level 0 biodegradation [1, 2].

IOP Conf. Series: Earth and Environmental Science 1047 (2022) 012032 doi:10.1088/1755-1315/1047/1/012032

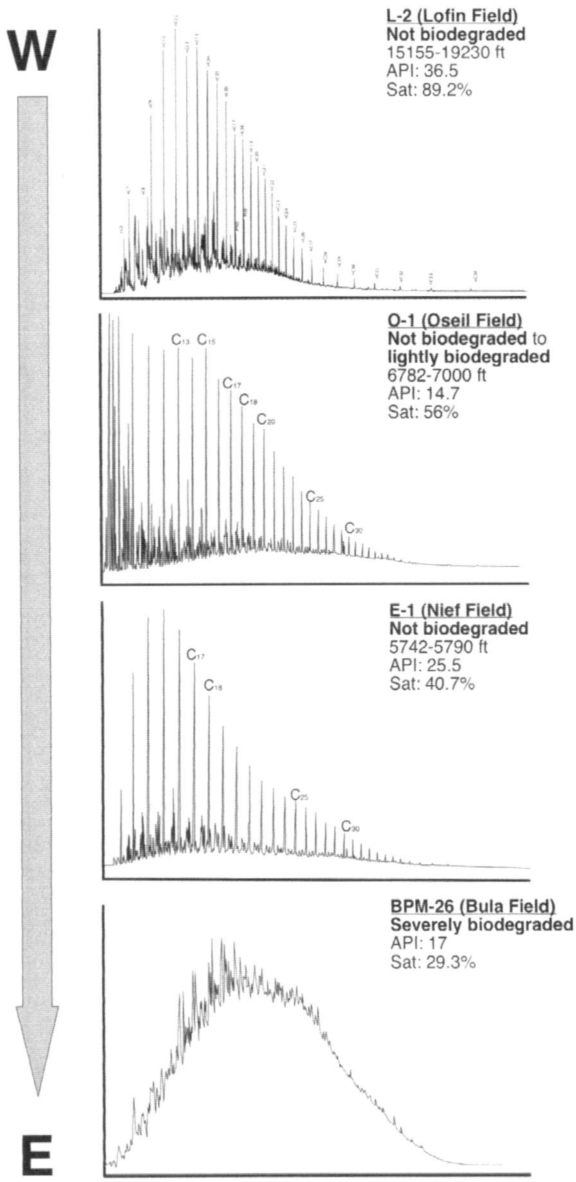

Figure 5. Whole oil chromatogram of represented oil samples from each field (referenced on figure 1). The biodegradation tends to be more extensive in the eastern oil fields since no peaks are observable from Bula Oil (represented by oil retrieved from BPM-26 well).

Further east, oils from Oseil Field have the lowest API values in Seram, for example, oil retrieved from O-1 have API that is only 14.5° with a moderate saturate fraction that is 56%. This low API could indicate that oil from Oseil Field, represented by oil from O-1, has been altered, especially biodegraded [2, 7]. However, the chromatograms tell differently with the nC_6-nC_{12} alkanes of Oseil oil is still intact (figure 5). Based on Wenger et al. [2] biodegradation scales, the oil must not be biodegraded or at least lightly biodegraded (level 1) as unresolved complex mixtures (UCM) is slightly observed. Therefore, the low API of Oseil oil is not related to biodegradation but to other hydrocarbon alterations. The possible explanation is the occurrence of water-washing since it has been suggested that there is possible fluid movement in Manusela reservoir [17]. This process also lowered the API values, saturate and aromatic fractions, and increased the polar fraction [3, 21].

IOP Conf. Series: Earth and Environmental Science 1047 (2022) 012032 doi:10.1088/1755-1315/1047/1/012032

In the most southeast, Nief oils share similar characters to Oseil oils with a moderate saturate fraction (Sat: 40.7%) but with slightly higher API values that reached 25.5°. The API is still categorized as low indicating that Nief oils, represented by oil from N-1, are also possibly altered. Nief oil chromatogram, however, suggests that no biodegradation has occurred since no UCM is observed and nC_6-nC_{12} alkanes are still intact (figure 5) [1, 2]. Therefore, the low API is also interpreted as the result of water-washing. This interpretation is based on the position of Nief and Oseil fields that is in the same lineaments with an NW-SE trend (figure 1) possibly related to the trend of major thrust-belt in Seram [15, 16, 17, 18, 20]. Both oil fields also produced the oils from Manusela reservoir [17].

Bula oils have been regarded by several authors to be extensively biodegraded [11, 12, 13]. The earliest work on Seram oils [11] shows that not limited to oil seeps, oils from the shallow Fufa reservoir tend to be biodegraded, making crude oils characterization in Seram Basin difficult. It is the fact that could not be neglected since our data also suggest that Bula oils have been extensively altered. Bula oils, represented by oil from BPM-26 and B-1, have a low API, respectively 17° and 23.7°. The most distinctive characters of Bula oils are the high sulphur content (S%> 2%), which reached 2.69% in BPM-26. Biodegradation in the Bula Field is also shown by the chromatogram that shows a prominent amount of UCM. Moreover, the n-alkanes have been completely unrecognized together with the isoprenoids (figure 5). Such characteristics indicate that oil from BPM-26 is severely biodegraded [1, 6] at least level 4 biodegradation [2]. To be more specific, other biomarkers such as the bicyclic alkanes (on m/z 123), terpanes (on m/z 191), and steranes (on m/z 217) are still intact (figure 6) in the Bula oils suggesting level 4 biodegradation [2]. Moreover, the 25-norhopanes that usually occurred in level 5-6 biodegradation [1, 2] are not observed in Bula oils (on m/z 177; figure 6) in line with Price et al. [11] observation confirming that biodegradation in Bula Field is level 4 biodegradation.

5.4 Geological Validation

The biodegradation discussed before must be in line with its geological condition. Lofin oils are retrieved from the deep reservoir that is more than 14,000 ft in depth (figure 7). The reservoir temperatures would be high enough to inhibit microbial activities in such depth. For example, in the L-2 well, the reservoir salinity is only 14 PSU that still opens up possibilities for biodegradation to occur (bacteria died at 100 PSU [3]). However, the bottom hole temperature (BHT) measurement in the reservoir suggests that the temperature reached 269°F which will inhibit microbial activities as bacteria only could withstand temperature only to 160°F [1, 2, 3]. Therefore, the geological conditions validate the reason why Lofin oils are unaltered.

Oseil and Nief oils are retrieved from the Manusela reservoir which is at 5500 ft to 7000 ft in depth (figure 7). No reservoir fluid analysis was performed in this field. Temperature data is only available in the Oseil Field, so that the reservoir temperature in Nief Field is estimated from the previous study [17]. Reservoir temperature measured in Oseil Field reached 165°F which is slightly higher than the threshold for bacteria to live [1, 2, 3]. Such condition still provides a little possibility for biodegradation to occur. Reservoir temperatures in Nief Field are calculated from the geothermal gradient that is 4.6°C/100 m (40.28°F/328 ft) [17]. With that geothermal gradient, the reservoir temperature in Manusela reservoir observed in Nief field (~5700 ft) should be at 175°F which will also inhibit microbial activities. These reservoir conditions confirm that biodegradation should not be occurred also in Oseil and Nief Fields. To confirm the possibilities of water-washing that is responsible for the low API of Oseil and Nief oils, more adequate data such as pressure data is needed to confirm the occurrence of hydrodynamic activities in the subsurface. Moreover, more proxies must be utilized such as C^{13} isotope and dibenzothiophene (DBT) [3, 21] to detect water-washing.

The severe biodegradation in Bula Field is clearly the result of the different reservoir targets in the Bula Area, which is the Quaternary Fufa and Tertiary Pre-Fufa reservoirs [11] while the other fields targeted Pre-Tertiary Manusela reservoir. Fufa and Pre-Fufa reservoirs are shallow reservoirs (< 1500 ft; figure 7) and have been targeted for exploration since the 1900s [11]. While no reservoir temperature and salinity data are available in this area, this study estimates the temperature from the given geothermal gradient [17] in the Fufa reservoir is only 80°F to 90°F. A similar observation in Oseil Field, the temperature measured in the ~1000 ft is 104°F. Therefore, the reservoir conditions of Fufa and Pre-Fufa Formations provide a suitable condition for bacteria to biodegrade the accumulated crude oils [1-3].

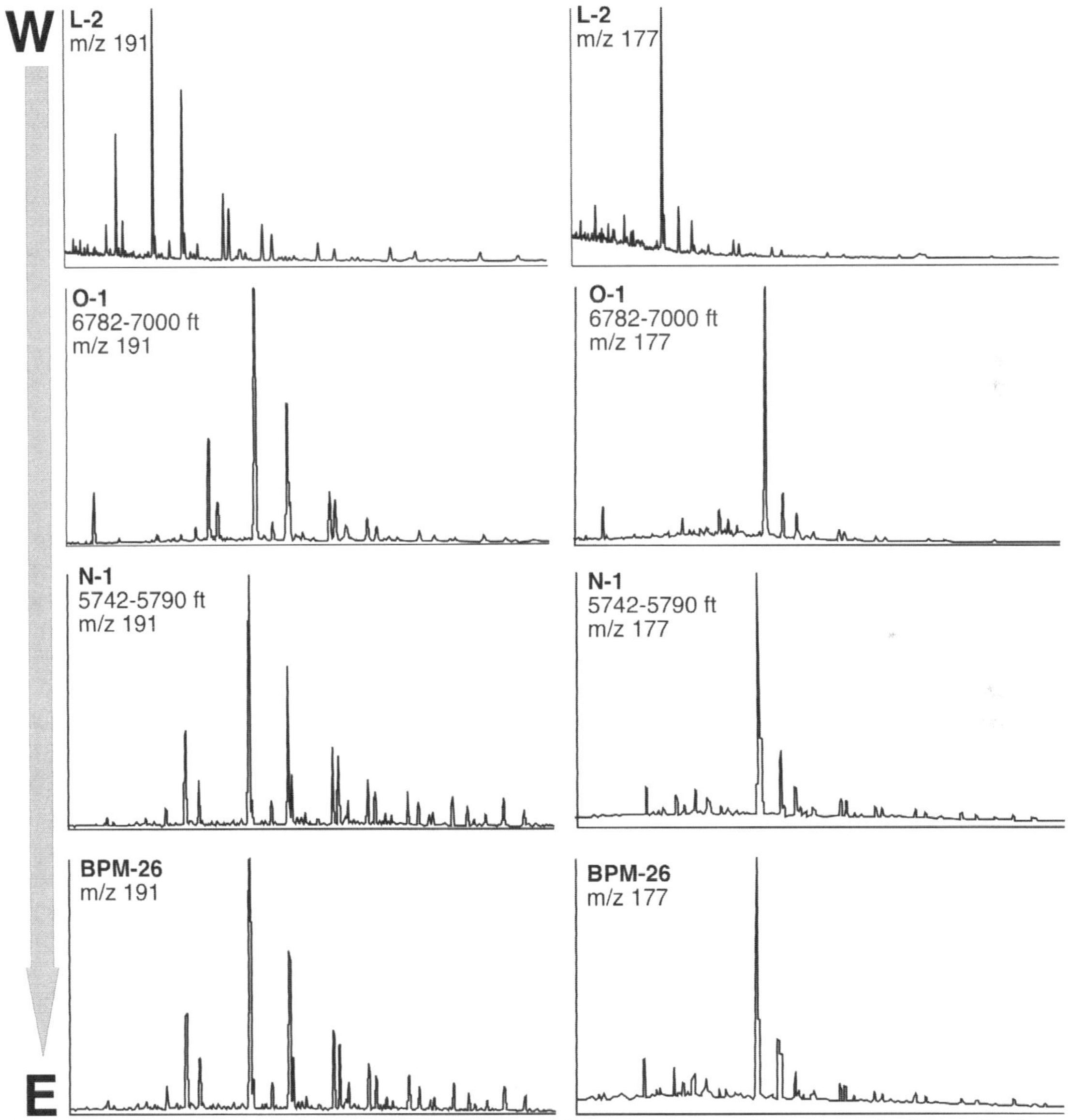

Figure 6. Partial m/z 191 (left) and m/z 177 (right) fragmentograms of represented oil samples from each field. The severely biodegraded Bula oil that is represented by sample from BPM-26 displays a similar character with other oils suggesting that the biodegradation is no more than level 4 biodegradation.

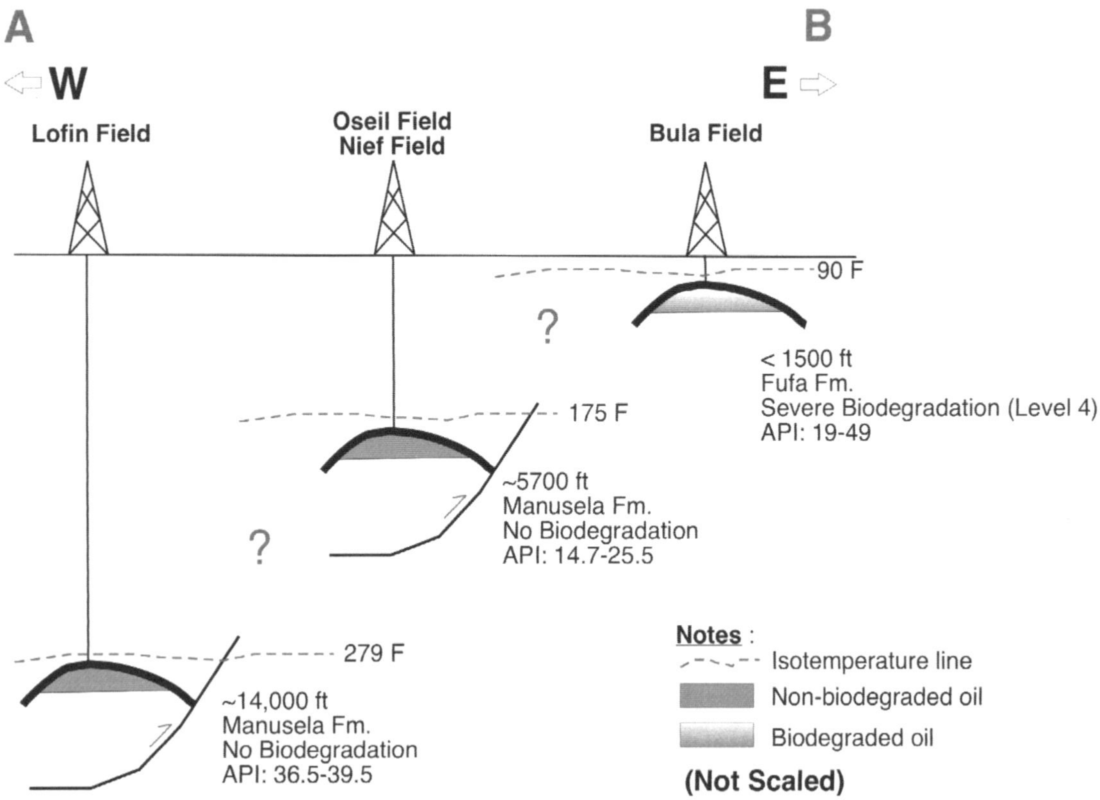

Figure 7. Schematic model of biodegradation processes in each field in Seram Basin that tends to be more extensive in the eastern part (cross-section A-B referenced on the red line in figure 1). The different reservoir depths are directly related to temperatures that control the degree of biodegradation.

5.5 Implication to the Petroleum System

From the biodegradation study, it has been understood that the oils retrieved from relatively west and south, Lofin, Nief, and Oseil Fields are intact to the biodegradation compared to the Bula Field that is situated more north (figure 7). This pattern is similar to the new petroleum system model proposed by Hartono et al. [14] that has several highlights, one of which is the timing of migration. The first migration in Seram Basin involved Saman-Saman Formation that expelled Family A oils and these oils are accumulated in the Manusela Formation. The collision of Banda microplate during Mio-Pliocene [16, 17, 18, 20] leaked this accumulation so that the previously accumulated oils are migrated to shallow Fufa and Pre-Fufa reservoirs, while oils expelled from the Manusela Formation, Family B oils, accumulated in the same Manusela intervals but with good porosity and permeability [14]. If this model is correct, then the biodegradation in the Fufa and Pre-Fufa reservoirs should be intensive. This study confirms the previous model as oils retrieved from Fufa reservoirs in Bula Field, which belongs to Family A, are severely biodegraded, while Family B oils are intact in the Manusela reservoirs.

6. Conclusions

To summarize, this study successfully demonstrates how oil biodegradation provides essential information regarding the petroleum system. In this case, the different degree of biodegradation in the Seram Area is in line with the newly proposed petroleum system model of Seram Basin. The extensive biodegradation of hydrocarbon in the Bula Field is indirectly the result of the Neogene deformation which leaked the previous accumulation so that the oils migrated to the shallower reservoir. Therefore, it is crucial to understand biodegradation in a basin together with its supporting geological factor to improve the knowledge of one petroleum system, especially for the critical time.

7. Acknowledgement

The authors would like to acknowledge CITIC Seram Energy Limited and Pusat Survei Geologi that has provided the data and allowed this research to be presented and published for International Seminar on Earth Sciences and Technology (ISEST) 2021.

References

[1] Volkman J K, Alexander R, Kagi R I, and Woodhouse G H 1983 *Organic Geochemistry* **47** 785-794
[2] Wenger L M, Davis C L, and Isaksen G H 2002 *SPE Reservoir Evaluation & Engineering* 375-383 SPE 80168
[3] Palmer S E 1984 *AAPG Bulletin* **68** 137-149
[4] Curiale J A and Bromley B H 1996 *Organic Geochemistry* **24** 1097-1113
[5] Curiale J A 2002 *Organic Geochemistry* **33** 1389–1400
[6] Peters K E, Walters C C, and Moldowan J M 2005 *The Biomarker Guide vol.2: Biomarker and Isotopes for Petroleum Exploration and Earth History* (Cambridge: Cambridge University Press)
[7] Larter S, Huang H, Adams J, Bennet B, Jokanola O, Oldenburg T, Jones M, Head I, Riediger C, and Fowler M 2006 *AAPG Bulletin* **90** 921-938
[8] Dembicki Jr. H 2017 *Petroleum Geochemistry for Exploration and Production* (Amsterdam: Elsevier Science)
[9] Scotchman I C, Griffith C E, Holmes A J, and Jones D M 1998 *Organic Geochemistry* **29** 671-700
[10] Zhou S, Huang H, and Liu Y 2008 *Petroleum Science* **5** 87-94
[11] Price P L, O'Sullivan T, and Alexander R 1987 *Proceedings Indonesian Petroleum Association, 16th Annual Convention and Exhibition* (Jakarta: IPA)
[12] Peters K E, Fraser T H, Amris H, Rustanto B, and Hermanto E 1999 *AAPG Bulletin* **83** 1927-1942
[13] Wahyudiono J, Adlan R, Permanadewi S, and Gibran A K 2018 *Jurnal Geologi dan Sumberdaya Mineral* **19** 233-241
[14] Hartono B M, Subroto E A, Kesumajana A H P, Andrianto R, Malvinas G, Wahyudiono J, and Priyanto B 2021 *Proceedings of the International Meeting on Organic Geochemistry* (Virtual: EAGE-EAOG)
[15] Tjokrosapoetro S, Rusmana E, and Suharsono 1993 *Geology of the Masohi quadrangle, Maluku scale 1:250,000* (Bandung: Geological Research and Development Centre)
[16] Charlton T R 2004 *AAPG Bulletin* **88** 565-585
[17] Kemp G and Mogg H 1992 *Proceedings Indonesian Petroleum Association, 21st Annual Convention and Exhibition* (Jakarta: IPA)
[18] Charlton T 2003 *IPA Newsletter*
[19] Livsey A R, Duxbury N, and Richards F 1992 *Proceedings Indonesian Petroleum Association, 21st Annual Convention and Exhibition* (Jakarta: IPA)
[20] Audley-Charles M G, Carter D J, Barber A J, Norvick M S, and Tjokrosapoetro S 1979 *Journal Geological Society of London* **136** 547-568
[21] Kuo L 1994 *Organic Geochemistry* **21** 465-479
[22] Palacas J G, Anders D E, and King J D 1984 South Florida Basin-prime example of carbonate source rock of petroleum in *Petroleum Geochemistry and Source Rock Potential of Carbonate Rocks* ed Palacas J G (Tulsa, Oklahoma: AAPG) pp 71-96
[23] Hughes H B 1984 Use of thiophenic organosulphur compounds in characterizing crude oils derived from carbonate versus siliciclastic sources in *Petroleum Geochemistry and Source Rock Potential of Carbonate Rocks* ed Palacas J G (Tulsa, Oklahoma: AAPG) pp 181-196

The Projections of Climate Change using Conformal Cubic Atmospheric Model (CCAM) in Bali - Indonesia

L Toersilowati[1], B Siswanto[1], E Maryadi[1], I Susanti[1], M Suhermat[1], A Witono[1], S B Sipayung[1], S A Rahayu[1], F 'Adany[1], J Aminuddin[2] and R F Lubis[3]

[1] Center for Atmospheric Research and Technology, Research Organization of Aeronautical and Space, National Research and Innovation Agency, Jl. dr. Djunjunan 133, Bandung, Indonesia
[2] Department of Physics, Faculty of Mathematics and Natural Sciences, Soedirman University, Jl. Dr. Suparno 61, Purwokerto, Jawa Tengah, 53122, Indonesia
[3] Geotechnology Research Center, Research Organization of Earth Science, National Research and Innovation Agency, Jl. Cisitu Sangkuriang, Bandung, Indonesia

larastursilowati@gmail.com

Abstract. Research on climate change was conducted in Bali, Indonesia, using the Conformal Cubic Atmospheric Model (CCAM), which works with the Representative Concentration Pathway (RCP) 4.5. The history data means average data from 1971 to 2000, climate projections with RCP4.5 scenario means average data from 2040 to 2069, and anomaly (climate change) is RCP4.5 minus history. The results represent the history of temperature between 22.5-27.5 ^{0}C and the projection between 25.5-29.5 ^{0}C. Temperature anomalies can be observed in much of northern Bali, which has increased from about 1.6 to 2.9 ^{0}C. There is a trend towards reduced (drier) humidity in most parts of Bali, particularly in the northern part of Bali. In contrast, a small amount in the south increases the (wetter) humidity. The comfort index of the Bali region in history is still relatively comfortable (20-26 ^{0}C). The projection condition RCP 4.5, there is no comfortable area with an index of more than 26 ^{0}C (hot and dry).

1. Introduction

Climate is the average weather where the weather is the state of the atmosphere at a certain point in time. Climate is defined as the average measure and the variability of the relevant quantity of certain variables (such as temperature, precipitation, or wind), at a certain period, which ranges from months to years or millions of years. Climate changes continuously due to interactions between its components and external factors such as volcanic eruptions, variations in sunlight, and factors caused by human activities such as changes in land use and use of fossil fuels [1, 2, 3].

Many cities are at risk of climate change impact because of unplanned urbanization. Climate change in cities will ultimately also affect local and global climate change [4, 5]. The largest emitters of greenhouse gases located in the cities, which contribute to climate change [6, 7]. Reference [8] investigated the magnitude of these temperature differences (up to 10 K) can be much larger than the expected temperature changes due to climate change.

The combination of relative humidity in the atmosphere and air temperature could indicate a comfort index in the region [9, 10, 11]. Temperature Humidity Index (THI) or Heat Humidity Index is a method to observe comfort index. The Heat Index combines high temperature and humidity that can reduce the body's ability to cool itself. THI is related to the body's ability to cope with the severity of summer in the monsoon period [12]. THI is also used to measure the comfort index of eggs in laying eggs [13, 14] and to estimate the comfort condition of dairy cows in producing milk [15, 16].

This paper investigates the climate change projections using Conformal Cubic Atmospheric Model (CCAM) that runs with the Representative Concentration Pathway (RCP)4.5. CCAM is a global atmospheric model developed for climate modeling and regional climate downscaling [17, 18]. This model applies a quasi-uniform grid or the principle of projecting the earth's surface into a cube panel [19]. Hence, this model does not require lateral boundaries in the process of running it. In CCAM, grid stretching is governed by Schmidt's transformation.

The RCP4.5 is a scenario that stabilizes radiative forcing at 4.5 Wm^{-2} in the year 2100 without ever exceeding that value due to the presence of greenhouse gas [20, 21]. Radiative forcing is the difference between the radiation variables received by the earth and the energy reflected outside the earth. The more significant radiative forcing will affect the energy that enters the earth, so that the earth is getting hotter and vice versa. The RCP4.5 is equivalent to the increase in CO_2 by 650 ppm [22, 23, 24, 25]. RCP4.5 itself is widely used in climate projection modeling to determine future climate conditions, with intermediate stabilization pathways that more reliable, representative, and closer to the Paris agreement. RCP 2.6 is one pathways where too low radiative forcing peak (3 Wm^{-2}) before 2010, while RCP 8.5 too high (8.5 Wm^{-2}) by 2100 and continues to rise.

2. Methodology

The data used as input model is data from the Global Forecast System (GFS) – National Climatic Data Center (NOAA) produced by National Centers for Environmental Prediction (NCEP). GFS data covers the whole world with a horizontal resolution between grid points of about 18 miles (28 kilometers). Weather forecaster use GFS data to predict the weather for the next 16 days [26]. The spatial resolution of CCAM runs 14 km. References [27, 28] stated this CCAM uses the Lagrangian system advection approach with horizontal bi-cubic interpolation. The CCAM grid can be focused over a region of interest using a Schmidt transformation [29, 30]. CCAM was part of the first generation of stretched grid atmospheric models. The stretched grid approach avoids lateral boundary conditions and can be smoothly deformed into a global grid. The Stretched Grid Model (SGM) allows feedback between regional and global circulation, avoiding the need for two-way nesting [17, 18, 19, 31].

The output data from CCAM runs used for climate change analysis are air temperature (T_a), surface temperature (T_s), and relative humidity (*RH*). *THI* estimation could be derived from *Ta* and *RH* using equation [32]:

$$THI = 0.8Ta + \frac{(RH \; x \; Ta)}{500} \tag{1}$$

The history data means average data from 1971 to 2000, climate projections with RCP4.5 scenario means average data from 2040 to 2069, and anomaly (climate change) is RCP4.5 minus history. Data processing is done by downscaling and re-gridding. The data processing flow chart is shown in figure 1.

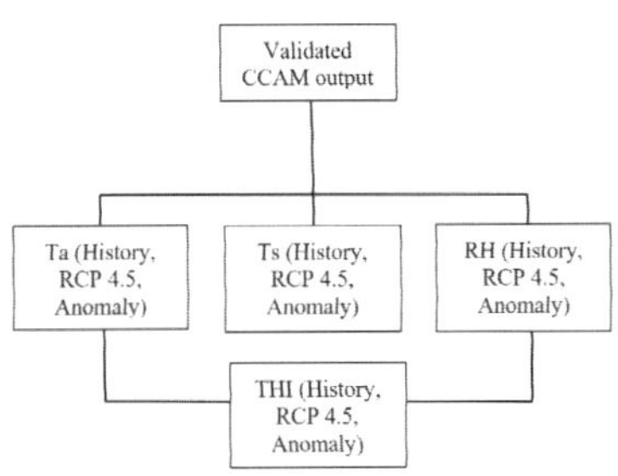

Figure 1. Data processing flow chart

IOP Conf. Series: Earth and Environmental Science 1047 (2022) 012033 doi:10.1088/1755-1315/1047/1/012033

The post-processing output data from CCAM runs use Grids Analysis Display System (GrADS) [33]. GrADS is an easy-to-use, interactive desktop tool. GrADS is used to manipulate and visualize earth science data, and it is freely distributed. GrADS has two data models for handling gridded and station data. GrADS can read data in various formats. In this study using the Net CDF data format.

The research location is the province of Bali, Indonesia, the longitude (114.4 – 115.8) ^{0}E, and latitude (8 – 8.9) ^{0}S (figure 2). Climate Change in Bali is interesting to study because Bali is one of the famous islands holiday destinations in the Indonesia archipelago, also known as the Island of the Gods and Thousand Temple Island. Bali has a tropical climate with severe environmental degradation in recent decades due to land use and lifestyles changes. The temperature rise in Bali caused by the tourist facility development has contributed to the microclimate warming. The rapid growth population has also influenced the microclimate change in Bali [34].

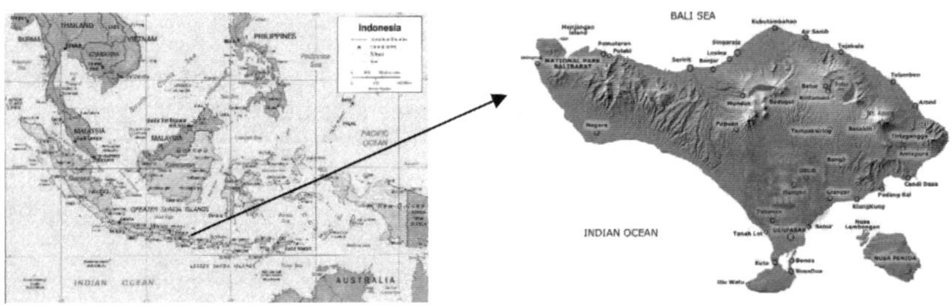

Figure 2. Bali Island, from https://www.wonderfulbali.com/bali-map/

3. Result and Discussion

3.1. *Ta history, RCP4.5, and anomaly*

Air temperature (Ta) is measured at a standard height of 1.2-2 m above the ground surface, while surface temperature is the temperature near the surface of the earth. Air temperature in Bali from the result of running CCAM is shown in figure 3. The *Ta* history (figure 3a) represents air temperature between 25-28 ^{0}C. The low temperature is seen in parts of Southern Bali, then further north is higher. The lowest temperatures (purple) are in the area around Ubud, Bangli, and Gianyar.

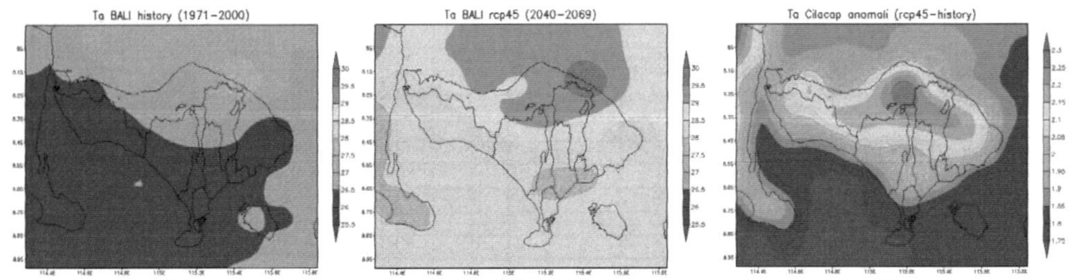

(a) (b) (c)

Figure 3. Air Temperature (*Ta*) Bali in ^{0}C, (a) History, (b) RCP4.5, and (c) Anomaly

Figure 3b illustrates the projection of Bali's air temperature (RCP4.5) with the temperature range between 27 until 30 ^{0}C with the pattern almost the same as figure 3a (the temperature getting north is higher. The low air temperature (25-26 ^{0}C), like in figure 3a, is no longer found in figure 3b. The magnitude of air temperature rise (anomaly) in Bali is shown by figure 3c, which ranges from 1.7 to 2.3 ^{0}C. The highest air temperature rise (around 2.3 ^{0}C) occurred in the Kintamani area (northern Bali).

3.2. *Ts history, RCP45, and anomaly*

Figure 4 represents the surface temperature of Bali a result of running CCAM. The history of surface temperature between 22.5 - 27.5 ^{0}C is shown in figure 4a. Low temperatures (22.5 ^{0}C) are located in the mountainous areas (rural areas in central Bali) and gradually rise to the beach (urban areas) (27.5 ^{0}C).

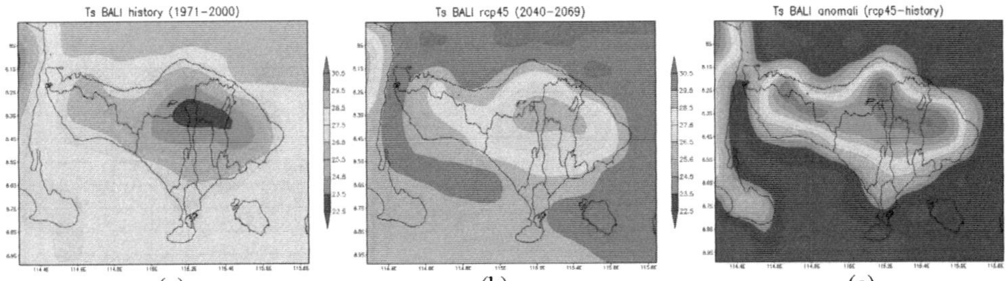

(a) (b) (c)

Figure 4. Surface Temperature (*Ts*) Bali in ^{0}C, (a) History, (b) RCP4.5, and (c) Anomaly

Figure 4b shows the spatial pattern of projected surface temperature with a range between 25.5 – 29.5 ^{0}C, with the lowest temperature in rural areas (mountains) and high in urban areas (beaches). There is a tendency to increase in surface temperature between history and projection. The low temperatures around 22.5 – 25.0 ^{0}C in figure 4a are no longer seen in the projection (figure 4b). Figure 4c shows the magnitude of the increase in surface temperature rise, which is around 1.6 to 2.9 ^{0}C.

Figure 4a (Ts Bali history) shows that the area with high temperatures is wider than figure 3a (Ta Bali history). Likewise, the high temperature in figure 4b (Ts Bali RCP4.5) is wider than figure 3b (Ta Bali RCP4.5). Air temperature is caused by the combination of air, water and land, hence surface temperature is hotter than air temperature.

3.3. *RH history, RCP4.5, and anomaly*

Relative humidity in Bali as a result of CCAM runs is shown in figure 5. The mean of relative humidity corresponding to the optimum comfort temperature was close to 73% [35]. *RH* history in Bali (figure 5a) represents the optimum comfort temperature close to 73% on the northern coast of Bali. The most humid area is in the eastern central part of Bali (Mt. Agung).

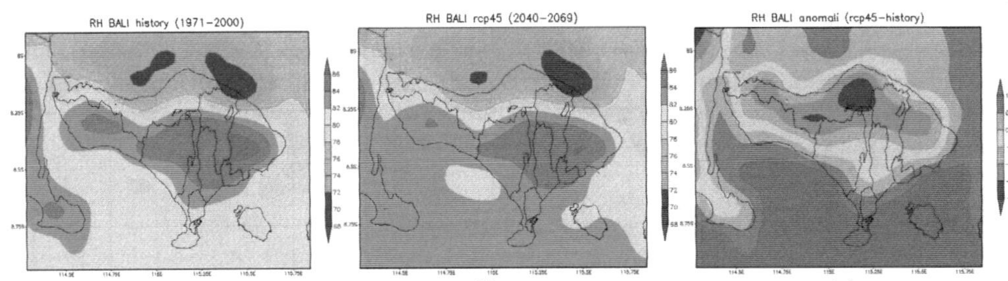

(a) (b) (c)

Figure 5. Relative Humidity (*RH*) Bali in %, (a) History, (b) RCP4.5, and (c) Anomaly

The projection of RH in figure 5b shows that there is a tendency of humidity decreases (drier) in most areas of Bali, especially the northern part of Bali, while a small part in the south of moisture increases (wetter) (figure 5c).

3.4. *THI history, RCP4.5, and anomaly*

Temperature Humidity Index is a single value representing the combined effects of air temperature and humidity associated with the level of thermal stress. Most people are quite comfortable (cool and wet)

when the index is below 20, comfortable when the index is between 20 – 26, and very uncomfortable (hot and dry) when the index is above 26. Figure 6 represents the history of the Comfort index Bali area still relatively comfortable (20-26), but on the condition, RCP4.5 year is no comfortable area for greater than 26 (hot and dry).

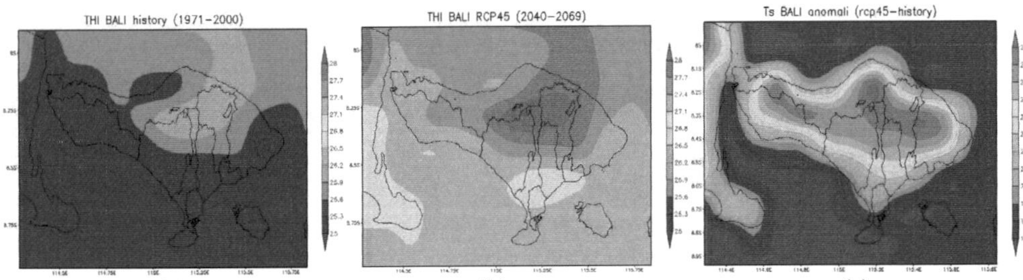

(a) (b) (c)

Figure 6. Temperature Humidity Index (*THI*) Bali, (a) History, (b) RCP4.5, and (c) Anomaly

3.5. Ta, Ts, RH, and THI analysis

Quantitatively, the Bali climate change projection can be shown in Table 1. Air Temperature (Ta) is measured at a standard height of 1.2 m above the ground surface. In general, air temperatures above the surface temperatures. This can be shown in the table 1, the history of Ta (25 ^{0}C – 28 ^{0}C) higher than Ts (22.5 ^{0}C – 27.5 ^{0}C), and the projection (RCP4.5) of Ta (27 ^{0}C – 30 ^{0}C) higher than Ts (25.5 ^{0}C – 29.5 ^{0}C). The anomaly of Ta around 1.7 ^{0}C to 2.3 ^{0}C, Ts (1.6 ^{0}C to 2.9 ^{0}C). RH history and projections are around 68% - 86%, but the changes of RH vary from -2% to 1%

Table 1. Projection of Climate Change in Bali.

	HISTORY	RCP4.5	Anomaly
Ta (^{0}C)	25 - 28	27 - 30	1.7 to 2.3
Ts (^{0}C)	22.5 - 27.5	25.5 - 29.5	1.6 to 2.9
RH (%)	68 - 86	68 - 86	-2 to 1
THI	25 - 26.1	26.8 - 28	1.75 to 2.05

The highest anomaly of temperature and RH is concentrated in a mountainous area. This case similar with the result of [36, 37], the surface temperature changes vary spatially, western Europe and parts of Asia shows the strongest hight altitude warming during the period of record. The seasonal and diurnal variability of RH is enhanced toward the mountain summit [38].

3.6. Statistical Analysis

To observe the trend of increasing or decreasing of surface temperature (Ts) and relative humidity (RH), a statistical study was carried out for 30 years (climatologically) from 1991 to 2021. The climatology of Ts and RH was observed in 9 district/municipalities of Bali. Each regency/municipality is represented by 1 city, except for the districts of Buleleng and Bangli which are represented by 2 cities/regions. Jembrana regency was represented by Negara, Tabanan by Tabanan, Badung by Badung/Mangupura, Denpasar by Denpasar, Gianyar by Gianyar, Klungkung by Semarapura, Bangli by Bangli and Batur, Karangasem by Amlapura, Buleleng by Singaraja and Gilimanuk (figure 2).

3.6.1. Ts statistical analysis. The trend of Ts in Bali is shown in figure 7, and the linear equation is shown in table 2. Here it can be seen that all regions have positive trends, meaning that there is a tendency to increase of temperatures. The temperature increase in all areas of Bali is in accordance with the Ts projection and anomaly as shown in the figure 4b and 4c.

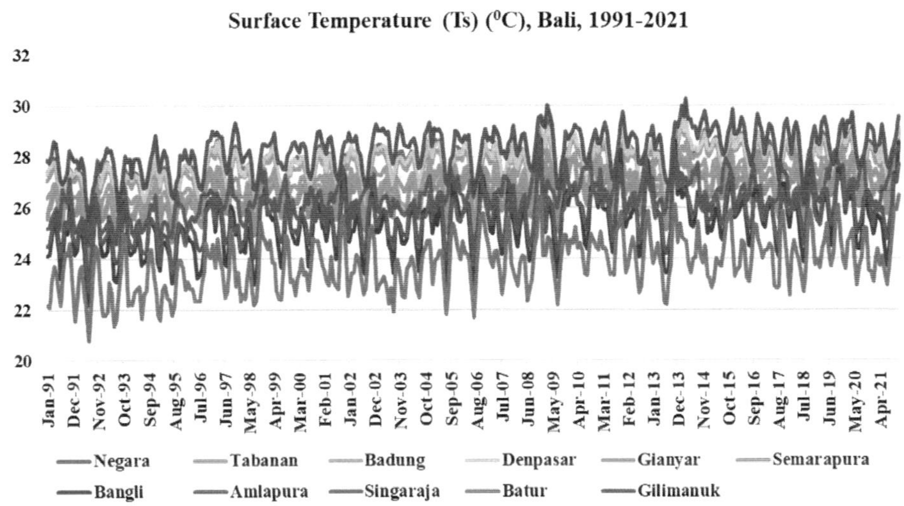

Figure 7. Climatology of Surface Temperature (Ts) Bali (^{0}C), 1991-2021

Table 2 represent the average of Ts from highest to lowest are Gilimanuk (28.323 ^{0}C), Negara (27.712 ^{0}C), Denpasar (27.715 ^{0}C), Tabanan (27.435 ^{0}C), Semarapura (26.907 ^{0}C), Gianyar (26.576 ^{0}C), Singaraja (26.477 ^{0}C), Badung (26.255 ^{0}C), Amlapura (25.626 ^{0}C), Bangli (25.348 ^{0}C), and Batur (23.801 ^{0}C).

Table 2. Ts equation trends and Ts Average (^{0}C), Bali, 1991-2021

	District	Capital/City	Ts equation trends	Ts average
1	Jembrana	Negara	y = 0.0001x + 23.55	27.712
2	Tabanan	Tabanan	y = 0.0001x + 23.171	27.435
3	Badung	Badung/Mangupura	y = 0.0001x + 22.009	26.255
4	Denpasar	Denpasar	y = 0.0001x + 23.501	27.715
5	Gianyar	Gianyar	y = 0.0001x + 21.902	26.576
6	Klungkung	Semarapura	y = 0.0001x + 22.527	26.907
7	Bangli	Bangli	y = 0.0001x + 20.072	25.348
8	Bangli	Batur	y = 0.0002x + 18.199	23.801
9	Karangasem	Amlapura	y = 0.0001x + 20.613	25.626
10	Buleleng	Singaraja	y = 0.0001x + 21.343	26.477
11	Buleleng	Gilimanuk	y = 0.0001x + 24.236	28.323

Temperature falls 6.5 ^{0}C for each 1 km altitude increase. Hence, Gilimanuk as the beach area in the west tip of Bali (altitude about 4 m above sea level) have the hottest temperature. Otherwise in the Batur mountain as the highest study area (altitude about 1000 m above sea level) located in east of Bali have the coldest temperature.

3.6.2. RH analysis. The trend of RH in Bali is shown in figure 8, and the linear equation is shown in table 3. A positive RH trend means the area is wetter than before, otherwise if the RH trend is negative, the area is drier. The positive trends that there is a tendency to increase of RH can be seen in Negara, Tabanan, Denpasar, and Semarapura. On the other hand, the negative trends that means the tendency to decrease of RH can be seen in Badung, Gianyar, Bangli, Batur, Amlapura, Singaraja, and Gilimanuk.

The tendency for an increase or decrease in RH is in accordance with the spatial pattern shown in figure 5b (RH projection) and figure 5c (RH anomaly).

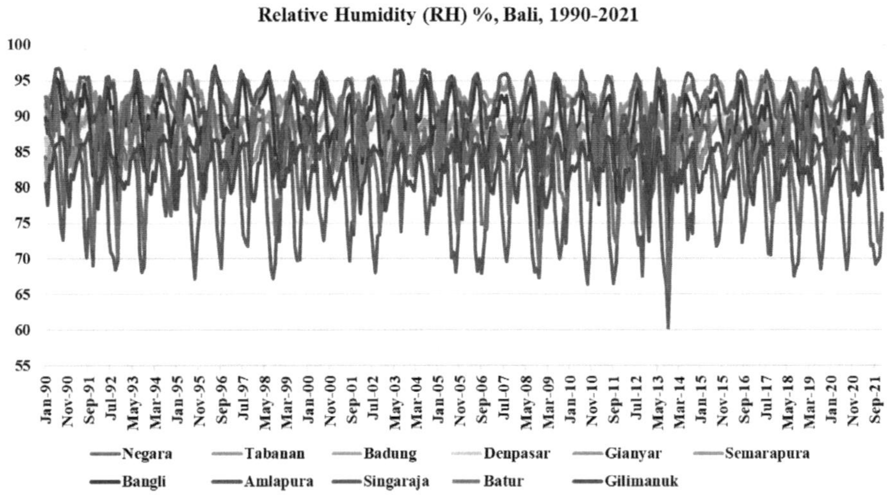

Figure 8. Climatology of Relative Humidity (TH) Bali (%), 1991-2021

Table 3 represent the average of RH from highest to lowest are Badung (92.862%), Gianyar (92.1802%), Amlapura (92.388%), Semarapura (90.032%), Bangli (89.180%), Denpasar (88.312%), Negara (87.072%), Tabanan (87.016%), Batur (83.925%), Gilimanuk (83.244%), and Singaraja (80.769%).

Table 3. Regression Linear Equation and Average of RH (%) Bali, 1991 - 2021

	District	Capital/City	RH equation trends	RH average
1	Jembrana	Negara	y = 7E-06x + 86.796	87.072
2	Tabanan	Tabanan	y = 5E-06x + 86.806	87.016
3	Badung	Badung/Mangupura	y = -2E-05x + 93.531	92.862
4	Denpasar	Denpasar	y = 1E-05x + 87.789	88.312
5	Gianyar	Gianyar	y = -1E-05x + 93.302	92.802
6	Klungkung	Semarapura	y = 2E-06x + 89.936	90.032
7	Bangli	Bangli	y = -9E-05x + 92.815	89.180
8	Bangli	Batur	y = -0.0002x + 90.273	83.925
9	Karangasem	Amlapura	y = -4E-05x + 94.026	92.388
10	Buleleng	Singaraja	y = -0.0002x + 86.663	80.769
11	Buleleng	Gilimanuk	y = -2E-06x + 83.315	83.244

Relative Humidity reflects the percentage of saturation of the atmosphere, it is basically an amount of water vapor present in the air. In the higher altitude, the atmosphere becomes thinner (less pressure), this means that the total amount of water vapor of the atmosphere can potentially hold is decreased. The higher percentage of relative humidity in the air says that it is more humid. It is not dependent on the height but on the capacity of the air to hold moisture. Evaporation in the southern part of Bali more effective than in the northern part because it is influenced by Indian ocean evaporation. Therefore, the southern area of Bali (including Badung) has a high water vapor content which represent a high RH.

4. Conclusions

The Climate Change in Bali has been investigated using CCAM. Based on history IPCC AR5 as baseline climatology average (1971 - 2000), climate projections with RCP4.5 scenario (2040 - 2069), while the anomaly (climate change) is the difference between RCP4.5 and history. The air temperature (*Ta*) history temperature is between 25-28 ^{0}C. The low *Ta* is seen in parts of Southern Bali, then further north is higher. The projection of Bali's air temperature (RCP4.5) with the temperature range between 27 until 30 ^{0}C has a pattern similar to *Ta* history. The low temperature (25-26 ^{0}C), like Ta history, is no longer found in *Ta* projection. The magnitude of air temperature rise (anomaly) in Bali ranges from 1.7 to 2.3 ^{0}C. The highest air temperature rise (around 2.3 ^{0}C) occurred in the Kintamani area (northern Bali). The history of surface temperature in Bali is between 22.5 - 27.5 ^{0}C, and RCP4.5 between 25.5 - 29.5 ^{0}C. The temperature anomalies can be seen in most of northern Bali, which increased by about 1.6 to 2.9 ^{0}C. Ts Bali history and RCP shows that the area with high temperatures is wider than Ta Bali history and RCP. Air temperature is caused by the combination of air, water and land, hence surface temperature is hotter than air temperature. There is a reduced humidity tendency (drier) in most parts of Bali, especially the north part of Bali, while a small portion in the south increases moisture (wetter). The highest anomaly of temperature and RH is concentrated in a mountainous area. The comfort index of the Bali region in history is still relatively comfortable (20 - 26), but on the condition RCP4.5 there is no comfortable area with an index more than 26 (hot and dry).

References

[1] National Research Council 2020 *Climate Change: Evidence and Causes: Update 2020.* Washington, DC: The National Academies Press, https://doi.org/10.17226/25733

[2] Aubry T J, Staunton-Sykes J, Marshall L R, Haywood J, Abraham N L and Schmidt A 2021 Climate change modulates the stratospheric volcanic sulfate aerosol lifcycle and radiative forcing from tropical eruptions. *Nature Communications* **12**:4708 https://doi.org/10.1038/s414617-021-24943-7

[3] El-Ramady H, Belas A A and El-Marsafawy S M 2012 *Contemporary Environmental Readings Volume 1 Climate Change A blessing or a Curse fo Agriculture.* Lambert Academic Publishing, https://www.researchgate.net/publication/256457778

[4] Tursilowati L, Sumantyo J T S, Kuze H and Adiningsih E S 2012 The integrated WRF/urban modeling system and its application to monitoring urban heat island in Jakarta, Indonesia. *Journal of Urban and Environmental Engineering* (JUEE) **6**(1) pp 1-9 ISSN 1982-3932 doi: 10.4090/juee.2012.v6n1.001009

[5] McCarthy M P, Best M J and Betts R A 2016 Climate change in cities due to global warming and urban effects. *Geophysical Research Letters* **37** L09705 doi: 10.1029/2010GL042845

[6] Hoornweg D, Sugar L and Gomez C L T 2014 Cities and greenhouse gas emissions: moving forward, *Environment & Urbanization* **23**(1) 207-227 doi: 10.1177/0956247810392270

[7] IPCC, Climate Change 2007 Impacts, Adaptation and Vulnerability *Forth Assessment Report* Cambridge University Press New York.

[8] Yow D M 2007Urban heat island: Observations, impacts, and adaptation, *Geography Compass.* Wiley **1**(6) pp 1227-1251 https://doi.org/10.1111/j.1749-8198.2007.00063.x

[9] Tursilowati L, Sumantyo J T S, Kuze H and Adiningsih E S 2012 Remote sensing technology for estimation of surface energy balance components relate with land use and land cover in Semarang-Indonesia *Research Journal in Engineering Sciences* (RJEAS) **1**(5) pp 291-298

[10] d'Ambrosio Alfano F R, Palella B I and Riccio G 2011 Thermal environment assessment reliability using temperature – humidity indices *Industrial Health* **49** pp 95-106

[11] Epstein Y and Moran D S 2006 Thermal comfort and the heat stress indices *Industrial Health* **44** pp 388-398

[12] Mohan M, Gupta A and Bhati S 2014 A modified approach to analyze thermal comfort classification *Atmospheric and Climate Sciences* **4** pp 7-11 http://dx.doi.org/10.4236/acs.2014.41002

[13] Kilic I and Simsek E 2013 The effects of heat stress on egg production and quality of laying hens. *J. of Anim. and Vet. Adv.* **12**(1) pp 42-47

[14] Barbosa Filho J A D, Silva M A N, Silva I J O and Coelho A A D 2005 Egg quality in layers housed in different production systems and submitted to two environmental conditions *Brazilian J. of Poultry Sci.* **8**(1) pp 23-28

[15] Gantner V, Mijić P, Kuterovac K, Solić D and Gantner R 2011 Temperature-humidity index values and their significance on the daily production of dairy cattle *Mljekarstvo* **61**(1) pp 56-63

[16] Dikmen S and Hansen P J 2009 Is the temperature-humidity index the best indicator of heat stress in lactating dairy cows in a subtropical environment? *J. Dairy Sci.* **92** pp 109-116 American Dairy Science Association doi:10.3168/jds.2008-1370

[17] McGregor J L 2005 C-CAM: Geometric Aspects and Dynamical Formulation (electronic publication). *CSIRO Atmospheric Research Technical Paper No. 70*

[18] McGregor J L and Dix M R 2008 An updated description of the Conformal-Cubic Atmospheric Model. In: Hamilton, K., Ohfuchi, W. (Eds.) *High Resolution Simulation of the Atmosphere and Ocean* Springer pp 51–76

[19] McGregor J L 1993 Economical determination of departure points for Semi-Lagrangian models. *Mon. Wea. Rev.* **121** pp 221-230

[20] Thomson A M, Calvin K V, Smith S J, Kyle G P, Volke A, Patel P, Dlegado-Arias S, Bond-Lamberty B, Wise M A, Clarke L E and Edmonds J A 2011 RCP4.5: a pathway for stabilization of radiative forcing by 2100 *Climatic Change* **109** pp 77-94, doi:10.1007/s10584-011-0151-4

[21] van Vuuren D P, Edmonds J, Kainuma M, Riahi K, Thomson A, Hibbard K, Hurtt G C, Kram T, Krey V, Lamarque J-F, Masui T, Meinshausen M, Nakicenovic N, Smith S J and Rose S K 2-11 The representative concentration pathways: an overview *Climatic Change* **109**(1) pp 5-31

[22] Zhang D and Zheng W 2004 Diurnal cycles of surface winds and temperatures as simulated by five boundary layer parameterizations *J. Appl. Meteor.* **43** pp 68–75

[23] Smith S J and Wigley T M 2006 Multi-gas Forcing stabilization with the MiniCAM *The Energy Journal Multi-Greenhouse Gas Mitigation and Climate Policy* Special Issue #3 373–391 doi: 10.5547/ISSN0195-6574-EJ-VolSI2006-NoSI3-19

[24] Clarke L, Edmonds J, Jacoby H, Pitcher H, Reilly J and Richels R 2007 Scenarios of the greenhouse gas emission and atmospheric concentrations. *Sub-report 2.1A of Synthesis and assessment product 2.1 by the U.S. climate change science program and the subcommittee on global change research* Washington USA Department of Energy Office of Biological & Environmental Research p 154

[25] Wise M A, Calvin K V, Thomson A M, Clarke L E, Bond-Lamberty B, Sands R D, Smith, S J. Janetos A C and Edmonds J A 2009 Implications of limiting CO_2 concentrations for land use and energy *Science* **324** pp 1183–1186

[26] NOAA, Global Forecast System (GFS), Formerly the National Climatic Data Center, https://www.ncdc.noaa.gov/data-access/model-data/model-datasets/global-forcast-system-gfs, last accessed 2020/03/05.

[27] National Science Foundation, Solving the Puzzle 2009 Researching the impacts of climate change around the world *Jupiter images Corporation* p 114

[28] Prasad N, Ranghieri F, Shah F, Trohanis Z, Kessler E and Sinha R 2009 Climate resilient cities: A primer on reducing vulnerabilities to disasters *The World Bank*, Washington DC p 186

[29] Meyers R A 2012 Mathematics of complexity and dynamical systems *Encyclopedia of Complexity and Systems Science* Springer New York

[30] Wimp J 1981 Sequence Transformations and Their Applications *Mathematics in Science and Engineering* Vol. 154 Academic Press Inc. New York

[31] McGregor J L 1966 Semi-Lagrangian on conformal-cubic grids. *Mon. Wea. Rev.* **124** pp 1311-1322

[32] Niewolt S 2013 Tropical Climatology, New York: John Wiley & Sons, 1977, (in Singh R P, Parr J F and Steward B A): Advances in Soil Science: Dryland Agriculture: Strategies for Sustainability Springer *Science* 373 pages
[33] GrADS, Grid Analysis and Display System *George Mason University* http://cola.gmu.edu/grads/, last accessed 2020/04/10
[34] Rajendra A 2020 Climate change impacts on the coastal tourist resorts of Bali *IOP Conf. Series: Earth and Environmental Science* **423** pp 1-8 doi: 10.1088/1755-1315/423/I/012044
[35] Djamila H, Chu C M and Kumaresan S 2014 Effect of humidity on thermal comfort in the humid tropics *Journal of Building Construction and Planning Research* Scientific Research **2** pp 109-117
[36] Diaz H F and Bradley R S 1997 Temperature variations during the last century at hight elevation sites *Climate Change* **36** pp 253-279
[37] Sambuco E N, Mark B G, Patrick N, DeGrand J Q, Porinchu D F, Reinemann S A, Baker G M and Box J E 2020 Mountain temperature changes from embedded sensors spanning 2000 m in Great Basin National Park, 2006-2018 *Frontiers in Earth Science* **8**(292) pp 1-18 doi: 10.3389/feart.2020.00292
[38] Duane W J, Pepin N C, Losleben M L and Hardy D R 2018 General Characteristics of Temperature and Humidity Vaiability on Kilimanjaro, Tanzania *Archtic, Antarctic, and Alpine Research* **40**(2) pp 323-334 ISSN: 1523-0430 (Print) 1938-4246 (online) doi: 10.1657/1523-0430(06-127)[DUANE]2.0.CO;2

Acknowledgment
The author would like to thank the Head of LAPAN Center for Atmospheric Science and Technology who has provided the facilities I needed to conduct research. The author would also like to thank the Climate Change group who assisted in the analysis and discussion.

Contributor Statements
Laras Toersilowati as a primary contributor (author), devised and directed the research, data processing, and result analysis. Other contributing authors are Bambang Siswanto, and Edy Maryadi, run the CCAM and computational framework. Indah Susanti, Mamat Suhermat, Adi Witono, and Sinta Berliana Sipayung, visualization data and assisted methodology development. Fildzah 'Adany, Jamrud Aminuddin, Soni Aulia Rahayu, and Rachmat Fajar Lubis are assisted in data processing, and verification of results. All authors provided feedback, discussed results, and contributed to the final manuscript.

Characteristics of volcanic rocks and associated intrusions based on petrography analysis in Jari-Krondonan Area, Bojonegoro Regency, East Java

J Marin[1], Y M Muhammad[1] and T Winarno[1]

1 Department of Geological Engineering, Faculty of Engineering, Universitas Diponegoro, Jl. Prof. Soedarto, SH, Tembalang, Semarang, Jawa Tengah, Indonesia 50275

Corresponding author: jenianmarin@gmail.com

Abstract: Pandan volcano is one of the dormant Quaternary volcanoes in East Java characterized by several volcanic centers which extend to the northern part. This study is aimed to identify the volcanic rocks and intrusions of Jari - Krondonan area, Bojonegoro as well as determine their petrology and mineralogical characteristics. Geological observation was conducted to obtain primary data and to collect samples. Thin section of samples was prepared to analyze the petrographic aspects. Field observation shows several volcanic hills at Mount Jati, Mount Puru, Mount Watu, and Mount Lawang sites. Based on field mapping, lithologies of the research area are andesitic lava and intrusion with columnar joint or sheeting joint structure, surrounded by andesitic breccia. Andesite characterized by porphyritic texture with visible phenocrysts and volcanic glass groundmass. Samples from Mount Lawang and Mount Watu are composed of plagioclase and hornblende as main minerals. Samples from Mount Jati and Mount Puru are composed of plagioclase and pyroxene. Quartz, sanidine, and olivine present as accessory minerals. Plagioclase and pyroxene occasionally formed glomerocryst and poikilitic texture, which play an important role in the fractionation and crystal settling processes. Sieve and regular zoning in plagioclase suggest magma mixing. Gabbro and metamorphic xenoliths found in Mount Lawang and Mount Watu indicates an interaction with country rock during magma rising.

1. Introduction

Pandan Volcano is a Pleistocene volcano [1], part of the Quaternary Sunda-Java volcanic arc magmatism [2]. The volcano is located in the Wilis – Lasem arc segment, East Java Province (Figure 1). Currently, the volcano is not active but still shows magmatic activity as indicated by the presence of geothermal manifestations such as hot springs and travertine on the northern slope of the volcano [1, 3]. Stratigraphy of Pandan Volcano is composed of the Pandan Breccia Formation consisting of andesite breccia and andesite lava as well as several intrusions were found in the northern part around Jari and Krondonan, Gondang District, Bojonegoro Regency, East Java. Pandan Volcano, which is Lower Pleistocene, situated in the Kendeng Zone. Kendeng Zone is Neogene sedimentary rock formations consist of volcaniclastic and pelagic marine successions [4] (Figure 2).

Petrology and volcanostratigraphy studies are discussed in several sites of igneous rocks from Pandan Volcano, particularly associated with geothermal manifestations. Several volcanic centers of Pandan Volcano on the northern slope are produced andesite lava, pyroclastic fall deposit, and pyroclastic flow breccia. There is also porphyritic andesite intrusion of Selo Gajah Hill which cause contact metamorphism effect on its surrounded sedimentary rocks [5]. Previous study on the petrological and geochemical characteristics of Pandan Volcano stated that magmatic activity is the source of heat in the surrounding geothermal system. The study found that the igneous rocks in the area are hornblende andesite and basaltic andesite originating from basaltic-andesitic magma with calc-alkaline affinity [3]. A long history of volcanoes and their evolution are related to magmatic processes which produce igneous rocks that built them. Igneous texture recorded in volcanic rocks and intrusions is an essential key in recognizing these processes. The aims of this study were to identify the distribution of volcanic rocks and intrusions of the research area, determine their petrographic and mineralogical characteristics, and interpret the magmatic processes. The results were expected to provide more detailed information as an integrated igneous petrology study of magmatism in Java Island and the evolution of Pandan Volcano as well.

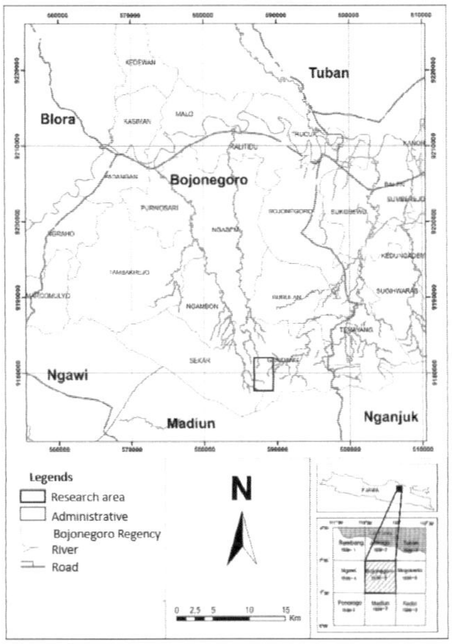

Figure 1. Administrative location of the research area in Bojonegoro Regency, East Java

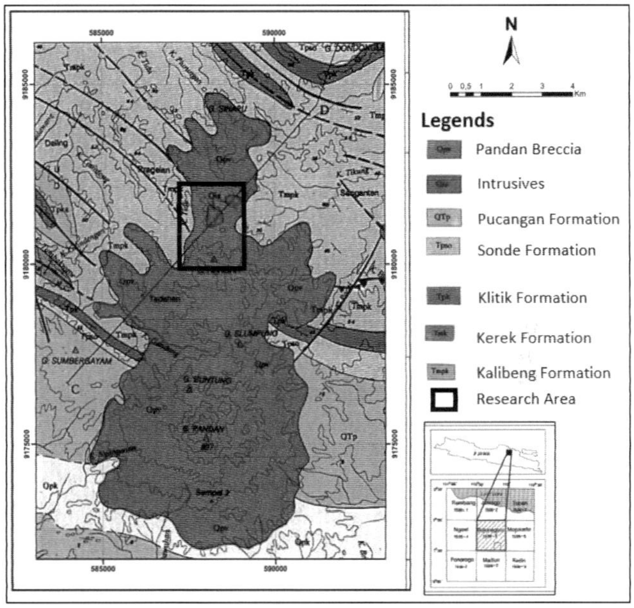

Figure 2. Geological map of Pandan Volcano and its surrounding [4]

2. Methods

Field observation and geological mapping were conducted specifically at Jari-Krondonan village area to collect geological field data consist of geomorphology, lithologic description, and geological structure measurement. To delineate the geological unit of the volcanic area, it is noteworthy to correlate the geomorphological feature with lithological unit. Rock sample were taken from igneous rocks of lava and intrusion, as volcaniclastic rock cannot be used to determine primary magmatic processes due to its nature as secondary product of volcanic activity.

Eight selected rock samples were prepared as thin section with 30 μm thickness to represent each volcanic rocks and intrusions sites found in the field. To obtain the best results, the rock samples had to be as fresh as possible with minimum weathering. Petrographic analysis was performed by observing thin section under polarization microscope to determine mineralogical composition and igneous textures.

3. Results and Discussion

3.1. Geological setting of the study area

Based on the geological mapping, there are four sites of volcanic rocks and intrusions in the study area, namely hornblende andesite from Mount Watu and Mount Lawang and pyroxene andesite from Mount Jati, and Mount Puru (Figure 3). Morphology of the study area shows moderate – steep slope volcanic hill (Figure 4). The lithology units can be distinguished based on geomorphological and physical characteristics of the outcrop, the latter confirmed further by petrography analysis. In addition to igneous rocks from the four main locations as mentioned above, several other units are volcanic breccia from Pandan Breccia Formation, also sandstone and claystone that belong Kalibeng Formation.

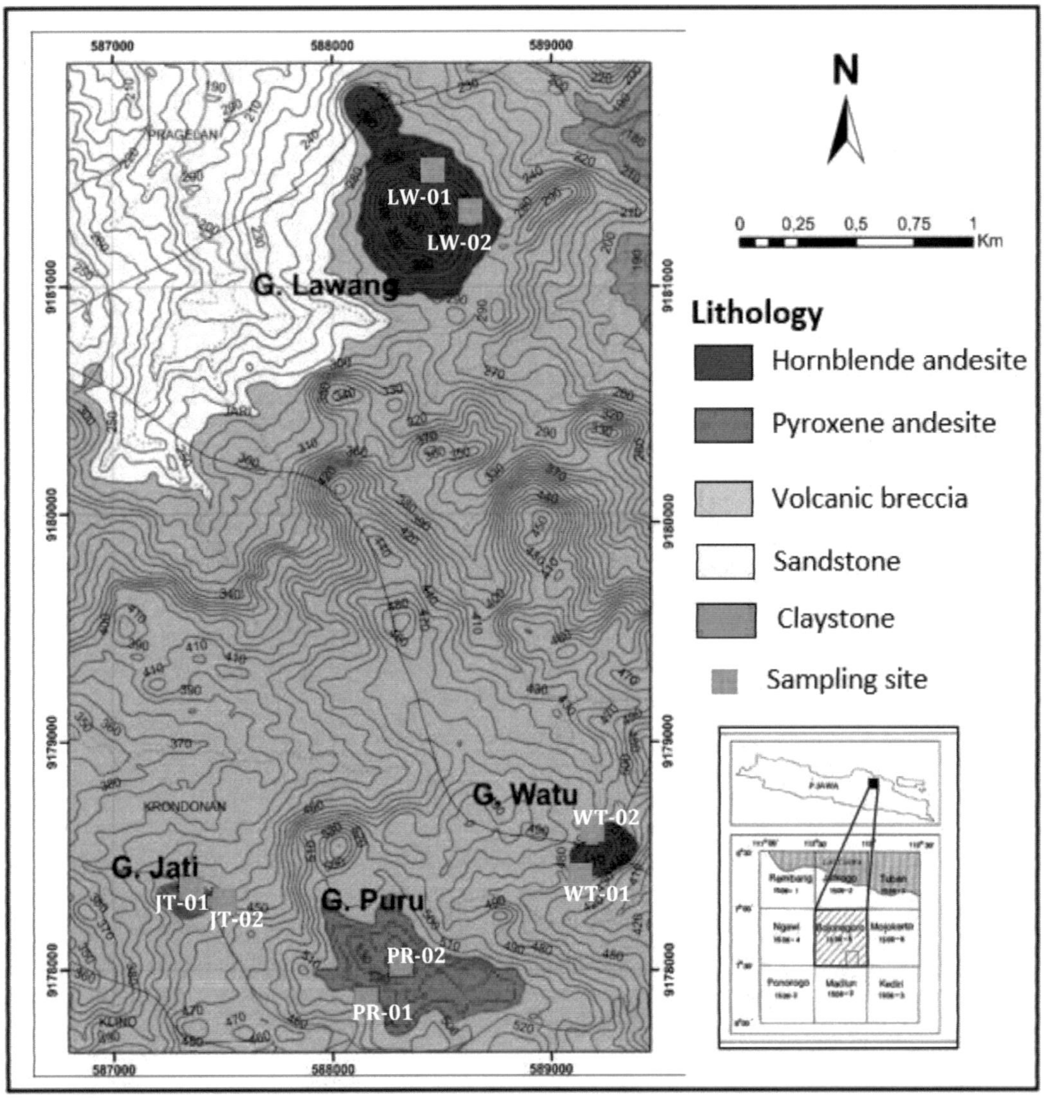

Figure 3. Geological map of the study area

3.1.1. Mount Lawang. This site is located in the western part of Jari Village, Gondang District. The morphology is characterized by steep intrusive hill with maximum elevation of 390 m and slope 60° - 80°. Igneous rocks are exposed by outcrops along the cliffs and the distribution of fresh boulders throughout the hill. Irregular columnar joint can be observed as the result of magmatic solidification process when the rock cooled and contracted during its formation. Lithology of this site is hornblende andesite characterized by light grey color, hypocrystalline, porphyritic texture, the main phenocrysts are plagioclase and hornblende surrounded by finer groundmass. Xenoliths are found in rock outcrops, consisting of mafic igneous rocks and metamorphic rocks which size is approximately 2-3 cm up to 10 cm. Igneous xenoliths are identified as gabbro by the dark color and coarse crystal size, while metamorphic xenoliths are greenish gray color and have phyllitic structure. Xenoliths are firmly embedded in rock bodies making them difficult to sample separately (Figure 5).

IOP Conf. Series: Earth and Environmental Science 1047 (2022) 012034 doi:10.1088/1755-1315/1047/1/012034

3.1.2. Mount Watu. Mount Watu is located in the Krondonan Village, Gondang District. This site is a lava dome with 250 m x 300 m dimension, maximum elevation of about 490 m and the slope is around 40° - 50°. On the western part of Mount Watu, sheeting joints were found with the main orientation of NE-SW that developed parallel to the hill slopes. Rather than tectonic related process, this structure is formed as extensive fracture in response to unloading in the shallow part of rock bodies. Lithology of MountWatu is hornblende andesite with characteristics of dark-grey color, hypocrystalline, medium to strongly weathered except in the fresh outcrop. The megascopic appearance and rock composition is similar to Mount Lawang with plagioclase and hornblende as main phenocrysts as well.

3.1.3. Mount Jati. Located in the southern part of the study area, Mount Jati has maximum elevation of 460 m and forms a volcanic hill in the northern part of Pandan Volcano. Based on the geomorphological appearance, Mount Jati is an intrusive hill that easily distinguished apart its surrounding. Several outcrop sites can be observed with fresh and slightly weathered rock condition. Field observation shows that this site is composed of pyroxene andesite, gray color, hypocrystalline, porphyritic texture with phenocrysts visible to the naked eye consisting of pyroxene and plagioclase and the aphanitic groundmass.

Figure 4. Geological feature of the four volcanic rocks and intrusion sites, show geomorphology of moderate – steep slope volcanic hill, columnar joints at Mount Lawang (red lines), sheeting joints at Mount Watu

3.1.4. Mount Puru. This site is located in Krondonan Village, also in the southern part in between Mount Jati and Mount Watu. Elevation at the top of Mount Puru is 540 m, while the slope is 15° - 25°, relatively gentler than other sites and it occupies an area of 400 m x 500 m. The lithology is pyroxene andesite lava which characteristics are similar to Mount Jati that can be observed from numerous igneous rock boulders scattered throughout the hill slopes. No geological structure was observed at this location.

IOP Conf. Series: Earth and Environmental Science 1047 (2022) 012034 doi:10.1088/1755-1315/1047/1/012034

Figure 5. Megascopic appearance of hornblende andesite and pyroxene andesite from the outcrops show porphyritic texture and xenoliths

3.2. Mineralogy of rocks

Pandan Volcano was formed in an island arc tectonic environment with medium – high K calc-alkaline magma affinity. The morphology of the volcano is constantly evolving, where this strato volcano consists of several volcanic centers and the youngest eruptive center is the main edifice of Pandan Volcano itself. Spatially, this Pleistocene (1.2 Ma) volcanic complex tends to move younger to the south [3][6].

These volcanic activities produce igneous rock with basaltic-andesite characteristics from intrusive, extrusive, and pyroclastic mechanisms. In this study, the mineralogical characteristics are focused on intrusive and extrusive rocks or lava which is the result of direct magma crystallization from eight thin section samples as summarized in Table 1.

Table 1. Composition of rock samples from Mount Lawang, Watu, Jati dan Puru

Site	Sample	Type	Composition (%)									Xenolith
			Plg	Hbl	Cpx	Opx	Opq	Qz	Sa	Ser	V	
Mount Lawang	LW-01	Hornblende andesite	55	16	3	-	2	2	4	-	15	Gabbro, Phyllite
	LW-02	Hornblende andesite	53	15	2	-	5	2	5	-	17	
Mount Watu	WT-01	Hornblende andesite	53	15	3	-	4	-	-	6	19	Gabbro
	WT-02	Hornblende andesite	50	15	5	-	2	-	-	5	23	
Mount Jati	JT-01	Pyroxene andesite	58	-	14	4	5	-	-	-	19	
	JT-02	Pyroxene andesite	55	-	14	3	5	3	1	-	19	
Mount Puru	PR-01	Pyroxene andesite	58	-	15	5	5	-	-	-	14	
	PR-02	Pyroxene andesite	55	-	15	5	7	2	-	-	16	

Notes:
Plg = Plagioclase, Hbl = Hornblende, Cpx = Clinopyroxene, Opx = Orthopyroxene, Qz = Quartz, Sa = Sanidine, Ser = Sericite, Opq = Opaque mineral, V = Volcanic glass

Observation in the rock outcrops show the general appearance of light gray to slightly dark gray color which is typical of intermediate igneous rocks. Igneous rocks in the study area are grouped into two types based on the mineral composition observed from petrographic analysis, which are hornblende andesite and pyroxene andesite. The main composition of the rock consists of phenocryst and groundmass with hypocrystalline and porphyritic texture (Figure 6a.). Volcanic glass as groundmass is counted for about 15-23% with its appearance in hornblende andesite is relatively lighter in color than in pyroxene andesite, probably due to slightly more acidic composition. The main phenocryst is plagioclase with abundance > 50% in all analyzed samples. The appearance of plagioclase under a polarizing microscope is colorless, low relief, subhedral-euhedral, 0.1 – 2.5 mm in size, prismatic and lathlike shape, order 1 interference color, and shows Carlsbad-albit and albite twins. The mineral has extinction angle of 22°-34°, hence the plagioclase composition is andesine to labradorite (An40 – An60). Other than as larger phenocryst, plagioclase is abundant as finer grained microlite as well, which is the result of later and more rapid crystallization.

(a) Plane polarized light (b) Cross polarized light

Figure 6. (a) Photomicrograph of thin section rock shows porphyritic andesite with groundmass and phenocryst, (b) Plagioclase (Plg) and pyroxene (Px) are the main phenocryst in pyroxene andesite from Mount Puru (PR-1)

Pyroxene is the next abundant phenocryst in pyroxene andesite samples. This mafic mineral presents as clinopyroxene (14-15%) and orthopyroxene (3-5%) in Mount Jati and Mount Puru (Figure 6b.). The characteristics of clinopyroxene are colorless to pale green, moderate relief, 0,1-2 mm in size, subhedral-euhedral, prismatic shape, order 2 interference color, inclined extinction, and shows simple contact twinning. While orthopyroxene has nearly similar characteristics, it tend to be lighter in color, lower order of interference color, and has parallel extinction. Some pyroxene phenocrysts show inclusion of opaque minerals.

Samples from Mount Watu and MountLawang contain significant amount of hornblende as main phenocryst beside of plagioclase with the average of 15% (Figure 7a., 7b.). In the plane polarized light, this phenocryst is easily distinguishable among the others. It has noticeable brown color and some crystal show pleochroism, which absent in other associated minerals. Other characteristics are moderate relief, 0,1-3 mm in size, subhedral-euhedral, prismatic and bladed shape, two-directional cleavage, order 2 interference color, inclined extinction, and shows simple contact twinning. Opaque mineral is commonly featured as inclusion inside the larger hornblende crystals. While pyroxene is present as accessory mineral in hornblende andesite (Figure 7b), hornblende is rather absent in pyroxene andesite from Mount Jati and Mount Puru.

Since metamorphic rocks and mafic igneous rocks are generally found deeper in the subsurface, the xenoliths presence indicates the interaction of magma with country rocks during its rising to the surface. According to regional study [7], the basement of Kendeng – Rembang Zone is most likely composed of volcanic arc and ophiolithic rocks similar to the rock complexes in Karangsambung. Therefore, it is suggested that the xenoliths found in Mount Lawang and Mount Watu is part of the Kendeng Zone basement.

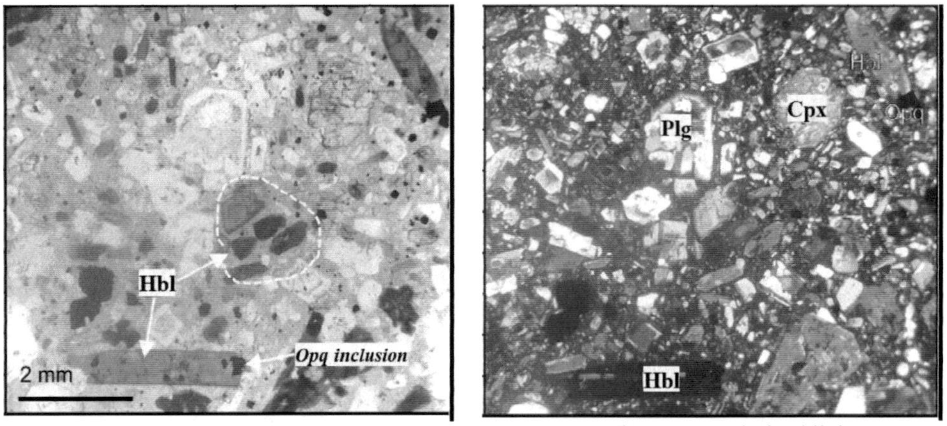

(a) Plane polarized light (b) Cross polarized light

Figure 7. (a) Photomicrograph of thin section rock of WT-1 sample light-dark brown hornblende (Hbl) phenocryst featured opaque mineral (Opq) inclusion, (b) Clinopyroxene (Cpx) as accessory mineral in hornblende andesite

Besides opaque minerals as mentioned above, samples contain other minerals as minor composition which accounted for less than 5%. Quartz presents in both pyroxene andesite and hornblende andesite with characteristics colorless, low relief, anhedral, and 0.1-0.2 mm size. Sanidine is more frequent in Mount Lawang, colorless, low relief, size less than 0.25 mm, subhedral-euhedral, and shows Carlsbad twinning. Sericite is alteration product replacing plagioclase which can be found in Mount Watu.

3.3. Igneous Textures and Magmatic Processes Interpretation

Formation of crystal through cooling of magma can be straightforward and simple or can be highly complex. In igneous petrogenesis study, textures of igneous rock help to understand what magmatic processes have formed them [8]. Volcanic rocks and intrusions in the study area are porphyritic andesite with more than 50% phenocrysts embedded in volcanic glass and microlite groundmass. Petrographic analysis of thin section has identified more detailed textural relationship that is essential to understand the origin of rocks (Table 2.).

Table 2. Igneous texture found in the sample

Site	Sample	Textures						
		Porphyritic	Intergranular	Pilotassitic	Glomerophyric	*Zoning*	*Sieve*	Poikilitic
Mount Lawang	LW-01	++	-	-	++	++	-	*
	LW-02	++	-	-	++	++	-	*
Mount Watu	WT-01	++	-	-	++	++	*	*
	WT-02	++	-	-	++	++	*	*
Mount Jati	JT-01	++	++	-	++	++	-	*
	JT-02	++	++	-	++	++	*	*
Mount Puru	PR-01	++	-	++	++	++	++	++
	PR-02	++	-	++	++	++	++	++

Symbol: ++ = frequent, * = less common, - = absent

Thin section from Mount Puru lava shows an additional texture involving the arrangement of microlites in the groundmass. Plagioclase microcrystalline laths exhibit pilotassitic texture, the sub parallel arrangement amongst them (Figure 8). The phenocryst in igneous rock tend to formed by slower crystallization in the magma chamber that the modification processes are inevitable. Whereas, microlite crystallizes quicker so it directly records the magma flow during rock emplacement, typically lava flow. Particular texture also can be identified from Mount Jati, namely intergranular texture which exhibits space filling amongst plagioclase phenocrysts by other crystals commonly pyroxene (Figure 9). This texture formed when early formed crystals trapped magma and it crystallized later [9].

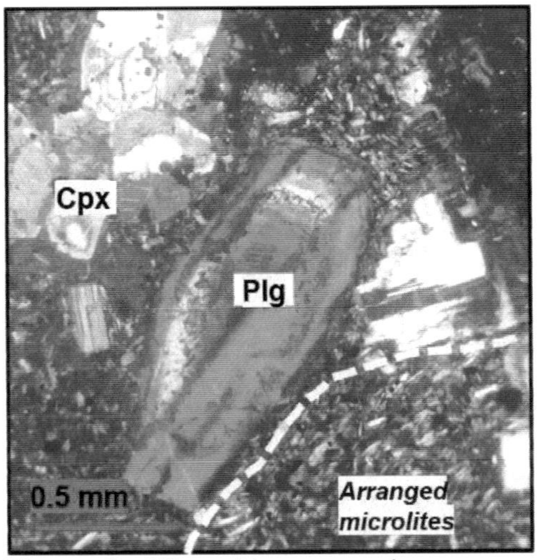

Figure 8. Pilotassitic texture in pyroxene andesite from Mount Puru (PR-2) shows plagioclase microlite arrangement in sub parallel pattern

Figure 9. Intergranular texture in JT-1 sample from Mount Jati shows angular space between plagioclase laths is filled by another crystal grains

Plagioclase as the most common phenocryst shows regular zoning pattern which is an indication to changes in magma composition during the crystal growth. This phenomenon can be identified using cross polarized microscope as the crystal changes color and angle of extinction during rotation due to the difference in mineral composition from core to rim (Figure 10.). Despite the fact that petrographic observation is insufficient to determine whether the variation of chemical composition is normal or reversed, zoned plagioclase shows continuous reaction of initial crystal to the magma melt. This reaction can be the result of the difference in crystal composition to the evolved magma or influx of fresh magma to the residual magma inside the chamber. Furthermore, plagioclase with sieve texture is also found in several samples (Figure 11.), which also can be the effect of influx of the more calcic magma [10]. Another possible cause of sieved plagioclase is the magma decompression during its rise towards the surface [11], in which the crystal corroded in melt while it continued to grow.

The phenocryst cluster composed of plagioclase and pyroxene often form crystal aggregates, also called glomerocryst, as known to be glomerophyric texture (Figure 12.). This intergrowth occurs as indication of simultaneous crystal growth, in account for heterogenous nucleation rates amongst crystals. In the fractionation process, glomerocryst is heavier than the individual crystals so it is more prone to sinking. In addition, another form of crystal aggregate occasionally presents as poikilitic texture when the phenocryst contains other minerals. The order of crystallization can be determined, which the enclosed crystal (chadacryst) must be crystallized earlier than the enclosing crystal (oikocryst). Figure 13 shows smaller grain clinopyroxene enclosed in larger clinopyroxene.

Figure 10. Plagioclase crystals exhibit concentric pattern or regular zoning texture from Mount Jati (JT-2)

Figure 11. Sieved plagioclase from Mount Watu (WT-2)

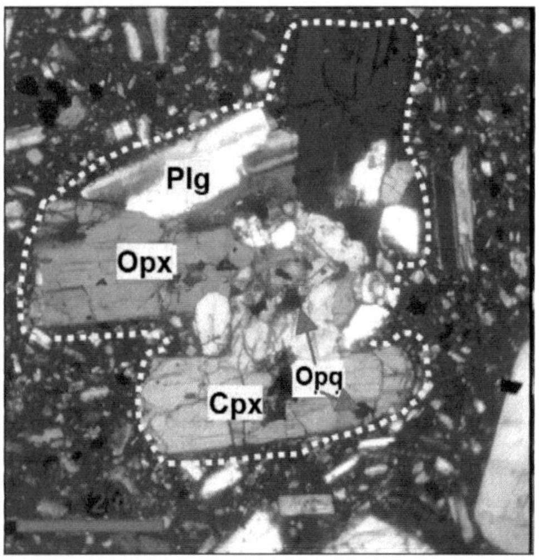

Figure 12. Plagioclase (Plg), orthopyroxene (Opx), and clinopyroxene (Cpx) cluster from MountJati (JT-2)

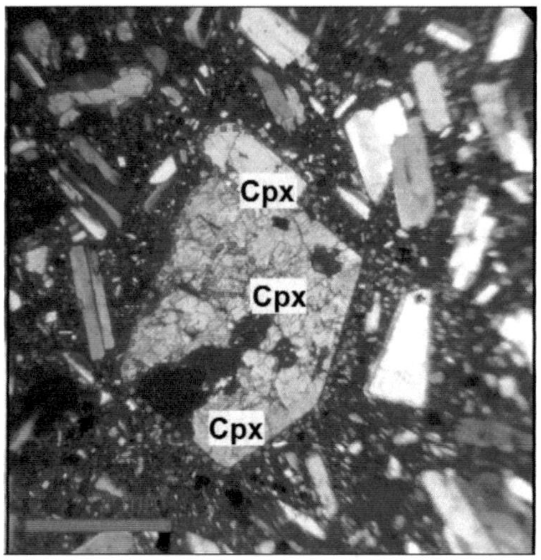

Figure 13. Poikilitic texture in which clinopyroxenes (Cpx) occur both as enclosed and enclosing crystal from MountJati sample (JT-2)

4. Conclusions

Volcanic activity in Pandan Volcano formed volcanic rocks and intrusions extends to the northern part/foot slope. Rock type in the research area is porphyritic andesite with fine grained groundmass and main phenocrysts of plagioclase, hornblende, pyroxene. Rock textures indicate magmatic processes such as crystal fractionation, basement interaction, magma mixing, and change in magmatic composition. Future research suggestion is geochemistry analysis to determine magmatic affinity and to construct petrogenetic model

5. References

[1] Thoha M, Parman P, Pratistho B, Yudiantoro D F, Hati I P, Jagranata I B 2014 Geology and geothermal manifestations of Mount Pandan *Proc. Int. ITB. Geoth.Workshop* **3**

[2] Lunt P, Netherwood R, Huffman F 1998 Guidebook of IPA Field Trip to Central Java *Indon. Petrol. Assoc. Field Trip* **65**
[3] Yudiantoro D F, Takashima I 2018 Magmatism and geothermal potential in Pandan Volcano East Java Indonesia *Jur. Min. Ener. LinMount* **2** 50-60
[4] Pringgoprawiro H, Sukido 1992 *Peta Geologi Lembar Bojonegoro Jawa Timur* (Jakarta: Pusat Penelitian dan Pengembangan Geologi)
[5] Winarno T, Marin J, Jati W W 2020 Effect of porphyritic andesite intrusion on the formation of contact metamorphism aureole in Selo Gajah Hill Bojonegoro Regency East Java *Jour. Geosc. EnMount Env. Techn.* **5** 101-108
[6] Lunt P 1991 The Neogene geological history of East Java, some unusual aspects of stratigraphy *Proc. Ann. Conv. Indon. Assoc. Geol* **20** 26-36
[7] Smyth H, Hall R, Hamilton J, Kinny P 2005 East Java: Cenozoic basins, volcanoes, and ancient basement *Indon. Petrol. Assoc. Ann. Conv.* **13**
[8] Jerram D A, Dobson K J, Morgan D J, Pankhurst M J 2018 The petrogenesis of magmatic systems: using igneous textures to understand magmatic process *Volcanic and Igneous Plumbing Systems* Steffi Burchardt (Amsterdam: Elsevier) chapter 8 pp 191-229
[9] Sen G 2014 *Petrology: Principles and Practice* (Berlin: Springer) p 359
[10] Seaman S J 2000 Crystal clusters, feldspar aglomerocrysts, and magma envelopes I the Atascosa Lookout Lava Flow Southern Arizona, USA: recorders of magmatic events *Journal of Petrology* **41** issue 5 693-716
[11] Nelson S T, Montana A 1992 Sieve-textured plagioclase in volcanic rocks produced by rapid decompression *American Mineralogist* **77** 1242-49

Acknowledgments
This research is supported by research grant from Faculty of Engineering, Diponegoro University 2022. Finally, special thanks to our colleagues for the endless help and encouragement during the course of this study.

Balancing nature and landscape conservation practices of Leukon Community in Simeulue Island

T Lubis[1*], A F Abus[2], Ramlan[3], T Nasution[4], N Saputra[5]

[1]Universitas Sumatera Utara, Medan, Indonesia
[2]Aceh Landscape Development Centre, Langsa, Indonesia
[3]Universitas Jabal Ghafur, Sigli, Indonesia
[4]Universitas Simalungun, Pematangsiantar, Indonesia
[5]Sekolah Tinggi Ilmu Tarbiyah Al-Hilal Sigli, Sigli, Indonesia

*Email: tasnimlubis@usu.ac.id

Abstract. This study described the Leukon Community's daily interactions with the natural environment to preserve nature's balance and landscape conservation in social spaces conducive to sustainable growth. The natural environment's behavior features can be studied through the interaction of interacting areas such as fields, flora, and the sea that are habitable for the Leukon Community. Natural resources and the environment play a critical role in the Leukon Community's survival system due to the community's availability and ability to utilize raw materials. This research applies a qualitative strategy, an investigative procedure used to comprehend social or human problems based on a holistic description. The ethnographic technique is employed as a paradigm, with in-depth interviews and participant observation. The results indicate that the Leukon Community practices environmental stewardship by avoiding mechanization and chemical use in farming, capturing fish with environmental friendly technology, and relying on woody tree vegetation for basic requirements. As a result, the delicate balance between nature and environmental preservation is preserved. Avoiding soil and water damage promotes landscape conservation, resulting in a more natural and beautiful landscape. This is because the Leukon Community manages the area sustainably by utilizing natural resources in accordance with local wisdom values and belief systems.

1. Introduction

Balancing nature deals with the effort to treat the environment well. The environment is everything that surrounds or affects human life in their living space and time. The environment can be natural or man-made. Balancing environment deals with maintaining the component in nature [1]. Working at different scales is one of the greatest challenges of large landscape conservation. Management of landscapes at increasingly larger scales is proving critical points to understand the large-scale stressors, such as climate change. Landscape conservation is an approach that brings people together across geographies, sectors, and cultures to collaborate on conserving our important landscapes and the myriad ecological, cultural, and economic benefits they provide [2]. This study focused on the natural environment regarding the human interaction and effort to maintain the natural balance. The human interaction with their environment covered all of their life activities to survive and exist for a long time.

Leukon Community is a community that lives only on a small island located on the westernmost of Aceh Province, Indonesia. They live in remotes areas namely Langi and Lafakha villages [3]. Almost all of them work as both farmers and fishermen. Leukon farmers and fishermen are expert in determining the particular time to grow rice and fishing. This becomes their local knowledge, which is transmitted from their ancestors. The local knowledge is delivered through verbal communications. Oral tradition as cultural practice contains words, phrases, and expressions as media to inform the knowledge and experience. It has become local wisdom to manage the activity, including food security [4].

From the preliminary observation, the Leukon Community still relies on their ancestor's ways to manage their work as farmers and fishermen. One of the ways they receive the knowledge is from verbal communications. Oral literature is the result of creative thinking from the ancestor in expressing experiences, observations, and feelings. It comprises policies that include values related to human and their environment [5]. This oral literature arose in the community and lasted for generations. The Leukon Community had been aware that their living area bears a greater challenge, since their living area is surrounded by the Indian Ocean and only have a few barrier islands compared to other locations in the Simeulue Island.

Faced with this situation, the Leukon people tend to stay in their territory and work like their ancestors did to manage their activities related to landscape conservation. This study aimed to describe the Leukon Community's daily interactions with the natural environment to preserve nature's balance and landscape conservation in social spaces conducive to sustainable growth.

2. Methods

This study used the ethnography method and anthropolinguistic approach [6][7]. The data were audio-video recording of the Leukon Community narrative on several topics regarding managing their growing rice, fishing, and sagoo processing activites. In addition, in-depth-interview and participant observation on how Leukon Community management, procedure, and also the value in their activities growing rice, pick up the clove, coconut, and catch the fishes. According to Spradley (1979), there are twelve stages covered data collection technique and analysis. To find out the relationship between Leukon Community effort or landscape conservation practices and balancing nature, the semantic relationship 'reason' was chosen. It means all of the activities relating to the environment, must be based on the reason that derived from their indigenous knowledge. The relationship is shown in Figure 1.

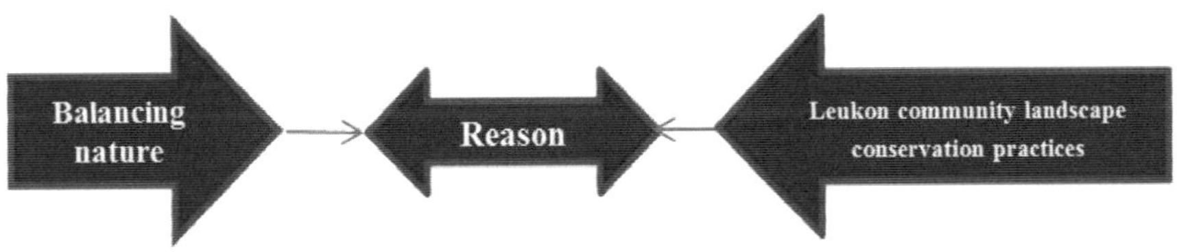

Figure 1. Diagram of Domain analysis with semantic relationship 'reason'.

The diagram above described the reasons for the Leukon Community's effort in landscape conservation along with their activities to fulfill their needs as well as maintaining the balance of the environment due to its natural sources.

3. Result and discussion

In Lafakha and Langi village, most coastal areas have endemic vegetation such as coconut, mangrove, and sagoo. In common, clove trees also grow well in Simeulue Island and make this island became popular as a clove production. The clove planting began since Netherland imperialism came to Indonesia including on this island. The are several main foods of the Leukon Community; rice, sagoo, taro (taleuk), and *taeb* (mangrove fruit processed). But nowadays, the Leukon Community does not consume mangrove anymore, even the younger generation do not know about them. These main food grow well on this island and also can be cultivated due to the natural location support [4]. There is only one round trip transportation (minibus) from Sinabang (the capital of Simeulue Island) to Lafakha and Langi villages in a good weather. When the windstorm, sometimes, the wood along the road fell to the street and obstruct the trip. The heavy rain for long period caused the flood, also able to obstruct the trip, so it has to be canceled until the normal restored. The situation is shown in Figure 2.

Figure 2. The Fallen tree blocked the trip.

Recently, they only consume rice and sagoo for the main food. Leukon Community knew how to grow rice from Minangkabau nomade. To meet all the rice needs of the Leukon residents, those who have large lands to plant rice will be assisted by other Leukon residents who do not own land. Therefore, all Leukon people will have rice as their main meal. If the crop is sufficient, the owner will give charity to the poor. This concept relies on Islam because all of Leukonese are Muslim. Besides oral literature, Islamic concept also implemented in Leukon Community life. Based on Islam's role, when the crop or harvest reaches nisab (minimum limit of assets that must be subject to *zakat*/alms), Muslims have to give charity to the poor. The law for zakat on agricultural products, including zakat on rice is the Qur'an letter Al Baqarah (2: 267) [8].

Based on this local wisdom, the Leukon Community can suffice their need for rice as one of the main food. The rice that they have is managed to consume for six months. To keep the rice from rice lice, they do not grind all rice (only as needed) and keep the rest in a specific container made of wood. They also have sagoo and taro in case of lack of rice. To have sagoo, usually, Leukonese conduct a mutual work. It begins by cutting the sagoo tree and processing it into sagoo powder. Leukonese knows what to do and survive in this area regarding its bad weather and condition. One of Leukonese characters is their obedience listen to their ancestors advice through their oral literature such as *nandong* (sung poetry), *nanga-nanga* (sung story), and *inafi-nafi* (story). Those oral literature are useful because it is consist of advice, story, and also the way how to survive and save them from disaster [9]. The landscape in this area also supports the maintaining ecosystem [10].

Besides being farmers, Leukonese are fish catchers. It is known that Simeulue Island has rich marine natural resources. Simeulue Island is a strategic water area that has a lot of marine resources. Simeulue Island, Aceh Province is one of the islands located in the Indian Ocean and 150 km apart from the mainland of the island of Sumatra. Simeulue Island is rich in marine and coastal resources including beautiful beaches, extensive mangroves, abundant reef fish, clean and clear waters as well as other marine biota such as seaweed, crabs, sea cucumbers and lobsters which are the icons of this island [11]. It covers fishes that are tremendously rich of number and type [12]. The popular marine resources are lobster, fresh fish, sea cucumber, and octopus. Leukonese fishermen still use traditional tools to catch fish, lobster, and octopus. Some of them also have the skill to dive for minutes to use tools, namely *tembak* (shot), to catch lobster and octopus. Those who have a canoe or boat, can go farther to catch bigger fish. Usually, the big fishes are sold to the *muge* (the person who buys fish from the fishermen), and they only take as their need. That is the way it goes. The life of the Leukon Community follows the natural cycle in utilizing natural resources for their daily needs so that landscape conservation is practiced The landscape use supporting the ecosystem [13].

In making boats, the Leukon Community uses wood as raw materials from the surrounding forest (Figure 3). Forest vegetation can also create a microclimate in one area to creates a comfortable nuance [14]. The transportation used to traverse the waters is a boat from mostly wood. The forest inhabited by the Leukon Community is a tropical rain forest that is rich in tree species. Apart from being a means of transportation, boats made of wood are also widely used by the Leukon Community to catch fish. The activity of making robin is shown in Figure 3.

Figure 3. The activity of making *robin/bokge* (boats) in the Leukon Community.

Renewable coastal area resources consist of mangrove forests, coral reefs, seaweed fields, marine fishery resources, and bioactive materials that contribute to landscape conservation. In the management of coastal areas, the concept of balancing nature and landscape conservation are two interrelated things that cannot be separated. The concept of development in coastal areas ideally can support increased growth and development as well as landscape conservation.

The Leukon Community lives on the coast and most of their lives depend on marine products, therefore they need *robin/bokge* (boats) as a means of transportation and at the same time to catch fish (Figure 4). This is done to achieve the target in the form of a synergy of economic and social activities in the area concerned while minimizing the negative environmental impacts resulting from these economic and social activities. Besides that, it also increases efforts to preserve the coastal area's environment in addition to increasing the ability to meet development needs through optimal utilization of owned resources towards natural balance. The integrated manner through the collaboration of the community, scientists, and the government to find the right coastal management strategy should be conducted entirely [15]. The robin as a transportation is shown in Figure 4.

Figure 4. *Robin/bokge* (boats) as a means of transportation and at the same time to catch fish.

Balancing nature and landscape conservation practices provides support for improving the quality of life of the Leukon Community today and is expected to continue to be implemented in the future. This ideal sustainable landscape management concept focuses on the characteristics of the Leukon Community ecosystem which is managed by taking into account aspects of environmental parameters, conservation, and quality of life.

The balance of nature should be maintained for the sake of the preservation of living things that can use nature in moderation and according to needs. Meanwhile, landscape conservation is maintaining the living environment or all the contents of the earth and sky that are in a balanced condition or state. Balance of nature is a form of balance between biotic life and the environment. Humans often destroy the balance of nature. For example by exploiting natural resources, cutting down trees carelessly and not using environmentally friendly materials. If the balance of nature is damaged and resources become extinct, this will affect the life of living things and the ecosystem they live in. In order not to damage the balance of nature and the extinction of natural resources, humans must be willing to play a major role in caring and preserving both flora and fauna in Lafakha and Langi villages.

4. Conclusion

The Leukon Community has strong reasons to protect their environment through their efforts towards landscape conservation. All of their activities on how they use natural resources are based on their years of experience. They believe that managing the way they extract from the environment will help them to achieve food security and prevent disasters. In addition, obedience to the advice of their ancestors in carrying out life is a good attitude that they must maintain and pass on to the next generation in order to survive on this island.

References

[1] Effendi R, Salsabila H and Malik A 2018 Pemahaman tentang Lingkungan Berkelanjutan *MODUL* **18** 75–82
[2] Lubis T and Purba A 2020 Metaphor and parallelism in political advertisements *Cogency* **12** 71–88
[3] Lubis T and Williams N 2019 *Preliminary Documentation of Leukon Language* (London)
[4] Zulkarnain, Lubis T, Ramlan, Dardanila, Hasrul S, Shaumiwaty and Saputra N 2021 Nandong as a culture-based effort to overcome food security toward COVID-19 pandemic situation in Simeulue Island *IOP Conf. Ser. Earth Environ. Sci.* **807** 022007
[5] Lubis T 2019 Learning Nandong in schools as a medium to inform the Simeuluenese local wisdom: An anthropolinguistics approach *Stud. English Lang. Educ.* **6** 262–72
[6] Spradley J 1979 *The Ethnographic Interview* (USA: Rinehart and Winston)
[7] Shaumiwaty S, Lubis M A, Lubis T, Dardanila, Purba A, Nasution T, Ramlan and Hasrul S

2020 Teacher performance toward students' mathematical literacy in teaching linear program mathematical models *J. Phys. Conf. Ser.* **1663** 012066

[8] Rosele M I, Ramli M A, Isamail M Z and Arifin M F 2014 Zakah on Agriculture Reformation: An Analysis in Malaysia *Sains Humanika* **3** 47–53

[9] Lubis T 2019 Oral Tradition Nanga-Nanga in Simeulue Island *J. Oral Tradit.* **1** 28–36

[10] Lubis T, Dardanila, Nasution T, Zulkarnain, Hasrul S, Ramlan and Abus A F 2021 Tradition lubuk larangan as a local wisdom for ecocultural tourism river management through landscape anthropolinguistic approach in Mandailingnese *IOP Conf. Ser. Earth Environ. Sci.* **926** 012029

[11] Edwarsyaha, Raudhatia N, Djumantob, Hendric A, Zurbaa N and Burhanis 2021 Karakteristik Suhu, Salinitas dan Klorofil-A di Pulau Simeulue Provinsi Aceh untuk Mengestimasi Zonasi Ikan *JFMR-Journal Fish. Mar. Res.* **5** 499–507

[12] Wisha U J, Ondara K, Gemilang W A, Rahmawan G A and Dhiauddin R 2018 Monitoring corals and fish abundance in Simeulue Island, Aceh Province before mass bleaching phenomenon caused by enso in 2016 *Prosiding Seminar Nasional Kelautan dan Perikanan IV* (Surabaya) pp 103–14

[13] Abus A F, Lubis T and Abus N A A 2021 The landscape concept of environment in Taman Gajah Mada Medan *IOP Conf. Ser. Earth Environ. Sci.* **922** 012035

[14] Lubis T, Perangin-angin A B, Dardanila, Nasution T, Hasrul S, Ramlan and Abus A F 2021 The role of environment restoration on microclimate sustainability in Perumahan Cemara Asri Medan *IOP Conf. Ser. Earth Environ. Sci.* **922** 012039

[15] Fabianto M D and Berhitu P T 2014 Konsep Pengelolaan Wilayah Pesisir Secara Terpadu dan Berkelanjutan yang Berbasis Masyarakat *J. Teknol.* **11** 2044–60

Assessing the optimal location for Sipora wave energy power generator site placement from a geological perspective

A W Pertala[1], I H Suherman[1], A H P Kesumajana[2], H Kurnio[1]

[1]Marine Geological Institute P3GL KESDM, Jl. Dr. Djunjunan No.236 Bandung, Indonesia.
[2]Department of Geology, Institut Teknologi Bandung, Jl. Ganesa No.10, Bandung, Indonesia.

Abstract. One of the potential sources of renewable energy in the Mentawai region is ocean wave energy. The western Sumatra sea has a wave energy potential of up to 20 kW/m. The wave energy potential modeling conducted by the Marine Geological Institute (P3GL) ESDM revealed that western Mentawai waters are locations that produce annual average wave energy of around 20 kW/m. The geological survey of the seabed for power generating wave energy turbine sites using acoustic equipment including the Echosounder Odec, SBP Stratabox, and Star Fish Side Scan Sonar. The mapping of the seabed allows for the identification of undersea morphology, slopes, structural elements, and seafloor objects. Ocean depth sounding and bottom surface imaging are used in this research to determine the placement of the turbine site from the wave energy generator connection grid. The existence of underwater corals and fluvial plain units, continental shelf with shallow seabed and a gentle slope with an average depth of 30 meters was recognized as the most important factor in determining the site's location in this research. Based on the weighting and scoring results, optimal locations for potential power plant sites are in coastal areas located in offshore near Batu putih coast.

Keywords: Mentawai, seafloor, wave, energy, Underwater, exploration, Geology

1. Introduction

The Bali action plan Joint Implementation (JI) was established at the 2007 COP-13 meeting in Bali, Indonesia, as an emission reduction mechanism through which developed countries could transfer emission reductions through joint projects to reduce greenhouse gas emissions and the rate of climate change in the country. The meeting's protocol became the government's basis for responding to global issues, with the Minister of Energy and Mineral Resources issuing Ministerial Regulation Concerning the Regulation of the Electric Power System Network, which requires PT PLN (Persero) to be able to optimize the potential for power generation from new and renewable energy sources. Offshore wave energy farms provide one of the best, new possibilities for clean, abundant, environmentally benign, and reliable energy generation. Wave energy is produced close to coastal population and does not require significant additional build out of transmission. The Sipora islands are known as the destination of endless surf breaks. Waves tend to be heavy hitters with some exposed sections of shallow reef. Consistent Indian Ocean swell to make it one of the most sought on earth. There are at least 22 villages that are part of the non-electrified village in Mentawai Islands [1]. The wave energy potential modeling conducted by the Marine Geological Institute (P3GL) ESDM revealed that western Mentawai waters are locations that produce annual average wave energy of around 20 kW/m[2]. Based on ESDM research and Sipora natural conditions, the Sipora offshore are most suitable for the use of wave energy.

1.1 Location

Tuapejat is the main village on the island, with a population of around 20,000 people, PT PLN still relies on inadequate diesel generators. Despite the fact that the electricity supply is limited, Tuapejat's electrical network installation is well connected. In order not to require a significant additional transmission, we decided to conduct a survei in Tuapejat, Sipora Utara (Figure 1).

IOP Conf. Series: Earth and Environmental Science 1047 (2022) 012036 doi:10.1088/1755-1315/1047/1/012036

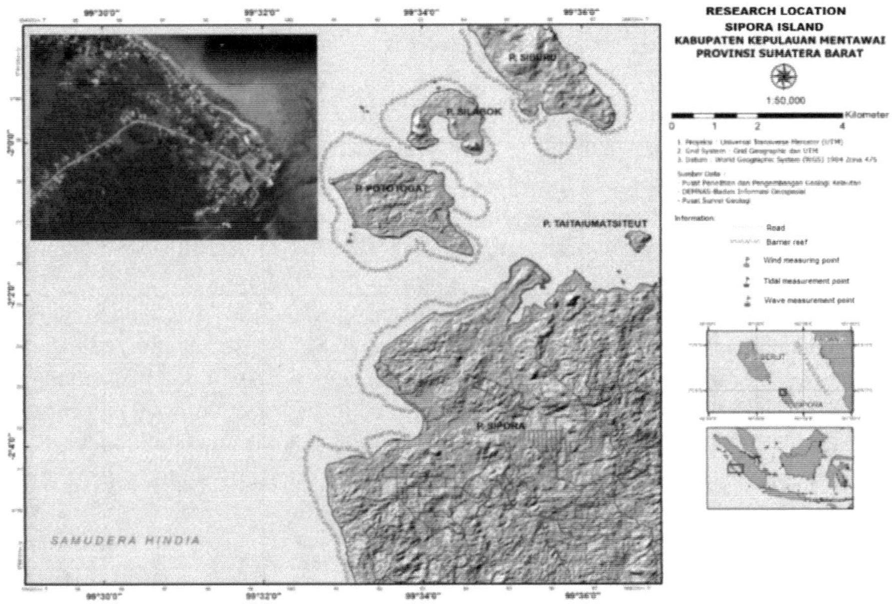

Figure 1. Study area are the world-famous surf destination located in Sipora island. As the largest island, the town and administrative center is also located on the island, precisely in Tuapejat city that located on the North side.

1.2 Equipment plan

The waves originally blow far from the coastline, due to the wind blowing on the surface of the water inducing the water particles to rotate and that rotation extends deep down below the surface. This rotational energy turns into elliptical energy and eventually backward and forwards. The mechanism was actually developed by a company called WaveRoller. The system was inspired by a professional diver, Rauno Koivusaari. While diving in a shipwreck and watching shipwreck hatch moves back and forth with the waves. This prototype approximately has about 10 square meters of steel panel fixed to the seabed via the floatable foundation. The equipment is planned to be deployed at around 15-20 m of water depth, where there is still strong wave energy coming in. It is submerged at a depth of around 20 m, so it is protected from extreme waves (Figure 2).

The system generates electricity by capturing the movement with a hydraulic circuit in a machinery room underneath the surface. That hydraulic energy we turn into electricity with a hydraulic motor, accumulator, and generator[3].

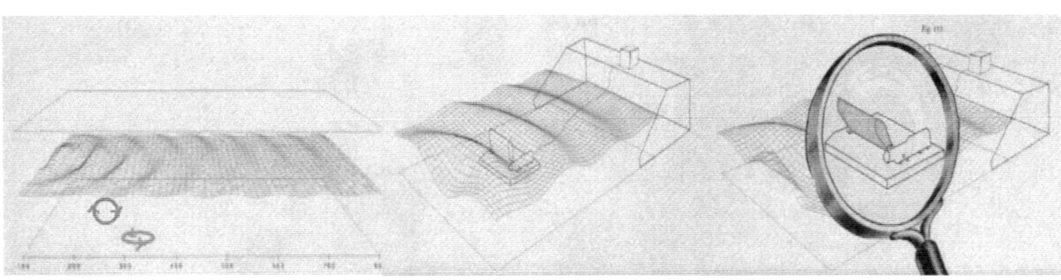

Figure 2. The device is installed underwater at depths of approximately 15 - 20 meters, which connects to a hydraulic motor that drives an electricity generator. The wave energy power plant's electrical output is subsequently connected to the electric grid through a subsea connection. [3]

IOP Conf. Series: Earth and Environmental Science 1047 (2022) 012036 doi:10.1088/1755-1315/1047/1/012036

1.3 Regional Geology

The basement rock of the sipora island is constituted of accreted oceanic crustal rocks mixed with with scoured metamorphic rock and the Paleogene deep-sea flysch sediments (Bancuh). The geodynamic framework suggests the presence of a magmatic arc province that migrated as a result of the compressional and strike-slip tectonic forces. During late Miocene, extensive stretching of the crust that was related to the eastwards migration of the subduction zone. This transtensional rifting brought about the input of sea water in the Mentawai basin, leading to the development of both continental and shallow-marine depositional environments. The sedimentary rock of the Mentawai Islands comprises clastic and carbonate rocks with lignite intercalation [4]. In the Sipora Islands, the sedimentary rock of the Mentawai Group was included into the Tolopulai, Maonai, and Batumonga Formations. The Early Miocene Tolopulai Formation was deposited within the shallow marine environment, probably as fore-reef, and overlain by the Maonai Formation. The later interfingers with the Batumonga Formation of Middle-Late Miocene - Pliocene age deposited within an inner littoral environment [5](Figure 3).

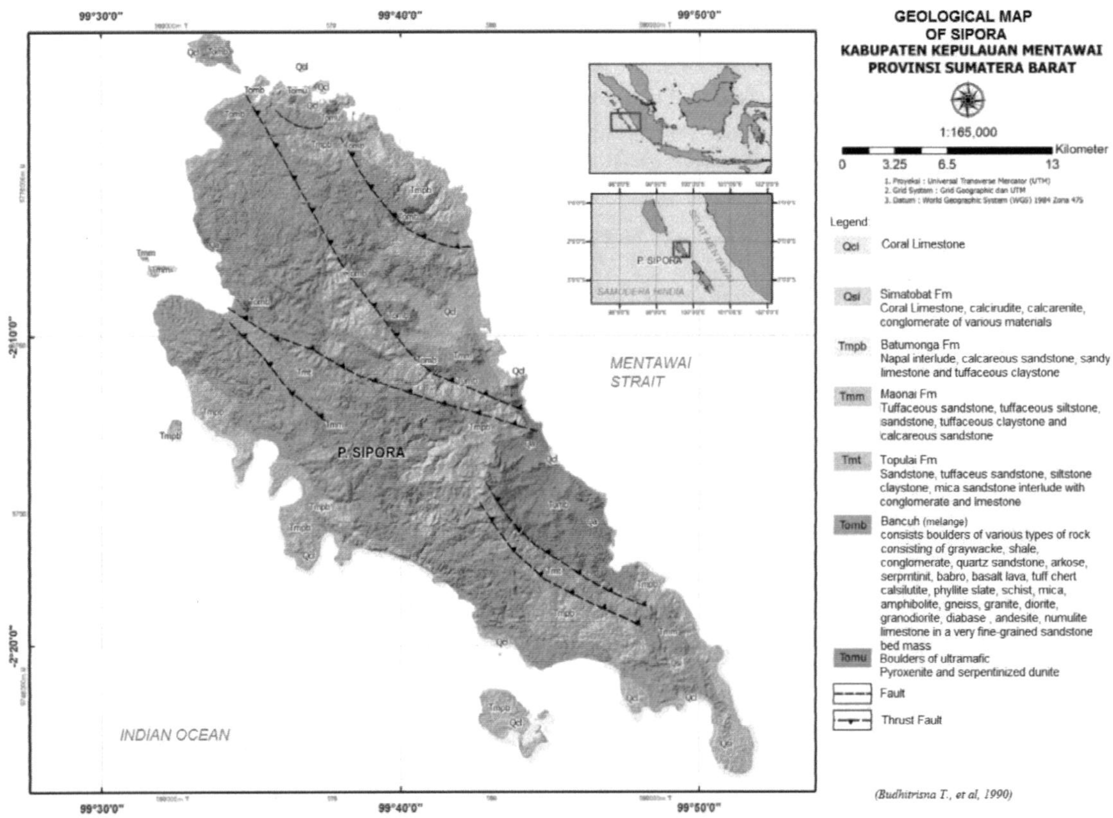

Figure 3. Geological Map of Pagai and Sipora[5]

2. Method

2.1 The coastal characteristics Mapping

The coastal characteristics were mapped out qualitatively according to the classification proposed by Dolan [6], which included lithology, beach slope, shoreline characteristics, and dominant processes. The objective of mapping the coastal characteristics is to understand in detail the characteristics of the beach, the process of accretion and abrasion in the study area. The results of the- mapping of coastal characteristics are used for determination of the landing point of the cable route transmission.

2.2 Sounding

Sounding is a process of gaining sea depth measurements simply using an Echosunder Single Beam Navison Reson 210. Data is retrieved graphically and digitally utilizing a recorder installed into the equipment. A signal transducer tool is put on the ship's left hull, at a depth of 1 (meter) meter below sea level.

The echosounder is triggered at 1 second intervals with a 10 cm reading accuracy. The depth data are corrected for the transducer position and the tides of the survey area, and the outputs are projected on a map of the research area's location, ultimately result in a map of sea depth.

2.3 Side Scan Sonar

Side Scan Sonar Starfish 900F used for imaging seabed objects and providing information on the shape and size of underwater objects. This method can provide quantitative information in the form of the amplitude value of the object or the bottom of the water. The purpose of this research is to visualize or map the bottom of the waters and classify sediments or objects on the seabed qualitatively and quantitatively, so that the unique signal characteristics of various substrates can be known.

2.4 seismic reflection (SBP)

Sub-bottom profilers Stratabox are usually composed of a single channel source that transmits sound pulses into the shallow sediments on the seabed. The sound pulses bounce off the sea floor and subsequent buried sediment layers according to differences in their acoustic impedance (hardness). Acoustic impedance is related to the density of the object and the speed at which sound travels through this material. The different times taken for this signal to be returned and recorded by the SBP indicate how deep the layers are below the seabed. The result of SBP data processing is a representation of the geological structure.

2.5 Seabed surface sediment sampling

Sampling of seabed sediments was carried out to determine the distribution of seabed sediments and as a side scan sonar image correction in the study area. This method is very important to determine the distribution of sediment as a sediment carrying capacity for wave power plant. Around Tuapejat offshore, Sampling of seabed sediment was carried out at 16 locations using a grab sampler.

2.6 Data processing

Data from sounding and SBP were corrected for transducer positions and tidal before being projected onto a map of the study area location. Depth data from sounding is processed using Surfer and Global Mapper to create contour shapes and three-dimensional morphology. Data from SBP were analyzed qualitatively using petrel software and side-scan sonar data is interpreted and classified qualitatively using the SonarPro software to determine the geometric shape and uniqueness of objects. Ocean depth sounding and bottom surface imaging also used in this study to determine the cable route transmission from the wave energy generator connection grid [7].

The main purpose are GIS analysis with raster-based spatial analysis is used to determine the location of the generator, assuming several parameters to get the best weight for determining the location.

3. Result

3.1 Coastal characteristics

Based on the type of lithology in the coastal area can be divided into four groups, reef limestone (Qcl), Tufaceus sandstone (Tmm), Ultramafic Boulder (Tomb) and Bancuh Melange (Tomb) (Figure 3)

Most of the mineral that composes the beach is from coral limestones (Qsi). The grain size of the coral limestones sand found on the beach is generally medium-coarse, greyish white in color, and contain few quartz minerals, lithic fragments, and mafic minerals. This type of sediment dominates the Pototogat island coast and the northwest coast of Sipora. Around the Jati beach exposed few melange rocks that underlie the island of Sipora, with a very weathered outcrop. The north side of the Sipora island is dominated by mangrove forests (Figure 4). The sediment that dominates this area is grey mud which is believed to have a high organic matter content. This beach is also strongly influenced by the tides of sea water which periodically stirs the sediment. The interaction of organisms with sediment and the high effect of water evaporation also greatly affects this coastal environment. The wave conditions on the north side mangrove beach is relatively calm so they are unable to carry the sediment to the deep waters on the high seas so the suspended sediment spread over a wide area of water. The slope and morphology of the study area can be devided into two categories. The north side of the Sipora island is dominated with hilly beach approximately occupies 30% of the study area. The slope is between 20 - 30 degrees, with high cliffs in some sections. The lowlands occupy about 70% of the Sipora island are coastal plains units with concave slopes, exogenous morpho-genesis (marine), 0–2% slopes, and land use for settlements, agriculture, and smallholder plantations. This type of morphology dominates the Pototogat island coast and the northwest coast of Sipora (Figure 4).

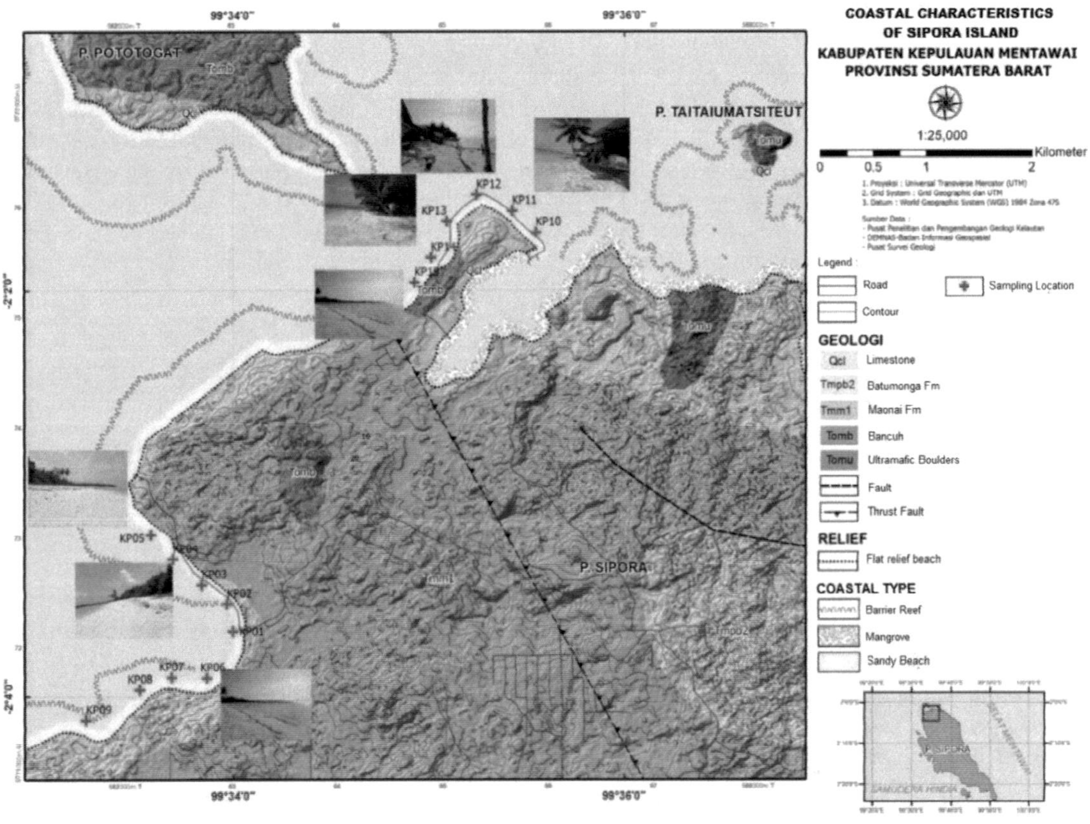

Figure 4. Coastal characteristics map of Sipora Island (Tuapejat-Pantai Jati-Mapadegat)

The dominant factor affecting the configuration most of the north Sipora and Patotogan island coastline is marine processes. This is because this area is an overwash area, which has higher energy. The slope of the beach ranges from 2–6%. This shows that the wave energy in this area is quite high, because the steepness of the beach slope is directly proportional to the wave energy [8]. Currents and waves are estimated to be coming from the west, southwest, south, and southeast. Waves can generate energy to form beaches, creating currents and sediment transport in a perpendicular direction and along the coast and cause forces acting on coastal structures [9]

Overall, we discovered two types of beaches as a result of the coastal characteristics mapping, there are sandy beaches and mangrove beaches. The sandy beaches are characterized by the presence of coral reef terraces near shore with fairly coarse white loose sand consisting of coral debris and shell fragments and few mafic minerals from reworked melange in front of the beach, while the the north side of the Sipora island is dominated with hilly beach and mostly covered by mangroves (Figure 3). This type of beach has shallow, clear, and inundated calm water conditions so that coral reefs can develop well, as well as several other marine biotas (locally and in small quantities), especially fish and seaweed. This is due to the presence of a wide and shallow coral terrace in front of the beach, which serves as a barrier against the crashing waves. Based on the data above, the best proposal for the placement of the cable line transmission landing point is the northwest coast of Sipora.

3.2 Seabed surface sediment

Seabed sediments were taken from 15 locations using sample grabs at a depth of 10 meters to 40 meters. Sampling of seafloor sediments was performed at 16 sample locations and only one location contained sand samples at M20-08. the other samples were only fragments of coral reefs and mollusk shells fragments with grain size granules to gravel. The results of megascopic descriptions can be seen in Table 1. Samples M20-8 consisted of medium-coarse sand, brownish white, poorly sorted, moderately angled, contain few quartz minerals, lithic fragments, and mafic minerals.

As a result, nearly the entire feasible area is covered with coral. In this case, only 20 percent of the study area is covered with sand (Figure 5)."

Table 1. List of Samples, and description

No. Samples	Coordinate (DD)	Description
M20-01	02.061352 /99.566660	Coral, red, brown and white coral reefs and mollusk shells fragments, granules to gravel
M20-02	02.058995 /99.566172	Coral and mollusk shells fragments, granules to gravel
M20-03	02.057558 /99.563961	Coral and mollusk shells fragments, granules to gravel
M20-04	02.055352 /99.561528	Coral, red, and mollusk shells fragments granules to gravel
M20-05	02.052489 /99.559621	Coral, and mollusk shells fragments, granules to gravel
M20-06	02.065159 /99.564449	Coral and mollusk shells fragments, granules to gravel
M20-07	02.065149 /99.561445	Coral and mollusk shells fragments, granules to gravel
M20-08	02.066131 /99.558701	Medium-coarse sand, brownish white, poorly sorted, moderately angled, contain few quartz minerals, lithic fragments, and mafic minerals
M20-09	02.068680 /99.554173	Coral and mollusk shells fragments, granules to gravel
M20-10	02.028561 /99.592453	Coral and mollusk shells fragments, granules to gravel
M20-11	02.026810 /99.590458	Coral and mollusk shells fragments, granules to gravel
M20-12	02.025491 /99.587373	Coral and mollusk shells fragments, granules to gravel
M20-13	02.027655 /99.584880	Coral and mollusk shells fragments, granules to gravel
M20-14	02.030637 /99.583553	Coral and mollusk shells fragments, granules to gravel
M20-15	02.032701 /99.582147	Coral and mollusk shells fragments, granules to gravel

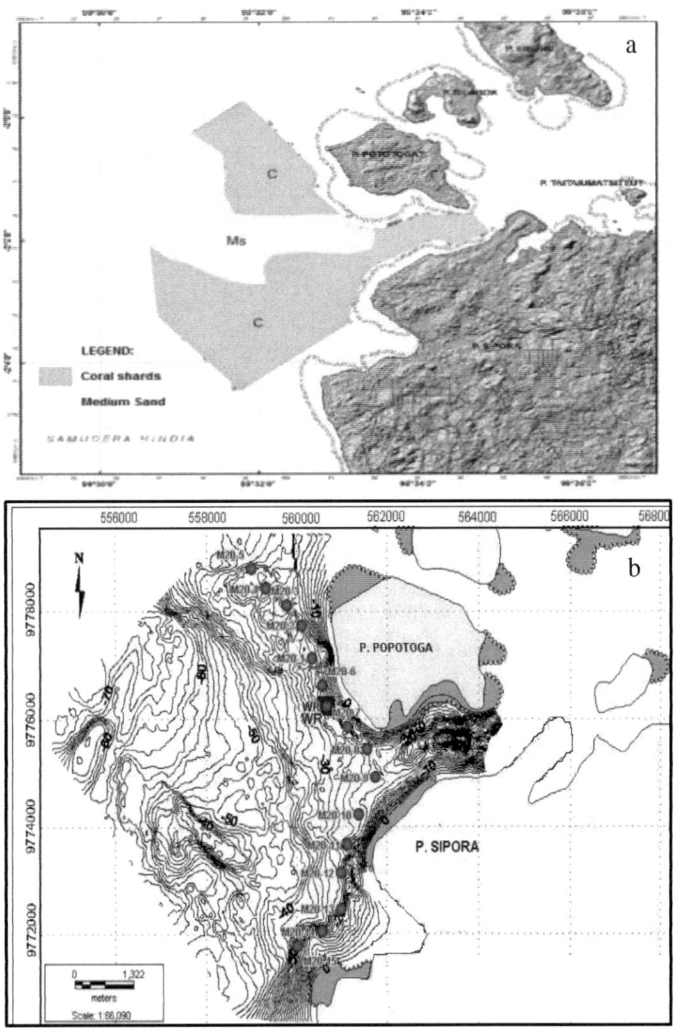

Figure 5. a.) Sediment distribution and b.) sampling location

3.3 Seafloor Morphology & Bathymetry

The integrated sounding data produces a sounding fix point in the form of coordinates, latitude (X) and longitude (Y), as well as the depth value (Z). The result of bathymetric measurements was corrected with tidal MSL.

Data selection for Interpretation is related to the target location which aims to find out the bottom profile of the seabed. The basis of picking on stratabox data is based on the analysis model seismic facies[10]. In this study, the direction of the trajectory underwater morphological data collection is more focused on the location of the tool placement static ocean wave meter with the direction of trajectory southwest - northeast and vice versa.

The coastal parts of Sipora Island and Pototagat Island have relatively high subsurface relief, reaching a depth of approximately 40 m. From the SBP records, the rock that forms the base of the island and seabed sediments cannot be separated. This indicates that the stratabox acoustic waves cannot penetrate the layers on the seabed surface, making it difficult to identify the sediment layers below the seabed surface. The shape of the underwater morphology is a result of tectonic formations working in the study area. The sediment layer at the bottom of the subsea surface is rework material from the bedrock where the relative thickness is not visible in the stratabox recording results because the accoustic energy does not penetrate the hard rock.

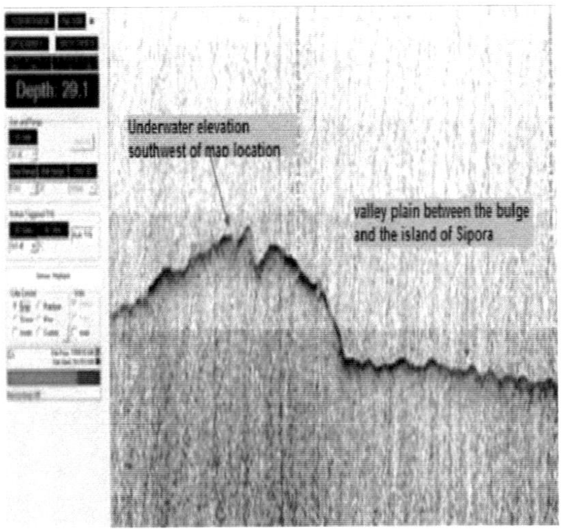

Figure 6. SBP records of seafloor the morphology show subsurface elevation in the Southwest of the study area.

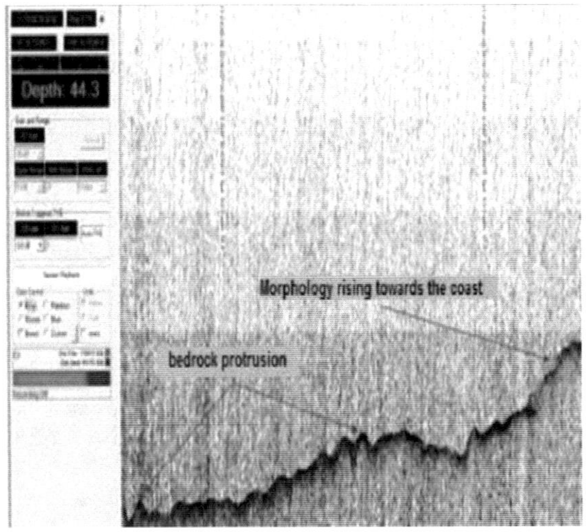

Figure 7. SBP records in the Northwest of study area, the appearance of the slopes towards features the coast with a depth of 35 meters and shows the bedrock protrusion morphology.

Bathymetric data show that the morphology of the Sipora seabed is graded starting from a depth of -15 m near the shoreline and getting deeper into the open sea to a depth of -60 m (Figure 4). The coastline pattern relatively follows the pattern of the north to south shoreline, the contour pattern begins to change and there is a wide valley plain up to a depth of 7 meters. In the northeastern area, the contour pattern mapping forms an oval circle with a depth of -35 meters which is the morphology of the protrusion of the bedrock (Figure 7), on the southwest side the depth gets deeper to -60 meters and slopes back towards the coast. The west side coast of Mapadegat, there is height morphology in the middle of the sea with a height of about 10 m from the seabed which is estimated to be hard rock (Figure 6). The condition of the seabed surface in the study area shows a steep morphology in the form of coral reefs and shallow basins but it appears flat in other areas and only steep away from the coast. Based on these results, the submarine power line corridor is included in the topography of the continental shelf or continental shelf with a depth value of up to -30 meters. This matches up to [11] statement that the continental shelf is a shallow seabed with a gentle slope and a steep slope with an average depth of 30 meters.

Tectonic events are primarily responsible for differences in the topography of the seabed surface. Moreover, oceanographic factors such as waves, and tides also affect recent geological processes. Based on the measurement results, the distribution of the slope of the depth contour pattern is described in the form of color gradations. The redder the color, the steeper the slope (Figure 9). On the map can be shown from east to west, to a depth of -30 meters from the shoreline. The contour pattern is relatively loose, showing the morphology of the seabed coastline with moderate relief, and a gentle slope to the open sea with a slope of 2° to 8°.

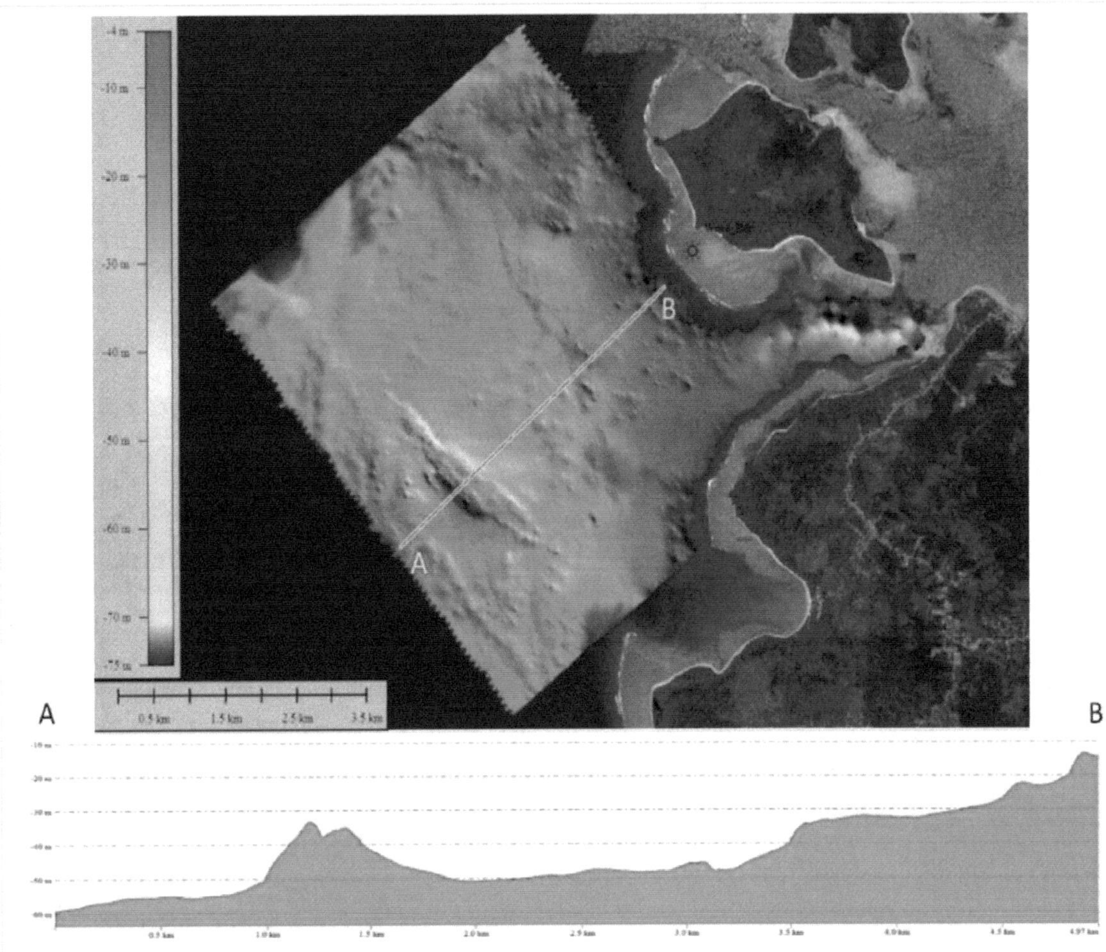

Figure 8. A-B cross-section of Sipora seabed morphology shows the presence of underwater elevations in the Southwest of the study area.

The slope of the seabed is visualized in the cross section line A-B'. The cross-sectional image of the seabed depth from A to B shows the morphology of the height in the middle of the seabed which is estimated to be hard rock (Figure 8). Based on the map cross-section the morphology of the seabed slopes off the coast and gets higher towards the coast. The morphology can be interpreted that the wave run-up will be higher reaching coastal areas due to morphology that is increasingly rising towards the coast.

Furthermore, the current flow will be higher towards the beach due to the presence of straits or gaps narrow between the two islands of Sipora and Pototogat, as well as the area where the waves break in the bay, which will also be higher towards the shore. A slope of more than 6 degrees will pose a risk to the burial plow [12] The seabed slope is also of great concern in the cable burial installation (Figure 9).

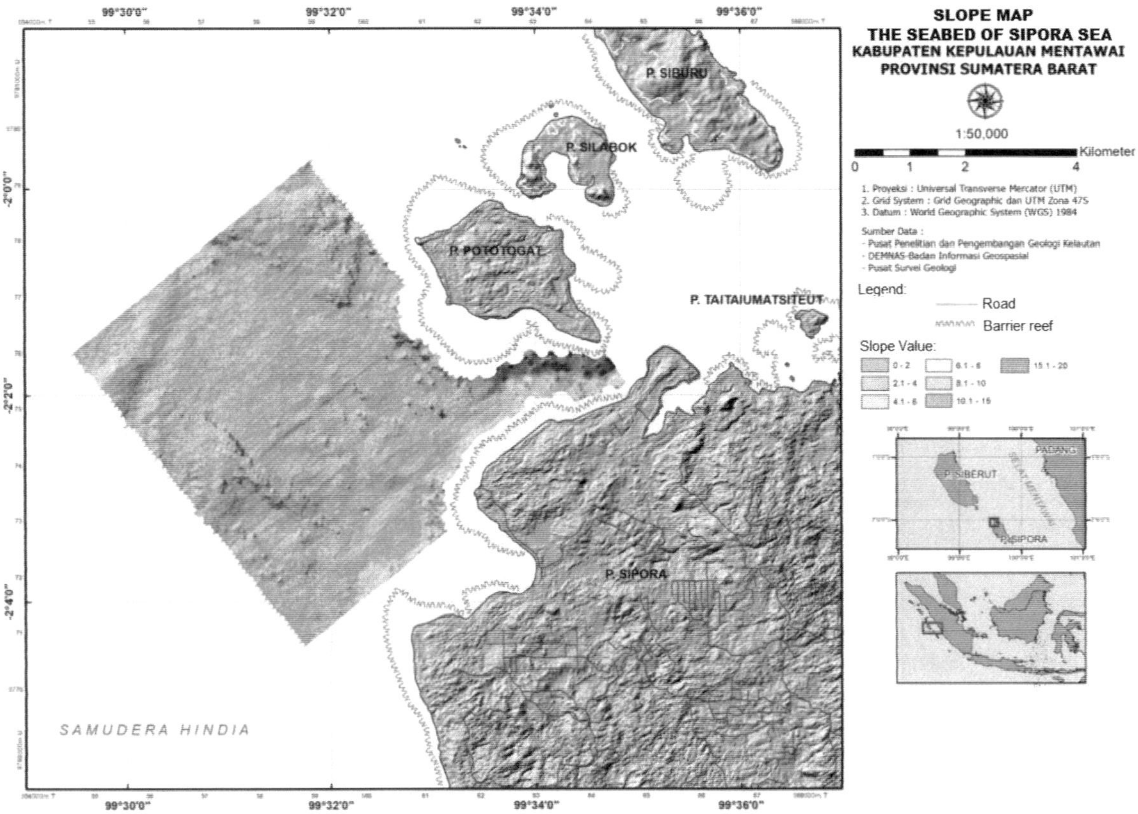

Figure 9. The slope map of the Sipora seabed shows the difference in the slope indicated by the color gradation

3.4 Seabed image

Side-scan sonar mapping obtained identification of rocks and sediments on the seabed. The side-scan sonar recording on the track in the southeastern region of the trudy area shows the presence of coral and sand sediments showing a ripple mark sediment structure.

The morphological characteristic of the seabed shows that the distribution of sand and coral sediments with diameters of up to 3 metres dominates the seabed of the study area in the strait (Figure 10). Coral colonizes or rests on a bed of hard rock that lies beneath the two islands.

Hard rock or coral features are interpreted as elongated boulder shapes with bright and shadowy saturation hues and a large and clustered texture pattern. The type of sand sediment is interpreted to be in the form of small grains with a fine texture pattern and has bright saturation hue.

The appearance of the seabed texture in the visualization of side-scan sonar data appears to have rougher and smoother parts. When compared to the appearance of a fine texture with a dark color intensity representing areas of low backscatter, coarse texture produces a bright backscatter signal.

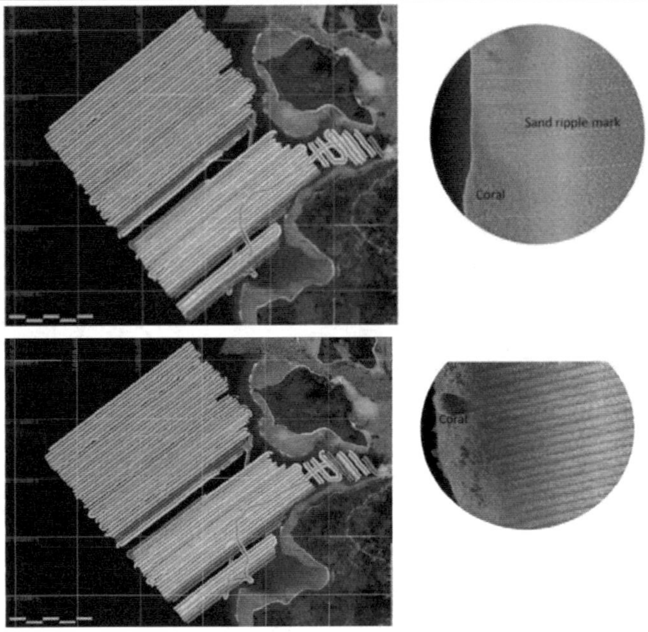

Figure 10. SBP and Side-scan sonar trajectory map showing the appearance of coral and sand ripple mark in the side scan sonar image

4. Analysis

GIS (Geographical Information System) modeling with raster-based spatial analysis is used to determine the location of the power plant, by assuming several parameters in order to obtain the best weight or assessment when deciding where to locate the power plant.

A total of four (criteria), including wave factor, lithological factor, slope factor, and coastal characteristics were adapted to evaluate sites for wave power generator (Table 2). The turbine should be installed in an area with a gentle slope and a topography of no more than 10%. Areas that have unstable topography should not be used as locations.

Table 2. List of constraint criteria, threshold values, and conformity mapping

Criteria	Application Model	Description
Annual Average Wave Energy Potential Map[2]	Constraint and suitability mapping	the places that received more than 15 kW/m per year are considered as potential sites. The higher the value, the higher the priority.[2]
Depth	Constraint and suitability mapping	The generator must be located between 15 to 25 meters [13].
Slope	Constraint and suitability mapping	The generator must be placed on a gentle slope in the range of 0-10% [14]
Types of seabed sediments	Constraint and suitability mapping	The recommended sediment as a base for water structures is sand (Soil parameters properties requirements for on-bottom

IOP Conf. Series: Earth and Environmental Science 1047 (2022) 012036 doi:10.1088/1755-1315/1047/1/012036

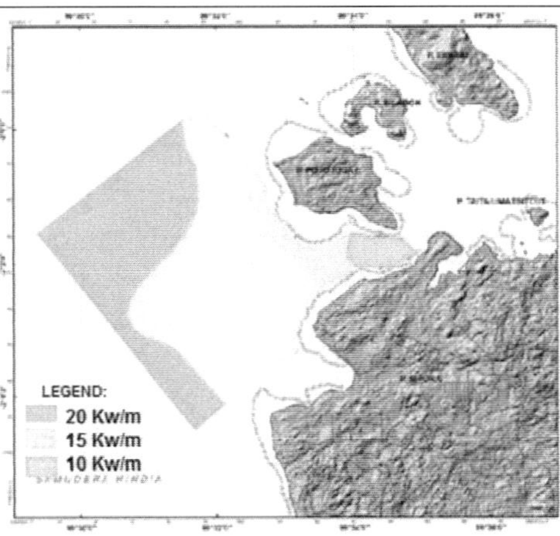

Figure 11 Annual Average Wave Energy Potential Map shows the highest average potential is 20 kW/m and the lowest is 10 kW/m. The higher the value, the higher the priority[2].

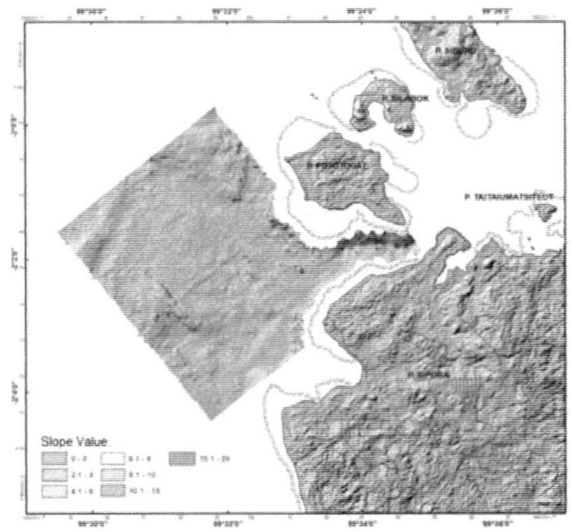

Figure 12 The slope map of the Sipora seabed shows the difference in the slope indicated by the color gradation. The generator must be placed on a gentle slope in the range of 0-10%.

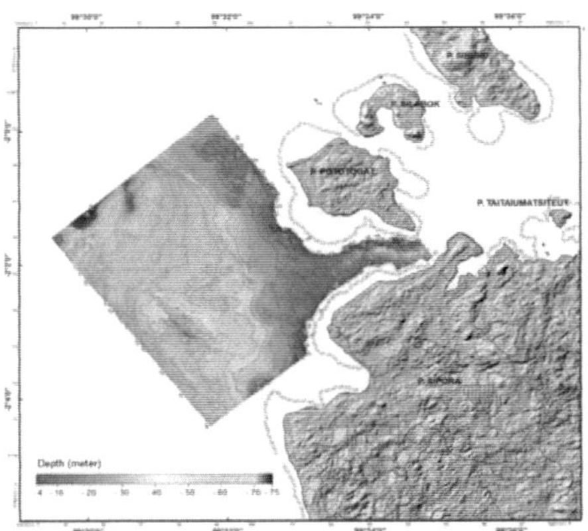

Figure 13 Bathymetric Map shows subsurface elevation features indicated by the color gradation. The generator must be located between 15 to 25 meters.

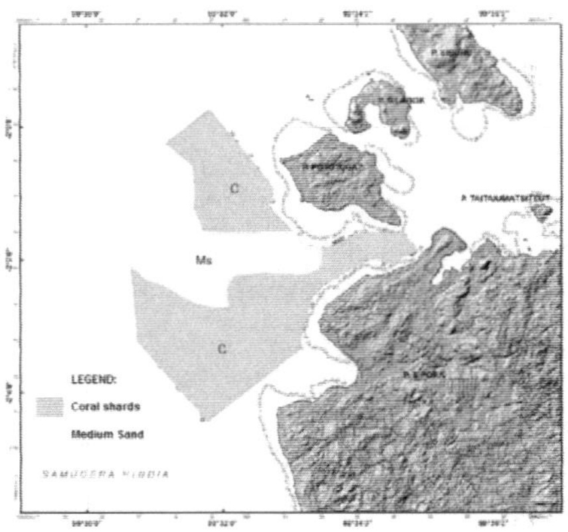

Figure 14 Sediment distribution map shows distribution of coral and sand. The recommended sediment as a base for water structures is sand.

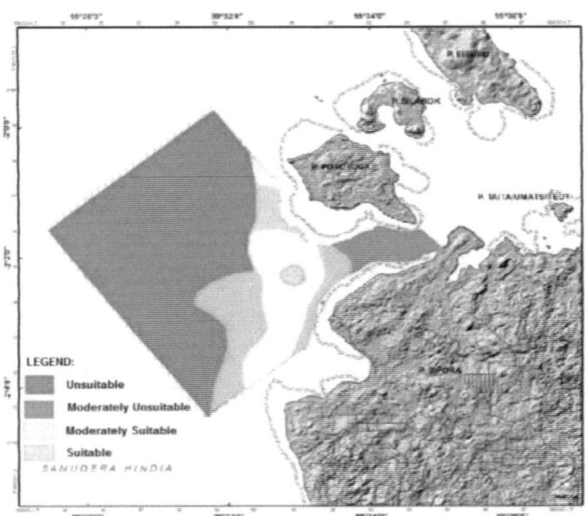

Figure 15 Suitability map based on weighting and scoring results.

The weights and standardized suitability maps are then combined to produce the overall suitability score. As a result, nearly the entire feasible area (77 percent of total area) is labeled "Unsuitable". In this case, 21.1 percent of the region is deemed "moderately suitable." Only less than 2% of the region is deemed "suitable”. Within this particular zone, the average depth of the coastal waters, the topography of the seabed, suitable sediment characteristics and the relatively reasonable wave energy potential (Figure 11) presuppose that the implementation of the wave energy system is practically achievable. Based on the weighting and scoring results, suitable locations for potential power plant sites are identified, including in coastal areas located in offshore near batu putih coast can be seen on the map above (Figure 15) the green color on suitability map shows the most suitable location with a depth of approximately 20 meters with estimation of annual average wave energy potential 15 kW/m.

5. Discussion and Conclusion

Subbottom profiling and side scan sonar data show that geological phenomena such as sloping plains on the high seas and a slight elevation of limestone exist at the study site.

Underwater morphology is a tectonic formation that works in the study area and is generally composed of hard and rigid rocks with a reflection pattern at the top and no reflection at the bottom. This is due to the fact that the rocks observed in the survey area are generally in the form of the Mentawai forearc's basement.

Bathymetry around the survey site is generally less than 100 meters. While the morphological contour pattern of the seabed beach with moderate relief is obtained with a slope of 10° to 20°, towards the open sea reaching 2° to 8°.

Based on the weighting and scoring results including wave factor, lithological factor, slope factor, and coastal characteristics suitable locations including in coastal areas located in offshore near batu putih coast with a depth of approximately 20m with estimation of annual average wave energy potential 15 kW/m. One of the most important value in the use of the GIS tools in that they constitute a distinctive information integration and management system that allows to gather, analyze, represent and facilitate the management and interpretation (qualitative and quantitative) of spatial and attribute information in a far more effective, rapid and integrated way. Additionally, it makes good use of the enormous capacity of spatial data in a single work environment for the fusion and review of various information layers, extraction of relevant data and ongoing information updating to enrich the system, compared to other types of manual methodologies and traditional map interpretation techniques used before. Such advantages allow us to handle several scenarios and generate map during the planning stage, before any decision is made on actions to take. Furthermore, these allows us to handle the assessment in quantitative way, providing us greater real-world approximation validity and less subjectivity in the analysis and selection of criteria.

6. Implication

Regarding the incorporation of overlay and weighting techniques, the advantage of this methodology is that it is integrated in GIS as a relatively common tool for a number of investigations as reported here is worth mentioning, where there are several factors and variables that influencing in the occurrence of a given fact, phenomenon or objective and there are several points of view in the decision-making process. Knowledge from these efforts will ensure that, as commercial-scale wave farms are developed, work is performed within a framework of understanding regarding associated effects on disaster risk, durability, and environmental sustainability at local and regional levels.

Acknowledgments
The author would like to thank supervisors A H P Kesumajana of ITB. Thank to I H Suherman for the opportunity to work with them in 2020 PPPGL team who provided me data used within this study.

References

[1] PT PLN Persero Wilayah Sumatera Barat 2018 *Laporan Musyawarah Perencanaan Pembangunan Daerah Kabupaten Kepulauan Mentawai (Pengembangan Listrik Pedesaan).*

[2] Suherman I H 2017 *Survey Report on Research on the Potential of Ocean Wave Energy for Electricity Development on Enggano Island*, Center for Research and Development of Marine Geology of Energy and Mineral Resources Bandung Indonesia

[3] Bloomberg Quicktake 2021 *Wave Power Could Be Energy's Next Big Leap.* https://www.youtube.com/watch?v=jahAum3zLsY [4] Mangga S A Kusnama K Suyono S 2006 *Stratigraphy and Tectonic Development of Mentawai Islands, West Sumatera, Based on Plate Tectonic Theory* Jurnal Geologi dan Sumberdaya Mineral 16 (3), 136-143

[5] Budhitrisna T & Andi Mangga S 1983 *Pagai and Sipora Geological Report* Sumatera in scale 1: 250.000 Geol Res Dev Centre Open file report.

[6] Dolan R B Hayden M Vincent 1975 Classification of coastal landform of the Americas. *Encyclopedia of Beaches and coastal Environment* P.72–88.

[7] EMEC, 2009, *Assessment of Wave Energy Resource and Guidlines for Project Development inthe Marine Energy Industry*, The European Marine Energy Centre Ltd.

[8] Darlan Y 1996 Geomorphology of Coastal Areas *Integrated Regional Planning and Management Training Paper* Ministry of Home Affairs and ADB Bogor.

[9] Cruz J 2008 *Ocean wave energy: current status and future perspectives* Green Energy and Technology Springer Berlin ISBN 978-3-540-74894-6.

[10] Mitchum R M Jr 1977 Seismic stratigraphy and global changes of sea level part 11 Glossary of terms used in seismic stratigraphy in C E Payton (ed) Seismic Stratigraphy Applications to Hydrocarbon Exploration Am Assoc Petrol Geol Memoir 26, pp. 205–212.

[11] Kennett J P 1982 *Marine Geology* Prentice-Hall Englewood Cliffs New Jersey 813pp

[12] Zuidam R A Van 1985 *Aerial Photo-Interpretation in Terrain Analysis and Geomorphology Mapping* Smith Publisher The Hague ITC

[13] AW-Energy Oy 2021 https://aw-energy.com/waveroller/#technology

[14] Republik Indonesia 2016 *Regulation of the Minister of Transportation of the Republic of Indonesia PM 129 shipping lanes at sea and buildings and/or installations in waters.*

[15] Lekkerkerk H J van der Velden R Roders J Haycock T de Vries R Jansen P and Beemster C 2006 *Handbook of Offshore Surveying* Book Two London Clarkson Research Services Limited ISBN 1-902157-74-5

Impact of natural conditions and anthropogenic activities on groundwater quality in Puntang volcanic area, West Java, Indonesia.

R. Maria [1,2]., T.Y.W.M Iskandarsyah[1], B.R. Suganda[1], A.F.Rusydi.[2] Hendarmawan Hendarmawan[1]

[1]Faculty of Geological Engineering, Universitas Padjadjaran, Jatinangor 45363, Sumedang, Indonesia

[2]Research Center for Geotechnology BRIN, Sangkuriang, Bandung 40135, Indonesia, 40135
rizk002@brin.go.id ; rizka_maria@yahoo.com

Abstract. The Puntang area in the southern mountains of West Java has potential water resources that are used to fulfill the community's raw water. The rapid development of the region and tourism will affect the quality of groundwater. This research was conducted on the slopes of Mount Puntang with the aim of analyzing natural and anthropogenic conditions on groundwater quality. Measurements and groundwater samples are taken from 5 springs, 10 dug wells and 8 drilling wells. The research method begins with a literature study followed by data collection in the field and laboratory analysis. Data analysis used statistical analysis and hydrogeochemical spatial analysis. Hydrogeochemical analysis using Piper diagrams. Based on the analysis results, it is known that the groundwater facies are CaCl, CaMgCl $CaHCO_3$, and $NaHCO_3$. The CaCl facies, enrichment of Cl^- is caused by contamination of volcanic domain. The CaMgCl facies has enrichment of Mg^{2+} cations in the process of groundwater and rock water interactions. The high Cl^- value in the proximal area is influenced by volcanic domain, while the Cl^- value in the medial and distal areas may be influenced by anthropogenic. The $CaHCO_3$ facies is influenced by local meteoric water, with a high structural control that shows water flowing through the fracture. The NaHCO3 facies is a wellbore in the intermediate groundwater flow and anthropogenic influences have occurred. The hydrochemical evolution that occurs is that Ca^{2+} becomes Mg^{2+} and then turns into Na^+. The $CaMgHCO_3$ facies dominates the proximal to distal area. The presence of HCO_3^- indicates that groundwater mixing has occurred and is influenced by the decomposition of organic matter. Indications of groundwater mixing can be seen in groundwater with $NaKHCO_3$ facies in the distal area. The medial and distal areas are densely populated areas which also affect groundwater. Geological and environmental conditions around the water resources are one of the supporting data to determine polluting sources.

Keywords: water resources, environmental, hydrogeochemical, rock water interactions.

1. Introduction

Groundwater is the main source of water needs for the domestic, industrial and agricultural sectors worldwide [1]. The physical and chemical quality of groundwater will be influenced by geological conditions and changes in the surrounding environment [2]. The decline in groundwater quality in several countries is caused by natural processes and anthropogenic activities [3]. Anion cations analysis serves as a natural tracer in aquifers originating from natural and anthropogenic influences [4]. Volcanic rocks are easily changed when in contact with hydrothermal solutions which affect the chemical processes of groundwater [5].

The Puntang area in the southern mountains of West Java has potential water resources that are used to fulfill community raw water. The rapid development of the region and tourism will affect the quality of groundwater. Regional developments have an impact on increasing demand for clean water [6]. Characteristics of landforms in volcanic areas have a fairly high potential for water resources, including from springs [7]. The compositional characteristics of volcanic aquifers are formed from loose material deposits to integrated pyroclastic materials [8]. The morphology of the volcanic cone encourages the formation of orographic rain so that the area around the volcanic mountains has a fairly high rainfall. The potential of water resources in volcanic areas is usually marked by the appearance of a row of springs at the bend of the slope on the slope and foot of the volcano [9]. This study was conducted on the slopes of Mount Puntang with the aim of analyzing natural and anthropogenic conditions on groundwater quality. The study of natural and anthropogenic conditions on groundwater quality is very important to know the interaction of water rock, geothermal, and anthropogenic domains that affect the potential of groundwater resources.

2. Methods

The research was conducted in the Puntang volcanic area, geographically located at 107°32'30" E–107°45'30" E and 7°02'30" S–7°07'30" S. Mount Puntang is located in Cimaung Districts, Bandung Regency, about 29 Km far from the center of Bandung City with an altitude of 2,224 meters above sea level. (Figure 1).

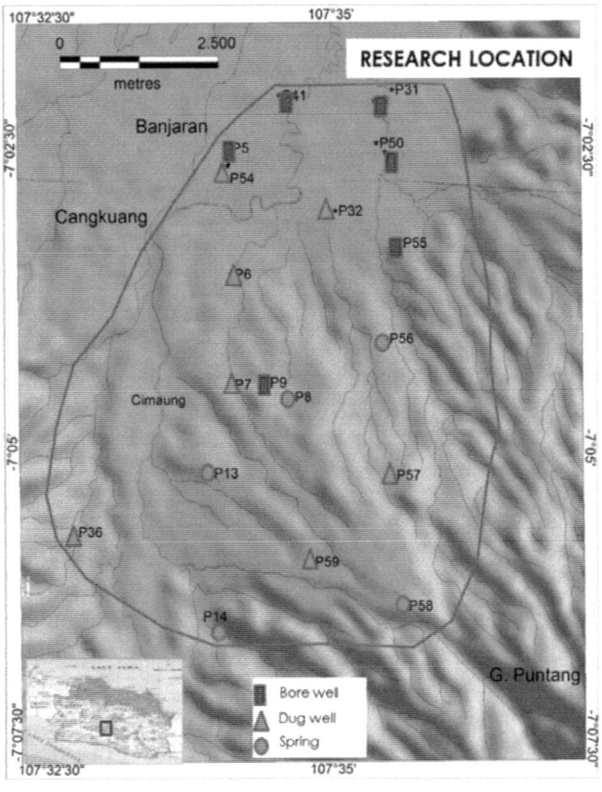

Figure 1. Research location in the Puntang mountain area

The research method begins with a literature study followed by data collection in the field and laboratory analysis. Data analysis used statistical analysis and hydrogeochemical spatial analysis. Statistical analysis was processed using Minitab 19 software. Hydrogeochemical analysis was processed using Scholler and Piper diagrams. (Figure 2). To determine the groundwater interactions, measurements and groundwater samples are taken from 5 springs, 10 dug wells and 8 drilling wells.

Collecting primary data for groundwater samples in the field including measurements of pH, conductivity and temperature in dug wells, boreholes and springs using the Horiba U 10 water quality checker measuring instrument. Laboratory analysis was carried out for the major ions: Ca^{2+}, Mg^{2+}, Na^{+}, K^{+}, SO_4^{2-}, Cl^{-}, HCO_3^{-}, CO_3^{-}. The analysis was carried out in the groundwater laboratory of the Research Center for Geotechnology BRIN. The analysis of Na^{+}, K^{+} was carried out using the atomic absorption flamephotometry (AAS) method. Analysis of Ca^{2+}, Mg^{2+} ions using complexometric method. The acidimetric method for the analysis of HCO_3^{-}, the turbidimetric method for the analysis of SO_4^{2-}, the argentometric method for the analysis of Cl^{-}, and the gravimetric method for the analysis of TDS.

Hydrochemical multivariate analysis was carried out on the physical and chemical properties of pH, EC, Ca^{2+}, Mg^{2+}, Na^{+}, K^{+}, SO_4^{2-}, Cl^{-}, HCO_3^{-}. Correlation analysis of hydrochemical parameter can be used to determine the relationship between groundwater quality parameters [10]. The hydrochemical parameters were analyzed using the correlation coefficient using the Pearson method [11]. Hierarchical group analysis was conducted to analyze homogeneous data with the principle of grouping data that tend to have the same properties based on the physico-chemical properties of groundwater [12].

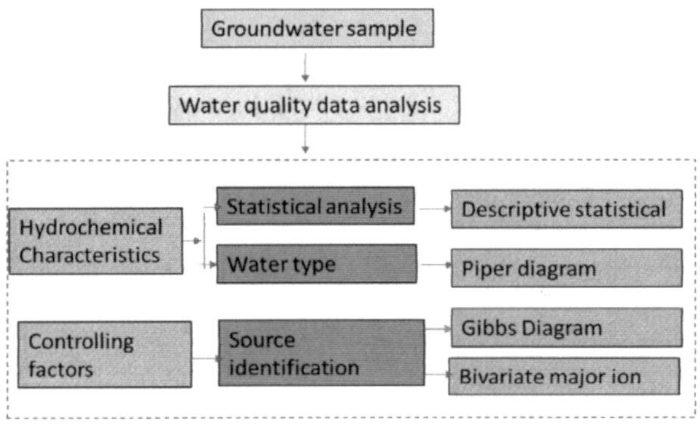

Figure 2. Research method

3. Data

The geological condition from the study area consist of: The Beser Formation (Tmb) is the oldest rock in the late Miocene, consist of tuffaceous breccias, lava, andesite and basalt. Quaternary volcanic rocks overlay Tertiary volcanic rocks. Old unraveled volcanic deposits (Qopu) of Pleistocene age, consisting of fine-coarse crystalline tuff, dacite, tuff breccia containing old pumice and andesite lava deposits. The youngest rock unit is lake deposits (Qd) of the Holocene age [13] (Figure 3).

Volcanic facies have characteristics based on the geological condition, sedimentary structures, and biological conditions [14;15]. The study area's volcanic facies consist of the central, proximal, medial, and distal facies [15]. The proximal facies on the upper slopes of Mount Malabar, the medial facies on the lower slopes, and the distal facies plain area of the lake deposits. The distribution of springs is located in the Malabar volcanic deposits (Qmt).

Landsat imagery analysis shows that the morphological lineaments are east-southeast-northwest in the direction of N $120^0 - 137.5^0$ E/N $300^0 - 317.5^0$ E (Figure 4). Morphological lineament analysis is needed to know the fracture aquifer system. The control structure height indicates the water flowing through the fracture and dilution process.

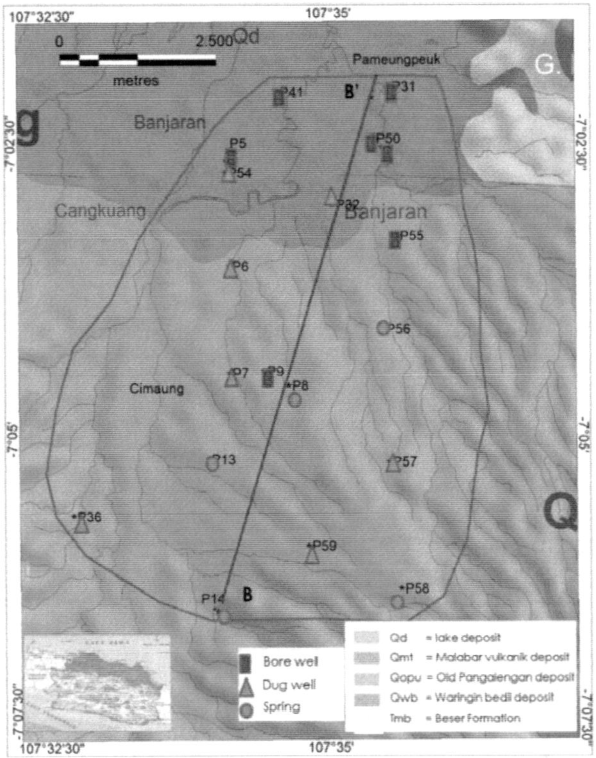

Figure 3. Geological map in the Puntang area (Modification from Alzwar [13])

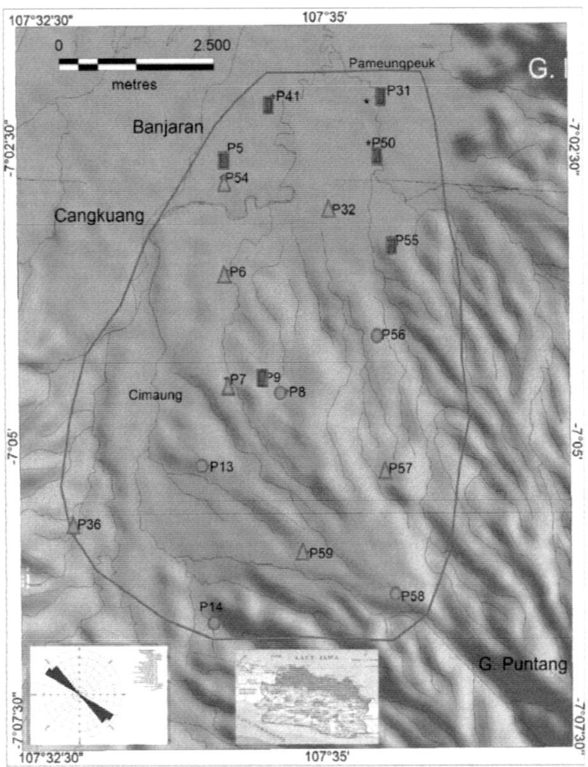

Figure 4. Morphological lineaments in the Puntang mountain area

The land use in the Gunung Puntang area consisting of primary forest, industrial forest plantations, plantations, settlements, dry land agriculture and rice fields (Figure 5). The locations for taking spring samples were mostly in settlements and paddy fields. The land cover of the research area affects the hydrochemical conditions of the research area. Anthropological influences from domestic and agricultural activities affect the concentration of the major ionic elements.

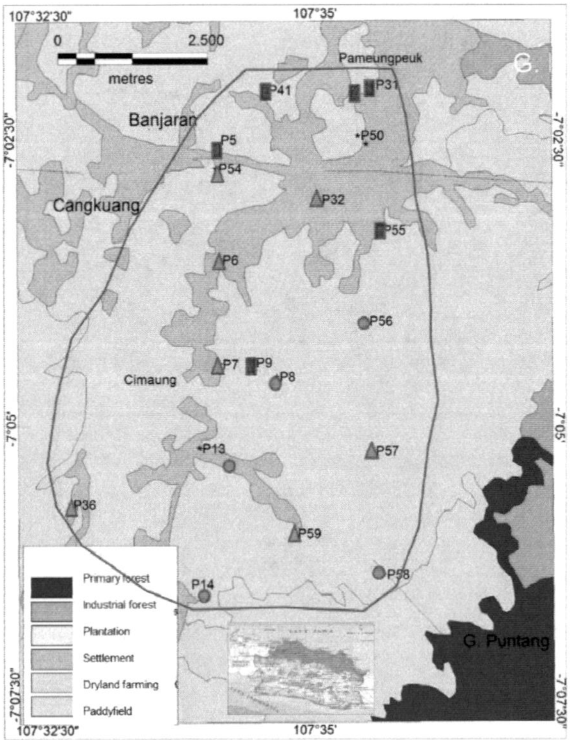

Figure 5. The land use in the Puntang area (Modification from Bakosurtanal [16])

4. Results and Discussion

The summary of the physical dan chemical properties is given in tables 1 and 2. The physico-chemical properties from the field data collection and the chemical properties are taken from laboratory analysis.

Table 1. The physico-chemical properties in the research area

No	ID	pH	Temp(°C)	EC(μS/cm)	TDS(mg/L)
1	P5	7.08	25.9	400	242
2	P6	5.25	26.2	15	84
3	P7	5.76	28.4	70	58
4	P8	6.65	26.3	60	22
5	P9	7.39	28.4	100	126
6	P13	6.1	26.5	120	116
7	P14	7.06	25.5	160	114
8	P30	7.27	25.1	270	198
9	P31	6.96	26.2	350	216

10	P32	7.18	26.4	430	260
11	P36	6.4	25.5	180	220
12	P38	6.96	28.9	380	198
13	P41	7.16	25.4	320	104
14	P42	7.43	29.3	460	186
15	P50	7.67	28.7	230	172
16	P51	7.77	29.9	280	324
17	P52	7.6	28.9	400	308
18	P54	7.27	28.2	330	252
19	P55	6.61	33	50	56
20	P56	7.46	28.8	80	136
21	P57	6.1	24.8	50	130
22	P58	6.53	24.2	60	20
23	P59	7.26	23.6	110	186

Table 2. Major ion's chemical analysis in the research area

No	ID	Na^+ (mg/L)	K^+ (mg/L)	Mg^{2+} (mg/L)	Ca^{2+} (mg/L)	Cl^- (mg/L)	SO_4^{2-} (mg/L)	HCO_3^- (mg/L)	Mn (mg/L)	Fe (mg/L)	NH_4^-N (mg/L)
1	P5	19.02	11.79	19.03	49.33	37.19	9.70	190.21	1.45	3.86	2.34
2	P6	11.33	0.71	5.19	6.98	27.55	4.12	20.81	0.25	0.05	0.16
3	P7	3.65	0.77	3.64	2.87	13.15	5.09	12.18	0.01	0.01	0.07
4	P8	2.03	0.38	2.31	2.77	6.77	5.10	9.45	0.01	0.02	0.08
5	P9	10.12	4.97	3.92	7.74	6.89	6.44	63.24	0.01	0.14	0.15
6	P13	7.69	0.45	5.19	7.93	22.04	10.59	18.45	0.01	0.01	0.02
7	P14	14.17	4.68	4.94	14.63	9.64	6.71	81.69	0.01	0.20	0.09
8	P30	7.37	1.41	8.36	37.19	17.91	23.89	108.04	0.01	0.02	0.06
9	P31	11.60	3.53	10.10	35.28	39.94	21.31	98.04	0.02	0.12	1.06
10	P32	26.68	2.47	6.70	52.48	46.83	28.77	142.84	0.07	0.16	0.53
11	P36	7.98	4.39	2.98	10.99	20.04	18.39	26.35	0.09	0.04	0.08
12	P38	15.83	1.86	16.37	48.80	27.55	17.56	187.10	0.65	5.51	1.89
13	P41	13.70	2.12	25.21	29.65	30.30	32.31	143.04	0.07	0.07	0.13
14	P42	27.55	8.59	18.11	29.82	13.77	4.77	258.25	0.46	3.82	6.46
15	P50	25.42	3.38	6.08	15.11	20.66	4.85	103.31	0.12	0.59	1.38
16	P51	24.00	3.20	7.06	16.73	20.66	5.59	112.94	0.21	1.43	3.62
17	P52	32.88	2.84	7.92	20.60	53.76	4.64	107.03	0.25	1.93	2.99
18	P54	20.81	2.48	21.09	25.45	12.40	8.34	191.83	0.81	6.56	1.78
19	P55	1.31	0.68	1.20	5.01	5.26	4.80	13.08	0.01	0.01	0.09
20	P56	3.90	0.68	2.38	6.30	9.02	7.08	21.35	0.01	0.05	0.11
21	P57	2.61	0.50	2.70	6.38	12.40	6.09	10.54	0.01	0.00	0.17
22	P58	1.96	0.50	2.37	4.93	6.89	6.13	16.35	0.00	0.02	0.12
23	P59	4.23	0.86	3.75	11.88	13.77	6.78	42.16	0.00	0.00	0.21

Table 3. The summary of the physico-chemical properties and Major ion in the research area

	Max	Min	Mean
pH	7.77	5.25	6.91
Temp	33	23.60	27.13
Cond	460	15	213.26
TDS	324	22	161.22
Na^+	32.88	1.31	12.86
K^+	11.79	0.38	2.75
Mg^{2+}	25.21	1.20	8.11
Ca^{2+}	52.48	2.77	19.52
Cl^-	53.76	5.26	20.63
SO_4^{2-}	32.31	4.12	10.83
HCO_3^-	258.25	9.45	86.01
Mn	1.45	0	0.20
Fe	6.56	0	1.07
NH_4^-N	6.46	0.02	1.03

Based on the summary of the hydrochemical parameters of Mount Puntang, the cation concentration is Ca^{2+}> Na^+> Mg^{2+} and the anion concentration is HCO_3^-> Cl^-> SO_4^{2-}.

Based on the physical and chemical data, the next analysis is statistical correlation with the Pearson method. Correlation coefficient of statistical analysis using the Pearson method was carried out to determine the relationship between several hydrochemical parameters. Correlation coefficient with a value of (r)> 0:7 indicates a strong correlation, a value between 0.5-0.7 indicates a moderate correlation, and a value <0.5 indicates a weak correlation [4]. The pearson correlation coefficient matrix for the Puntang mountains is described in table 3.

Table 4. Matrix of Pearson's correlation

	Ca^{2+}	Mg^{2+}	Na^+	K^+	HCO_3^-	SO_4^{2-}	Cl^-	pH	EC	TDS	Mn	Fe	NH_4^-N
Ca^{2+}	1												
Mg^{2+}	**0.68**	1											
Na^+	**0.56**	**0.51**	1										
K^+	0.49	0.49	**0.52**	1									
HCO_3^-	**0.81**	**0.85**	**0.76**	**0.65**	1								
SO_4^{2-}	**0.65**	0.42	0.09	0.01	0.30	1							
Cl^-	**0.65**	0.36	**0.63**	0.25	0.39	0.47	1						
pH	0.40	0.32	**0.56**	0.37	**0.57**	0.13	0.15	1					
EC	**0.87**	**0.74**	**0.81**	**0.59**	**0.91**	0.47	**0.67**	**0.59**	1				
TDS	**0.60**	0.37	**0.74**	0.43	**0.60**	0.23	**0.62**	**0.59**	**0.75**	1			
Mn	**0.57**	**0.67**	0.44	**0.68**	**0.67**	0.06	0.30	0.17	**0.55**	0.43	1		
Fe	**0.52**	**0.70**	**0.52**	0.44	**0.77**	0.07	0.15	0.30	**0.61**	0.47	**0.82**	1	
NH_4^-N	0.38	0.49	**0.73**	**0.60**	**0.75**	0.21	0.25	0.47	**0.68**	**0.55**	**0.51**	**0.63**	1

The Pearson correlation on groundwater at Mount Puntang are (Table 4);
Electrical conductivity (EC) has a strong correlation with TDS, Na^+, Ca^{2+}, Mg^{2+}, HCO_3^-, moderately correlated with pH, Cl^-, K^+, Fe, Mn and NH_4^-N. pH has a moderate correlation with Na^+, HCO_3^-, EC, TDS. Calcium (Ca^{2+}) is strongly correlated with EC, HCO_3^-, moderately correlated with Mg^{2+}, Na^+,

SO_4^{2-}, Cl^-, TDS, Fe, Mn. Magnesium (Mg^{2+}) has a strong correlation with EC, HCO_3^-, and Fe, moderately correlated with Na^+, Ca^{2+}, and Mn. Sodium (Na^+) has a strong correlation with EC, $HCO3^-$, TDS and NH_4^-N, moderately correlated with Cl^-, pH, K^+ and Fe. Potassium (K^+) is moderately correlated with EC, Na^+, HCO_3^-, Mn, and NH_4^-N. Bicarbonate ($HCO3^-$) has a strong correlation with EC, Na^+, Ca^{2+}, Mg^{2+}, Fe and NH_4^-N moderately correlated with TDS, Mn and K^+. Sulphate (SO_4^{2-}) is moderately correlated with Ca^{2+}. Chloride (Cl-) is moderately correlated with EC, TDS, Ca^{2+}, Na^+. Iron (Fe) is strongly correlated with $HCO_3^{-,}$ Mg^{2+}, Mn, moderately correlated with EC, Na^+, Ca^{2+} and NH_4^-N. Ammonium (NH_4^-N) is strongly correlated with Na^+, HCO_3^-, moderately correlated with EC, TDS, Mn and Fe. The strong correlation between EC and the main ion shows that the solubility of the major ion is quite high. The abundance of Ca^{2+} found in most of igneous rocks [17]. The abundance of Mg^{2+} comes from the dissolution of ferromagnesian alteration [18]. Na^+ can be adsorbed on clay and has a high cation exchange capacity [19]. NH_4^-N has a strong correlation with the major ion that heavily influenced by anthropogenic activities.

The next statistical process is Hierarchical Custer Analysis (HCA) to determine geological and environmental controls on groundwater conditions. Hierarchical statistical analysis groups based on the Euclidian distance in the Puntang mountain area is divided into two groups (A and B). Group A is groundwater with types of CaCl, MgCl and $CaHCO_3$. Group B is groundwater with types of $CaHCO_3$, $CaMgHCO_3$ and $NaHCO_3$. some locations in this group B have high NH_4^-N concentrations and indicate the influence of anthropogenic activities. Group A shows the hydrochemical grouping of groundwater that is influenced by groundwater-rock interactions and Group B which is influenced by anthropogenic interactions (Figure 6).

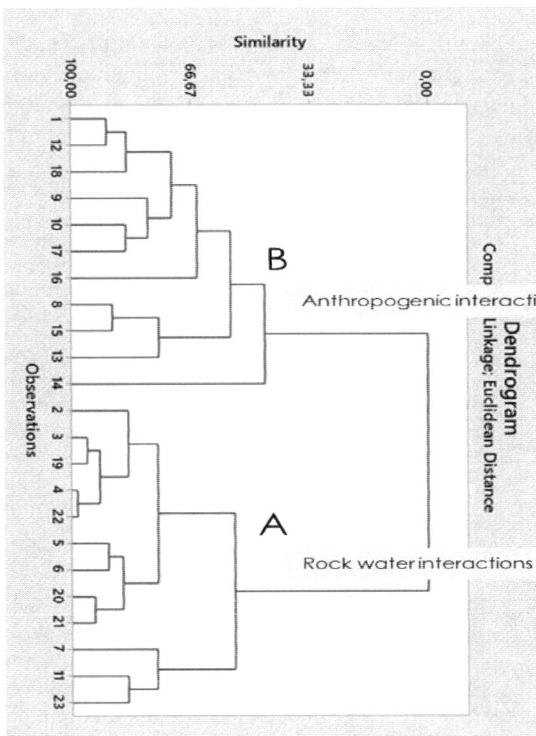

Figure 6. Hierarchical groups based on the Euclidian distance in the Puntang mountain area

Hydrochemical value will be more clearly described by spatial modeling of water quality parameters to show the distribution pattern of hydrochemical variations. (Figure 7).

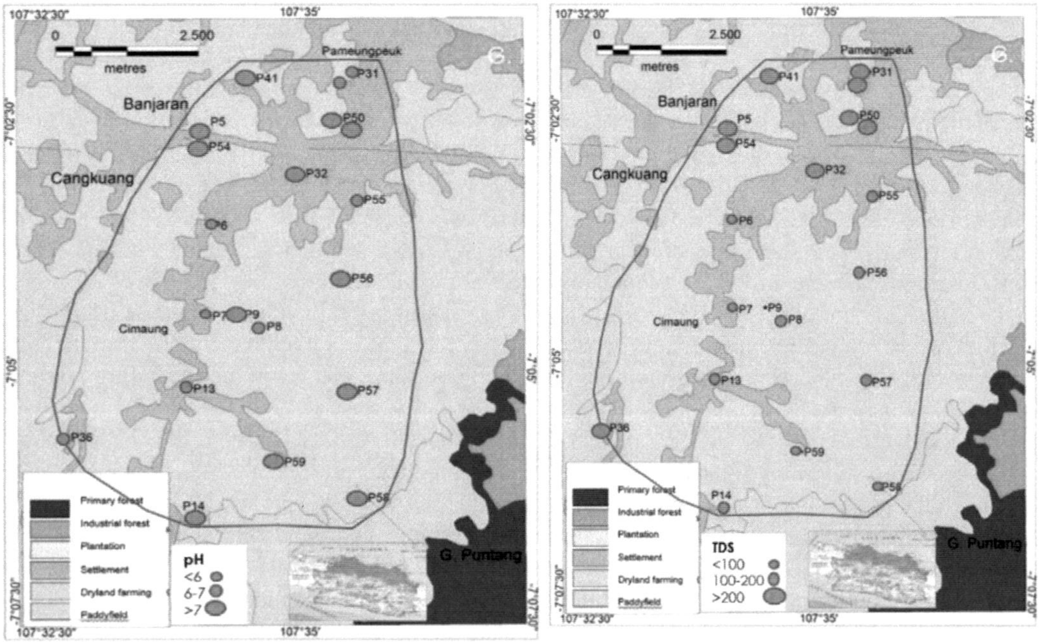

Figure 7a.b. Spatial distribution of pH and TDS in the Puntang mountain area

Alkaline pH (pH>7) indicates loss of CO_2 as alkaline mineral pH deposition (pH>7) caused by loss of CO_2 and mineral deposition. The pH value is evenly distributed, not affected by topographic conditions (Figure 7a). Sources of material availability from TDS and conductivity come from geological conditions and the influence of human activities such as household and agricultural waste [20]. Variations in TDS values can also characterize groundwater catchment areas. High TDS value indicates groundwater travel time (Figure 7b).

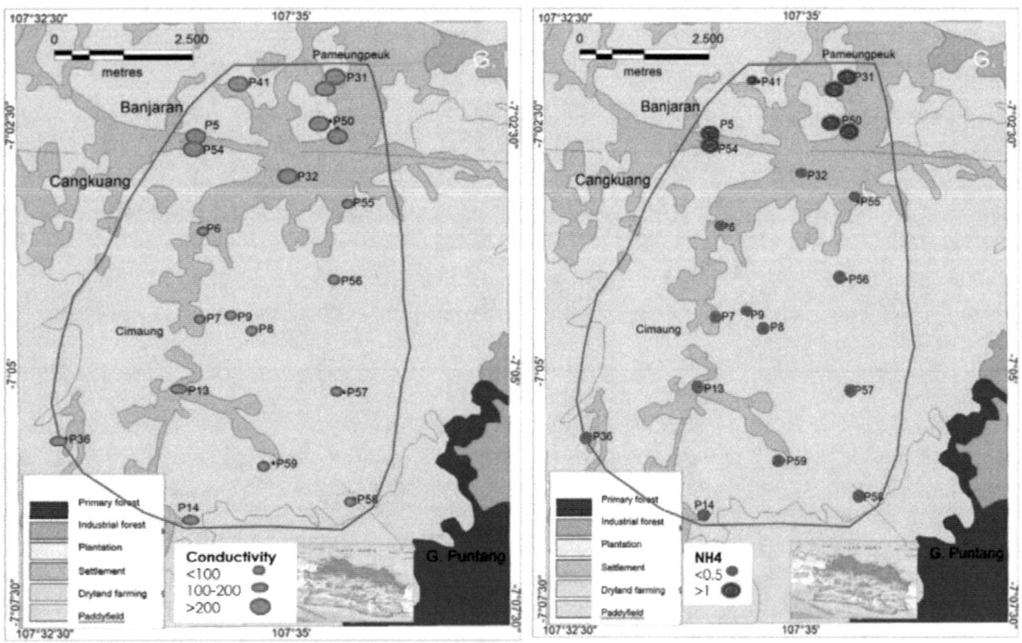

Figure 7c.d. Spatial distribution of EC and NH_4^-N in the Puntang mountain area

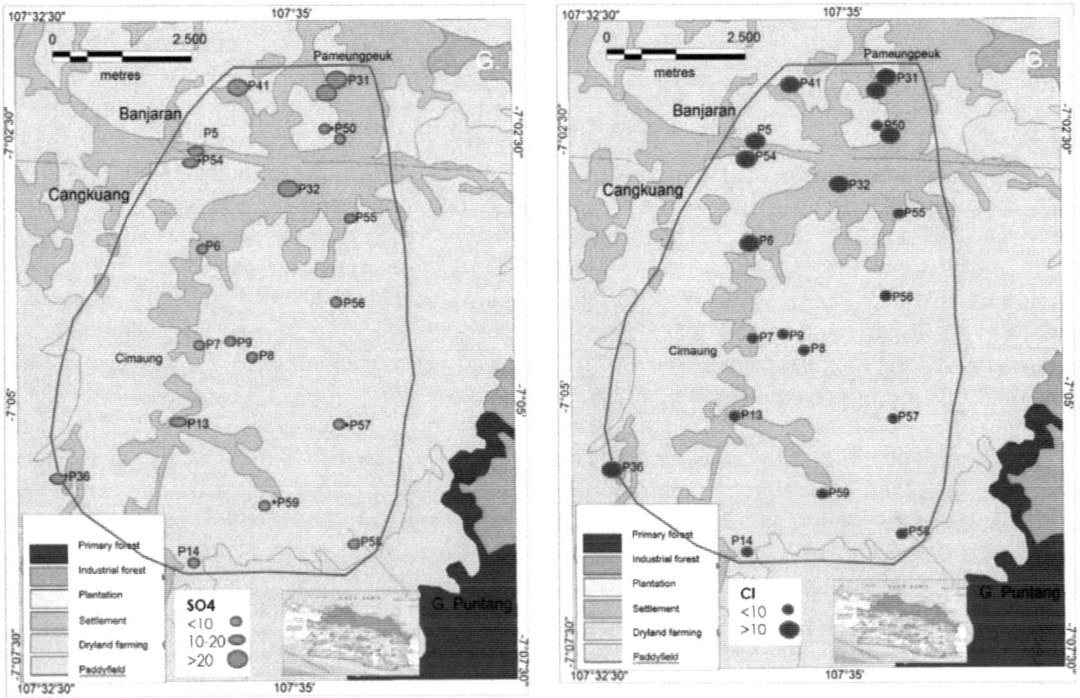

Figure 7e.f. Spatial distribution of Mn and Fe in the Puntang mountain area

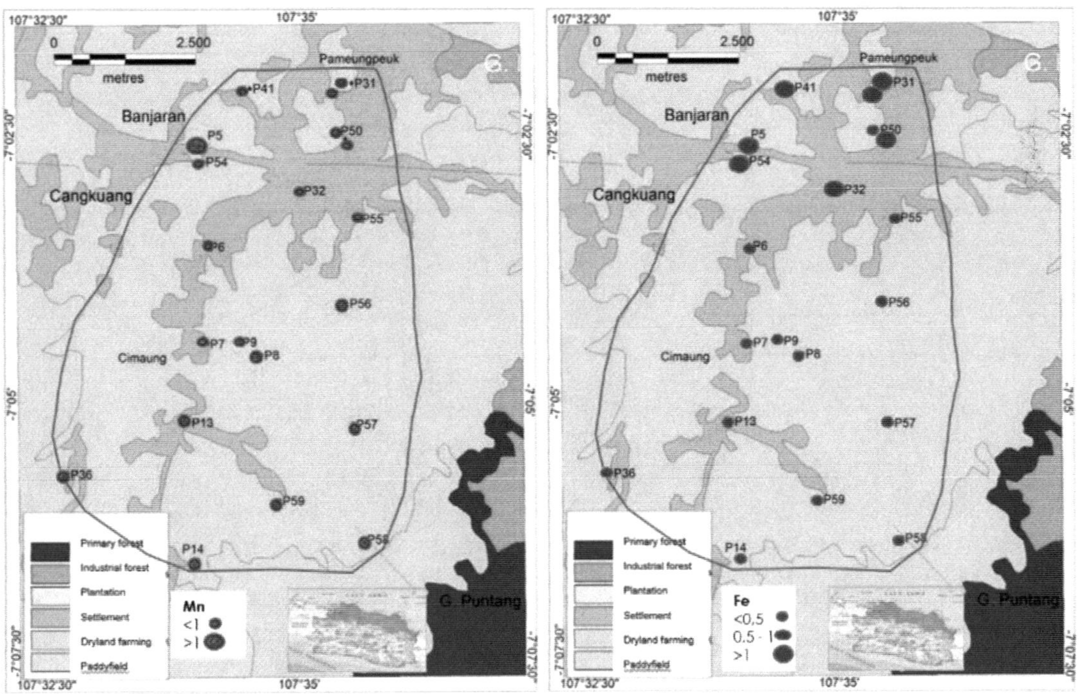

Figure 7g.h. Spatial distribution of SO_4^- and Cl^- in the Puntang mountain area

Electronic conductivity (EC) can be sourced from geological conditions and the influence of anthropogenic activities from households and agriculture [21]. Groundwater with low EC values have shorter contact times with rocks so that variations in EC values can indicate recharge areas. The low EC value is in the Quarter Malabar rock unit in the proximal and medial zones, while the high EC value is in the distal zone (Figure 7c). Ammonium (NH_4^-N) pollution comes from agricultural residues and domestic waste [22] [23]. The concentration of NH_4^-N that occurs in the rainy season is higher due to the effect of dissolution by rainwater that accumulates in the distal area with lake sediment lithology (Qd) (Figure 7d). The value of Manganese (Mn) in the study area is still below the threshold. only a few locations in the distal region exceeded the threshold (Figure 7e). The value of iron (Fe) in the proximal and medial areas is smaller than in the distal area. The process of dissolving rocks affects the concentration of iron in groundwater (Figure 7f).

Residues of agricultural activities will produce high ionic values for Cl^-, NH_4^-N, SO_4^{2-} and K^+ [24]. Anthropogenic activities such as agricultural fertilizers and backflow irrigation also affect groundwater conditions[25] . The concentration of SO_4^{2-} in the proximal area tends to be lower than in the distal area, which occurs due to the effect of dissolution by rainwater that accumulates in the distal area with lake sediment lithology (Qd) (Figure 7g). The Cl^- ion is the most abundant element in nature. The longer the groundwater circulation and the distance traveled to reach the discharge area [26]. The concentration of Cl^- ions is influenced by geothermal activity [27], and agricultural residues will produce high ion levels for Cl^- [24]. The high Cl^- value in the proximal area is influenced by volcanic domain, while the high Cl^- value in the distal area is influenced by anthropogenic activity (Figure 7g).

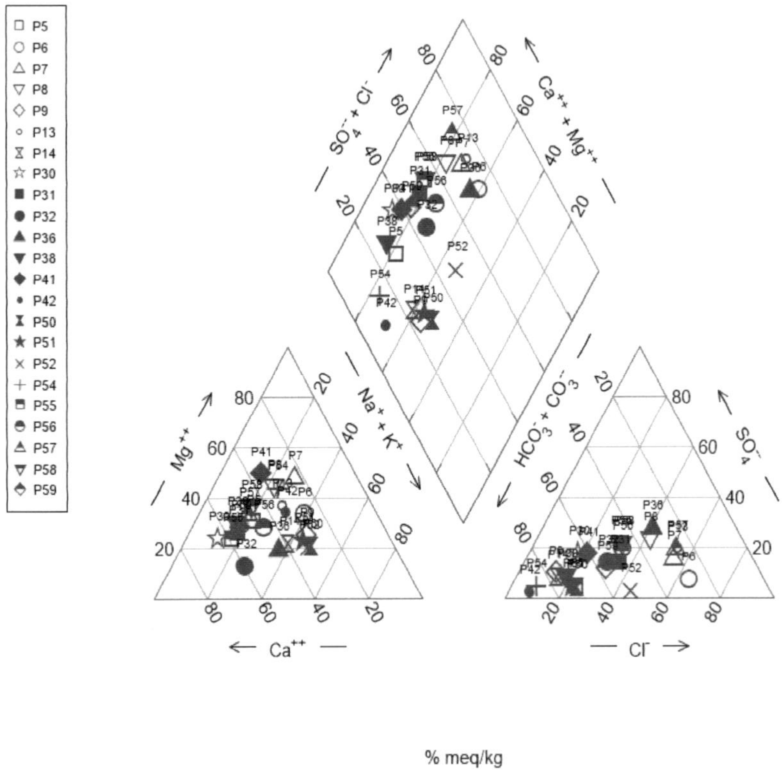

Figure 8. Piper Diagram in the Puntang mountain area

To determine the hydrochemical facies and characteristics of groundwater were plotted on a Piper diagram. Based on piper diagram analysis, the dominant types of hydrochemical facies are CaCl, CaMgCl, $CaHCO_3$, and $NaKHCO_3$ (Figure 8). The CaCl facies, enrichment of Cl^- is caused by contamination of geothermal or volcanic domain. The CaMgCl facies has enrichment of Mg^{2+} cations in the process of groundwater and rock water interactions. The $CaHCO_3$ facies is influenced by local meteoric water, with a high structural control that shows water flowing through the fracture. The $NaHCO_3$ facies in the distal area indicated anthropogenic influences have occurred. The hydrochemical evolution that occurs is that Ca^{2+} becomes Mg^{2+} and then turns into Na^+. The presence of HCO^{3-} indicates that groundwater mixing has occurred and is influenced by the decomposition of organic matter. Indications of groundwater mixing can be seen in groundwater with $NaKHCO_3$ facies in the distal area.Groundwater in the distal area is dominated by the $NaKHCO_3$ facies influenced by silica weathering and cation exchange process, that shows the intermediate groundwater flow system.

The hydrogeochemical process of groundwater comes from rainwater, weathering of rocks, crystallization, or evaporation [28]. The Gibbs diagram distinguishes the interaction effects of rock weathering, evaporation, and the impact of rainwater on the chemical processes of groundwater [29]. The Gibbs diagram shows that the groundwater are dominantly influenced by rock weathering, while in P6, P7, P8, P55, P58 are influenced by atmospheric precipitation. Groundwater which is affected by rainwater comes from springs and dug wells which unconfined aquifer in the volcanic rocks. (Figure 9).

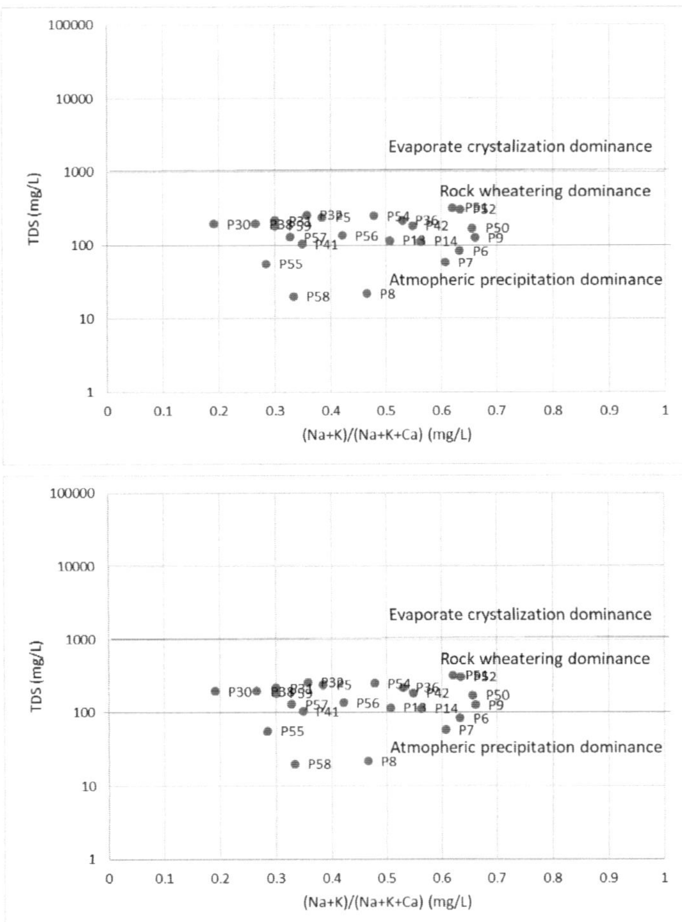

Figure 9. The Gibbs diagram in the Puntang mountain area

Anthropogenic activities such as agricultural fertilizers and irrigation affect groundwater conditions [29],[25]. Residues of agricultural activities will produce high ionic values for Cl^-, NH_4-N, SO_4^{2-} and K^+ [24]. Geological and environmental conditions around the water resources are one of the supporting data to determine polluting sources. High structural control allows water to flow through the fracture and facilitates the groundwater dilution process, thus helping the process of rock-water interaction. The high Cl^- value in the proximal area is influenced by volcanic domain, while the Cl^- value in the medial and distal areas may be influenced by anthropogenic from plantation and livestock activities. The medial and distal areas are densely populated areas which also affect groundwater quality. Sources of ammonium pollution come from septic tanks, residential and industrial wastewater, livestock and agricultural waste (Figure 10).

The results of bivariate Cl^- versus NH_4-N found that there were two groups of groundwater (Figure 10); Group A is groundwater located in the proximal to medial area. At Location P 32 has a low NH_4-N value but a high Cl^- value. This area is influenced by the volcanic domain. Meanwhile, the other group A had low Cl^- and low NH_4-N values which indicated the influence of groundwater and rock interactions. Group B is located in the medial to distal area. This area has a high Cl^- value and the NH_4-N value is also high above the threshold, so this area is influenced by anthropogenic activities. This condition shows the characteristics of lake deposits with thick sand and high porosity to support the movement of pollutants to groundwater

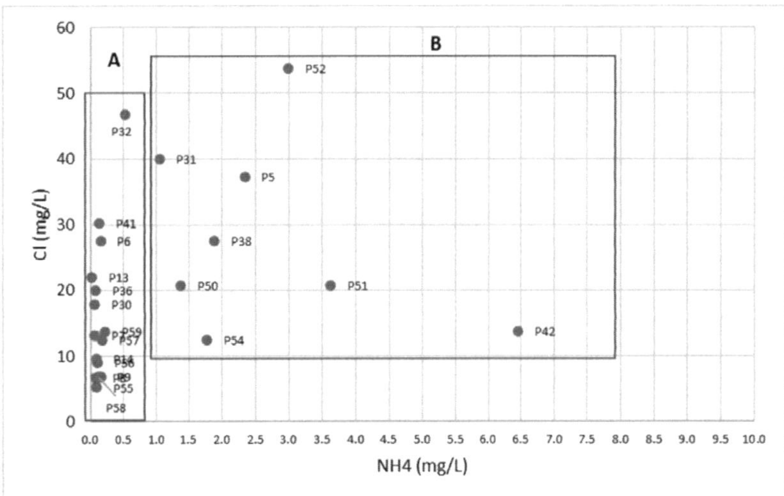

Figure 10. Groundwater hydrochemistry controlling factor from bivariate Cl^- versus NH_4-N

Conclusion

Geological and environmental conditions around the water resources are one of the supporting data to determine polluting sources. The high Cl^- value in the proximal area is influenced by volcanic domain, while the Cl^- value in the medial and distal areas may be influenced by anthropogenic. High structural control allows water to flow through the fracture and facilitates the groundwater dilution process, thus helping the process of rock-water interaction. The medial and distal areas are densely populated areas which also affect groundwater quality. The characteristics of lake deposits with thick sand and high porosity to support the movement of pollutants to groundwater. To understand the source of Cl^-, it is recommended to do more detailed research.

Acknowledgments
The author would like to thanks the Saintek Scholarship Program of the Ministry of Research and Technology, which has funded the education program. Thanks, are also given to the Head of the Research Center for Geotechnology, Indonesian Institute of Science, Dr. Eko Yulianto, who has given permission for providing laboratory facilities to carry out the water sample analysis. Mr. Dady Sukmayadi, and Mr. Wahyu Purwoko have helped a lot in laboratory analysis.

References

[1] B. Das and S. C. Pal, "Assessment of groundwater recharge and its potential zone identification in groundwater-stressed Goghat-I block of Hugli District, West Bengal, India," *Environ. Dev. Sustain.*, vol. 22, no. 6, pp. 5905–5923, 2020, doi: 10.1007/s10668-019-00457-7.

[2] G. Sakram and N. Adimalla, "Hydrogeochemical characterization and assessment of water suitability for drinking and irrigation in crystalline rocks of Mothkur region, Telangana State, South India," *Appl. Water Sci.*, vol. 8, no. 5, pp. 1–21, 2018, doi: 10.1007/s13201-018-0787-6.

[3] N. Aghazadeh, M. Chitsazan, and Y. Golestan, "Hydrochemistry and quality assessment of groundwater in the Ardabil area, Iran," *Appl. Water Sci.*, vol. 7, no. 7, pp. 3599–3616, 2017, doi: 10.1007/s13201-016-0498-9.

[4] S. Gaikwad, S. Gaikwad, D. Meshram, and V. Wagh, *Geochemical mobility of ions in groundwater from the tropical western coast of Maharashtra , India : implication to groundwater quality*, vol. 22, no. 3. Springer Netherlands, 2020.

[5] M. Chitsazan, N. Aghazadeh, Y. Mirzaee, and Y. Golestan, "Hydrochemical characteristics and the impact of anthropogenic activity on groundwater quality in suburban area of Urmia city, Iran," *Environ. Dev. Sustain.*, vol. 21, no. 1, pp. 331–351, 2019, doi: 10.1007/s10668-017-0039-1.

[6] M. L. Resubun, E. D. Wahjunie, and S. D. Tarigan, "ANALISIS KETERSEDIAAN DAN KEBUTUHAN AIR DI DAS Analysis Potency of Water Avaibility and Water Demand in Cisangkuy," vol. 2, no. 1, pp. 1–10, 2019.

[7] B. R. Suganda, "Kesesuaian Lahan Pada Gunung Api Vulkanik Kuarter Daerah Cicurug Dan Caringin Berdasarkan Kemampuan Lahan Serta Fasies Vulkanik Dengan Validasi Isotop Bagi Pengembangan Kawasan Permukiman dan Kawasan Industri," Universitas Padjajaran, 2016.

[8] H. Madrigal-Solís *et al.*, "Application of hydrogeochemistry and isotopic characterization for the assessment of recharge in a volcanic aquifer in the eastern region of central Costa Rica," *Isotopes Environ. Health Stud.*, vol. 56, no. 5–6, pp. 446–464, 2020, doi: 10.1080/10256016.2020.1814277.

[9] T. Hariyanto, S. A. Ekaprathama, and A. Kurniawan, "Analisis Potensi Mata Air Menggunakan Sistem Informasi Geografis (Studi Kasus : Wilayah Perbatasan Kabupaten Lumajang Dan Kabupaten Probolinggo)," vol. 12, no. 2, pp. 158–162, 2018, doi: 10.12962/j24423998.v12i2.3614.

[10] J. Mallick *et al.*, "Hydro-geochemical assessment of groundwater quality in Aseer Region, Saudi Arabia," *Water (Switzerland)*, vol. 10, no. 12, pp. 1–14, 2018, doi: 10.3390/w10121847.

[11] A. K. Mohamed, L. Dan, S. Kai, M. A. A. Mohamed, E. Aldaw, and B. A. Elubid, "Hydrochemical analysis and fuzzy logic method for evaluation of groundwater quality in the North Chengdu plain, China," *Int. J. Environ. Res. Public Health*, vol. 16, no. 3, 2019, doi: 10.3390/ijerph16030302.

[12] B. Sosi, A. Getabu, S. Maobe, and J. Barongo, "Relationship Between Aquifer Processes and Groundwater Quality: A Case of Olbanita Aquifer System, Lower Baringo Basin, Kenya Rift," *Air, Soil Water Res.*, vol. 12, 2019, doi: 10.1177/1178622119889910.

[13] M. Alzwar, N. Akbar, and S. Bachri, "Geological Map of the Garut and Pameungpeuk Quadrangle, Java," 1992.

[14] Hendarmawan, "Ketahanan Airtanah Wilayah Gunung Api," *Geomagz*, 2015.

[15] S. Bronto, "Fasies gunung api dan aplikasinya," *Indones. J. Geosci.*, vol. 1, no. 2, pp. 59–71,

2006, doi: 10.17014/ijog.vol1no2.20061.

[16] Bakosurtanal, "Digital Earth Map of Indonesia. Soreang Sheet. Scale 1:25,000.1208-633," 1999.

[17] N. Voutsis, E. Kelepertzis, E. Tziritis, and A. Kelepertsis, "Assessing the hydrogeochemistry of groundwaters in ophiolite areas of Euboea Island, Greece, using multivariate statistical methods," *J. Geochemical Explor.*, vol. 159, pp. 79–92, 2015, doi: 10.1016/j.gexplo.2015.08.007.

[18] H. Wu, J. Chen, H. Qian, and X. Zhang, "Chemical Characteristics and Quality Assessment of Groundwater of Exploited Aquifers in Beijiao Water Source of Yinchuan, China: A Case Study for Drinking, Irrigation, and Industrial Purposes," *J. Chem.*, vol. 2015, 2015, doi: 10.1155/2015/726340.

[19] J. Morán-Ramírez, R. Ledesma-Ruiz, J. Mahlknecht, and J. A. Ramos-Leal, "Rock-water interactions and pollution processes in the volcanic aquifer systemof Guadalajara, Mexico, using inverse geochemical modeling," *Appl. Geochemistry*, vol. 68, no. August 2019, pp. 79–94, 2016, doi: 10.1016/j.apgeochem.2016.03.008.

[20] B. J. Hibbs, "Factors in developing salt balances in groundwater basins," *Int. J. Hydrol.*, vol. 2, no. 2, pp. 231–238, 2018, doi: 10.15406/ijh.2018.02.00074.

[21] M. Sridharan and D. Senthil Nathan, "Groundwater quality assessment for domestic and agriculture purposes in Puducherry region," *Appl. Water Sci.*, vol. 7, no. 7, pp. 4037–4053, 2017, doi: 10.1007/s13201-017-0556-y.

[22] A. F. Rusydi, M. Saito, S. Ioka, R. Maria, and S. ichi Onodera, "Estimation of ammonium sources in Indonesian coastal alluvial groundwater using Cl- and gis," *Int. J. GEOMATE*, vol. 17, no. 62, pp. 53–58, 2019, doi: 10.21660/2019.62.4749.

[23] A. . Rusydi, W. Naily, and H. Lestiana, "Domestic and Agricultural Waste Pollution of Shallow Groundwater in Bandung Regency," *J. Ris. Geol. dan Pertamb.*, vol. 25, no. 2, p. 87, 2015, doi: 10.14203/risetgeotam2015.v25.201.

[24] N. J. Pawar and I. J. Shaikh, "Nitrate pollution of ground waters from shallow basaltic aquifers, Deccan Trap Hydrologic Province, India," *Environ. Geol.*, vol. 25, no. 3, pp. 197–204, 1995, doi: 10.1007/BF00768549.

[25] J. P. Deka, G. Tayeng, S. Singh, R. R. Hoque, A. Prakash, and M. Kumar, "Source and seasonal variation in the major ion chemistry of two eastern Himalayan high altitude lakes, India," *Arab. J. Geosci.*, vol. 8, no. 12, pp. 10597–10610, 2015, doi: 10.1007/s12517-015-1964-7.

[26] P. A. Domenico and F. W. Schwartz, *Physical and Chemical Hydrogeology*. John Wiley & Sons. Canada, 1990.

[27] N. Khatri and S. Tyagi, "Influences of natural and anthropogenic factors on surface and groundwater quality in rural and urban areas," *Front. Life Sci.*, vol. 8, no. 1, pp. 23–39, 2015, doi: 10.1080/21553769.2014.933716.

[28] S. Zereg, A. Boudoukha, and L. Benaabidate, "Impacts of natural conditions and anthropogenic activities on groundwater quality in Tebessa plain, Algeria," *Sustain. Environ. Res.*, vol. 28, no. 6, pp. 340–349, 2018, doi: 10.1016/j.serj.2018.05.003.

[29] A. K. Singh, B. Raj, A. K. Tiwari, and M. K. Mahato, "Evaluation of hydrogeochemical processes and groundwater quality in the Jhansi district of Bundelkhand region, India," *Environ. Earth Sci.*, vol. 70, no. 3, pp. 1225–1247, 2013, doi: 10.1007/s12665-012-2209-7.

AUTHOR INDEX

Institute of Physics
Dirac House, Temple Back
Bristol BS1 6BE UK

ISSN: 1755-1307
ISBN 978-1-7138-5825-6
9 781713 858256

CURRAN ASSOCIATES INC.
proceedings.com